AQA Mathematics

Unit 3 Higher

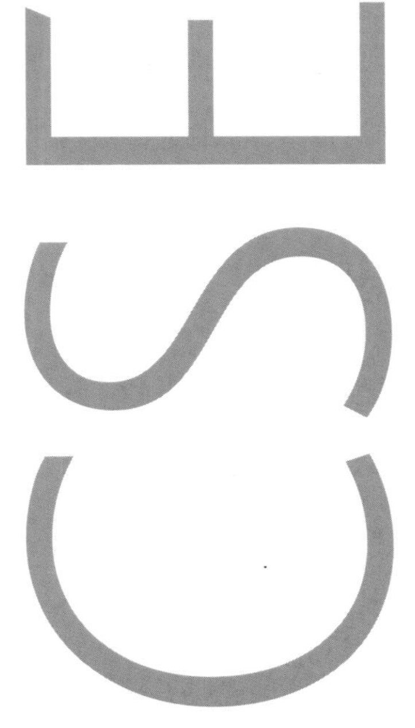

Series Editor
Paul Metcalf

Series Advisor
Andy Darbourne

Lead Authors
Sandra Burns
Shaun Procter-Green
Margaret Thornton

Authors
Tony Fisher
June Haighton
Anne Haworth
Gill Hewlett
Andrew Manning
Ginette McManus
Howard Prior
David Pritchard
Dave Ridgway
Paul Winters

Nelson Thornes

Published in 2010 by:
Nelson Thornes Ltd
Delta Place
27 Bath Road
CHELTENHAM
GL53 7TH
United Kingdom

10 11 12 13 14 / 10 9 8 7 6 5 4 3 2 1

A catalogue record for this book is available from the British Library

ISBN 978 1 4085 0630 1

Cover photograph: Photolibrary
Illustrations by Tech-Set, Rupert Besley, Roger Penwill and Angela Knowles
Page make-up by Tech-Set Limited, Gateshead

Printed and bound in Spain by GraphyCems

Photograph acknowledgements:
Alamy: 2.86, 4.7, 15.1, 18.1
Fotolia: 1.3, 1.4, 1.11, 1.13, 2.1, 4.8a, 8.26, 8.32, C1.8, C1.15, 9.1, 10.39, 12.1, 14.1, 14.7, 14.39, 16.1, 17.1, 19.1, 20.1, 23.1, C3.9
iStock: 3.1, 4.1, 7.1, 8.1, 9.32, 10.1, 10.21, 11.1, 13.1, 21.1, 21.11a, C3.4, C3.8
Science Photo Library: 5.1

Contents

Nelson Thornes and AQA

Nelson Thornes has worked in partnership with AQA to ensure that this book and the accompanying online resources offer you the best support for your GCSE course.

All AQA endorsed resources undergo a thorough quality assurance process to ensure that their contents closely match the AQA specification. You can be confident that the content of materials branded with AQA's 'Exclusively Endorsed' logo have been written, checked and approved by AQA senior examiners, in order to achieve AQA's exclusive endorsement.

The print and online resources together unlock blended learning; this means that the links between the activities in the book and the activities online blend together to maximise your understanding of a topic and help you achieve your potential.

These online resources are available on kerboodle! which can be accessed via the internet at **www.kerboodle.com/live**, anytime, anywhere.

If your school or college subscribes to kerboodle! you will be provided with your own personal login details. Once logged in, access your course and locate the required activity.

For more information and help on how to use these resources visit **www.kerboodle.com**.

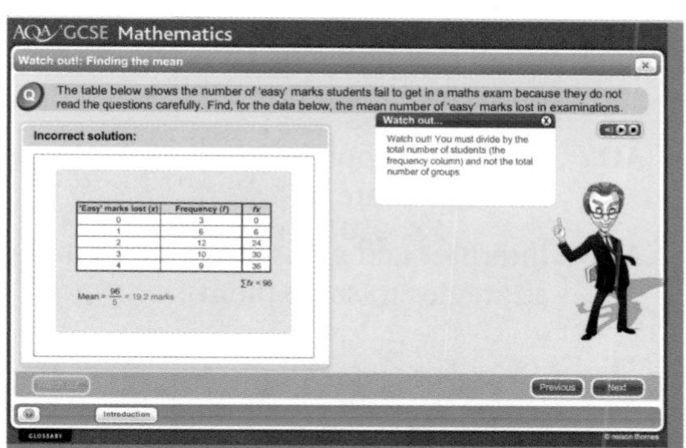

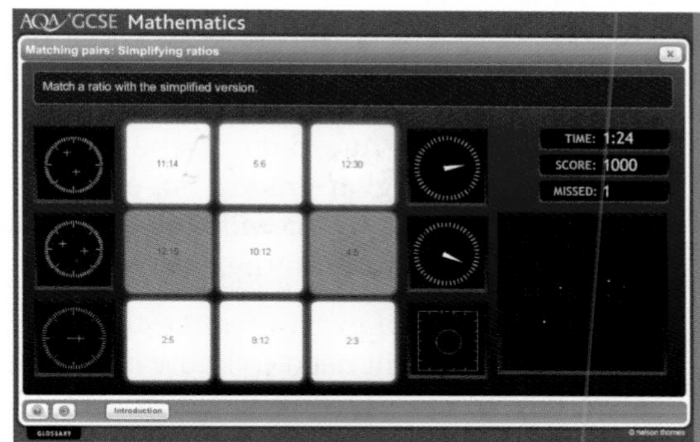

How to use this book

To help you unlock blended learning, we have referenced the activities in this book that have additional online coverage in *kerboodle!* by using this icon: 

The icons in this book show you the online resources available from the start of the new specification and will always be relevant.

In addition, to keep the blend up-to-date and engaging, we review customer feedback and may add new content onto *kerboodle!* after publication.

Welcome to GCSE Mathematics

This book has been written by teachers and examiners who not only want you to get the best grade you can in your GCSE exam, but also to enjoy maths. It covers all the material you will need to know for AQA GCSE Mathematics Unit 3 Higher. This unit allows you to use a calculator, so you will be able to use this most of the time throughout this book. Look out for calculator or non-calculator symbols (shown right) which tell you whether to use a calculator or not.

In the exam, you will be tested on the Assessment Objectives (AOs) below. Ask your teacher if you need help to understand what these mean.

AO1 recall and use your knowledge of the prescribed content

AO2 select and apply mathematical methods in a range of contexts

AO3 interpret and analyse problems and generate strategies to solve them.

Each chapter is made up of the following features:

Objectives

The objectives at the start of the chapter give you an idea of what you need to do to get each grade. Remember that the examiners expect you to do well at the lower grade questions on the exam paper in order to get the higher grades. So, even if you are aiming for a Grade A you will still need to do well on the Grade D questions on the exam paper.

On the first page of every chapter, there are also words that you will need to know or understand, called Key Terms. The box called 'You should already know' describes the maths that you will have learned before studying this chapter. There is also an interesting fact at the beginning of each chapter which tells you about maths in real life.

Learn...

The Learn sections give you the key information and examples to show you how to do each topic. There are several Learn sections in each chapter.

Practise...

These are questions that allow you to practise what you have just learned.

D The bars that run alongside questions in the exercises show you what grade the question is aimed at. This will give you an idea of what grade you're working at. Don't forget, even if you are aiming at a Grade A, you will still need to do well on the Grades D–B questions.

These questions are harder questions to test and challenge your mathematics.

These questions are Functional Maths type questions, which show how maths can be used in real life.

These questions are problem-solving questions, which will require you to think carefully about how best to answer.

These questions should be attempted **with** a calculator.

These questions should be attempted **without** using a calculator.

Assess

End-of-chapter questions written by examiners. Some chapters feature additional questions taken from real past papers to further your understanding.

Hint

These are tips for you to remember while learning the maths or answering questions.

AQA Examiner's tip

These are tips from the people who will mark your exams, giving you advice on things to remember and watch out for.

Bump up your grade

These are tips from the people who will mark your exams, giving you help on how to boost your grade, especially aimed at getting a Grade C.

Consolidation

Consolidation chapters allow you to practise what you have learned in previous chapters. The questions in these chapters can cover any of the topics you have already seen.

1 Fractions and decimals

Objectives

Examiners would normally expect students who get these grades to be able to:

D

- calculate with fractions and decimals
- find one quantity as a fraction of another
- solve problems involving fractions
- solve problems involving decimals

C

- add and subtract mixed numbers
- find the reciprocal of a number.

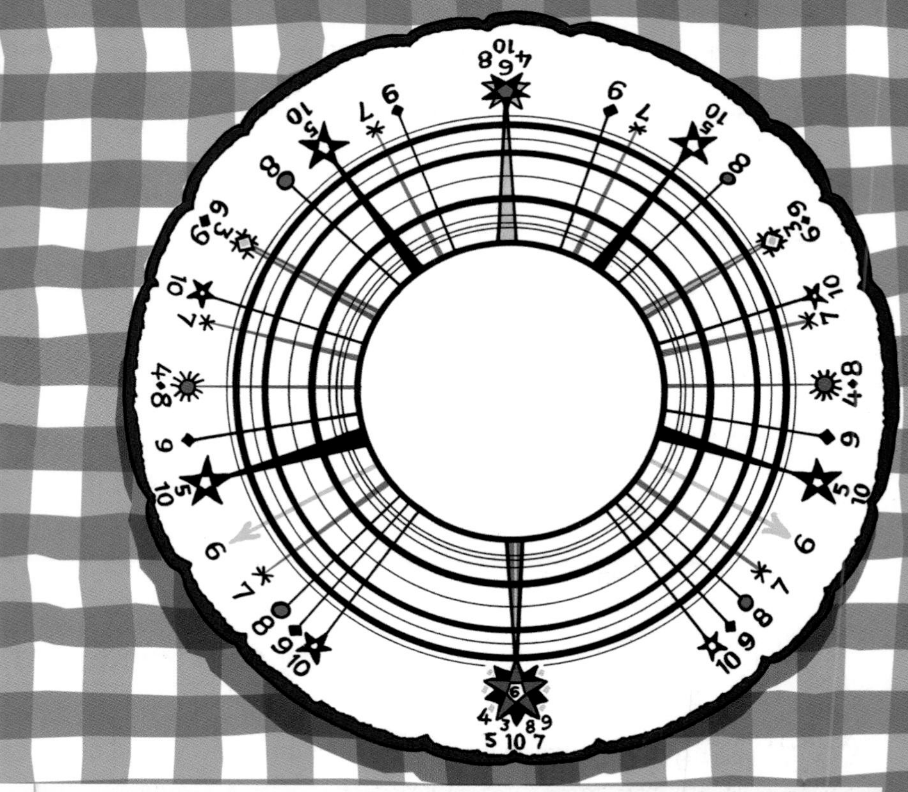

Did you know?

Plat diviseur

It is not easy to cut a pizza or a cake into pieces the same size, particularly if you need five or seven or nine pieces.

The French have a solution, the *plat diviseur*, or dividing plate, which has numbers marked round the edge to show where to cut for that number of pieces. Useful and beautiful!

You should already know:

- ✔ how to add, subtract, multiply and divide simple numbers
- ✔ the meaning of 'sum', 'difference' and 'product'
- ✔ how to work with simple fractions such as halves and quarters
- ✔ how to arrange decimals and fractions in order
- ✔ how to find equivalent fractions
- ✔ how to calculate fractions of quantities
- ✔ how to add and subtract fractions
- ✔ how to express fractions as decimals and percentages
- ✔ how to round numbers to the nearest whole number, ten, hundred and so on, and to different numbers of decimal places.

Key terms

numerator
denominator
rounding
integer
decimal place
significant figure
terminating decimal
recurring decimal
reciprocal

Learn... 1.1 One quantity as a fraction of another

To find one quantity as a fraction of another, make the first quantity the **numerator** and the second the **denominator**. Make sure that the units of the two quantities are the same. Use your calculator to simplify the fraction.

So 20 minutes as a fraction of 1 hour = 20 minutes as a fraction of 60 minutes $= \frac{20}{60} = \frac{1}{3}$

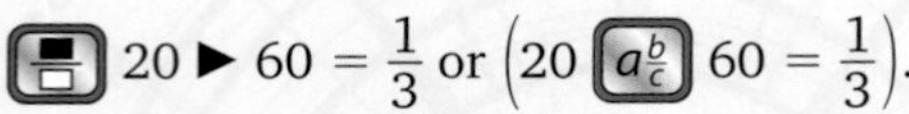

You need to use the fraction button on your calculator (▭ or $a\frac{b}{c}$) in this chapter.

You should not need a calculator to simplify the fraction $\frac{20}{60}$, but it is done like this:

▭ 20 ► 60 $= \frac{1}{3}$ or (20 $a\frac{b}{c}$ 60 $= \frac{1}{3}$).

Example: In a mathematics exam, there are 35 marks for Section A and 45 marks for Section B. What fraction of the exam's marks is for Section A?

Solution: Total number of marks for the exam is $35 + 45 = 80$

Fraction of marks for Section A is $\frac{35}{80} = \frac{7}{16}$

AQA Examiner's tip

Remember to give your answers in their simplest form.

Practise... 1.1 One quantity as a fraction of another

k! D C B A A*

D

1 Express the first quantity as a fraction of the second quantity.

a 25 minutes, $2\frac{1}{2}$ hours

b Half an hour, three hours

c 40p, £2.40

d 20 cm, 1.8 m

e 400 g, 1.2 kg

2 Gill has a dozen eggs. She uses three eggs for breakfast and two in a cake.

What fraction of her eggs does she have left?

3 Here are nine shapes.

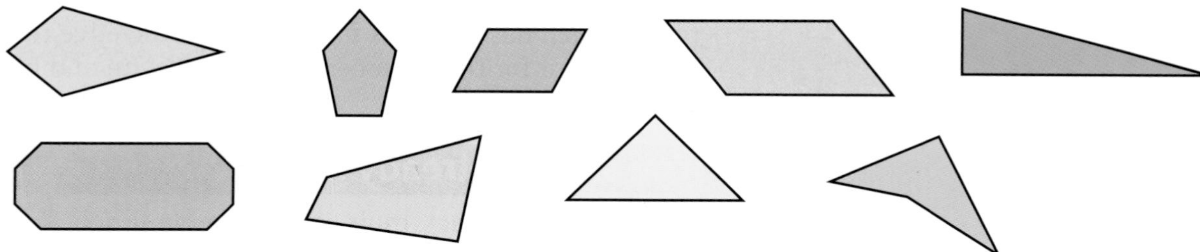

a What fraction of the nine shapes are quadrilaterals?

b What fraction of the nine shapes are parallelograms?

4 Pigs have an average weight of 150 pounds and a stomach weighing 6 pounds.

a What fraction of a pig's total weight is the weight of its stomach?

Sheep have a different digestive system. They have an average weight of 120 pounds and a stomach system weighing 30 pounds.

b What fraction of a sheep's total weight is the weight of its stomach system?

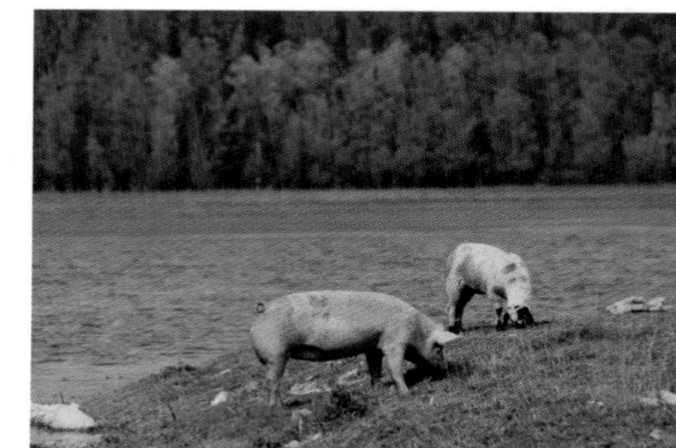

5 Which of these formulae can be used to find x minutes as a fraction of y hours?

i	$\frac{x}{y}$	**iii**	$\frac{60x}{y}$	**v**	$60xy$	**vii**	$\frac{x}{60y}$
ii	$\frac{y}{x}$	**iv**	$x \div 60y$	**vi**	xy	**viii**	$\frac{60y}{x}$

6 Nicola spends a quarter of her salary on the rent for her flat. Her salary goes up by £100 a month and she moves to a flat costing an extra £100 a month.

Is the fraction of her salary she spends on rent bigger, smaller or the same as before? Explain how you got your answer.

7 A pair of shoes costing £35 is reduced by £3.50 in a sale.

a What fraction of the original price is the reduction?

b What fraction is the sale price of the original price?

8 In New York one day in September, the sunrise time was 06:27 and the sunset time was 19:21. What fraction of the time from midnight on 5 September to midnight on 6 September was daylight?

9 The price of a dress is reduced by one fifth in a sale. The sale price is £25.80.

What was the original price?

Learn... 1.2 Working with fractions and decimals

Rounding

Numbers can be **rounded** to make them easy to work with and to understand. For example, if parents want to know how many students there are in a school, they probably want to know the approximate number, 900 for example, rather than the exact answer 936.

Rounded to the nearest 100, 936 is 900. It is between 900 and 1000 and nearer to 900.

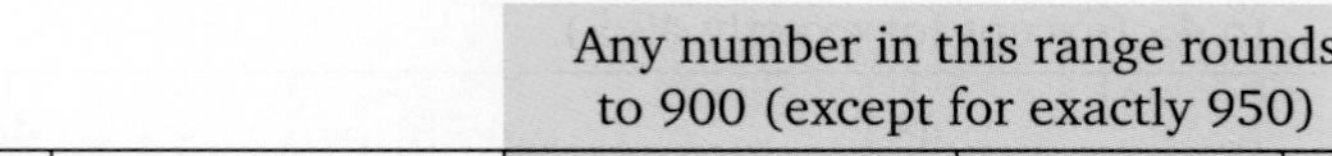
Any number in this range rounds
to 900 (except for exactly 950)
800
850
900
936
950
1000

Rounded to the nearest 10, 936 is 940. It is between 930 and 940 and nearer to 940.

Any number in this range rounds to 900 (except for exactly 945)

930 935 936 940 945 950

The diagrams also show that 936 to the nearest 50 is 950 and to the nearest 5 is 935. You need to be able to round numbers to different degrees of accuracy like this.

What is 950 to the nearest 100? It is exactly halfway between 900 and 1000 so it is not nearer to either. The usual rule is that numbers halfway between round up. So 950 is 1000 to the nearest 100.

You also need to be able to round decimal numbers to the nearest **integer** (whole number), to one **decimal place** (d.p.), two decimal places and so on.

Example: What is 7.846 to one decimal place?

Solution: 7.846 is between 7.8 and 7.9 and nearer to 7.8. So 7.846 to one decimal place is 7.8

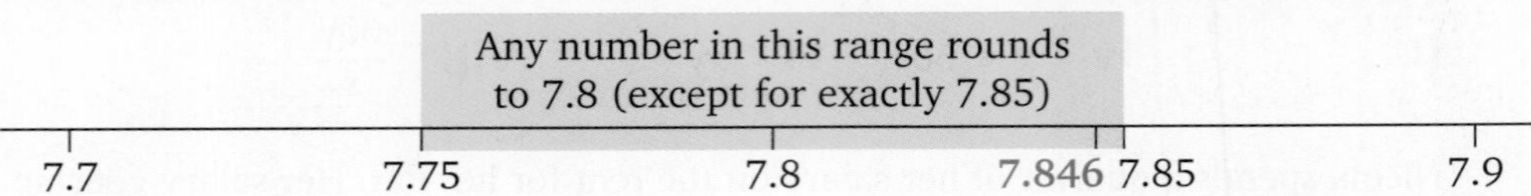

To two decimal places, 7.846 is 7.85; it is between 7.84 and 7.85 and nearer to 7.85

Significant figures

936 students in the school rounded to one **significant figure** (s.f) is 900.

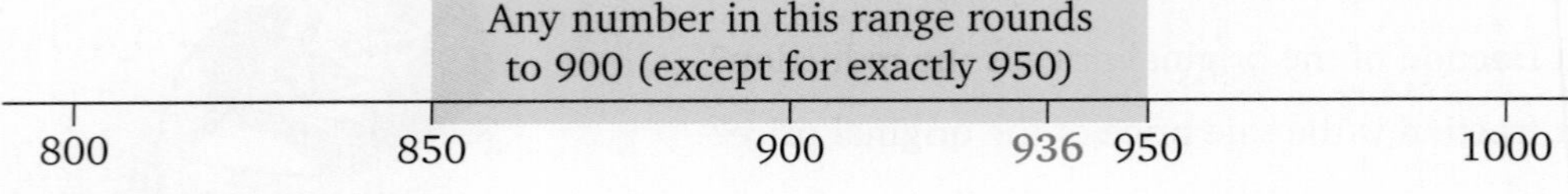

0.0936 rounds to 0.09

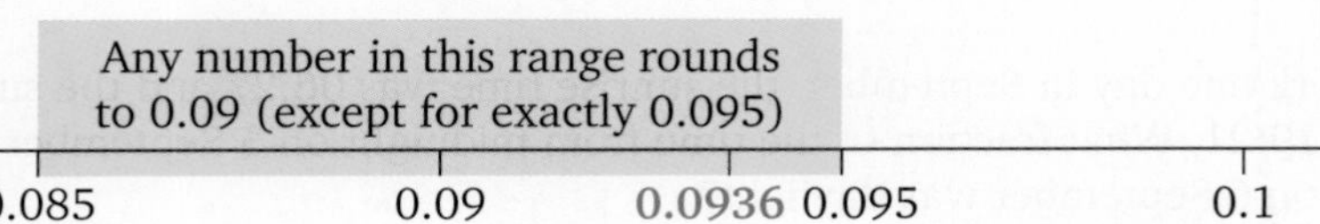

936 is 900 (one significant figure).
0.0936 is 0.09 (one significant figure).

The figure 9 in each number gives the approximate size.
Both numbers also need zeros to show the place value of the 9.
These zeros are not significant figures.

Note the difference between significant figures and decimal places:
936 is 940 (two significant figures)
0.0936 is 0.094 (two significant figures, three decimal places).

(Zeros **can** be significant; for example, the number 1020 has three significant figures, 1, 0 and 2. The zero in the units position is not significant. It is just there for place value.)

Example: Round these numbers to:

a the nearest integer **b** one decimal place **c** one significant figure.

i 44.79 **iii** 204.45
ii 0.5678 **iv** 0.0235

Hint
To round to the nearest integer, think which integers lie on either side of the number.

Solution: **a** 44.79 is between 44 and 45.

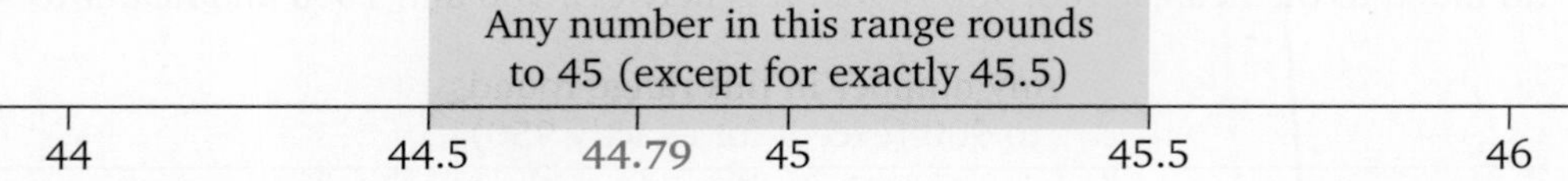

44.79 is more than 44.5 so is nearer to 45 than to 44. To the nearest integer, 44.79 is 45. The other numbers are rounded in a similar way.

a **i** 45 **ii** 1 **iii** 204 **iv** 0

To round to one decimal place, identify the numbers with one decimal place on either side: 44.79 is between 44.7 and 44.8

44.79 is greater than 44.75 so it is nearer to 44.8 than to 44.7. It rounds to 44.8

b **i** 44.8 **ii** 0.6 **iii** 204.5 **iv** 0.0

To round to one significant figure, look at the first two non-zero digits, which are 44. This is nearer to 40 than to 50, so to one significant figure 44.79 is 40.

c **i** 40 **ii** 0.6 **iii** 200 **iv** 0.02

Bump up your grade
To get a Grade C you have to be able to round to different numbers of significant figures.

Expressing fractions as decimals

Any fraction can be expressed as a decimal.
Some simple examples are $\frac{1}{2} = 0.5$, $\frac{3}{4} = 0.75$, $\frac{3}{10} = 0.3$

Make sure that you are familiar with simple examples like these.

To express a fraction as a decimal, use your calculator to divide the numerator by the denominator.

$\frac{3}{4} = 3 \div 4 = 0.75$ (Make sure that your calculator is set to give a decimal answer, not a fraction answer.)

0.75 is a **terminating decimal**. $3 \div 4$ works out exactly; the calculation comes to an end after two decimal places. Some fractions become **recurring decimals**; the calculation does not come to an end but one digit, or a group of digits, repeats forever.

$\frac{5}{11} = 5 \div 11 = 0.454545...$ This can also be written as $0.\dot{4}\dot{5}$

$\frac{2}{3} = 2 \div 3 = 0.666...$ This can also be written as $0.\dot{6}$

The calculator display shows that $\frac{2}{3}$ is 0.6666666667. The calculator can show only a limited number of digits and it rounds the final one in the display.

When decimals are used in calculations, you have to round them appropriately.

$\frac{2}{3}$ rounded to:

one decimal place is 0.7

two decimal places is 0.67

three decimal places is 0.667

four decimal places is 0.6667

and so on.

Expressing fractions as percentages

To express a fraction as a percentage, change the fraction to a decimal and then multiply the decimal by 100%.

$\frac{3}{4} = 0.75$

$0.75 \times 100\% = 75\%$

AQA Examiner's tip

In calculations, use all the figures already in your calculator whenever possible. Do not re-enter the number unless you really have to.

Example: **a** What is $\frac{5}{12}$ as a decimal? Write your answer correct to three significant figures.

b Write $\frac{5}{12}$ as a percentage.

Solution: **a** $\frac{5}{12} = 5 \div 12 = 0.41666... = 0.417$ to three significant figures.

b $\frac{5}{12} = 5 \div 12 = 0.41666...$

$= 0.41666... \times 100\%$

$= 42\%$, correct to the nearest whole percent.

AQA Examiner's tip

Make sure you always divide the numerator by the denominator when changing a fraction to a decimal.

Arranging fractions in order

When fractions are expressed as decimals or percentages, you can put them in order.

It is not easy to see which fraction, $\frac{3}{4}$ or $\frac{7}{9}$, is bigger when they are in fraction form, but in decimals $\frac{3}{4} = 0.75$ and $\frac{7}{9} = 0.77777...$ so you can see that $\frac{7}{9}$ is a bit bigger than $\frac{3}{4}$

Example: Arrange these in order of size, starting with the smallest.

$\frac{2}{3}$, $\frac{3}{5}$, 62%, $\frac{11}{18}$, 0.65, $\frac{7}{11}$

Solution: $\frac{2}{3} = 0.666...$, $\frac{3}{5} = 0.6$, $62\% = 0.62$, $\frac{11}{18} = 0.6111...$, $0.65 = 0.65$, $\frac{7}{11} = 0.636363...$

So the order of size is $\frac{3}{5}$, $\frac{11}{18}$, 62%, $\frac{7}{11}$, 0.65, $\frac{2}{3}$

Reciprocals

When two numbers multiply together to make 1, the numbers are the **reciprocals** of each another.
So 2 and $\frac{1}{2}$ are the reciprocals of each other because $2 \times \frac{1}{2} = 1$

The reciprocal of any fraction $\frac{a}{b}$ is $\frac{b}{a}$ because $\frac{a}{b} \times \frac{b}{a} = 1$

The reciprocal of any number x (integer, decimal or fraction) is $\frac{1}{x}$ because $x \times \frac{1}{x} = 1$

Your calculator can work out reciprocals both as fractions and as decimals; look for the button labelled  or x^{-1} and find out how to use it.

Example: Find the reciprocals of $\frac{1}{3}$, 0.2, $\frac{9}{20}$, 0.89

Solution: Reciprocal of $\frac{1}{3} = \frac{3}{1} = 3$

To find the reciprocal of 0.2, enter 0.2 into your calculator, press the x^{-1} key then =
The answer is 5

Reciprocal of $\frac{9}{20} = \frac{20}{9} = 2\frac{2}{9}$

Reciprocal of $0.89 = \frac{1}{0.89} = 1.12$ to three significant figures. (Enter 0.89 into your calculator, press the x^{-1} key, then =)

Bump up your grade

To get a Grade C you have to be able to work out reciprocals.

Practise... 1.2 Working with fractions and decimals

D

1 Round these numbers:

a to one significant figure **b** to two significant figures **c** to three significant figures.

i 12.89 **ii** 54.5 **iii** 109.87 **iv** 4.756 **v** 0.836

2 The number of wild tigers in India is 1500 to the nearest 500.
What is the biggest number of wild tigers there could be in India?
What is the least number?

3 Express these fractions as percentages.

a $\frac{4}{5}$ **b** $\frac{9}{10}$ **c** $\frac{11}{20}$ **d** $\frac{23}{50}$ **e** $\frac{67}{100}$

4 Arrange these in order of size, starting with the smallest.

$\frac{5}{12}$, 0.4, $\frac{4}{9}$, 45%, $\frac{3}{7}$, $\frac{7}{16}$

5 There are three girls and seven boys in the chess club. One more boy and one more girl join the club.

Is the fraction of girls in the club now more, less or the same?
Show how you found your answer.

C

6 **a** Which of these fractions are equivalent to recurring decimals?

i $\frac{3}{5}$ **ii** $\frac{9}{11}$ **iii** $\frac{5}{6}$ **iv** $\frac{7}{20}$ **v** $\frac{4}{15}$

b Can you say which fractions become recurring decimals without working them out?
Explain what you have found out, using examples.

C

7 **a** Match up each number with its reciprocal. The first has been done for you.

Number	Reciprocal
$\frac{5}{9}$	1
1	$\frac{1}{7}$
0.4	2.5
15	$1\frac{4}{5}$
7	0.066…

b What happens when you multiply a number by its reciprocal?

8 **a** Which number is the reciprocal of itself?

b Use your calculator to try to find the reciprocal of zero. What happens?

9 **a** Multiply 150 by $\frac{1}{4}$. What do you have to multiply the answer by to get back to 150?

b Choose another number. Multiply it by 2.5. What do you have to multiply the answer by to get back to the original number?

c What do your answers to parts **a** and **b** tell you about reciprocals?

Bump up your grade

To get a Grade C you have to understand that multiplying by a number and then by the reciprocal of the same number, has no effect.

10 Express the decimal 0.1296 as a fraction in its lowest terms.

11 Match up each number with its reciprocal. The first has been done for you.

Number	Reciprocal
$\frac{1}{x}$	$\frac{d}{c}$
$\frac{c}{d}$	$0.1x$
$\frac{10}{x}$	x
$15x$	$\frac{5}{2x}$
$0.4x$	$\frac{1}{15x}$

12 Jake got 12 marks out of 25 in his first maths test and 14 out of 30 in his second.

In which test did he do better?
Show how you found your answer.

13 Pigs have an average weight of 150 pounds and a stomach weighing 6 pounds. Sheep have a different digestive system. They have an average weight of 120 pounds and a stomach system weighing 30 pounds.

a What percentage of a pig's total weight is the weight of its stomach?

b What percentage of a sheep's total weight is the weight of its stomach system ?

1 Assess

D

1 One inch is approximately 2.54 cm. One yard is 36 inches.

a How many centimetres, correct to the nearest millimetre, are there in a yard?

b What percentage is one yard of one metre?

2 The price of a shirt is reduced by one-fifth and then the new price is increased by one fifth.

Is the final price less than, greater than or the same as the original price? Explain how you worked out your answer.

3 A skirt needs $1\frac{3}{4}$ yards of fabric and a jacket needs $2\frac{1}{8}$ yards.

How much fabric is needed altogether for four skirts and three jackets?

4 In a clinical trial of two new drugs, 2135 out of 3000 patients taking Drug A got better and 1855 out of 2500 patients taking Drug B got better.

Which drug appears to be the more effective?

5 In a butcher's shop, sausages are priced at £2.60 a pound. The butcher is legally required to price the sausages by the kilogram too.

What is the price of a kilogram of these sausages?

6 A number when rounded to two significant figures is 0.015

Write down three possible values of the number. What is its smallest possible value?

7 David buys a car costing £16 458. He has to pay one third of the price as a deposit. He borrows the rest on an interest-free loan to be paid back by equal monthly instalments for three years.

How much is each monthly instalment?

8 Asif earns £8.37 an hour for 35 hours a week and one and a third times this for each hour of overtime.

How many hours of overtime has Asif worked in a week when he earned £337.59?

C

9 Drew's annual salary is increased by $\frac{1}{4}$ and becomes £36 400.

What was his original salary?

10 Which of these numbers is/are the reciprocal of $\frac{5}{8}$?

0.625, $\frac{8}{5}$, $1\frac{3}{8}$, $1\frac{3}{5}$, $\frac{5}{8}$, 1.6

11 Show that the product of a fraction and its reciprocal is 1.

12 Kerry spent $34.60 New Zealand dollars in a souvenir shop. On her credit card bill, this appeared as £14.67.

What rate of exchange was this? How much in New Zealand dollars did she pay for a meal that appeared on her credit card bill as £57.60?

C

13 Write 0.8128 as a fraction in its simplest terms.

14 Which of these fractions cannot be expressed as a terminating decimal?

$\frac{2}{3}$, $\frac{3}{4}$, $\frac{4}{5}$, $\frac{5}{6}$, $\frac{7}{8}$, $\frac{8}{9}$, $\frac{9}{10}$, $\frac{10}{11}$, $\frac{11}{12}$

AQA Examination-style questions

1 Each fraction in this wall is the sum of the two supporting fractions in the wall below.
Complete the wall.

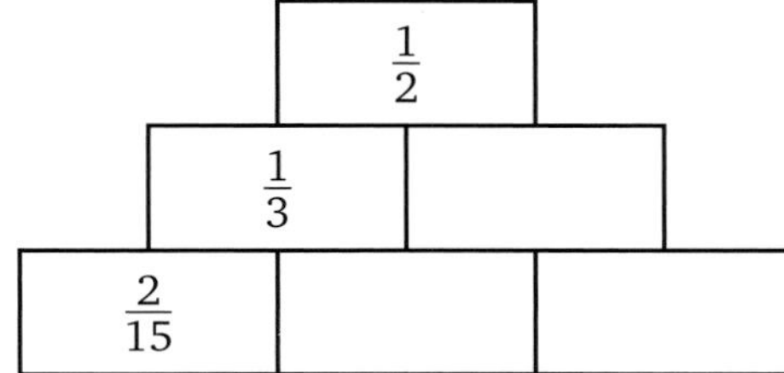

(3 marks)

AQA 2008

2 The cost of staying in a hostel is £13.50 for the first day.
The cost for each extra day is £10.50.
Eli has £75 to spend on accommodation.
He says that he can afford to stay in this hostel for seven days.

Is he correct?
You **must** show your working. *(3 marks)*

AQA 2009

2 Angles and areas

Objectives

Examiners would normally expect students who get these grades to be able to:

D

- recognise corresponding, alternate and interior angles on parallel lines
- understand and use three-figure bearings
- find the area of a triangle, trapezium and parallelogram
- find the area and perimeter of shapes made from triangles and rectangles
- calculate the circumference and area of a circle

C

- work out the perimeter and area of a semicircle and compound shapes made from parts of a circle.

Key terms

vertically opposite
parallel
transversal
alternate angle
corresponding angle
interior (allied) angles
bearing
base
perpendicular height

Did you know?

Wind and sails

If you want to sail a boat, you need to understand how the area of the sails and the angle the boat travels relative to the wind direction affect the boat's progress. If a sailor wants to travel 'upwind' (towards the wind), it's best to travel about 45° from the direction that the wind is coming from. This is called 'close-hauled' or 'on a beat'. In a storm sailors would minimise their sail area to prevent the sudden strong gusts of wind capsizing the boat. They would 'ride out the storm'!

You should already know:

- ✔ the terms clockwise and anticlockwise
- ✔ how to distinguish between acute, obtuse, reflex and right angles
- ✔ that angles at a point add up to 360°
- ✔ that angles on a straight line add up to 180°
- ✔ that vertically opposite angles are equal
- ✔ that angles in a triangle add up to 180°
- ✔ how to find the perimeter and area of rectangles
- ✔ the definition of a parallelogram and a trapezium
- ✔ the names of basic parts of a circle: centre, diameter, radius and circumference.

Learn... 2.1 Angles and parallel lines

Angles at a point add up to 360°.

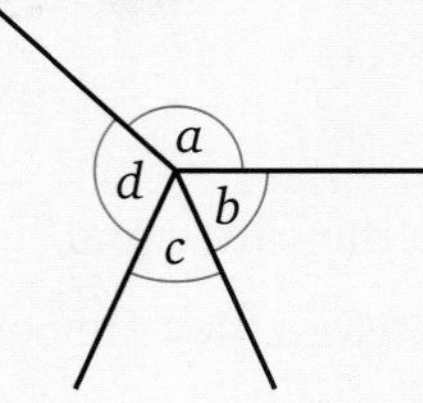

$a + b + c + d = 360°$

Angles on a straight line add up to 180°.

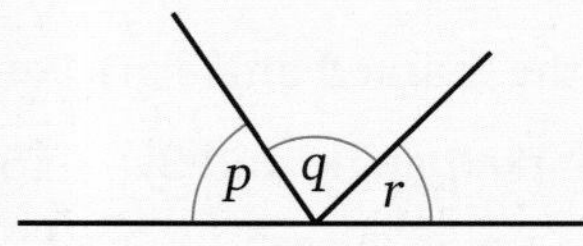

$p + q + r = 180°$

Where two lines cross, the **vertically opposite** angles are equal.

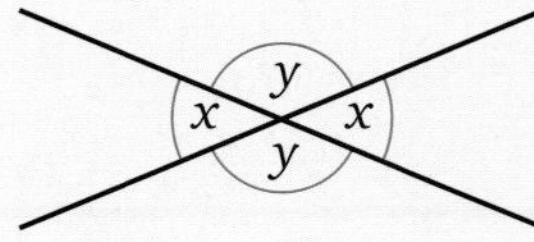

Two lines which stay the same perpendicular distance apart are called **parallel** lines.

The arrows show that the lines are parallel.

A line through the two parallel lines is called a **transversal**.

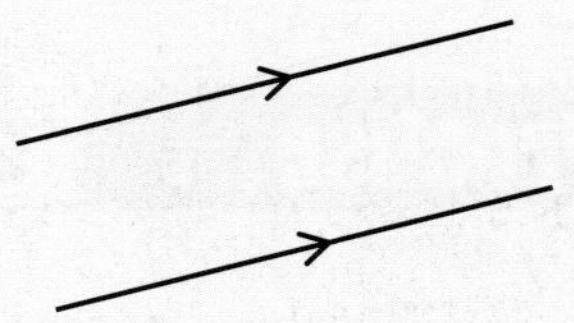

Several pairs of equal angles are formed between the parallel lines and the transversal.

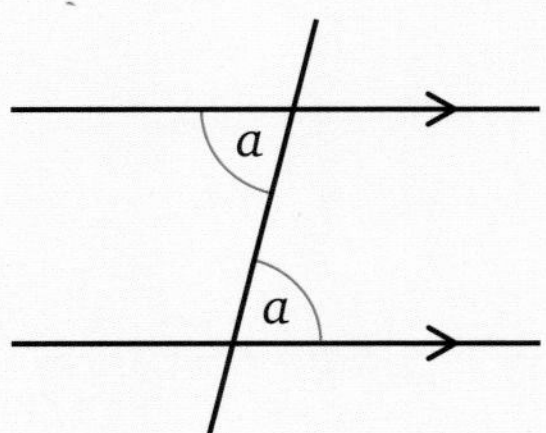

The angles marked *a* are **alternate angles**.

They are equal.

The angles are on opposite sides of the transversal.

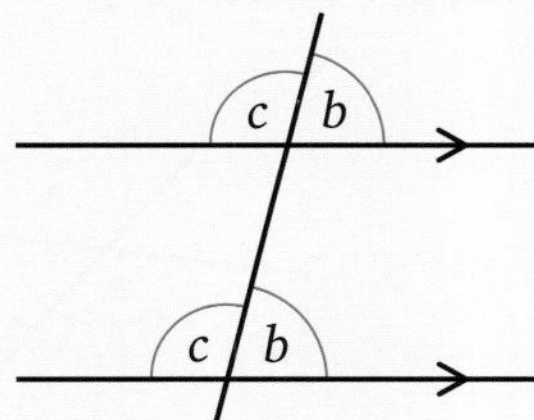

The angles marked *b* are **corresponding angles**.

They are equal.

The angles are in similar positions on the same side of the transversal.

The angles marked *c* are another pair of corresponding angles.

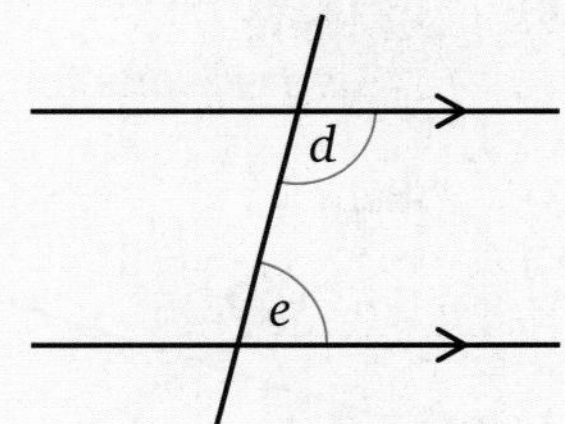

Angles *d* and *e* are **interior** (or **allied**) angles.

They always add up to 180°.

So $d + e = 180°$

Example: Work out *x*.

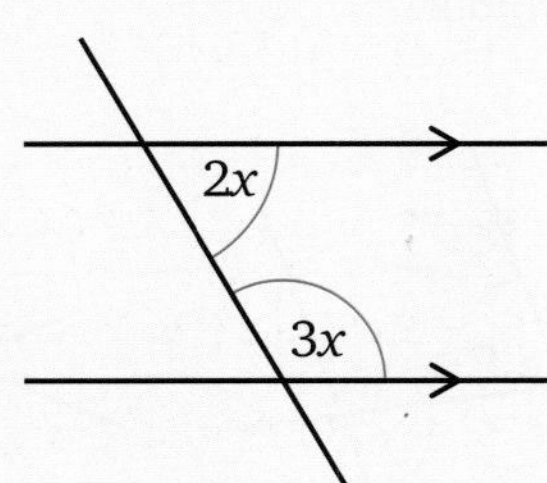

Solution: $2x$ and $3x$ are both interior angles and add up to 180°.

$$2x + 3x = 180°$$
$$5x = 180°$$
$$x = 36°$$

Example: Work out y.

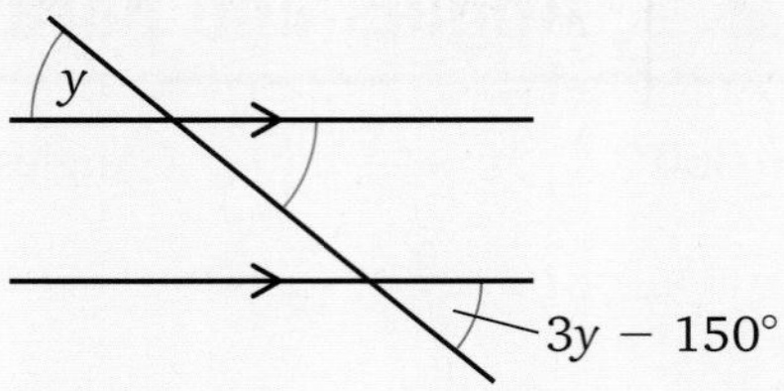

Solution: y and the marked angle are vertically opposite angles. Therefore they are equal.

The marked angle and $3y - 150°$ are corresponding angles. Therefore they are equal.

So y and $3y - 150°$ are equal.

$$3y - 150° = y$$
$$2y - 150° = 0$$
$$2y = 150°$$
$$y = 75°$$

It is often possible to find angles by more than one method.

Practise... 2.1 Angles and parallel lines 

D C B A A*

In this exercise the diagrams are not drawn accurately.

D

1 Work out the values of angles x, z, q, r and s.

Give reasons for your answers.

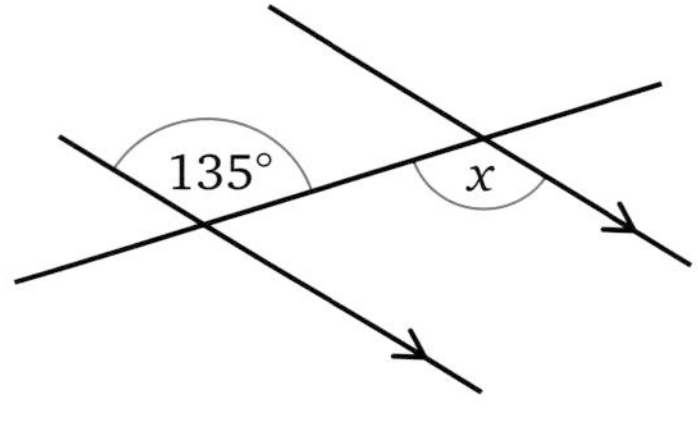

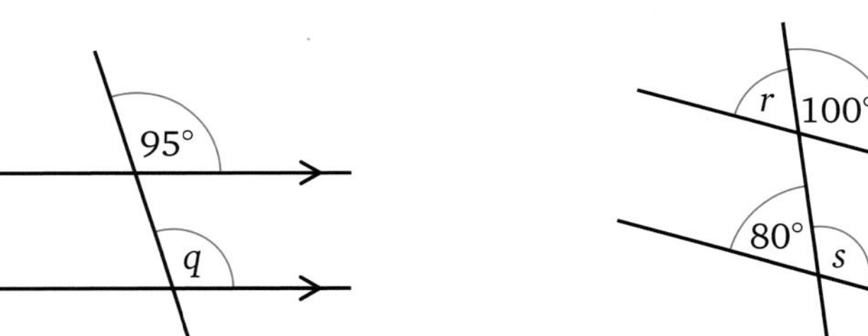

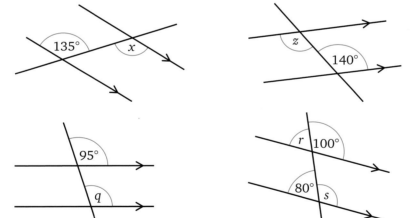

2 Work out the values of angles x, y and z.

Give reasons for your answers.

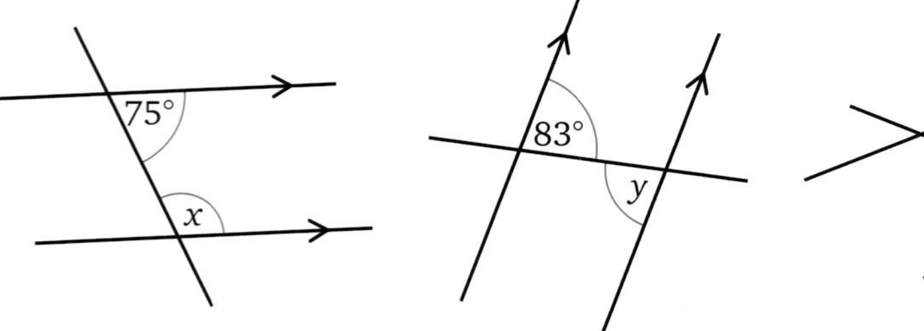

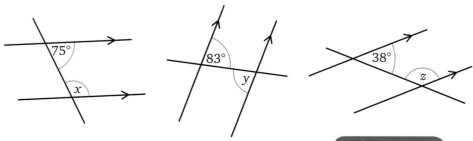

Hint

Remember that you can use facts about angles at a point or on a line as well as angle properties of parallel lines.

D

3 Work out the values of the angles marked with letters.
Give reasons for your answers.

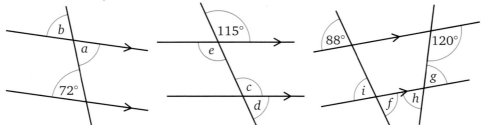

4 **a** Are lines AB and CD parallel?
Give a reason for your answer.

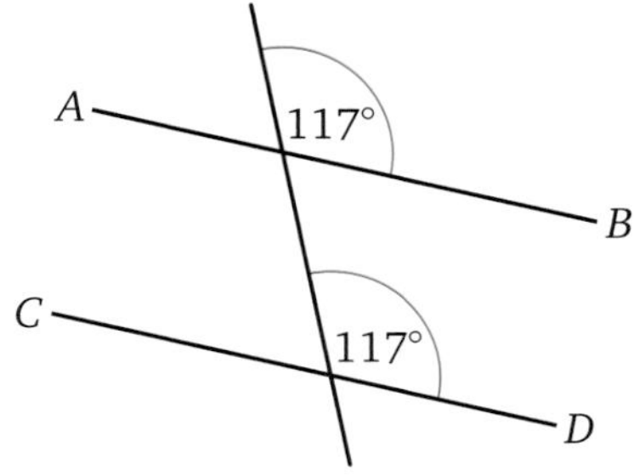

b Are lines EF and GH parallel?
Give a reason for your answer.

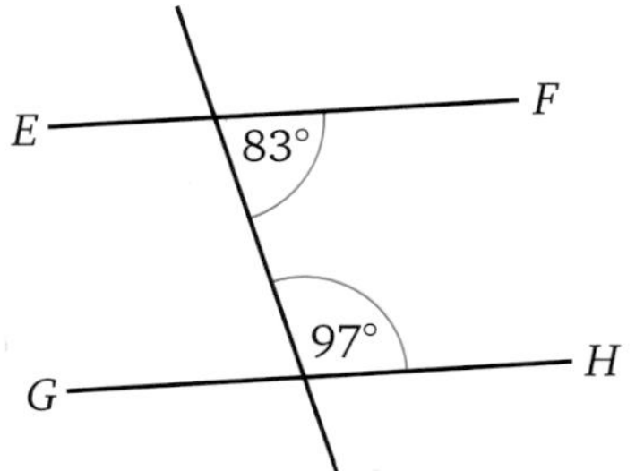

C

Bump up your grade

To get a Grade C you need to be able to prove that lines are parallel.

5 Calculate the size of the angles marked with letters.

a

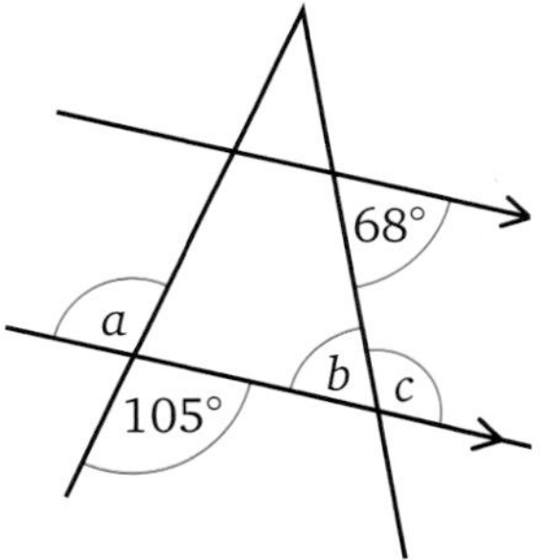

b

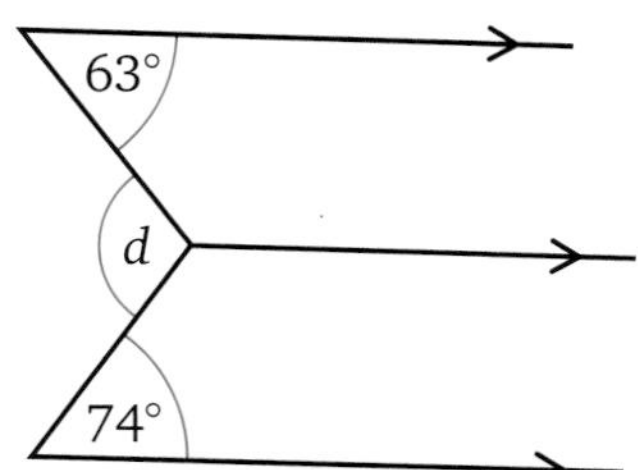

c

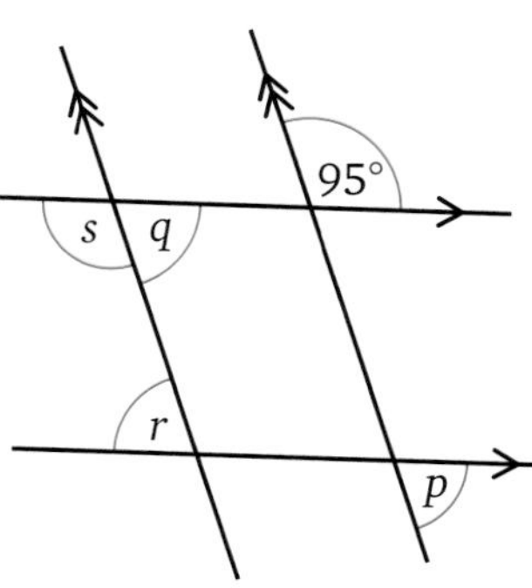

6 **a** Work out x and y.

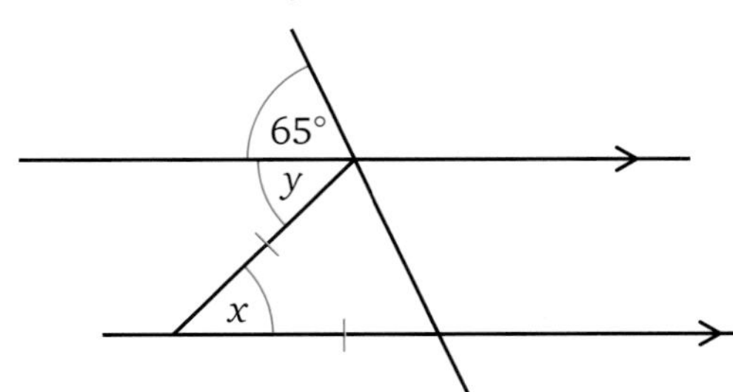

b Work out x and y.

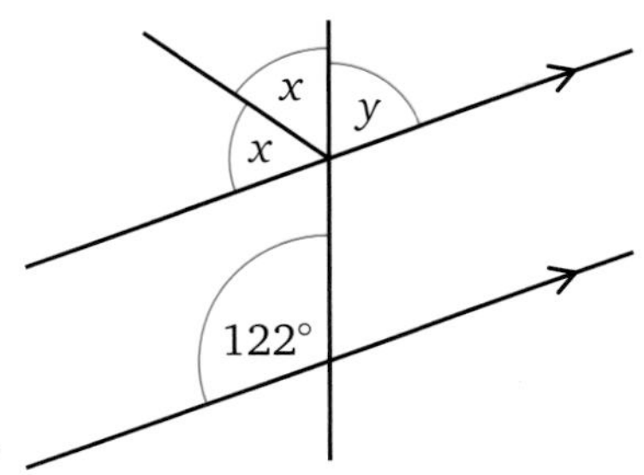

2.2 Bearings

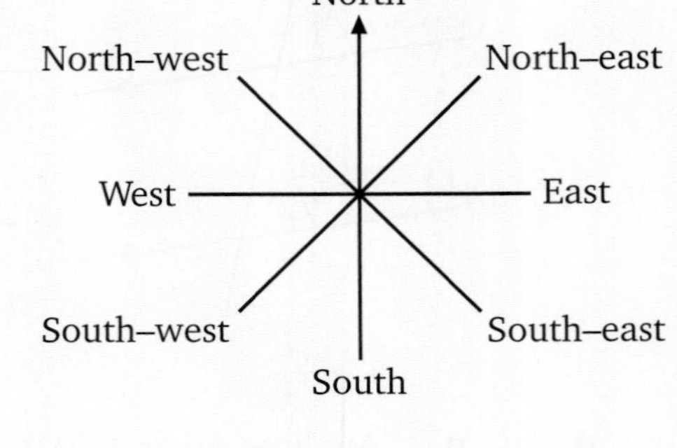

Directions can be described using the points of the compass such as south, north-west and so on.

Directions can also be described using **three-figure bearings.**

A three-figure bearing gives the angle measured in a clockwise direction from the north line.

Angles of less than 100° need a zero placed in front to make them three figures.

So south has a bearing of 180°, east has a bearing of 090° and west has a bearing of 270°.

Other directions can all be described using bearings.

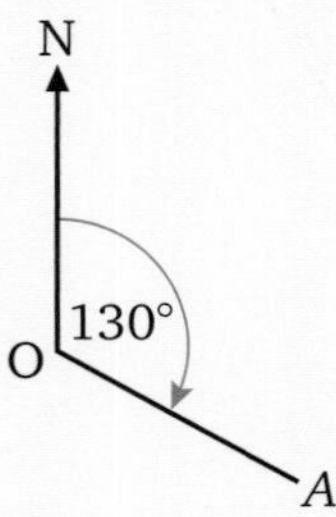

Bearing of 130°

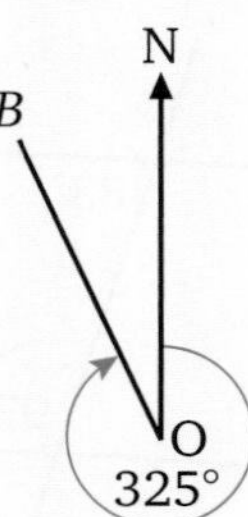

Bearing of 325°

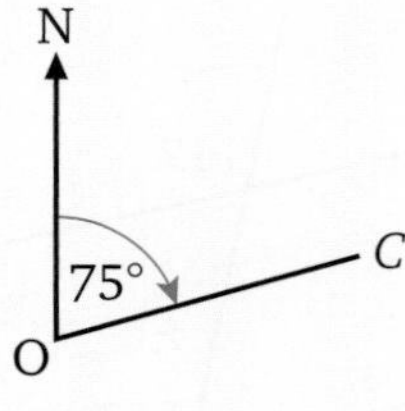

Bearing of 075°

Bearings **from** one place to another can be found by measurement or by calculation.

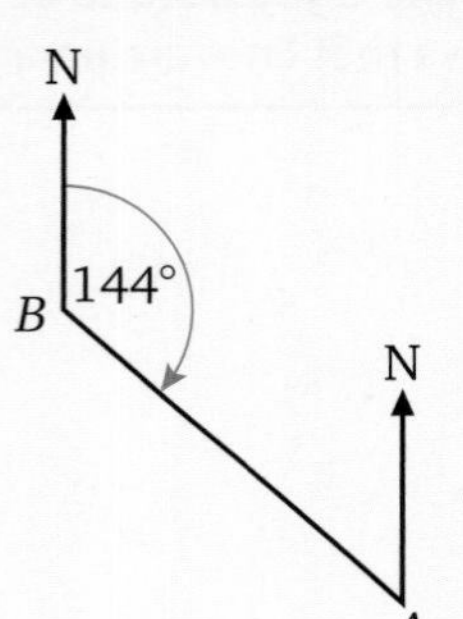

The bearing of A **from** B is 144°.

N P N 55° Q

The bearing of P **from** Q is 055°.

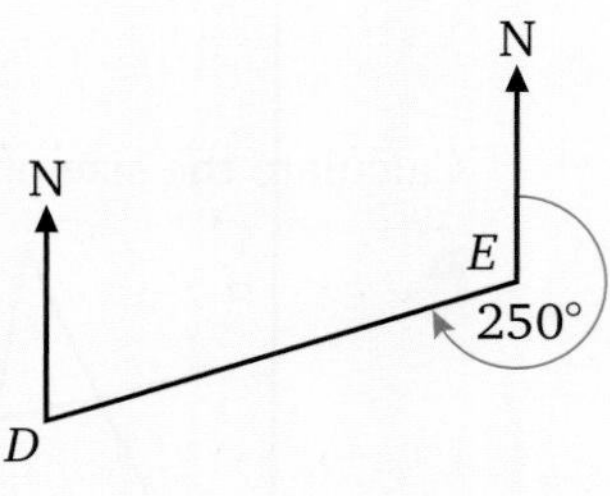

The bearing of D **from** E is 250°.

Example: Branton is on a bearing of 105° from Averby.

Work out the bearing of Averby from Branton.

Solution: Draw a sketch.

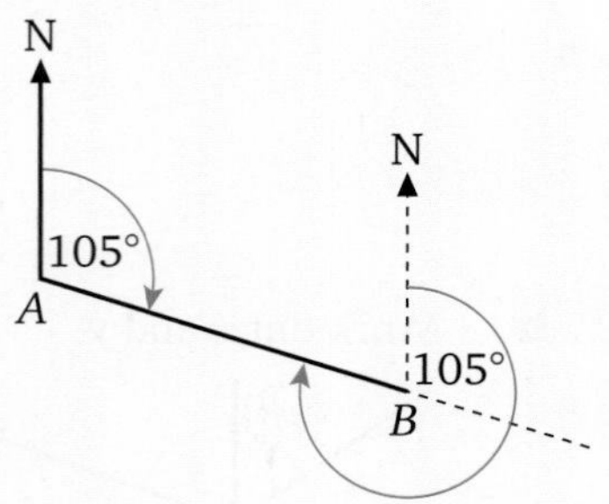

AQA Examiner's tip

Remember to put the north line at the place you are working the bearing out '**from**'.

Face towards Averby **from** Branton, so the north line is drawn at Branton, then the angle is measured clockwise from north around to the line joining B to A (as shown by the arrow). The angle measured clockwise from north is 105° + 180° = 285°.

So the bearing of Averby **from** Branton is 285°.

Practise... 2.2 Bearings

D

1 For each diagram, write down the three-figure bearing of D from E.

a

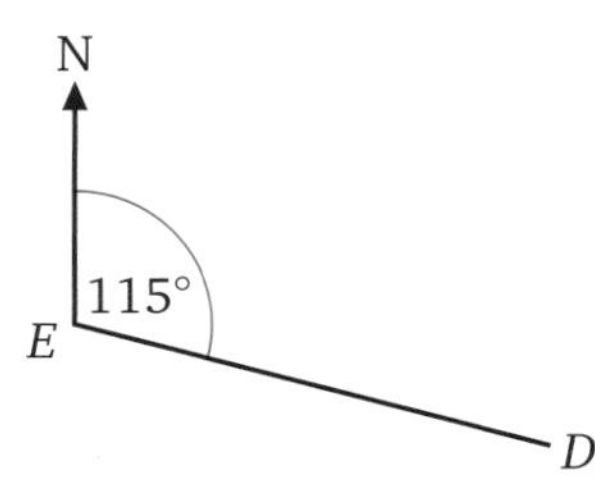

c

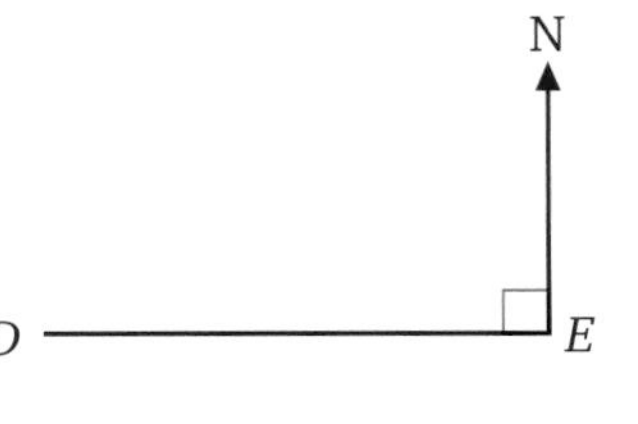

e

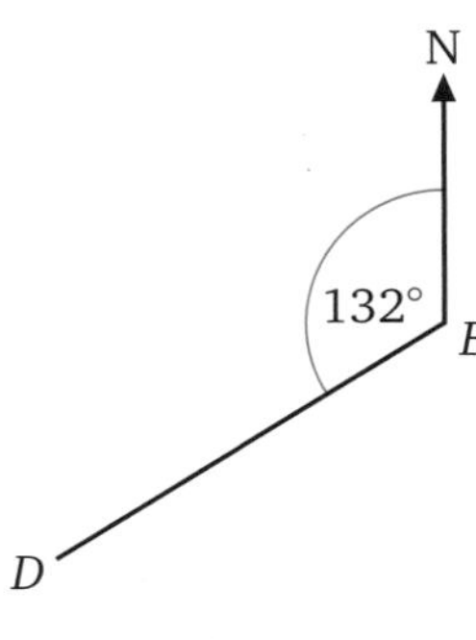

b

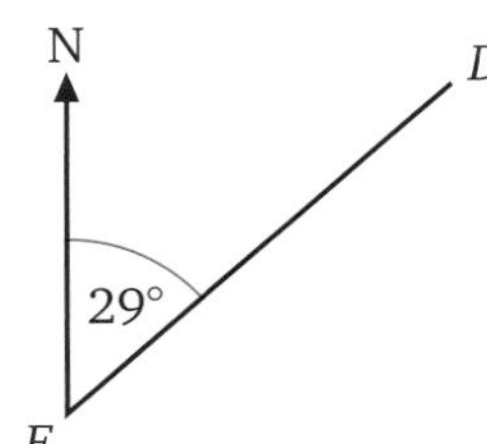

d

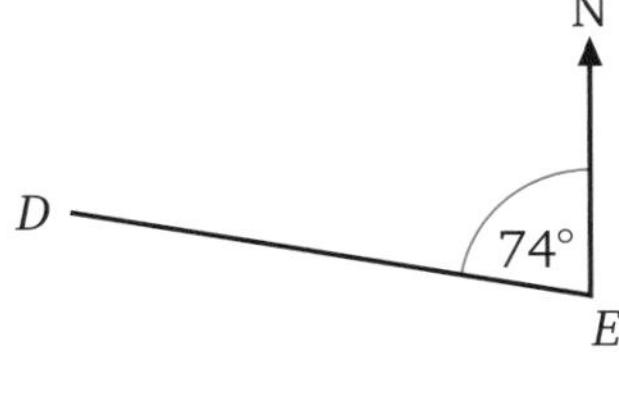

f

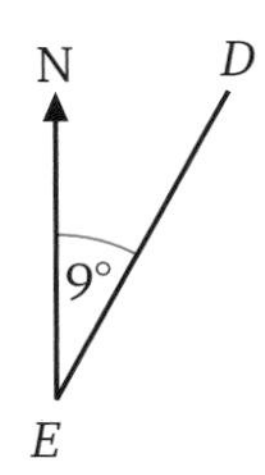

2 Use a protractor to draw accurate diagrams to represent these bearings.

a 140° **c** 210° **e** 85° **g** 163°

b 045° **d** 320° **f** 108° **h** 258°

3 Thatham is on a bearing of 78° from Benton.

Work out the bearing of Benton from Thatham.
Use a sketch to help you.

4 Newby is on a bearing of 250° from Reddington.

Work out the bearing of Reddington from Newby.

5 Jake leaves home and cycles for 6 kilometres on a bearing of 070°.
He then cycles for 5 kilometres on a bearing of 140°.

Make an accurate drawing of his route and use it to find the bearing and distance he needs to travel to return directly to his home.

6 Here is a map of an island.

P is a port. H_1 and H_2 are hotels.
T is a town and B is a beach.

a Which hotel is on a bearing of 055° from the port?

b Measure and write down the bearing of the beach from the port.

c Measure and write down the bearing of the port from the town.

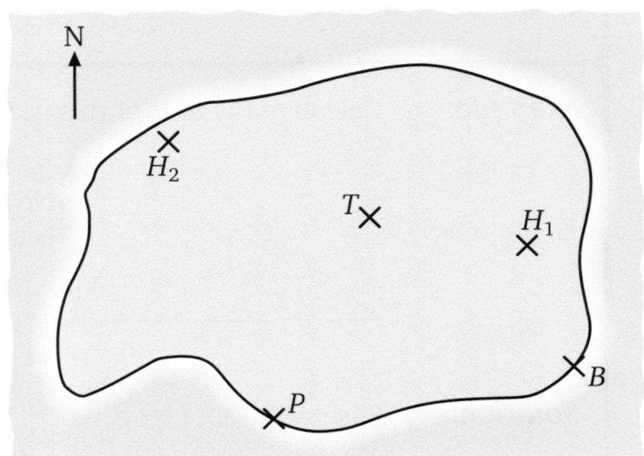

7 Bridgetown is 80 miles north and 50 miles east of Archway.

By making an accurate scale drawing, calculate the bearing of Archway from Bridgetown.

Learn... 2.3 Area of parallelograms, trapeziums and triangles

Here is a parallelogram drawn on squared paper.

If the shaded triangle is cut from one end of the parallelogram and put onto the other end, the shape becomes a rectangle.

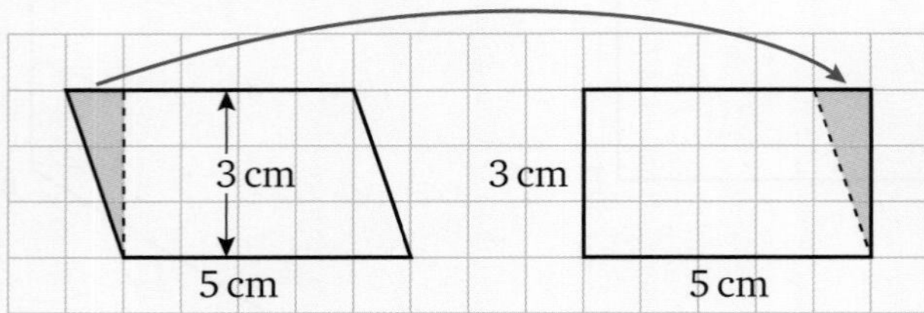

Not drawn accurately

The area of the parallelogram is the same as the area of the rectangle.

The **base** of the parallelogram is the same as the length of the rectangle.

The **perpendicular height** (at right angles to the base) of the parallelogram is the same as the width of the rectangle.

Area of rectangle = length × width = $5 \times 3 = 15\text{ cm}^2$

Area of parallelogram = $5 \times 3 = 15\text{ cm}^2$

The area of any parallelogram can be calculated using the formula:

area = base × perpendicular height

> **AQA Examiner's tip**
> Be careful that you do not use the sloped height. Examiners often put three measurements on a parallelogram.

A diagonal has now been drawn on the parallelogram.

The diagonal divides the parallelogram into two equal triangles.

The area of each triangle is half the area of the parallelogram.

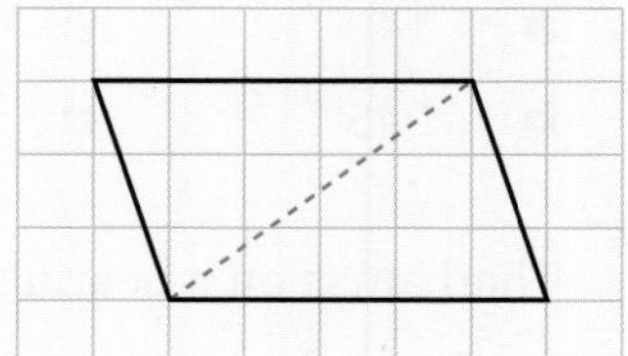

This gives the formula for the area of a **triangle**:

area = $\frac{1}{2}$ × base × perpendicular height

The shape in the example below is a **trapezium**.

The area of a trapezium can also be found by using the formula:

area = $\frac{1}{2}(a + b)h$

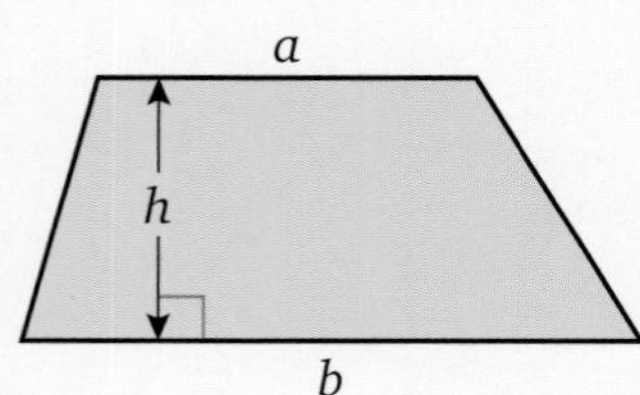

Example: Calculate the area of this triangle.

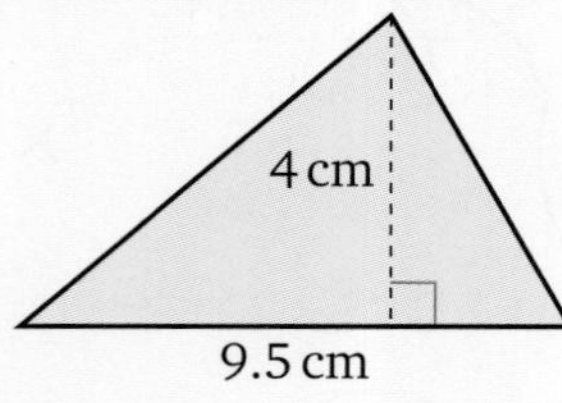

Not drawn accurately

Solution: Base = 9.5 cm

Perpendicular height = 4 cm

Area = $\frac{1}{2}$ × base × perpendicular height

Area = $\frac{1}{2} \times 9.5 \times 4 = 19\text{ cm}^2$

Example: A parallelogram has an area of 51 cm². The perpendicular height of the parallelogram is 3 cm. Work out the base of the parallelogram.

Solution:

Area = base × perpendicular height

$51 = \text{base} \times 3$

$\frac{51}{3} = \text{base}$ Divide each side by 3

$17 = \text{base}$

The base of the parallelogram is 17 cm.

Example: For this trapezium work out:

a the perimeter **b** the area.

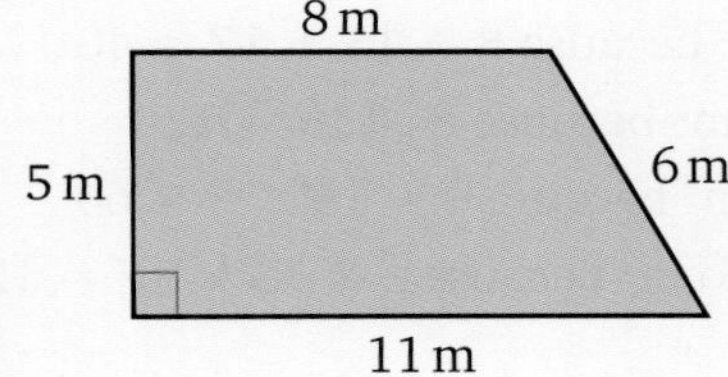

Not drawn accurately

Solution:

Perimeter = 8 + 6 + 11 + 5 = 30 m

Area $= \frac{1}{2}(8 + 11) \times 5 = \frac{1}{2} \times 19 \times 5 = 47.5\,\text{m}^2$

Practise... 2.3 Area of parallelograms, trapeziums and triangles

D C B A A*

The shapes in this exercise are not drawn accurately.

D

1 Work out the area of each parallelogram.

a

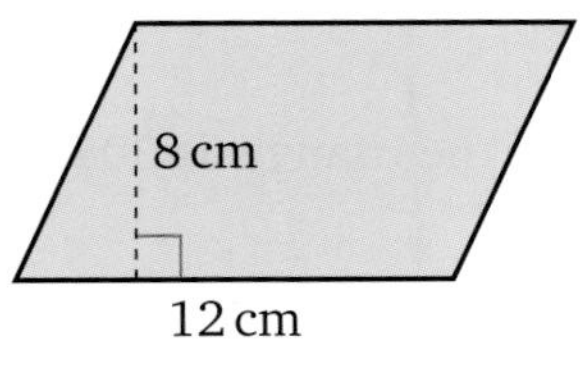

c

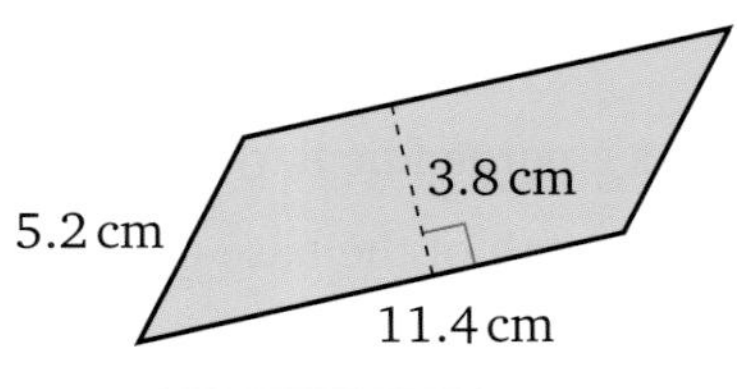

b

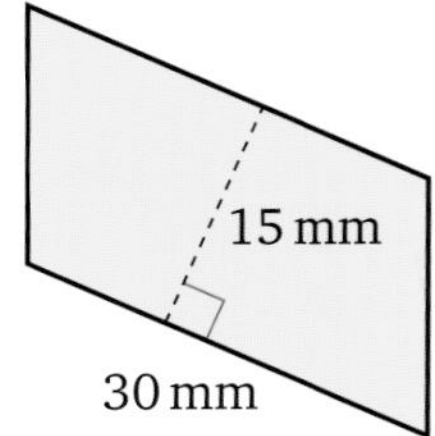

d

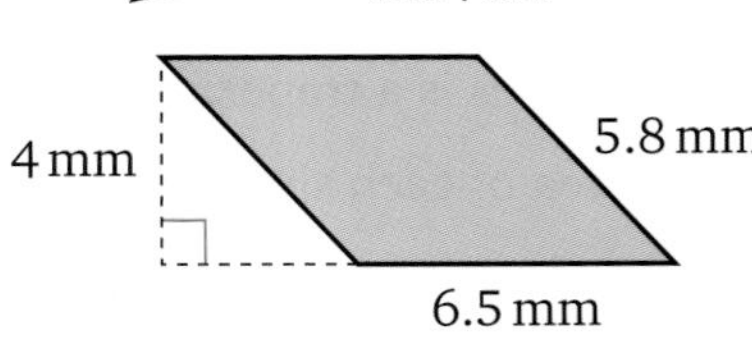

2 Work out the area of each triangle.

a

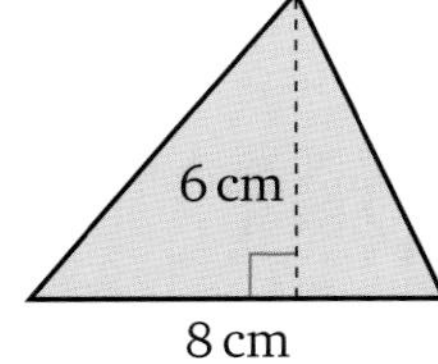

b

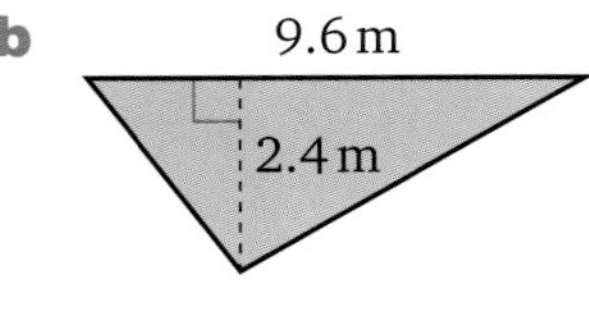

c

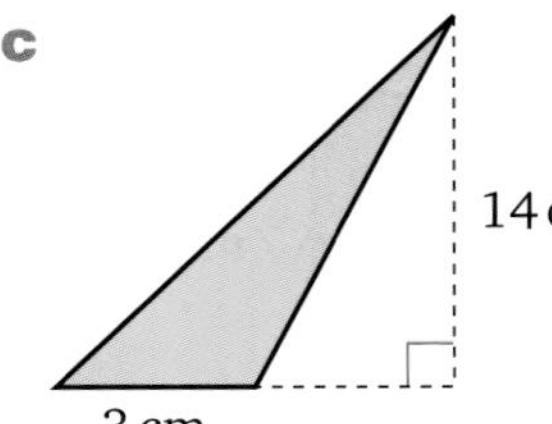

d

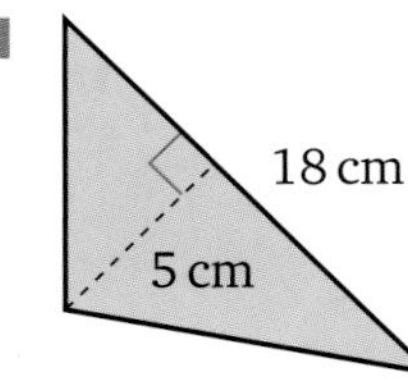

D

3 Which of the following shapes has the largest area?
Show your working.

a

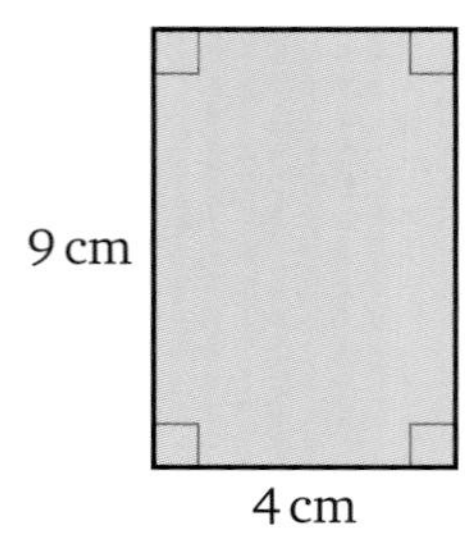

b

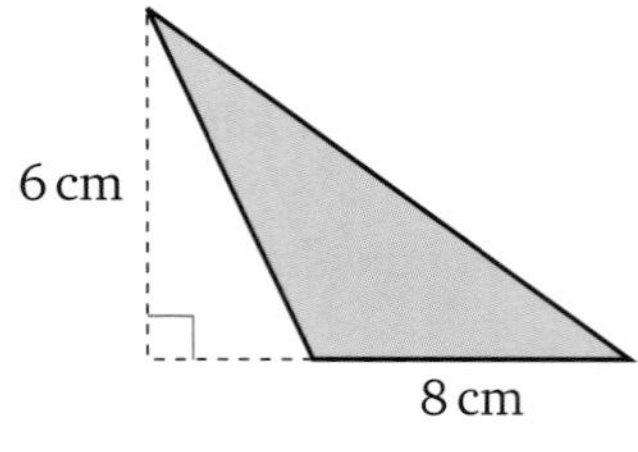

c

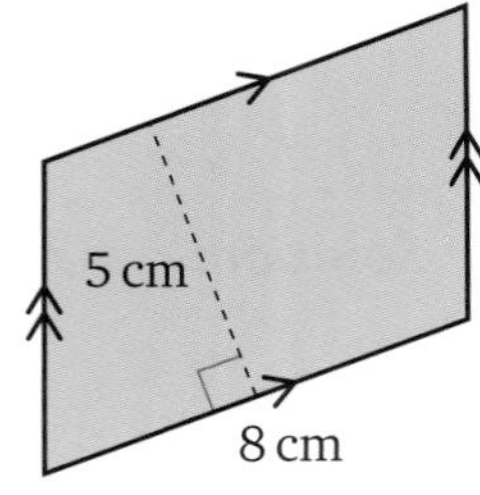

4 A parallelogram of base 7.2 cm has an area of 97.2 cm^2.

Work out the height of the parallelogram.

5 Four students are trying to find the area of this triangle.

Javed thinks the answer is 40 cm^2 because $8 + 15 + 17 = 40$

Kieran thinks the answer is 120 cm^2 because $8 \times 15 = 120$

Leanne thinks the answer is 60 cm^2 because $\frac{1}{2} \times 15 \times 8 = 60$

Megan thinks the answer is 127.5 cm^2 because $\frac{1}{2} \times 15 \times 17 = 127.5$

8 cm
17 cm
15 cm

Who is correct?

What mistakes have the other students made?

6 For these parallelograms, work out:

a the perimeter **b** the area.

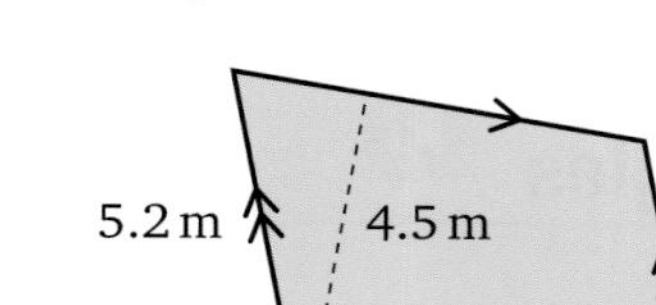

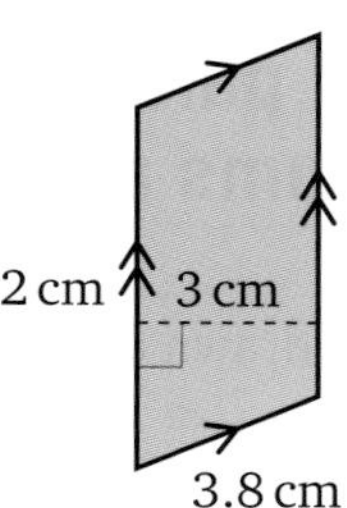

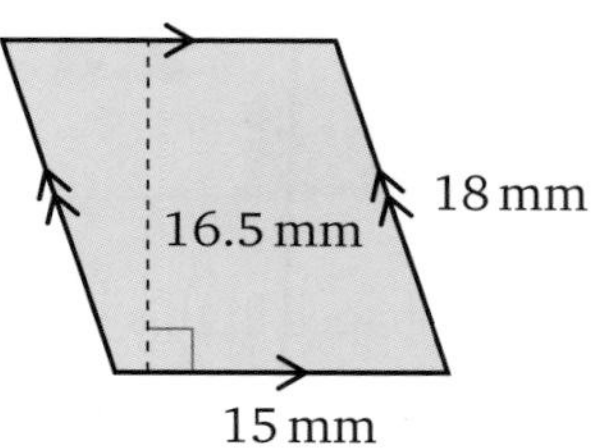

7 A triangle has a perpendicular height of 22 cm. The area of the triangle is 308 cm^2.

Work out the length of the base of the triangle.

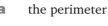

8 Each of these shapes is a trapezium.

Work out the area of each shape.

a

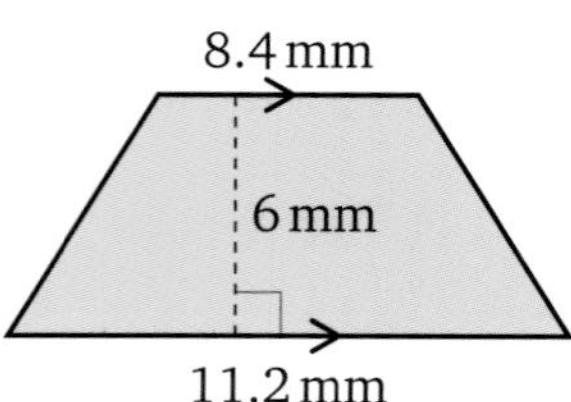

c

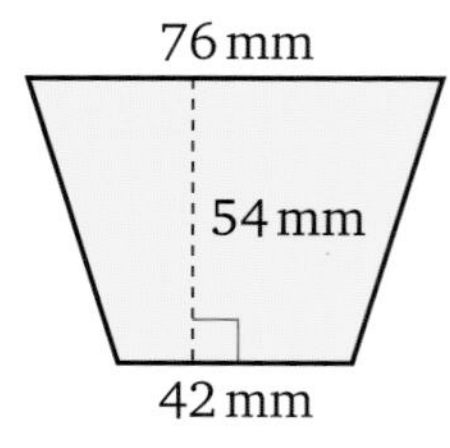

b

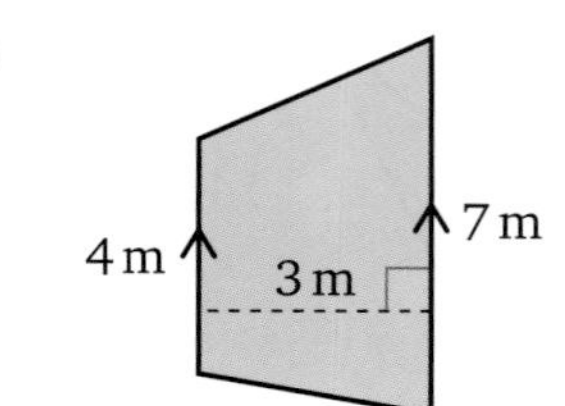

d

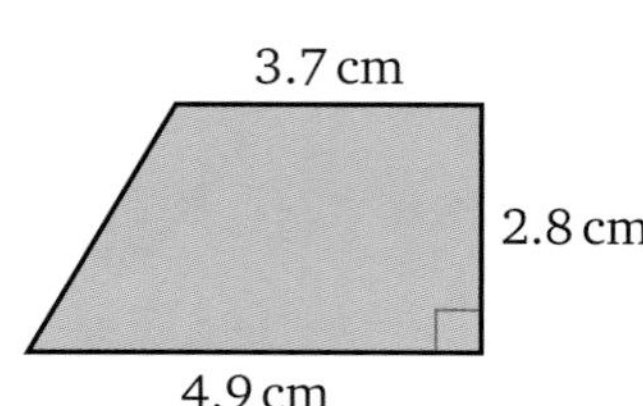

C

9 Work out the area of each of these shapes.

a

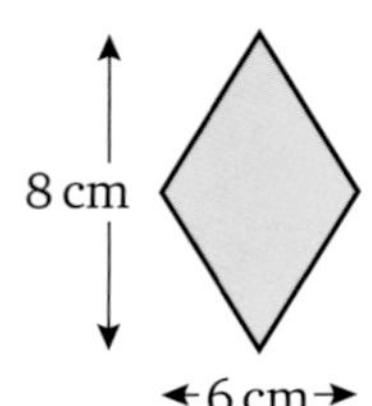

b

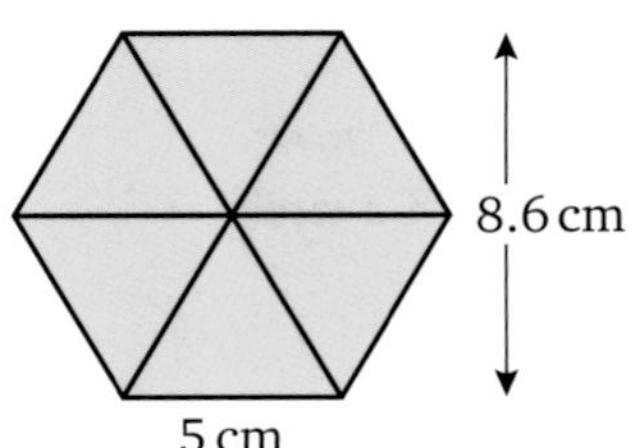

10 The triangle has the same area as the parallelogram. Work out h.

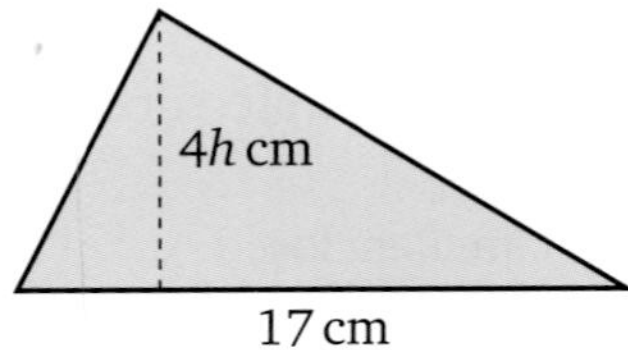

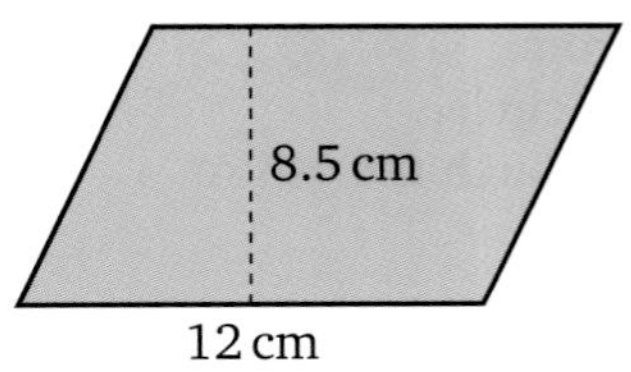

Bump up your grade

To get a Grade C you need to be able to find the height or base of a shape given the area.

11 A club logo is to be made out of metal.

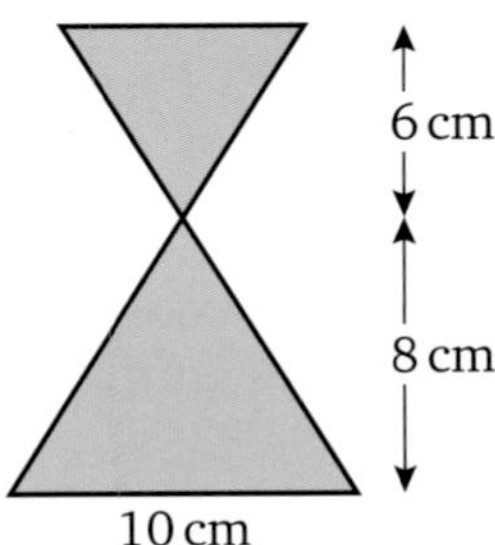

Work out the dimensions of the smallest rectangle that can be used to cut out a logo.

What percentage of metal is wasted?

12 A company makes kites. They cut triangles from yellow and blue silk as shown. The yellow silk costs £5 per square metre and the blue silk costs £7 per square metre. Find the cost of material used for each kite. Assume that the triangles for several kites can be cut from the material without wastage.

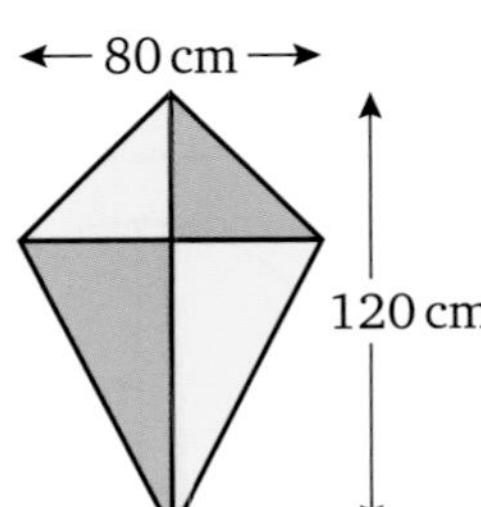

13 This trapezium is twice the area of the triangle. Work out x.

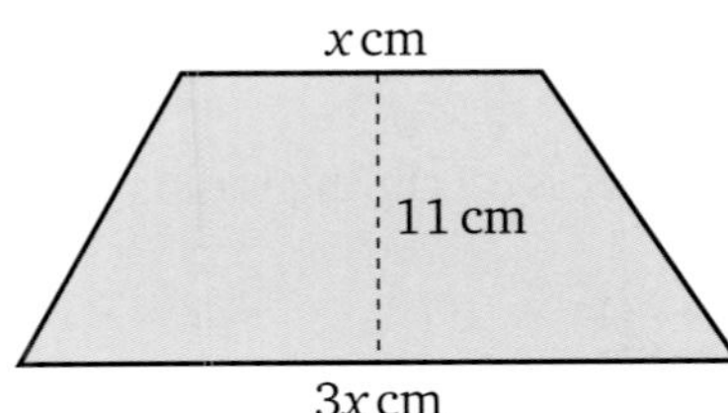

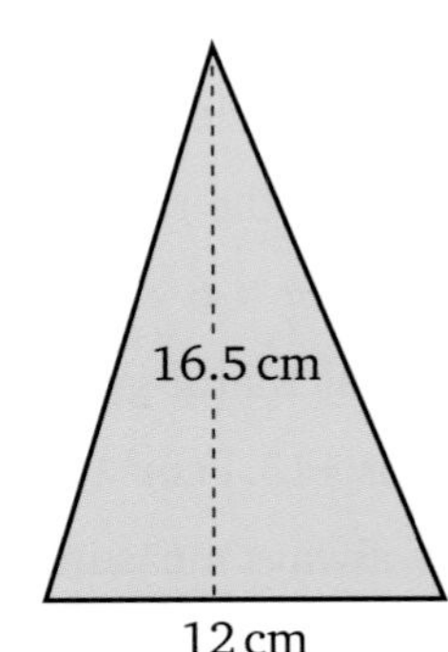

2.4 Circumference and area of a circle

Parts of a circle

Remember the following parts of a circle.

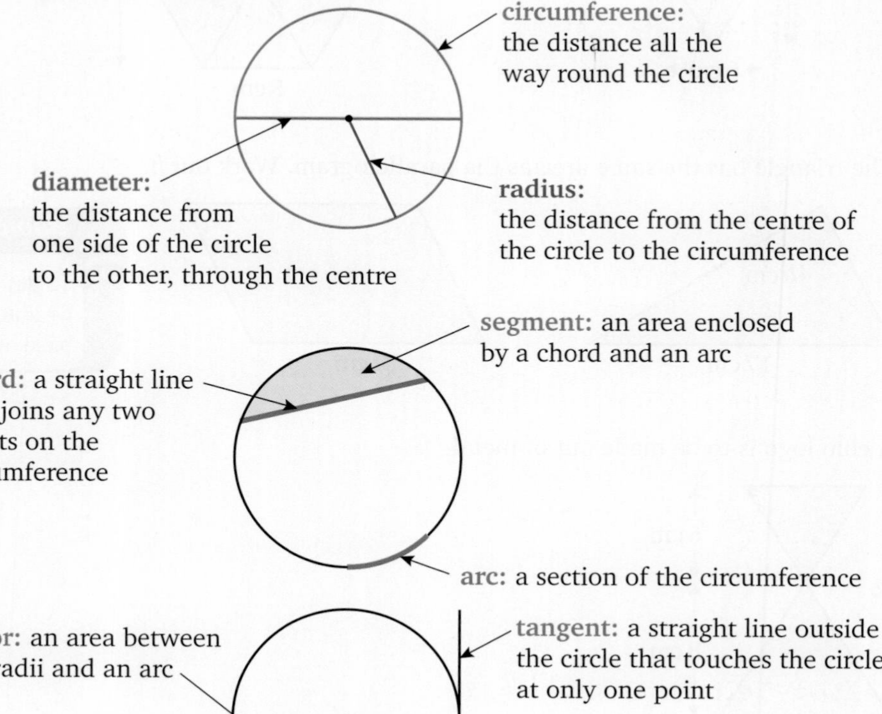
circumference:
the distance all the way round the circle
diameter:
the distance from one side of the circle to the other, through the centre
radius:
the distance from the centre of the circle to the circumference
segment: an area enclosed by a chord and an arc
chord: a straight line that joins any two points on the circumference
arc: a section of the circumference
sector: an area between two radii and an arc
tangent: a straight line outside the circle that touches the circle at only one point

Circumference of a circle

The circumference, C, of a circle can be calculated using $C = \pi d$

where d is the diameter of the circle and π is the Greek letter pi, and represents a value of 3.1415... (Look for the π button on your calculator.)

Area of a circle

The area, A, of a circle can be calculated using the formula $A = \pi r^2$ where r is the radius of the circle.

Example: Find the circumference and area of a circle of diameter 12 cm.

Give your answers to one decimal place.

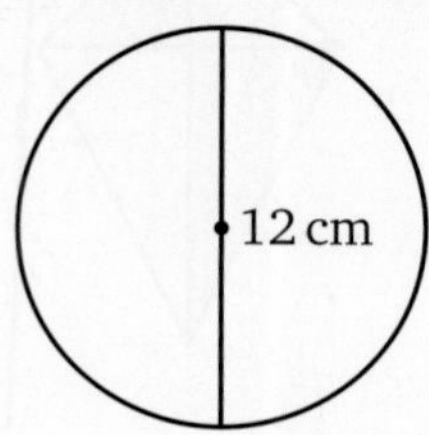
12 cm

Solution: $C = \pi d$ so $C = \pi \times 12$

Using the π button on a calculator gives $C = 37.69911...$

So the circumference of the circle is 37.7 cm (to 1 d.p.).

To find the area we must use the radius. If the diameter is 12 cm the radius is $\frac{12}{2} = 6$ cm

$A = \pi r^2$ so $A = \pi \times 6^2$ or $\pi \times 6 \times 6$

Using the π button on a calculator gives $A = 113.0973...$

So the area of the circle is 113.1 cm^2 (to 1 d.p.).

(Remember that area is always measured in square units.)

Example: Find the diameter of a circle of circumference 22.3 cm.

Give your answer to one decimal place.

Solution: $C = \pi d$ so $22.3 = \pi \times d$

Divide both sides by π.

$\frac{22.3}{\pi} = 7.0983...$

The diameter is 7.1 cm (to 1 d.p.).

AQA Examiner's tip

Always state the units of your answer and remember that area is always measured in square units.

Practise... 2.4 Circumference and area of a circle

D C B A A*

D

1 Calculate the circumference of each circle.
Give each answer to one decimal place.

a

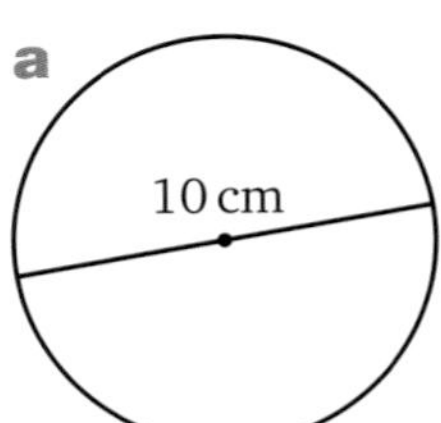

b

15 mm

c

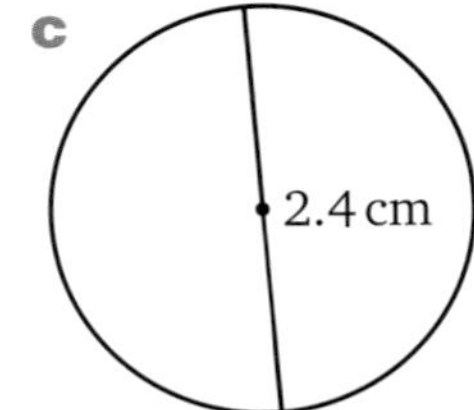

d

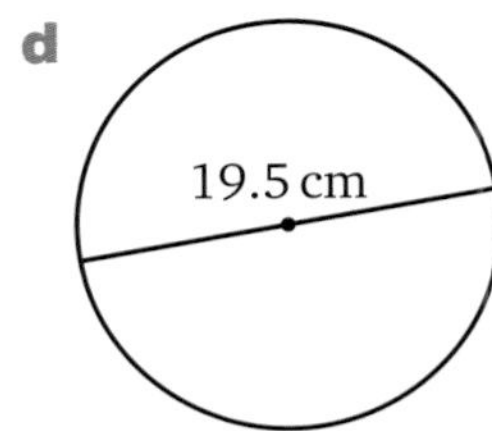

2 Calculate the circumference of each circle.
Give each answer to one decimal place.

a

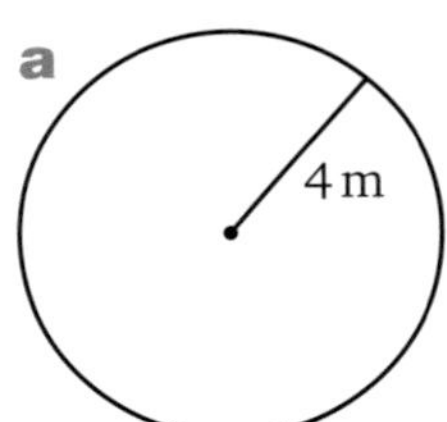

b

32 mm

c

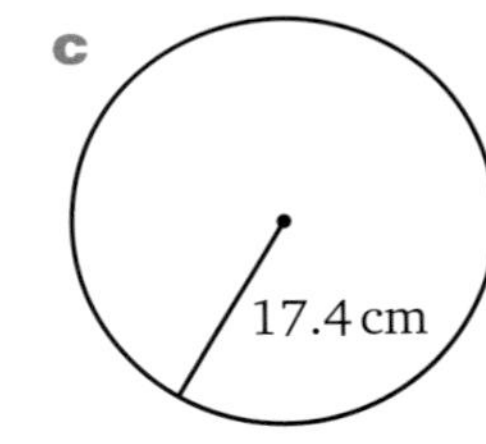

d

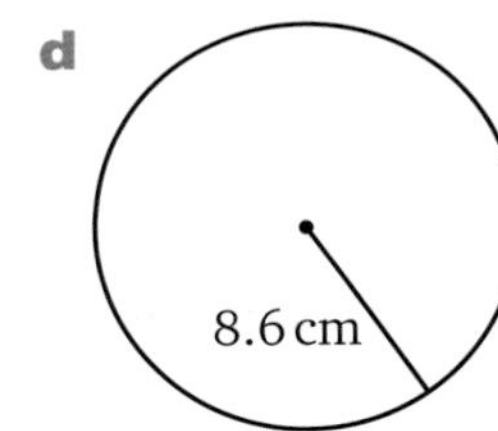

3 A circle has a circumference of 62.8 cm.

Work out the diameter of this circle.
Give your answer to the nearest whole number.

4 Calculate the area of each circle.
Give each answer to two decimal places.

a

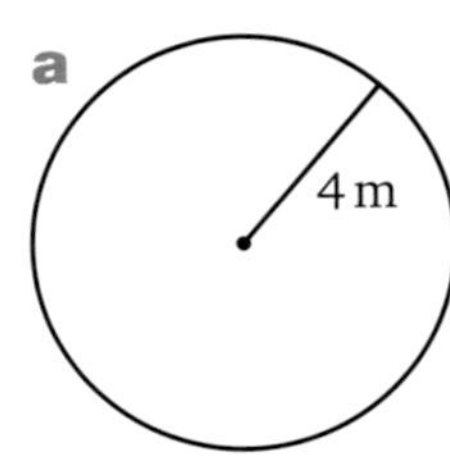

b

32 mm

c

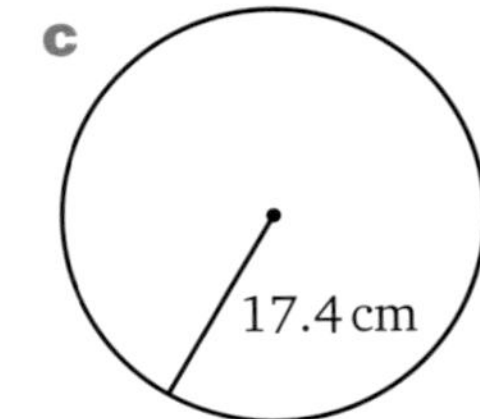

d

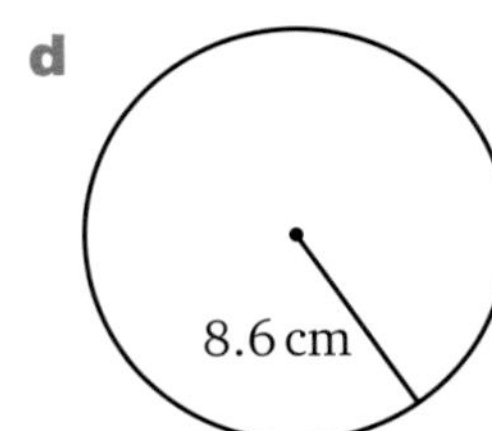

5 Calculate the area of each circle.
Give each answer to two decimal places.

a

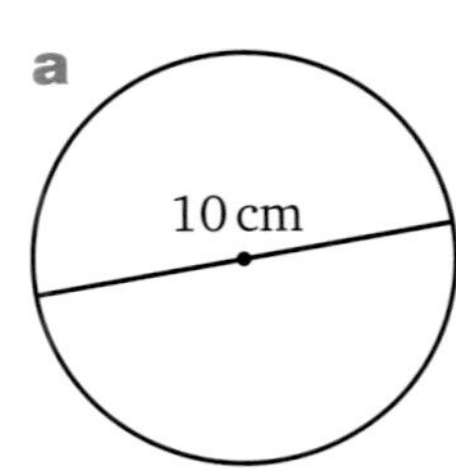

b

15 mm

c

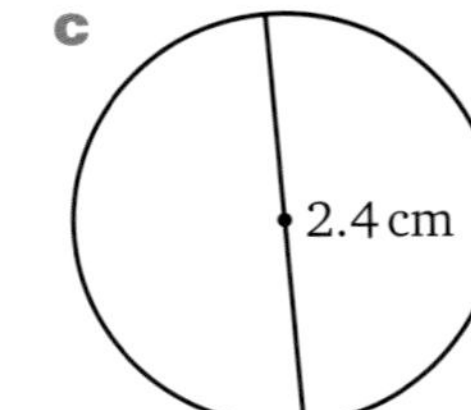

d

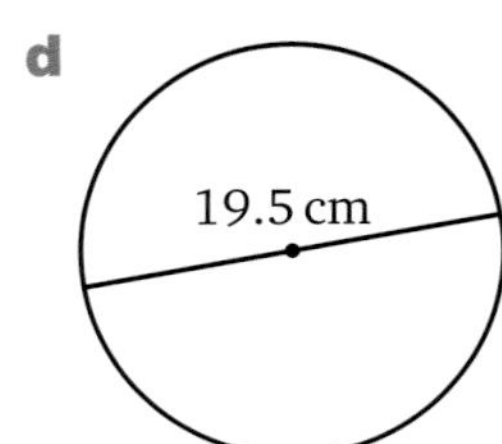

C

6 Find the perimeter and area of each semicircle.
Give each answer to one decimal place.

a

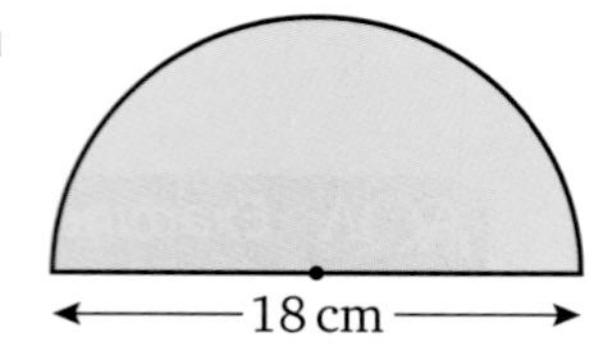

b

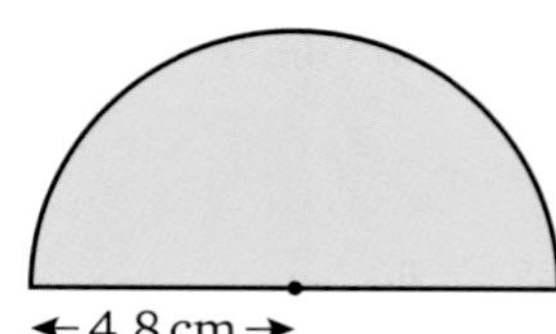

7 A semicircular rug has a diameter of 75 cm.
It is edged with fringing.

Calculate the length of fringing needed to go all the way round the rug.

Bump up your grade

To get a Grade C you need to be able to find areas and perimeters of semicircles.

8 A circle has an area of 201 mm^2.

Calculate the radius of the circle giving your answer to the nearest whole number.

9 Calculate the perimeter and area of each of these shapes.

a

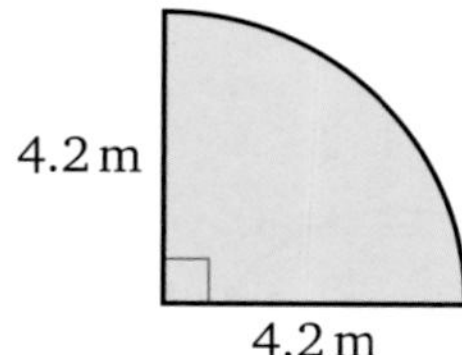

b

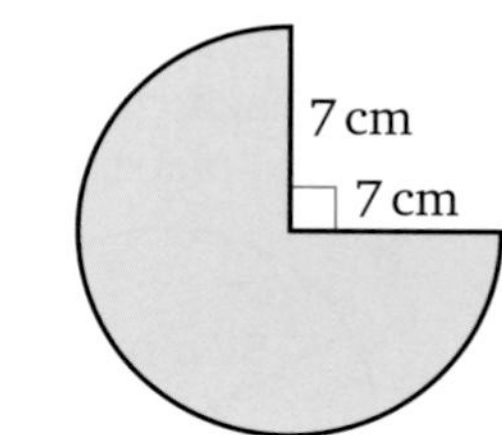

Not drawn accurately

10 Teri is training for a fun run.
She needs to run 10 000 m each week during her training.
The diagram shows the running track where she trains.
Teri says 'If I run five times round this track each day from Monday to Friday I will have run more than 10 000 m in a week.'

Is she correct?
Show your working.

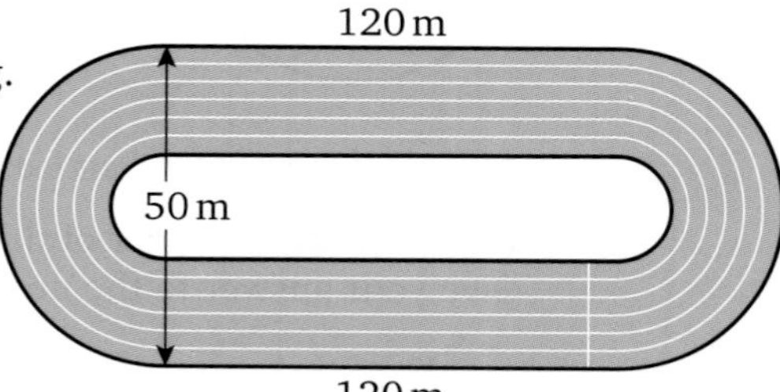

11 A circular cake tin has a diameter of 22.5 cm.
The lid is sealed with tape.

The ends of the tape overlap by 1.5 cm.

Calculate the length of tape needed to seal the tin.

? **12** The London Eye has a diameter of 135 m and takes approximately 30 minutes to complete one revolution.

How far does the base of a capsule travel every 5 minutes?

Learn... 2.5 Compound shapes

Many shapes are made up of triangles, semicircles and rectangles or squares. To find the area of more complicated or compound shapes divide the shape up into the shapes that you know about. Then find the area of the rectangle, then the triangle and so on, until you have found the areas of all the parts that make up the compound shape. The total area then equals the sum of all the smaller parts.

Example: Find the area of this shape.

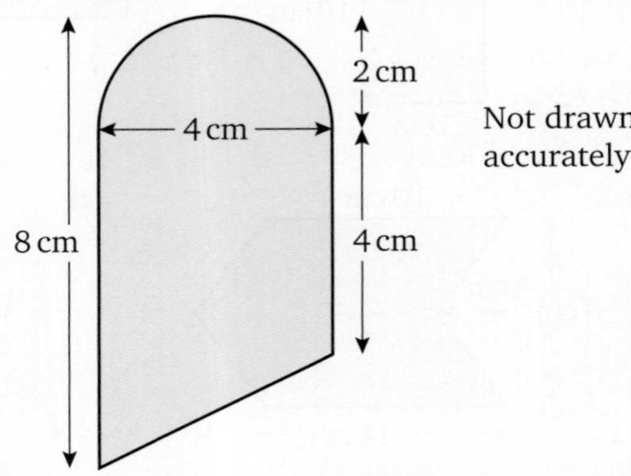

Solution: Divide the shape into a semicircle, a square and a triangle and work out the area of these smaller shapes.

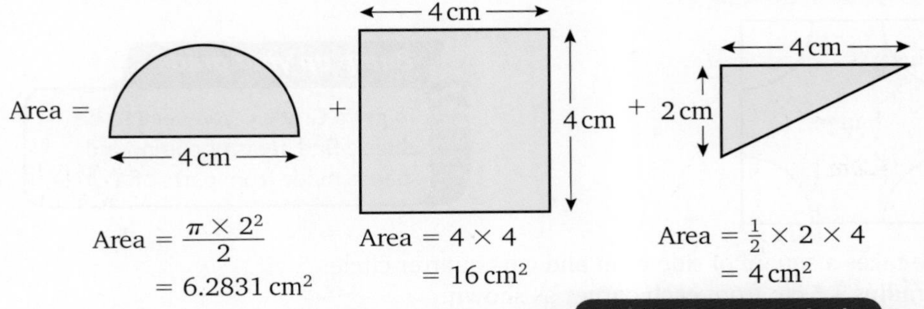

Area $= \frac{\pi \times 2^2}{2}$
$= 6.2831\ \text{cm}^2$

Area $= 4 \times 4$
$= 16\ \text{cm}^2$

Area $= \frac{1}{2} \times 2 \times 4$
$= 4\ \text{cm}^2$

Total area $= 4 + 16 + 6.2831 = 26.28\ \text{cm}^2$ to 2 d.p.

AQA Examiner's tip

Remember to state the units of your answer.

Practise... 2.5 Compound shapes

D C B A A*

The shapes in this exercise are not drawn accurately.

1 Work out the area of each shape.

a

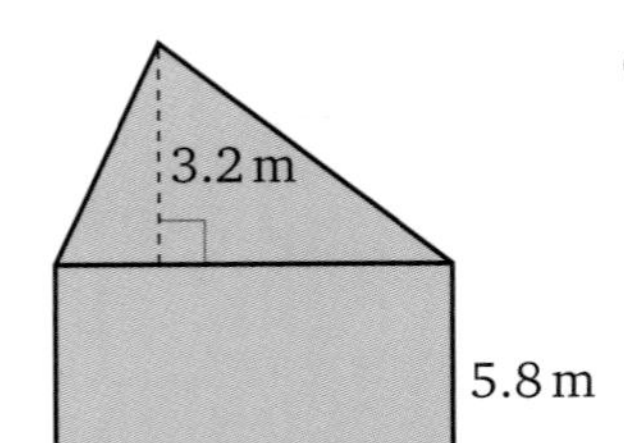

c

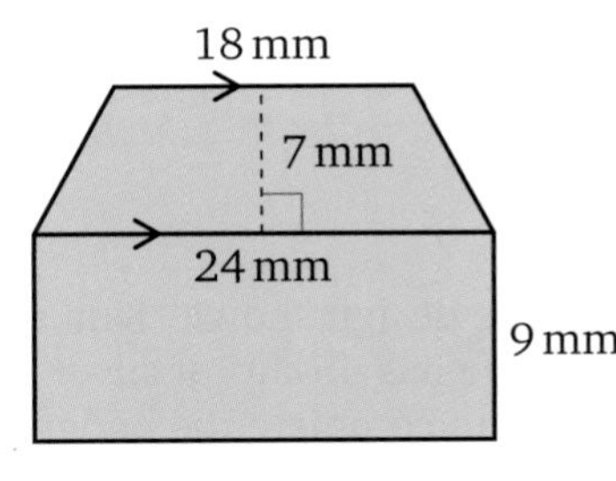

e

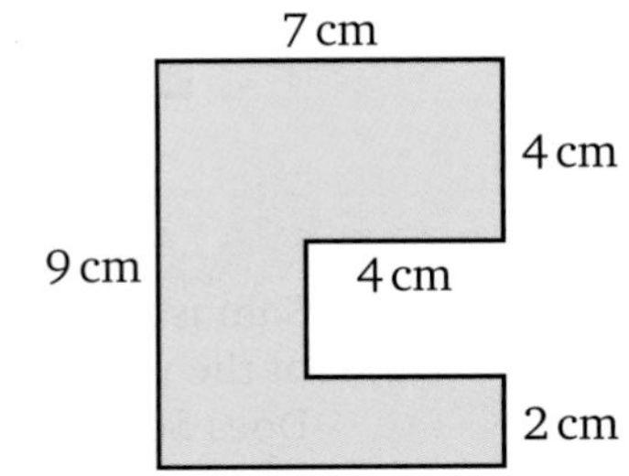

b

3.2 m
5.8 m
7.4 m

d

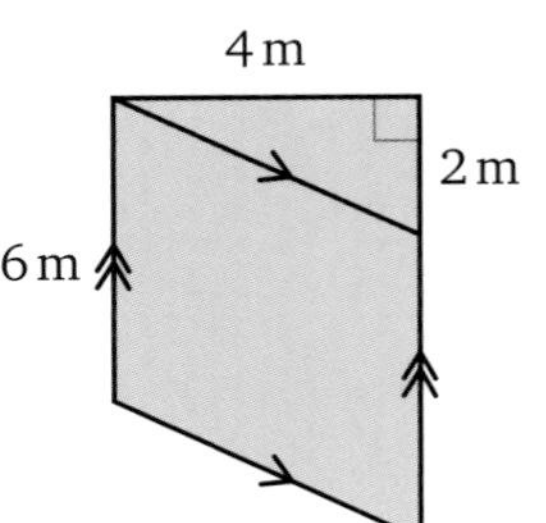

f

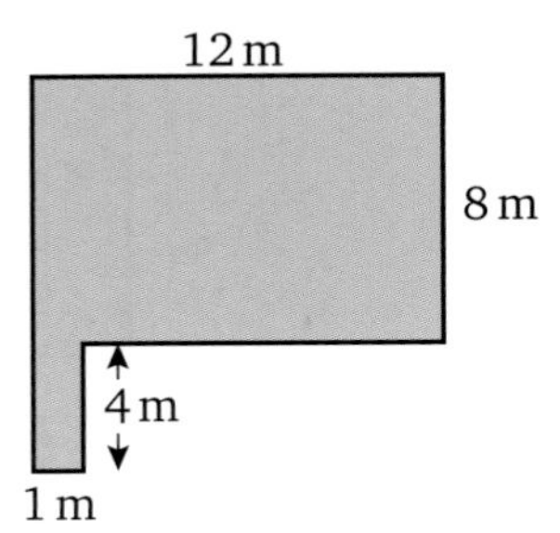

D C

D C

2 For each of these shapes find:

i the perimeter

ii the area.

a

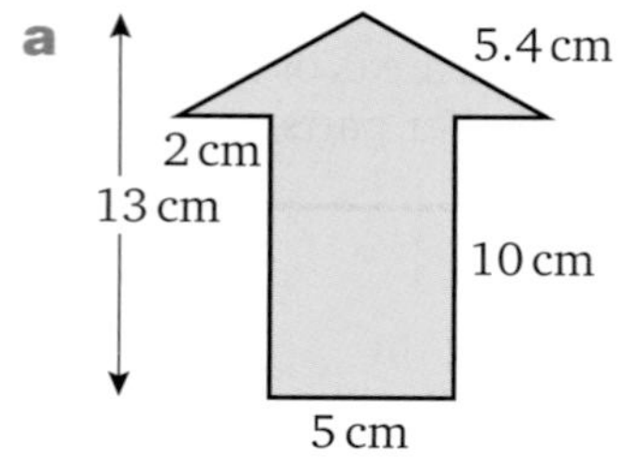

c

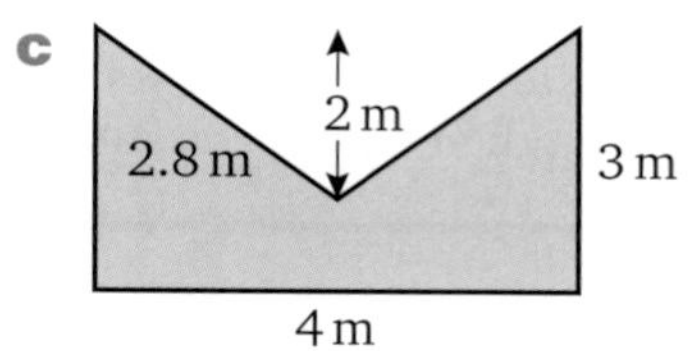

e

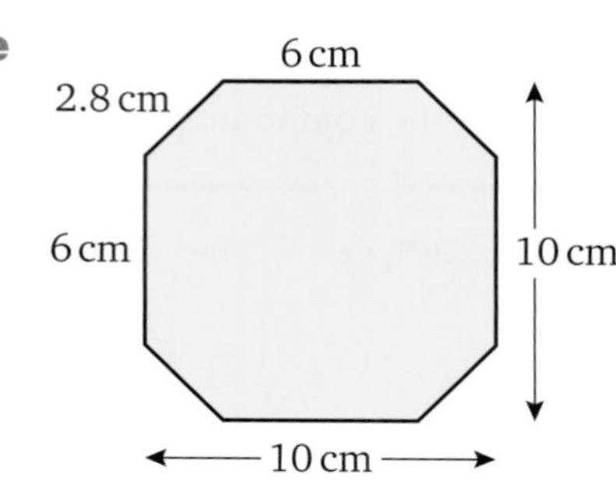

b

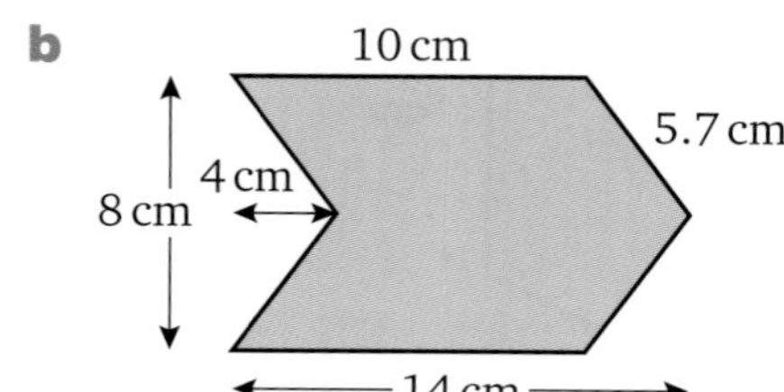

d

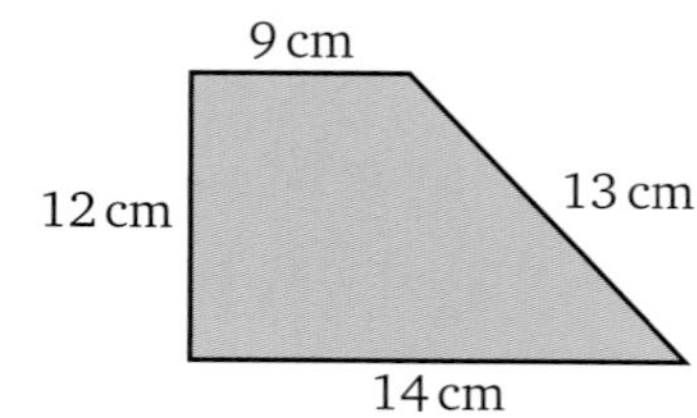

C

3 Alex is making badges out of card.

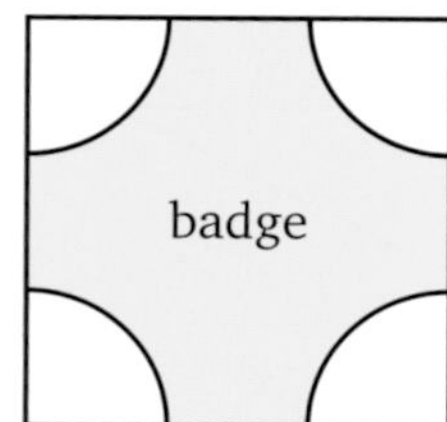

Bump up your grade

To get a Grade C you need to be able to find areas of compound shapes made from parts of a circle.

She takes a square of side 6 cm and cuts quarter circles of radius 1.5 cm from each corner as shown.
Calculate the area of the badge.
Give your answer to one decimal place.

4 Work out the area of the coloured part of each shape.

a

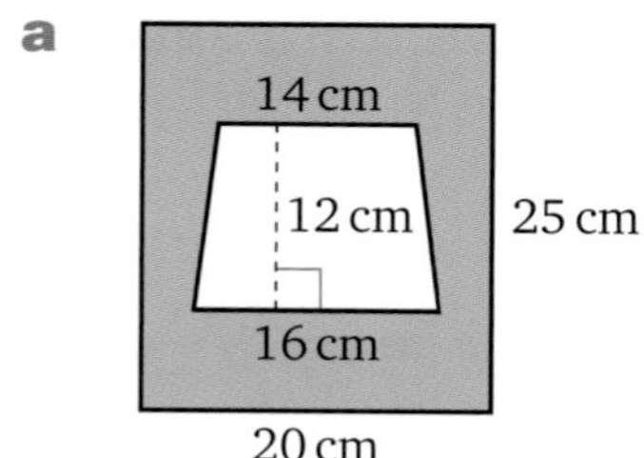

b

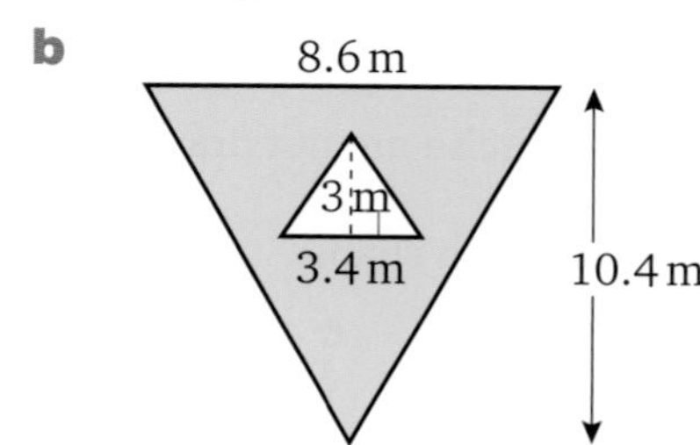

5 Sam is making badges to the design shown. Sam uses silver foil for the surround of the white centre shape. He has 45 cm^2 of silver foil left.
Does Sam have enough foil to complete the badge?
Show your working.

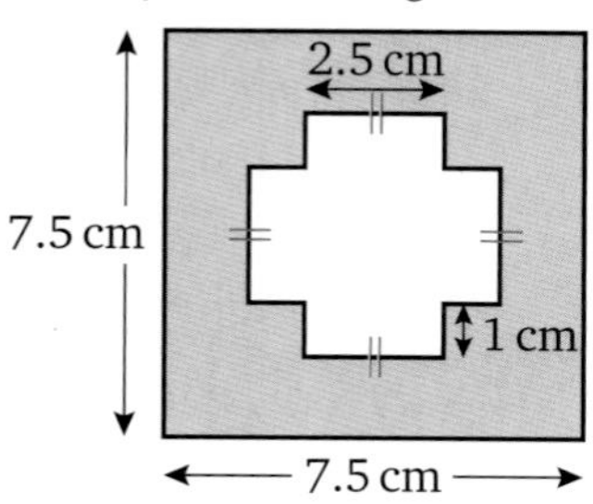

6 This is a sketch of Vikki's dining room.

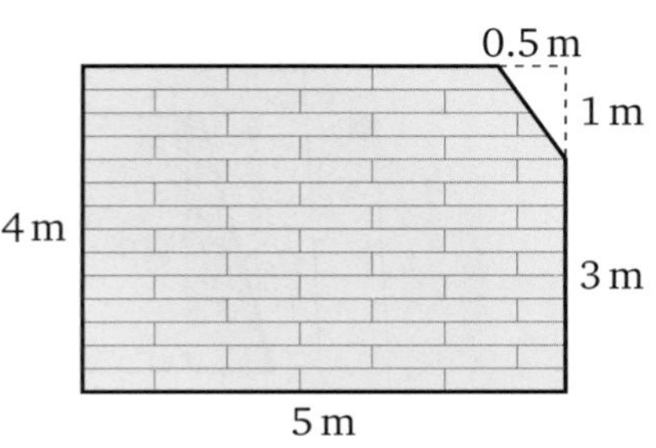

a Vikki wants to paint the floorboards which cover the whole of the floor.
One tin of floorboard paint will cover 9.5 m^2 of floor.
Vikki thinks that two tins will cover all her floor.

Is she correct?
You must show your working.

b Vikki fits new skirting board around all the room except the diagonal corner.
Skirting board is sold in 2.4 m lengths costing £7.99 each.

Work out the total cost of the skirting board.

2 Assess

D

1 Work out the size of each of the marked angles.
Give a reason for each answer.

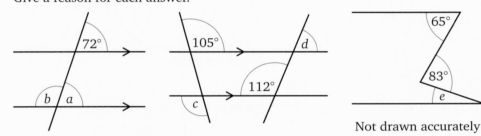

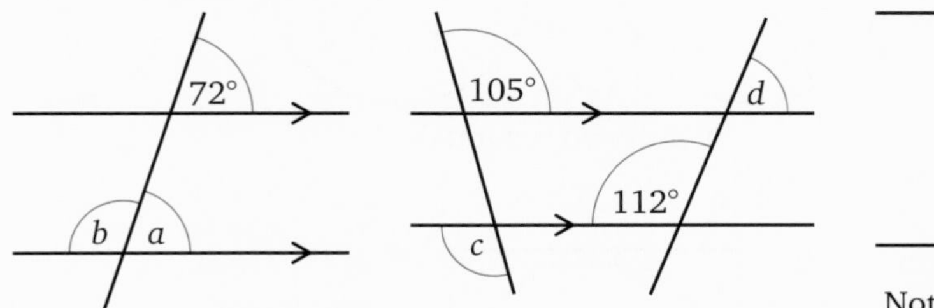

Not drawn accurately

2 For each diagram write down the three-figure bearing of P from Q.

a N, P, 64°, Q

b N, 168°, Q, P

c P, N, 46°, Q

Not drawn accurately

3 For each of the diagrams in Question 2, write down the bearing of Q from P.

4 The diagram shows the positions of three buoys A, B and C.

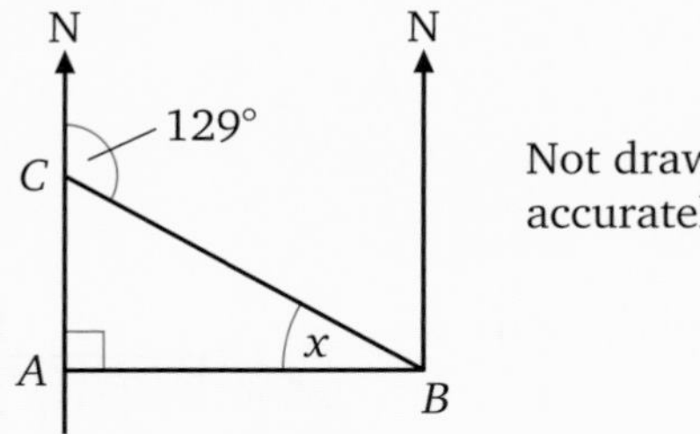

Not drawn accurately

B is due east of A and C is due north of A.

The bearing of B from C is 129°.

Work out angle x.

What is the bearing of C from B?

D

5 Find the area of each parallelogram.

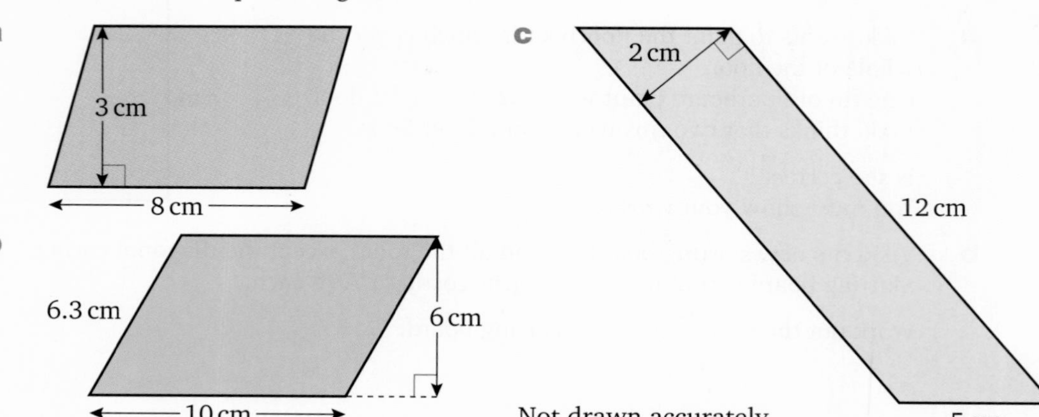

6 Find: **i** the circumference **ii** the area of each circle.

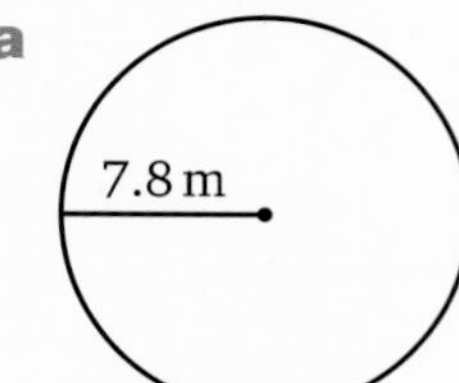

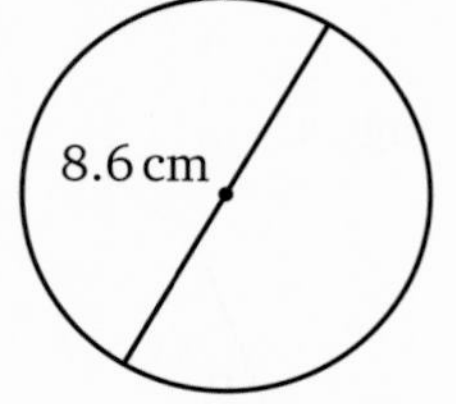

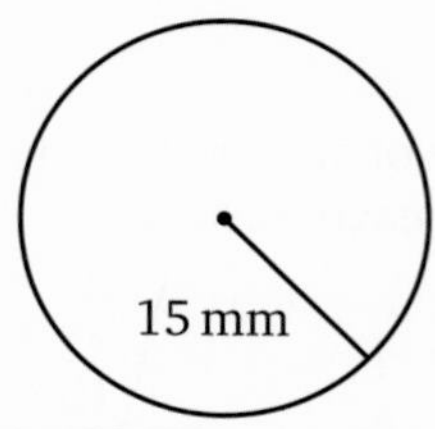

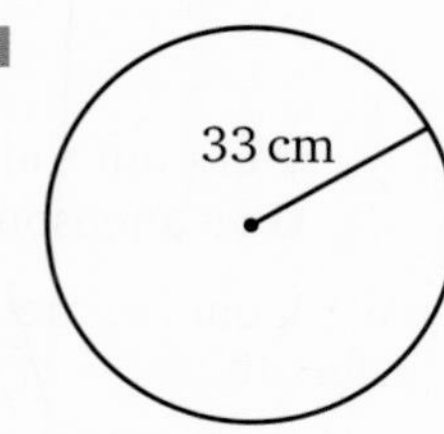

Not drawn accurately

C

7 A circle of diameter 45 cm is cut out of a square of side 50 cm.

Calculate the shaded area.

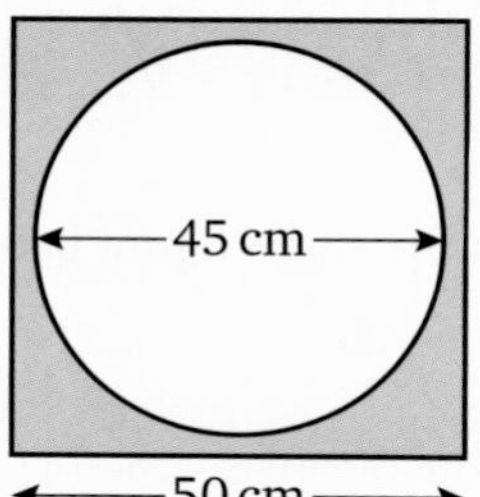

Not drawn accurately

8 Are lines *AB* and *CD* parallel? Give a reason for your answer.

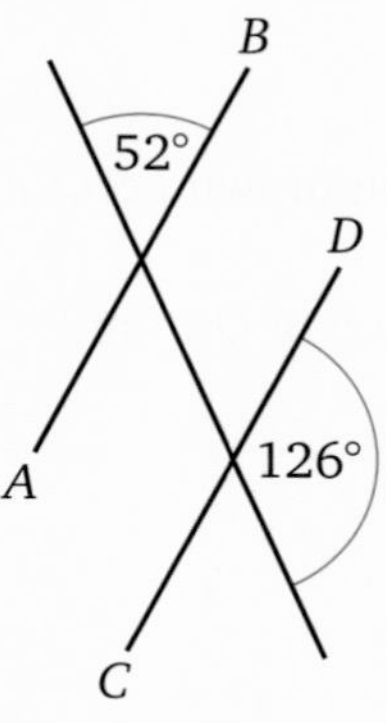

Not drawn accurately

9 The groundsman at the local sports field wants to spread lawn feed.
The field is the shape opposite.
One sack of lawn feed is enough to cover 100 m^2. He has bought 80 sacks.

Will he have enough lawn feed to cover the whole area?
Show your working.

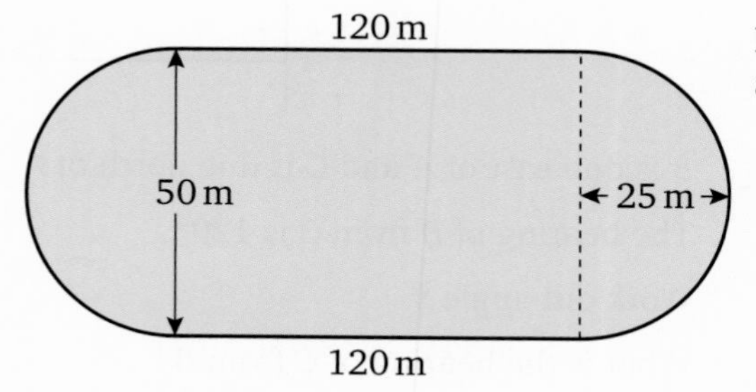

Not drawn accurately

AQA Examination-style questions

1 In the diagram AB is parallel to CD.

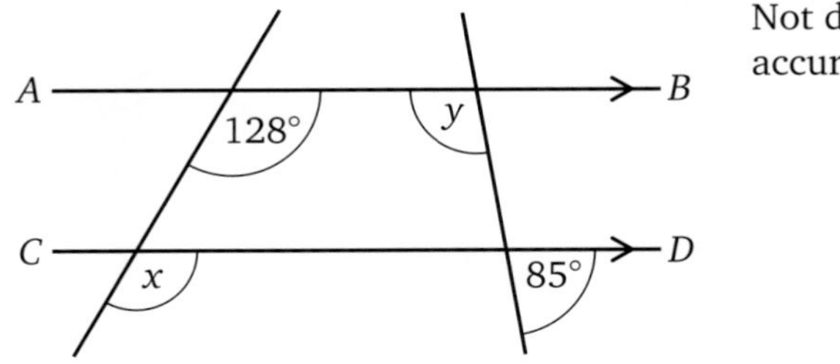

Not drawn accurately

a Write down the value of x. Give a reason for your answer. *(2 marks)*

b Work out the value of y. *(2 marks)*

AQA 2008

2 The radius of the semicircle is 10 cm.

Work out the area of the semicircle.
State the units of your answer.

Not drawn accurately

(3 marks)

AQA 2008

3 The shaded shape below is cut from a piece of rectangular card measuring 10 cm by 15 cm.

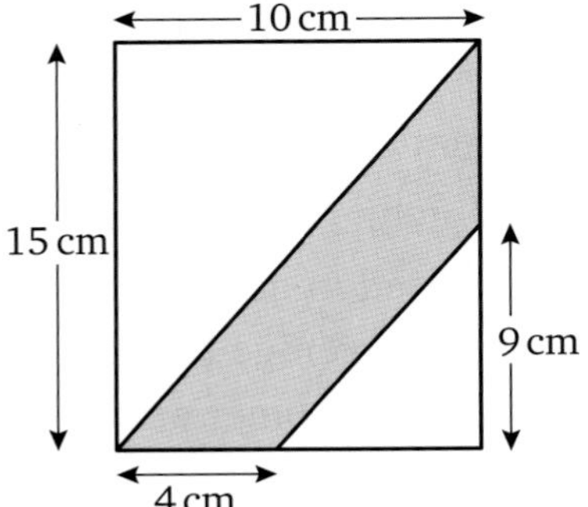

Not drawn accurately

a The larger right-angled triangle has two sides of 15 cm and 10 cm, as shown.
Show that the area of this triangle is 75 cm^2. *(1 mark)*

b Calculate the area of the shaded shape. State the units of your answer.
You must show all your working. *(4 marks)*

AQA 2008

3 Working with symbols

Objectives

Examiners would normally expect students who get these grades to be able to:

D

expand brackets in context such as $x(x + 2)$

factorise an expression such as $x^2 + 4x$

C

expand and simplify an expression such as $x(2x + 1) - x(2x - 3)$

B

expand and simplify two brackets in context such as $(x + 4)(x - 8)$

expand and simplify two brackets in context such as $(3x + 4)(2x - 8)$.

Did you know?

Queuing theory

Supermarkets use queuing theory to decide how many people they need to work at the checkout at any time.

There are mathematical formulae used in queuing theory.

This is Little's theorem: $N = \lambda T$

Little's theorem is used in queuing theory.

N stands for the average number of customers.

λ is the average customer arrival rate.

T is the average service time per customer.

Try testing out Little's theorem with real values to see if it makes sense.

You should already know:

- ✔ number operations and BIDMAS
- ✔ how to add and subtract negative numbers
- ✔ how to multiply and divide negative numbers
- ✔ how to find common factors
- ✔ how to add and subtract fractions
- ✔ the meaning of indices
- ✔ how to collect like terms.

Key terms

expand
expression
simplify
factorise
variable

Learn... 3.1 Expanding brackets and collecting like terms

When you **expand** brackets, **all** the terms inside the brackets must be multiplied by the term outside the brackets.

You will be given the instruction **expand** or **multiply out**.

If there is more than one bracket in the **expression**, each is done separately before collecting terms.

If there are terms not included in the bracket they are not included in the expansion.

These terms are collected once the brackets have been expanded.

Once you have the expanded expression, you should **simplify** it by collecting like terms.

Example: Expand $4(2y - 1)$

Solution: Everything inside the bracket is multiplied by the term in front of the bracket.

$$4(2y - 1) = 4 \times 2y - 4 \times 1$$
$$= 8y - 4$$

Example: Multiply out $5a(2a + 1)$

Solution: $5a(2a + 1) = 5a \times 2a + 5a \times 1$ Remember that $a \times a = a^2$

$$= 10a^2 + 5a$$

You may prefer to draw lines on the question to show which terms are multiplied.

Using this method saves a line in your working.

$$5a(2a + 1)$$
$$= 10a^2 + 5a$$

Example: Expand and simplify $9x - 2(x - 4)$

Solution: Note that $9x$ is not included in the bracket.

The part of the expression that needs to be expanded is $-2(x - 4)$

$9x - 2(x - 4) = 9x - 2x + 8$ Remember that the term in front of the bracket is -2 and not 2.

$$= 7x + 8$$

Leave $9x$ to be collected once the bracket has been expanded.

Example: Expand and simplify $4(3y + 2) - 5(y - 3)$

Solution: Separate the expression into two brackets to carry out the expansion.

Step 1 First bracket: Expand $4(3y + 2)$

$$4(3y + 2) = 12y + 8$$

Step 2 Second bracket: Expand $-5(y - 3)$

$$-5(y - 3) = -5y + 15$$

Step 3 Put the two answers together and collect like terms.

$$4(3y + 2) - 5(y - 3) = 12y + 8 - 5y + 15$$
$$= 7y + 23$$

Example: Write an expression for the area of this rectangle. Expand your answer.

$t - 5$

t

Solution: The area of a rectangle is width × length.

Area $= t(t - 5)$ Put the length in a bracket so both terms are multiplied by t.

Expand the brackets using your own method.

Area $= t^2 - 5t$ There are no terms to simplify because t^2 and t are different types of term.

AQA Examiner's tip

You can use any method to expand brackets as long as it works. Check your answer by substituting values.

Practise... 3.1 Expanding brackets and collecting like terms

D C B A A*

D

1 Expand:

a	$3(x + 4)$	e	$5(5d - 1)$	i	$3a(a - 2)$
b	$5(y - 2)$	f	$7(2 - 2f)$	j	$2d(3 - 5d)$
c	$8(2 - c)$	g	$3(10v + 7)$	k	$\frac{t}{2}(t - 4)$
d	$3(2p + 5)$	h	$11(7 + 3m)$	l	$6t(\frac{t}{2} - 2)$

2 a Tom expands $4f(2f - 3)$

Jade expands $6f(f - 2) + 2f^2$

Show that Tom and Jade get the same answer.

b Micky expands $8f(f - 1)$

What does Micky have to add to get the same answer as Tom and Jade?

3 Write an expression for the area of each of these rectangles. Expand your answer in each case.

a $5a + 1$, 3

c

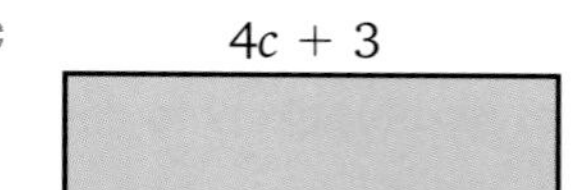

e

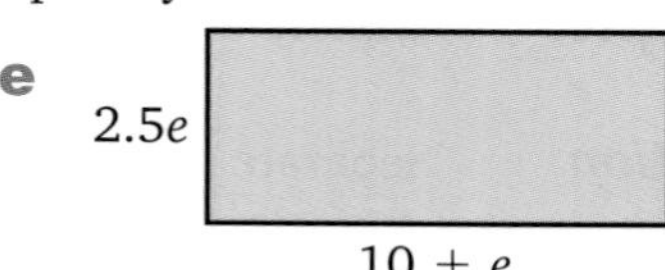

b $b - 5$, 2

d

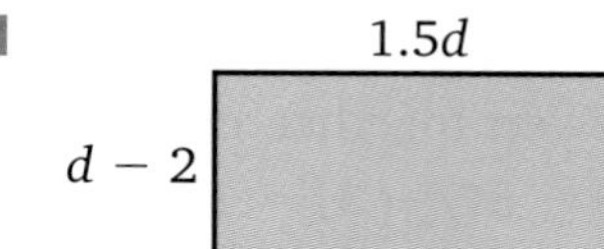

f

C

4 Sam and Jim are writing number puzzles using symbols. They use n to represent the missing number.

Sam says 'Think of a number, add two and then multiply the answer by 5.'

Jim writes his answer as $n + 2 \times 5$.

a Write down the mistake Jim has made and rewrite his answer correctly.

b Write each of these number puzzles as an expression, using n for the missing number.

i 'Think of a number, subtract 8 and then multiply the answer by 5.'

ii 'Begin with 10 and subtract the number, multiply the answer by 3 and then add 3 times the number.'

c Travis and his friends wrote these expressions for number puzzles. For each one, expand the brackets and collect like terms.

i $5(n + 2) - 10$

ii $3n - 2(n - 1)$

iii $5(n + 1) + 2(3 - n)$

iv $7(3 - 2n) + 10n$

v $3(2 + 3n) - 2(3 + 4n)$

vi $2(n + 1) + 5(n - 2) - 4n$

Bump up your grade

At Grade C, you may be asked to expand more than one bracket before simplifying.

5 Find an expression for the total area of each coloured shape. Expand and simplify your answer.

a

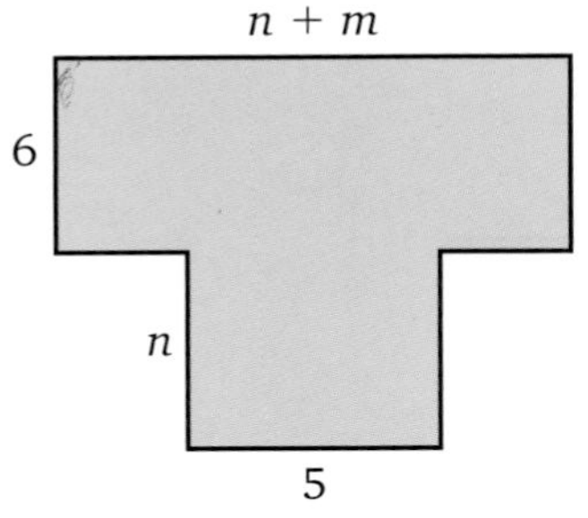

d

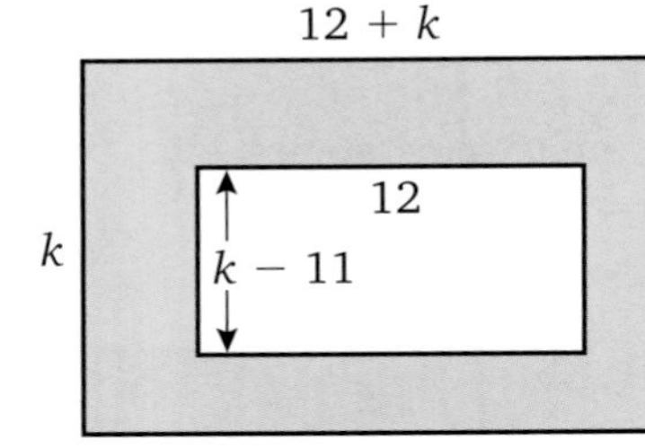

b

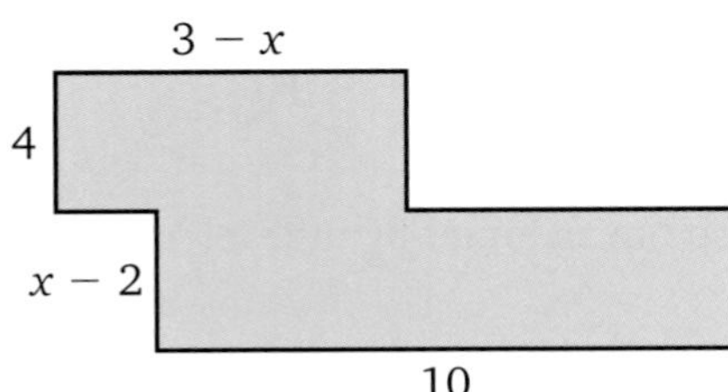

e

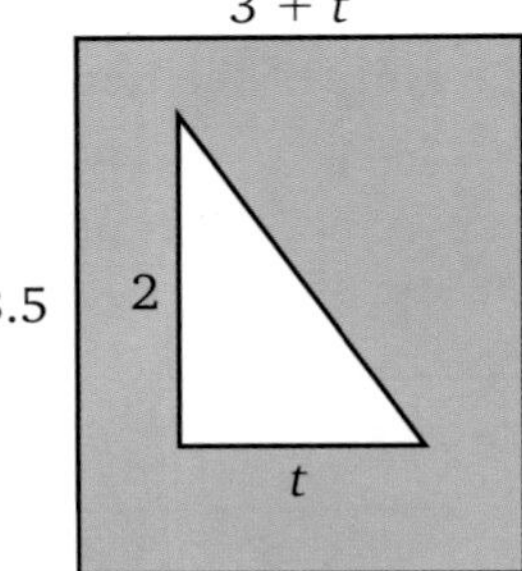

Hint

Remember the area of a triangle $= \frac{1}{2}$ base $\times$ height

c

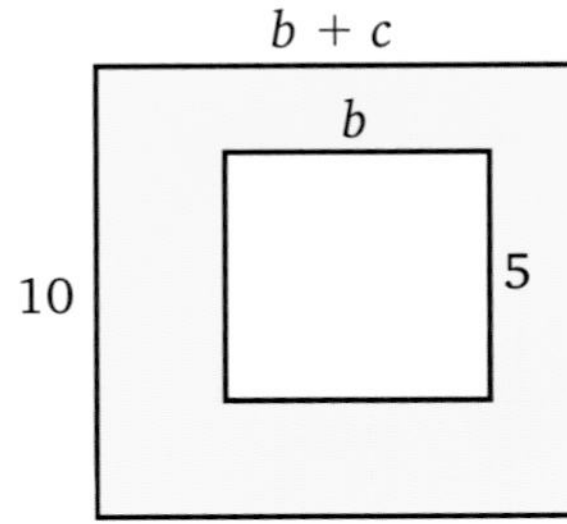

f

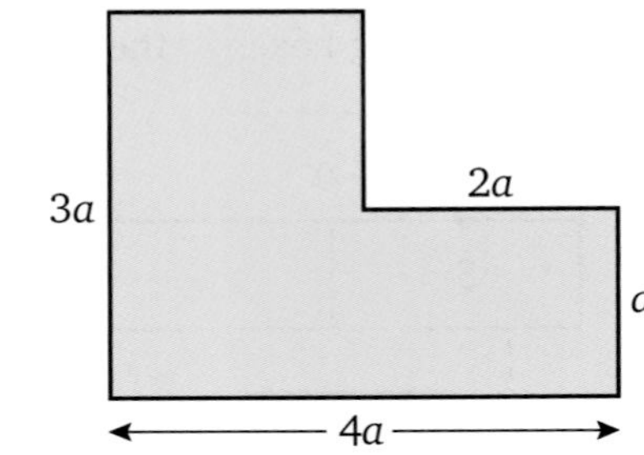

C

6 Find the missing numbers to make each pair of expressions equal.

a	$2(a + 4) + 3(a - 1)$	$?(a + 1)$
b	$5(1 - b) - 3(b - 3)$	$?(7 - 4b)$
c	$2(x + 1) - 5(x + ?)$	$-3(x + 6)$
d	$?(2x - 1) + 2(x - 3)$	$10(x - 1)$
e	$5(1 - x) + ?(4x + 5)$	$3(x + 5)$

? **7** In each diagram the value of the perimeter and the value of the area are the same.

In each question part, which value of x gives the same value for the perimeter and the area?

a

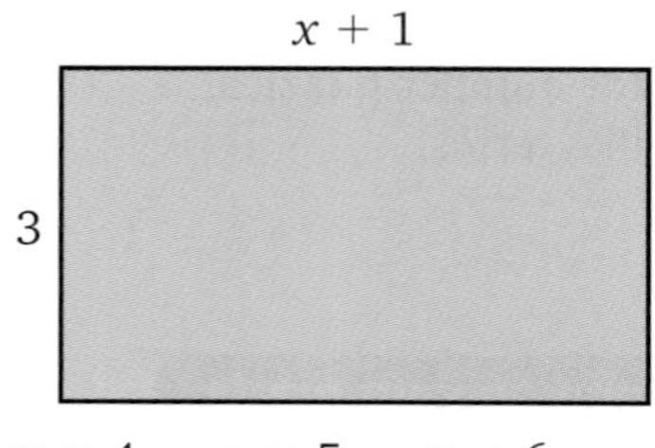

$x = 4$ $x = 5$ $x = 6$

b

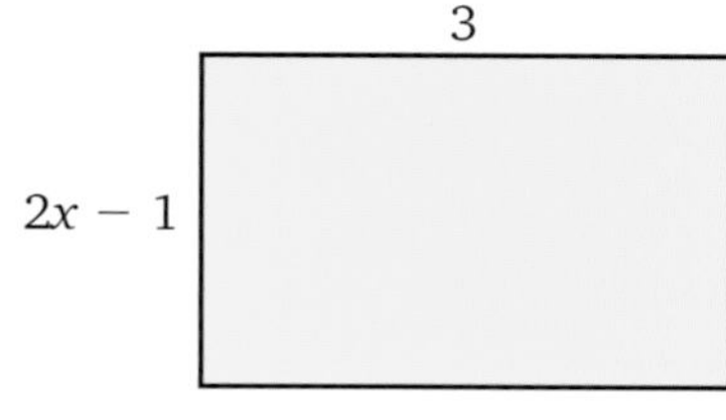

$x = 3.5$ $x = 4.5$ $x = 6.5$

c

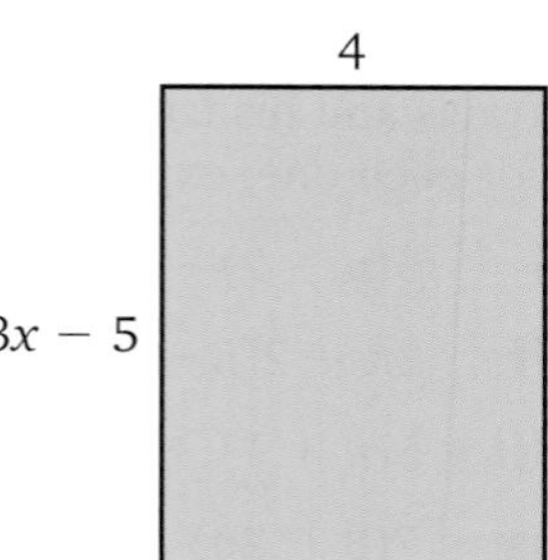

$x = 3.5$ $x = 1.3$ $x = 3.0$

Hint

Start by writing an expression for the perimeter and an expression for the area.

? **8** Here are two rectangles.

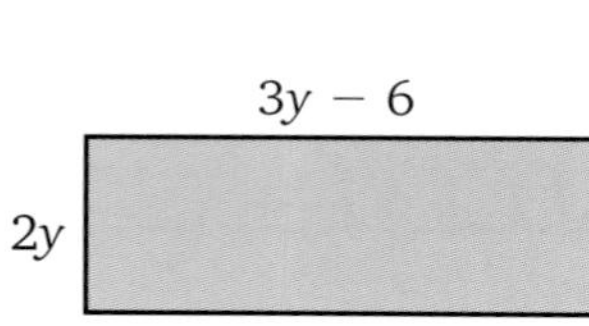

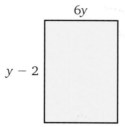

$6y$

$y - 2$

Show that the areas of the two rectangles are the same.

? **9** In the diagram below, the expression in the top box is found by multiplying the two adjacent boxes below.

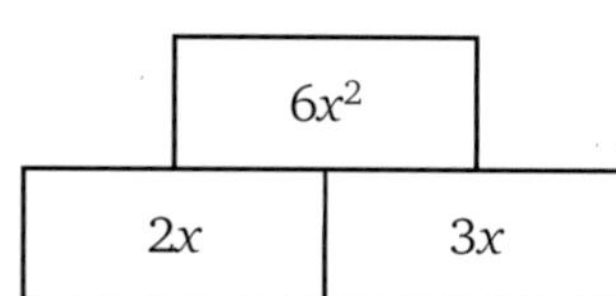

Fill in the missing boxes in the following diagrams.

a

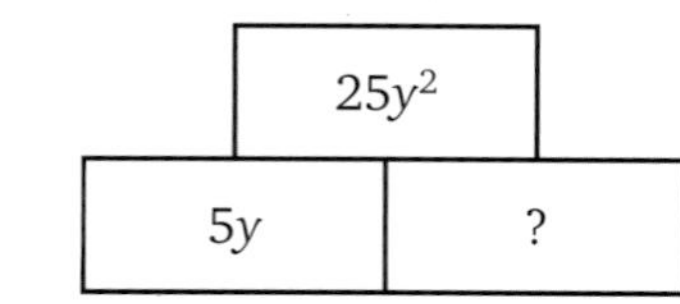

b

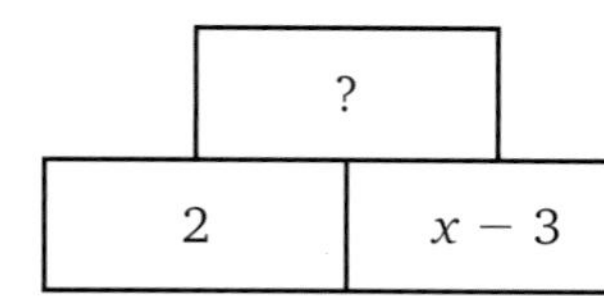

c

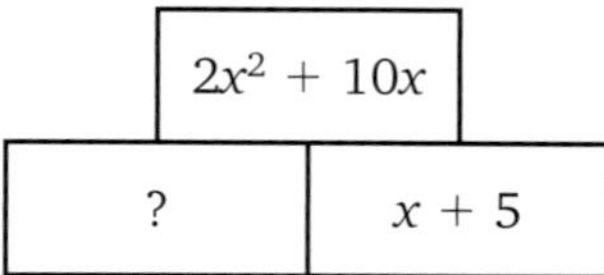

? **10**

Row A	3	$4x$	2	$3x$	$5x$	-5	$2x$	$-x$
Row B	$3 - 2x$	$x - 2$	$2x + 1$	$3 - x$	$x - 1$	$x + 2$	$2 - x$	$3x - 1$

Jess chooses two terms from row A and two terms from row B to make this expression:

$3(3 - x) + 3x(x - 2)$

Jess simplifies the expression in two steps:

$9 - 3x + 3x^2 - 6x$

$3x^2 - 9x + 9$

a Choose your own four terms to make an expression and simplify it in the same way that Jess has.

b The following shows Jess's working when simplifying three other expressions. Find the missing terms from the cards above, and copy and complete her working.

i $?(2x + 1) - 5(x + 2)$
$= 4x^2 + 2x - 5x - 10$
$=$

ii $4x(\ ?\ ) - x(2 - x)$
$= ? - ? - 2x + x^2$
$= 13x^2 - 6x$

iii $5x(x - 1) + ?(\ ?\ )$
$= 5x^2 - ? + 6 - ?$
$= 5x^2 - 7x + 6$

Learn... 3.2 Factorising expressions

Factorising is the opposite of expanding.

You will usually be given the instruction **factorise**.

Example: An expression for the area of this rectangle is $3xy + 6y$

Factorise fully the expression to find possible dimensions of the rectangle.

Link

You can find more on factorising in Unit 2, Chapter 5.

AQA Examiner's tip

If a questions asks you to factorise **fully** then it usually means there is more than one factor to take out. In this case, you can take out 3 and y.

Solution:

Step 1: Find the common factor of the two terms in the expression.

$3y$ is the common factor because 3 is a factor of 3 and 6

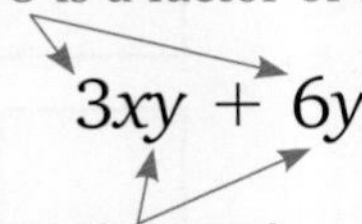

$$3xy + 6y$$

and y is a factor of xy and y.

Step 2: Divide each term by the common factor.

$3xy \div 3y = x \quad 6y \div 3y = 2$

Step 3: Write the common factor outside the bracket and the remaining terms inside the bracket.

$3xy + 6y = 3y(x + 2)$

Practise... 3.2 Factorising expressions 

D C

D

1 Factorise each expression.

a	$8c + 4$	**d**	$24 + 18k$	**g**	$12xy - 9y^2$	**j**	$15kl + 27k^2$
b	$12d - 15$	**e**	$20x + x^2$	**h**	$b^2 + 9ab$	**k**	$13f^2 - 65fg$
c	$20 - 10p$	**f**	$y^2 - 5y$	**i**	$4n + 18n^2m$	**l**	$36j^2k - 30jk^2$

AQA Examiner's tip

Multiply out your answers to check you have factorised correctly.

2 In each question, one side of the rectangle and the expression for the area has been given. Find the expression or value for the length of the other side.

a area $= 6x - 21$

?

3

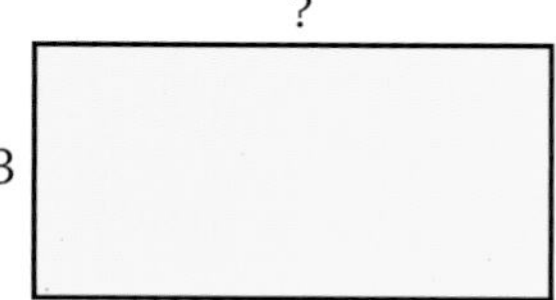

b area $= 20x^2 - 25x$

$5x$

?

c area $= 7p^2 - 5p$

p

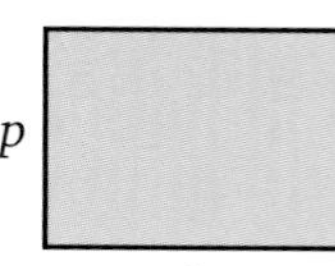

?

AQA Examiner's tip

Always factorise fully. The highest common factor should be taken outside the bracket.

d area $= 100t^2 + 125t$

$4t + 5$

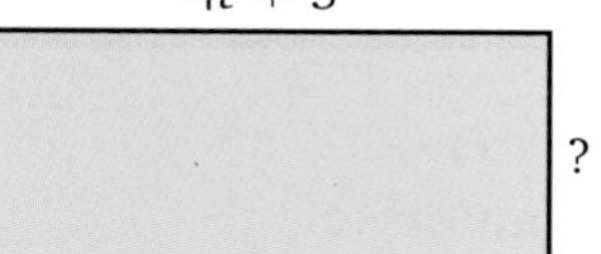

?

e area $= 2kq - 3kr$

k

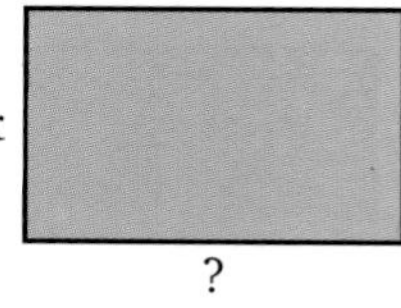

?

3 Factorise fully the expressions for the area of these rectangles.
Use your answers to give a term for the dimensions of each rectangle.

a area $= 5fg + 8fg^2$

b area $= 3x^2 + 21xy^2$

c area $= 6pq^2 - 2pq^3$

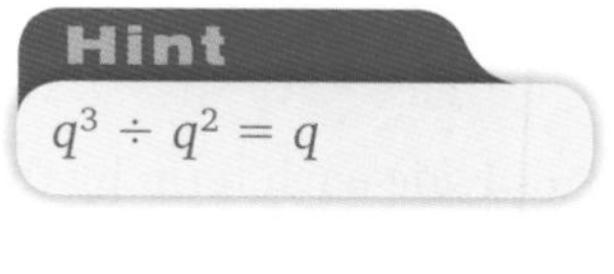

Hint

$q^3 \div q^2 = q$

d area $= 10a^2b + 15ab^2$

D

4 Chris and Sue both factorise the same expression but get different answers.

$12xy^2 - 18x^2y$

Chris's answer is $2x(6y^2 - 9xy)$

Sue's answer is $6xy(2y - 3x)$

Give the reasons why Chris is wrong and Sue is right.

5 These expressions have been factorised fully. Some are wrong and some are right.

For each question part, say whether the answer is wrong or right.

If it is wrong, give the correct answer.

a $3ap - 9p = 3p(a - 3)$

b $12f^2 - 18f = 3f(4f - 6)$

c $36 - 4t^2 + 12t = 4(9 - t^2 + 3t)$

d $15x^2y^2 - 20x^2y = 5xy(3xy - 4x)$

e $55k - 44klm^2 = 11k(5 - 4lm^2)$

C

6 In this factor puzzle the first expression is factorised. The factors are added. This is repeated until the expression cannot be factorised again.

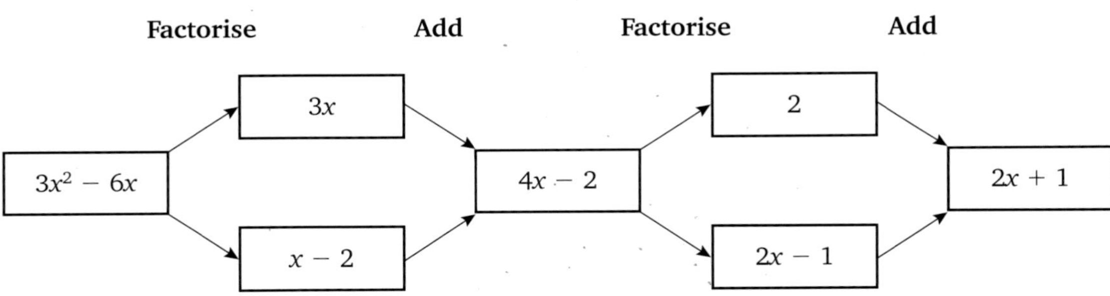

a Copy and complete this factor puzzle.

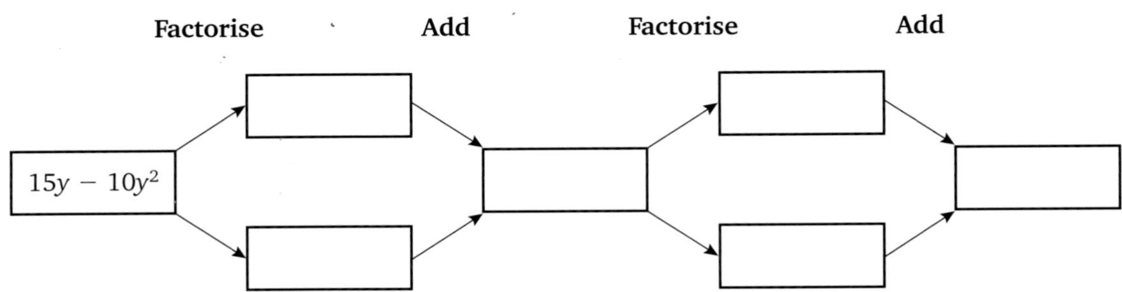

b Complete a factor puzzle for each of these starting expressions.

i $7a^2 - 14a$

ii $4p^2 - 12p$

iii $28c - 35c^2$

iv $20q - 25q^2$

Learn... 3.3 Multiplying two brackets together

There are different methods you can use to multiply two brackets together.

Use a method that works for you.

Everything in the first bracket is multiplied by everything in the second bracket.

They can be multiplied in any order.

Any like terms are then collected together.

$(2x - 1)(x + 2) = 2x^2 + 4x - x - 2 = 2x^2 + 3x - 2$

Link

See Unit 2, Chapter 5 for a reminder of the grid method of multiplying two brackets.

When there is a negative term in x an expression is often written with the constant before the **variable**.

e.g. $-3x + 8$ is written as $8 - 3x$

When the term in x^2 is also negative the expression is written in reverse order.

e.g. $1 + 3x - x^2$

Example: Expand and simplify:

a $(x + 1)(x - 5)$

b $(2x - 1)(x + 3)$

c $(5 - x)(2 - 3x)$

d $(5x - 1)^2$

Solution:

a $(x + 1)(x - 5) = x^2 - 5x + x - 5 = x^2 - 4x - 5$

b $(2x - 1)(x + 3) = 2x^2 + 6x - x - 3 = 2x^2 + 5x - 3$

c $(5 - x)(2 - 3x) = 10 - 15x - 2x + 3x^2 = 10 - 17x + 3x^2$

d $(5x - 1)^2 = (5x - 1)(5x - 1) = 25x^2 - 5x - 5x + 1 = 25x^2 - 10x + 1$

Don't make the mistake of just squaring the two terms in the brackets
e.g. $(5x)^2 = 25x^2$ and $-1^2 = 1$

Write the question as two brackets multiplied and then follow the same method.

Example: Find an expression for the area of this trapezium.

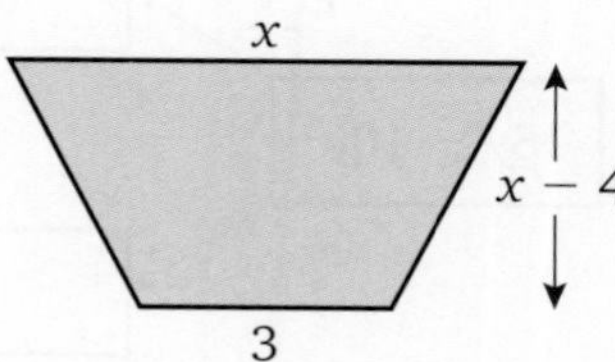

Solution:

Step 1. Form an expression using the formula for the area of a trapezium: $\frac{1}{2}(a + b) \times h$

$\frac{1}{2}(x + 3)(x - 4)$ Replace $(a + b)$ with $(x + 3)$ and h with $(x - 4)$

Step 2. Expand the brackets using your preferred method.

$\frac{1}{2}(x^2 - x - 12)$

The expression for the area of this trapezium is

$\frac{1}{2}(x^2 - x - 12)$ or $\frac{x^2 - x - 12}{2}$ or $\frac{x^2}{2} - \frac{x}{2} - 6$

AQA Examiner's tip

Remember that the formula for the area of a trapezium requires the perpendicular height and not the slant height of the trapezium. The formula will be provided in your exam.

Practise... 3.3 Multiplying two brackets together

D C B A A*

B

1 Expand and simplify:

a	$(x + 3)(x + 7)$	**e**	$(4 - f)(f + 1)$	**i**	$(5t - 1)(2t - 3)$
b	$(x - 2)(x + 4)$	**f**	$(j + 5)(6 - j)$	**j**	$(4 - k)(3k + 1)$
c	$(y + 8)(y - 3)$	**g**	$(5 - h)(10 - h)$	**k**	$(2a - 5)(5 - 2a)$
d	$(p - 3)(p - 5)$	**h**	$(3x + 4)(x + 1)$	**l**	$(3c + 5)^2$

2 Write an expression for the area of each polygon.
Expand and simplify your answer.

Hint
Put the expression for the length of each side in brackets.

a
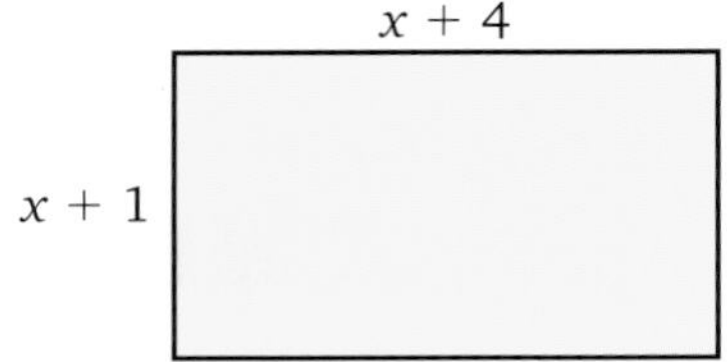

f
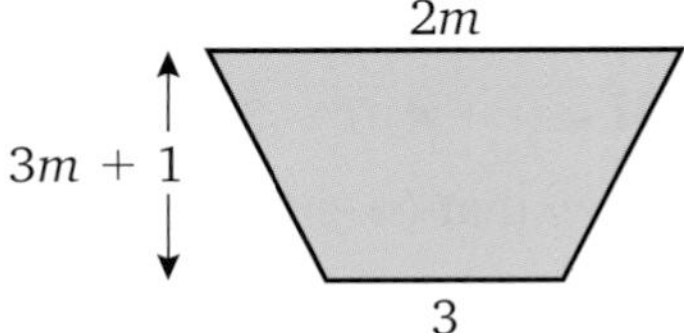

b
$x - 2$
$2x + 3$

g
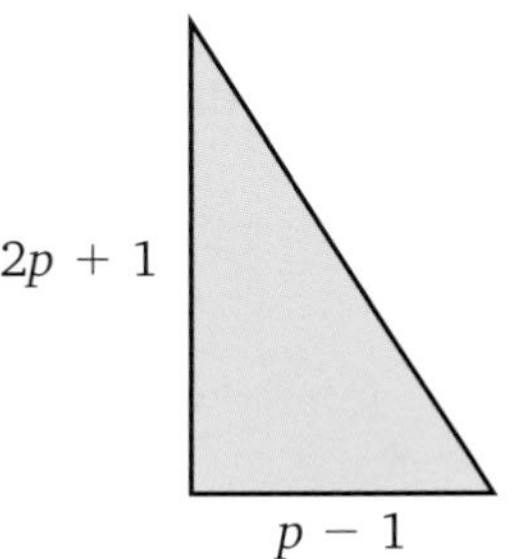

c
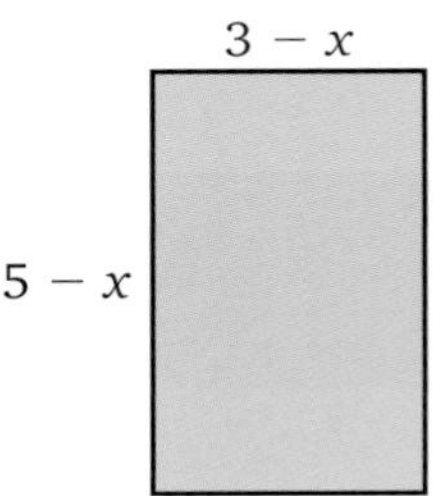

h
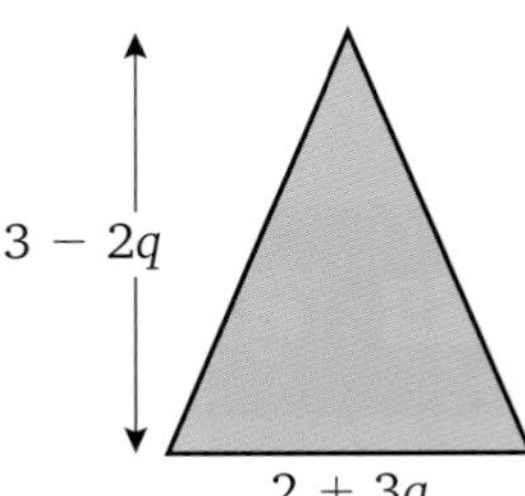

d
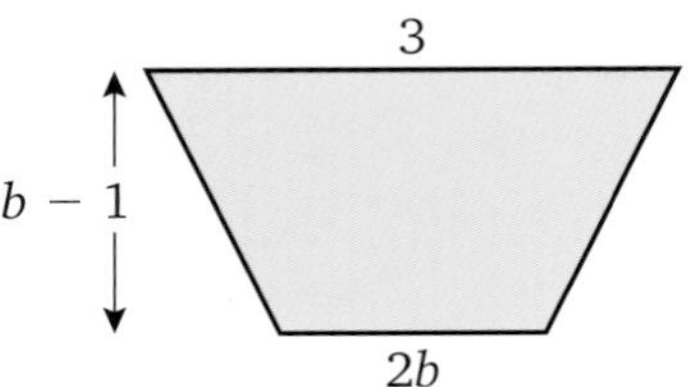

i
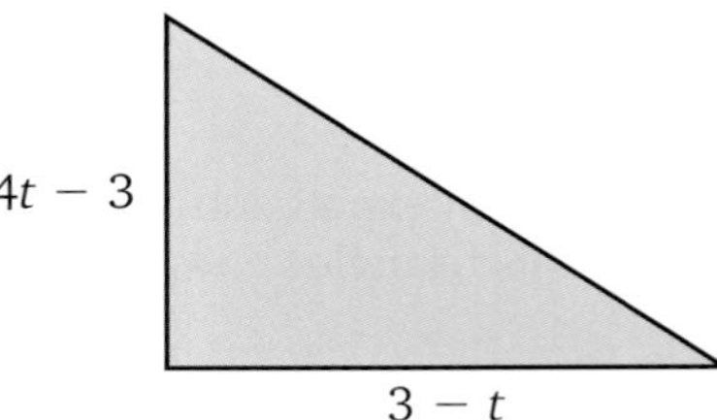

e
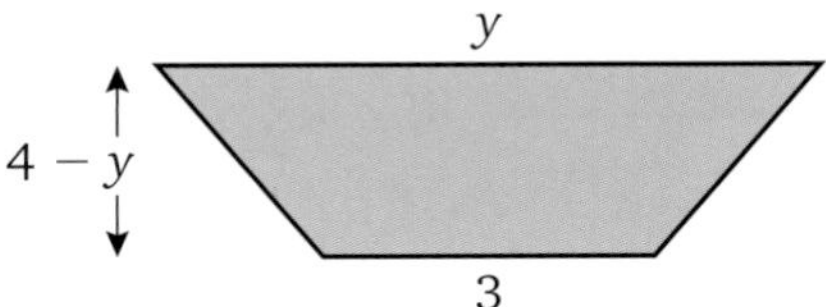

j A square with sides $g - 3$

k A square with sides $2 - 3w$

l A square with sides $3b + 7$

B

3 Cara and her friends formed expressions containing t^2.

Each student factorised their expression but some of them made mistakes.

By expanding the brackets in their answers, find out which students were right and which students were wrong.

Cara	$t^2 - 3t + 2$	$(t - 2)(t - 1)$
Tim	$2t^2 + 7t - 4$	$(t - 4)(2t + 1)$
Tariq	$t^2 - 5t + 6$	$(t - 6)(t - 1)$
Toby	$3t^2 + 2t - 1$	$(3t - 1)(t + 1)$
Emma	$6t^2 - 11t - 3$	$(2t - 3)(3t - 1)$
Sam	$4t^2 + 1$	$(2t - 1)^2$

4 **a** Expand and simplify:

i $(x + 1)(x + 2) + (x + 3)(x - 1)$

ii $(y - 3)(y + 1) + (y - 2)(y + 3)$

iii $(t + 4)(t - 3) - (t + 3)(t - 4)$

b Show that $(n + 2)^2 - n(n - 2) = 2(3n + 2)$

5 Any odd number can be written as $2n + 1$

Show that the product of two odd numbers is always odd.

Hint
The product of two numbers is the answer when they are multiplied.

6 Write an expression for the yellow area. Simplify your answer.

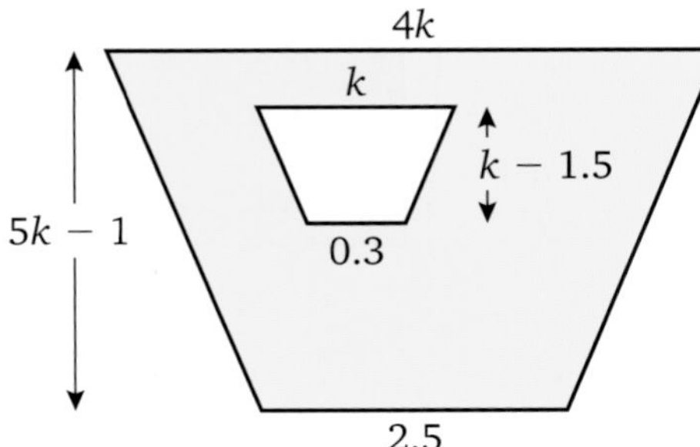

3 Assess

D

1 Expand:

a $4(x + 2)$ **b** $2(3y - 2)$ **c** $5(1 - 2g)$ **d** $2x(x + 5)$ **e** $\frac{t}{2}(t + 8)$

2 **a** Write an expression for the perimeter of this triangle.

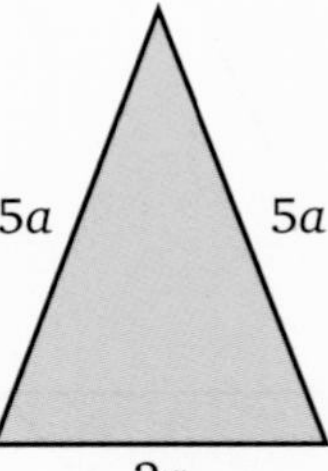

b Write an expression for the area of this rectangle.

$1.5d$

$d - 2$

D

3 Expand each expression and collect like terms.

a $6 + 2(x + 3)$ **b** $3(y - 1) + 3$

4 One side of the rectangle and the expression for the area has been given.

Find an expression or value for the length of the other side.

area $= 5v^2 - 10v$ $5v$

5 Factorise fully $15xy^2 - 24x^2y$

C

6 Simplify:

a $2(n + 1) + 3(n + 3)$ **b** $5(h - 1) + 4(h - 3)$ **c** $4(2y + 1) - 3(2y - 4)$

7 Find an expression for the total area of this shape. Simplify your answer.

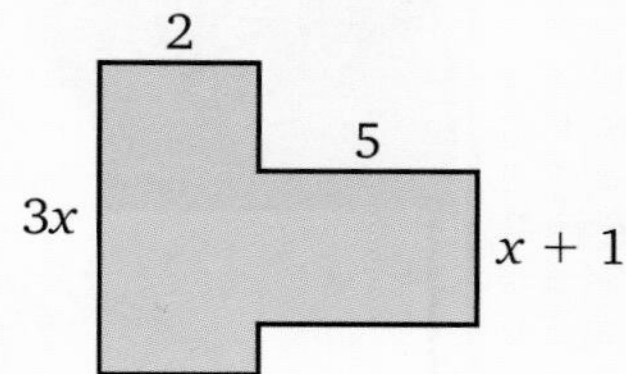

B

8 Expand and simplify:

a $(x + 3)(x + 2)$ **b** $(2y + 1)(y - 3)$ **c** $(4f + 3)(3f - 4)$ **d** $(3 - k)^2$

9 Write an expression for the area of each polygon. Expand and simplify your answer.

a

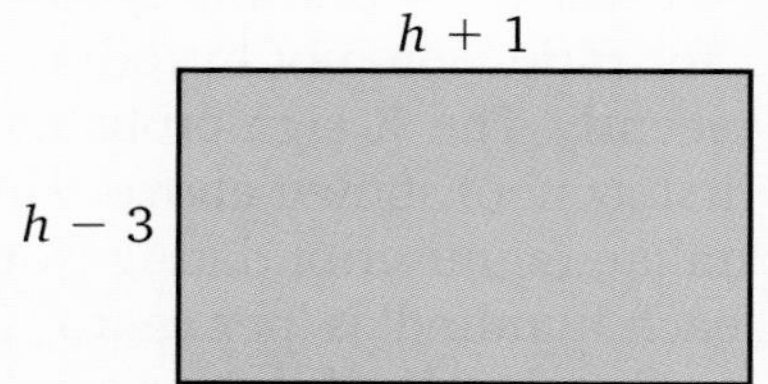

b

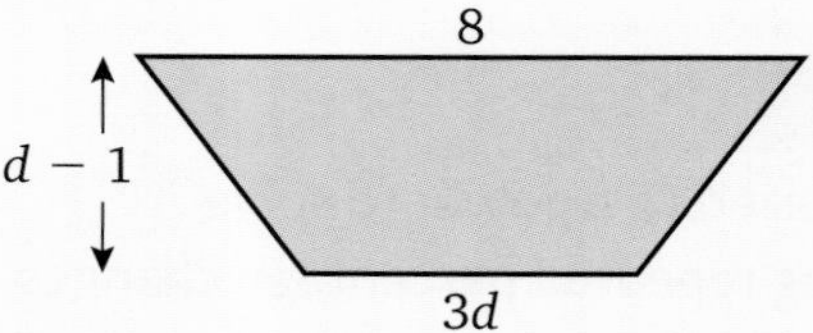

10 **a** Expand and simplify:

i $(x + 1)(x + 3) + (x + 4)(x - 1)$ **ii** $(t + 4)(t - 3) - (t + 3)(t - 4)$

b Show that $(n + 1)^2 - n(n + 2) = 1$

11 Any even number can be written as $2n$ and any odd number can be written as $2n + 1$.

Show that the product of an odd and even number is always even.

AQA Examination-style questions

1 Here is a flow chart.

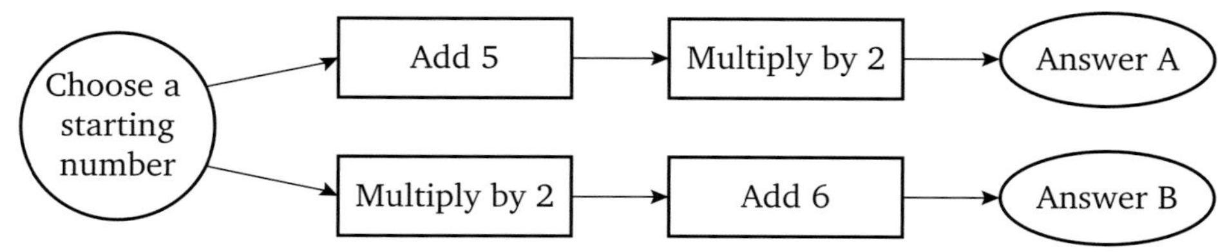

Show that for any starting number, the difference between A and B is always 4. *(4 marks)*

AQA 2009

4 Percentages and ratios

Objectives

Examiners would normally expect students who get these grades to be able to:

D

increase or decrease by a given percentage

express one quantity as a percentage of another

use ratio notation, including reduction to its simplest form and its links to fraction notation

solve simple ratio and proportion problems, such as finding and simplifying a ratio

C

work out a percentage increase or decrease

solve more complex ratio and proportion problems

solve ratio and proportion problems using the unitary method

B

use a multiplier raised to a power to solve problems involving repeated percentage changes and exponential growth

A

work out values and draw graphs in situations involving exponential growth. For example, draw a graph to show the way in which a population of bacteria grows

solve direct and indirect proportion problems

interpret graphs showing direct and indirect proportion problems.

Did you know?

The origin of mathematical symbols

15th century	17th century	Now
P$\frown$o	$\frac{o}{o}$	%

Did you know that the symbol **%** for per cent and **:** for ratio were not introduced until relatively recently. The % sign probably developed from the first symbol shown above. This appeared in an Italian manuscript dating from 1425. In Italian 'each hundred' is 'per cento', in French it is 'pour cent' and in English 'per cent'.

An Englishman, William Oughtred, was the first to use a colon : for ratios in his book, *Canones Sinuum,* in 1657.

Find out when other symbols such as +, −, ×, ÷ and = were introduced. What did people use before then?

You should already know:

- ✔ about place values in decimals and how to put decimals in order of size
- ✔ how to simplify fractions with and without a calculator
- ✔ how to write a fraction as a decimal and vice versa
- ✔ how to change a percentage to a fraction or decimal and vice versa.

Key terms

percentage	ratio
depreciation	unitary ratio
exponential growth	unitary method
Value Added Tax (VAT)	direct proportion
discount	indirect proportion

Learn... 4.1 Increasing or decreasing by a percentage

There are a variety of different ways to increase or decrease a quantity by a **percentage** but the most efficient way on a calculator is to use a multiplier. For example, suppose a new car costs £17 500, but its value **depreciates** (goes down) by 20% in the first year after it is bought.

The value of the car after one year will be £17 500 × 0.8 = £14 000

After a 20% decrease, the value is 80% of the original value, so the multiplier is 0.80 or 0.8

This is very quick to do on a calculator.
The method is summarised below.

To increase or decrease by a given percentage:

- work out the new percentage (add a percentage increase to 100%; subtract a percentage decrease from 100%)
- divide the new percentage by 100 to find the **multiplier**
- **multiply the original quantity by the multiplier.**

Work out the multiplier in your head if you can.

Suppose the value of the car depreciates by another 20% in the following year.

You can find the value of the car after 2 years in a single calculation:
$17\,500 \times 0.8 \times 0.8 = 17\,500 \times 0.8^2 = £11\,200$

AQA Examiner's tip

Multiplying multipliers together is the quickest way to combine percentage changes.

If you just want to know the overall percentage fall in the value of the car, multiply the multipliers together:

$0.8 \times 0.8 = 0.8^2 = 0.64$ so after 2 years the car is worth 64% of its original value.

Its value has depreciated by 36% because 100% − 64% = 36%

Percentage increases can also be combined. If the same percentage increase is repeated again and again, it leads to **exponential growth**. This is what happens in the following example.

Example: A population of bacteria increases by 60% every hour.

a Copy and complete this table of values.

Time (hours)	0	1	2	3	4	5
Number of bacteria	1000					

b Draw a graph of the number of bacteria against time.

Solution: **a** At the end of each hour, the number of bacteria will be 160% of the number at the beginning of that hour.

The multiplier = 1.60 = 1.6 for each hour

After 1 hour there will be 1000 × 1.6 = 1600 bacteria.

After 2 hours there will be 1600 × 1.6 = 2560 bacteria.

At each stage, continue the working on your calculator rather than starting again.

After 3 hours there will be 2560 × 1.6 = 4096 bacteria and so on.

The table gives the number of bacteria after each hour, with values rounded to the nearest whole number where necessary.

Time (hours)	0	1	2	3	4	5
Number of bacteria	1000	1600	2560	4096	6554	10 486

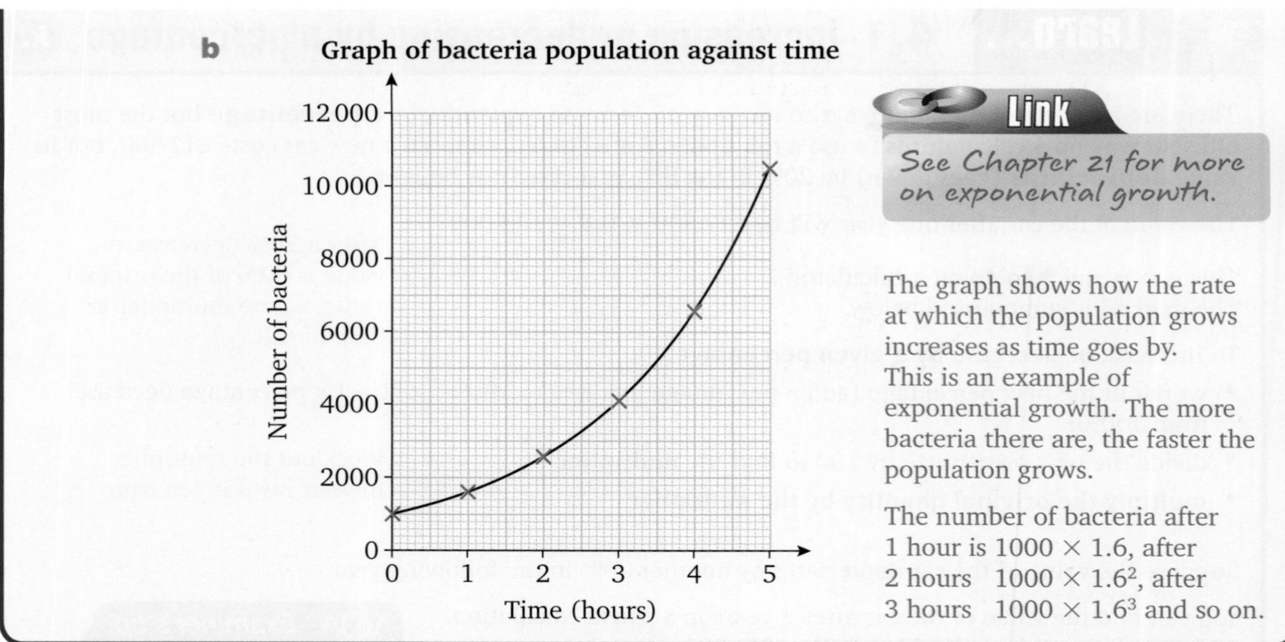

Link

See Chapter 21 for more on exponential growth.

The graph shows how the rate at which the population grows increases as time goes by. This is an example of exponential growth. The more bacteria there are, the faster the population grows.

The number of bacteria after 1 hour is 1000×1.6, after 2 hours 1000×1.6^2, after 3 hours 1000×1.6^3 and so on.

Practise... 4.1 Increasing or decreasing by a percentage

D

1 Use multipliers to:

a increase 250 m by 40%

b increase 80 kg by 70%

c decrease 24 miles by 5%

d decrease 37.5 litres by 12%

e increase £54.60 by 43%

f decrease £180 by 62.5%

Check your answers using a different method.

Hint

One way to check answers is to find the increase (or decrease) first, then add it to (or subtract it from) the original amount.

2 The prices of each of these items are given **excluding Value Added Tax (VAT)**.

Find the cost of each item **including** VAT at the rate given.

a

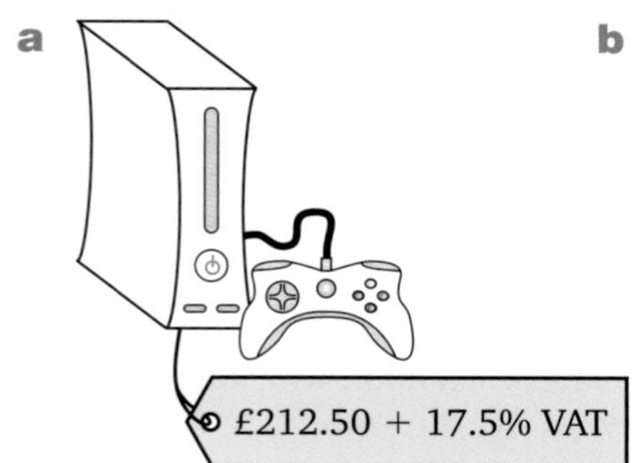

b Cost of electricity: £134.92 + 5% VAT

c

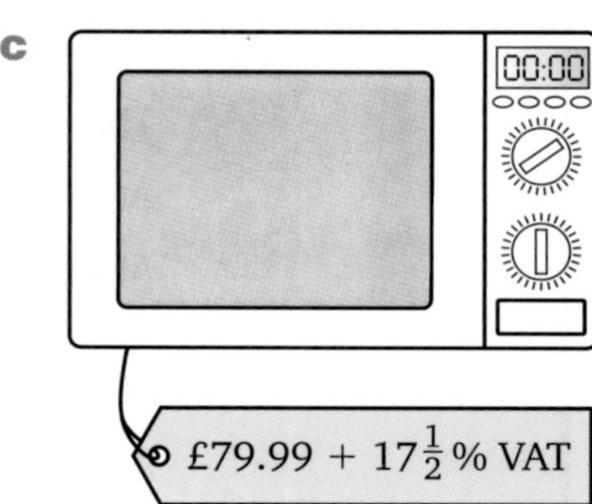

C

3 Zoe invests £2000 in shares. In the first year the shares go up in value by 9%

In the second year they go down in value by 4%

How much is Zoe's investment worth after 2 years?

Bump up your grade

For Grade C, you must be able to work out the result after more than one percentage change.

B

4 The rate at which bacteria increase depends on the conditions in the laboratory or in the outside world. Three populations of bacteria, *A*, *B* and *C*, all start with 1000 bacteria but are then kept in different conditions.

a Population *A* doubles every hour.

Find the number of bacteria in population *A* after two hours.

b Population *B* increases by 50% each hour.

Find the number of bacteria in population *B* after two hours.

c Population *C* increases by 25% every hour.

i Complete a table to show the number of bacteria in population *C* after 0, 1, 2, 3, 4, 5 and 6 hours.

ii Draw a graph of the number of bacteria in population *C* against time.

iii Describe the way in which this population grows.

5 There are estimated to be 5000 trout in a lake. This number is expected to decrease by 8% per year.

a How many trout are expected to be in the lake after:

i 2 years **ii** 5 years **iii** 10 years **iv** 20 years?

b Use your answers to part **a** to help you sketch a graph of the number of trout against time.

The way in which the number of trout decreases is called exponential decay. The next question gives another example of this.

6 A car worth £16 000 loses one quarter of its value every year.

a How much will the car be worth after 4 years?

b Sketch a graph of the value of the car against time.

7 A young manager has a job with a starting salary of £19 000 increasing by 2.4% per year.

a What is her salary after: **i** one year **ii** two years **iii** ten years?

b Your answer to part **a iii** may not be very realistic. Why?

A

8 The radioactive substance bismuth-210 decays at a rate of 12.9% every day. There is 1 kilogram of bismuth-210.

a How much bismuth-210 will there be after two weeks?

b How many more days will it take for the bismuth-210 to decay to less than one-tenth of its original mass?

9 **a** The organisers of a sponsored swim give 40% of all the money raised to a children's charity.
They give 25% of the remaining money to an animal sanctuary and the rest to a local hospice.

What percentage of all the money raised is given to the local hospice?

b The organisers aim to increase the amount raised by 20% each year.
The treasurer says that it will take 5 years to double the amount raised.

Do you agree? Explain your answer.

10 Scientists estimate that the area of the Arctic polar ice cap varies during the course of a year from about 7 million km^2 to about 15 million km^2. They also say that the ice cap is melting at a rate of 10% per decade.

Assuming that this is true, predict the way in which the area of the ice cap will vary over the course of a year at the end of this century.

11 Between 2000 and 2010 the world's population increased from 6.1 billion to 6.9 billion.

Use this information to make predictions about how the world's population may grow in each decade up to 2050. Describe any assumptions you make and any reasons why your predictions are likely to be inaccurate.

12 A shop usually sells sports equipment at prices that include a markup of 40% for profit. The manager decides to have a sale, but still wants to make at least 5% profit on the equipment sold.

What is the maximum percentage by which the manager can reduce the usual prices of the equipment in the sale and still make 5% profit?

Learn... 4.2 Writing one quantity as a percentage of another

To write one quantity as a percentage of another:

- divide the first quantity by the second; this gives you a decimal (or write the first quantity as a fraction of the second)
- then multiply by 100% to change the decimal or fraction to a percentage.

To write an increase or decrease as a percentage:

- find the difference between the old quantity and the new quantity; this gives the increase or decrease
- divide the increase (or decrease) by the **original** amount or write the increase (or decrease) as a fraction of the **original** amount
- then multiply by 100 to change the decimal or fraction to a percentage.

Sometimes you may need to write one part as a percentage of the whole amount (as in the first example below).

The quantities must always be in the **same units**.

You can also use this method to find a percentage profit or loss.

Example: Last season a school's football team won 18 matches, drew 4 matches and lost 10 matches.

a What percentage of the matches were won?

b What percentage were drawn?

c What percentage were lost?

Solution: You must write each number as a percentage of the total number of matches.

The total number of matches = 18 + 4 + 10 = 32

a percentage of games won $= \frac{18}{32} \times 100\% = 18 \div 32 \times 100\% = 56.25\%$

b percentage of games drawn $= \frac{4}{32} \times 100\% = 4 \div 32 \times 100\% = 12.5\%$

c percentage of games lost $= \frac{10}{32} \times 100\% = 10 \div 32 \times 100\% = 31.25\%$

or using the fraction key:

% won $= \frac{18}{32} \times 100\% = 56\frac{1}{4}\%$

% drawn $= \frac{4}{32} \times 100\% = 12\frac{1}{2}\%$

% lost $= \frac{10}{32} \times 100\% = 31\frac{1}{4}\%$

Add the percentages to check: 56.25% + 12.5% + 31.25% = 100% ✓

Example: Jack's pay rate has gone up by 45 pence to £8.55 per hour. Find the percentage increase.

Solution: The increase in Jack's hourly pay rate = 45 pence

His **original** hourly pay rate was 855 − 45 = 810 pence

To work in pence use £8.55 = 855 pence

% increase $= \frac{45}{810} \times 100\% = 45 \div 810 \times 100\% = 5.555...\%$

The % increase in Jack's hourly pay rate = 5.6% (to 1 d.p.)

or using the fraction key $\frac{45}{810} \times 100\% = 5\frac{5}{9}$

You get the same answer if you work in pounds:
% increase = 0.45 ÷ 8.10 × 100% = 5.6% (to 1 d.p.)

Practise... 4.2 Writing one quantity as a percentage of another

When answers are not exact, round them to one decimal place.

D

1 A school's athletic team is made up of 16 girls and 19 boys.

What percentage of the team are:

a girls **b** boys?

2 The table shows the amount of waste recycled by a local authority last year.

What percentage of the waste was glass?

Type of waste	Amount (tonnes)
Paper & cardboard	12 400
Glass	4100
Metal (including cans)	3500
Green waste	8200

C

3 A garden centre buys plants for 56 pence each. It sells them for 99 pence each.

Work out the percentage profit.

Hint

The percentage profit is the percentage increase in price.

4 Sharon buys a motor bike for £3400. She sells it a year later for £2900.

Work out her percentage loss.

Hint

The percentage loss is the percentage decrease in price.

5 A furniture shop reduces its prices in a sale. Dan buys a table and four chairs.

Work out the percentage reduction in the total price.

Bump up your grade

For Grade C you must be able to write an increase or decrease as a percentage.

6 The cost of Greg's car insurance has gone up by £60 to £480. Greg works out $60 \div 480 \times 100$. He says the cost has increased by 12.5%

a What mistake has he made?

b What is the actual percentage increase in the cost of the insurance?

B

7 Sara gets a 30% reduction on her car insurance because she has not made previous claims. She gets a further 10% **discount** on the reduced price by agreeing to pay the first £100 of any claim. Sara says the total reduction is 40%

Do you agree? Explain your answer.

8 A manufacturer makes a rectangular rug that is 160 cm long and 120 cm wide.
The manufacturer decides to increase both dimensions of the rug by 25%

a Find the percentage increase in:

i the perimeter of the rug **ii** the area of the rug.

b Would the percentage increase in the perimeter and area have been the same if the rug was square?
Explain your answer.

9 A manufacturer who sells muesli in 800 g bags wants to increase his profits.
One option is to increase the price of each bag by 5%
Another option is to keep the price the same but reduce the contents of each bag to 750 g.

Which option do you think would increase the profits more? Explain your answer.

10 Paul takes a maths test and a science test in December.
He takes another test in each subject the following June.
The table shows his scores.
Paul says 'I improved most in science.'
His teacher says 'Actually you improved most in maths.'
How did his teacher decide? How did Paul decide?
You must show your working to justify your answer.

	December	June
Maths	46%	62%
Science	53%	71%

Learn... 4.3 Using ratios and proportion

Ratios are a good way of comparing quantities. To do this, the quantities must be in the same units.

Two (or more) ratios that simplify to the same ratio are called equivalent ratios. For example, 8 : 12 and 100 : 150 are equivalent because they both simplify to 2 : 3.

To simplify a ratio, divide each part by the same number.

£3 = 300 pence
When the amounts are in the **same units**, you can omit the units.

For example, the ratio of £3 to 40 pence = 300 : 40 = 30 : 4 = 15 : 2

Each number is divided by 10, then 2

You can use the fraction key on your calculator to simplify ratios.

This is the **simplest form** of this ratio. It uses the smallest possible whole numbers.

Sometimes ratios are divided further until one side is 1. This gives the $1 : n$ (or the $n : 1$) form.
These are called **unitary ratios**.

Dividing 15 : 2 by 2 gives the ratio $7\frac{1}{2} : 1$ or 7.5 : 1 Dividing the original ratio 300 : 40 by 40 also gives this.

The scale of maps and models are often given as unitary ratios.

The **unitary method** is based on working out what happens with **one** unit of something.
It can be used to solve a variety of different problems involving ratio and proportion.

There are some links between ratios and fractions.

For example, suppose a brother and sister share an inheritance in the ratio 3 : 4

- This means that for every £3 the brother gets, the sister gets £4.
- The brother's share is $\frac{3}{4}$ of the sister's share. The sister's share is $\frac{4}{3}$ or $1\frac{1}{3}$ times the brother's share.
- The brother gets $\frac{3}{7}$ of the whole inheritance and the sister gets the other $\frac{4}{7}$

In the **multiplier method** you multiply the quantity by the fraction representing the ratio.

Example: A model of a car is made using a scale of 1 : 50

a The model car is 9 centimetres long.
How long is the real car in metres?

b The real car is 2 metres wide.
How wide is the model car in centimetres?

Solution:

a The ratio 1 : 50 means that 1 cm on the model car represents 50 cm on the real car.

To find the length of the real car, multiply the length of the model car by 50.

Length of the real car = 9 cm × 50 = 450 cm

Length of the real car in metres = 450 ÷ 100 = 4.5 m

1 metre = 100 centimetres

You can change the units before or after using the scale.

b The width of the real car = 2 m = 200 cm

To find the width of the model car, divide the width of the real car by 50.

Width of model = 200 ÷ 50 = 4 cm

Example: Liam earns £97.20 for working 15 hours in a supermarket.

How much does he earn for 24 hours at the same rate of pay?

The amount Liam earns is **proportional** to the time he works. (If he works twice as long, he gets paid twice as much. If he works 3 times as long, he gets paid 3 times as much… and so on.)

Solution: For 15 hours Liam earns £97.20.

For 1 hour he earns £97.20 ÷ 15 = £6.48. — Divide the pay for 15 hours by 15.

For 24 hours he earns £6.48 × 24 = £155.52. — Multiply the pay for 1 hour by 24.

You can do this all in one calculation:

£97.20 ÷ 15 × 24 = £155.52

(÷ 15: find the pay for 1 hour; × 24: then the pay for 24 hours)

Always check that your answer looks reasonable. Here the pay for 24 hours is more than that for 15 hours.

Multiplier method

£97.20 × $\frac{24}{15}$ = £155.52

Multiplying by $\frac{24}{15}$ is the same as dividing by 15 and multiplying by 24.

Example:

a Which jar of coffee gives the best value for money?

b Give a reason why you might decide to buy one of the other jars.

Solution: You can use the **unitary method** to solve 'best buy' problems.

a Find the cost of **1 gram** in each jar. Working in pence gives easier numbers to compare.

Small jar: Cost of 50 g = 156 pence
Cost of 1 g = 156p ÷ 50
= 3.12 pence

Medium jar: Cost of 100 g = 229 pence
Cost of 1 g = 229p ÷ 100
= 2.29 pence

Large jar: Cost of 200 g = 445 pence
Cost of 1 g = 445p ÷ 200
= 2.225 pence

The cost of 1 gram of coffee is **least** in the large jar. The large jar gives the best value for money.

There are sometimes other methods you could use. In this problem you could compare the cost of 100 g.

Small jar: 100 g (2 jars) costs 156p × 2 = 312 pence

Medium jar: 100 g costs 229 pence

Large jar: 100 g ($\frac{1}{2}$ jar) costs 445p ÷ 2 = 222.5 pence

This also shows that the large jar gives the best value for money.

Bump up your grade

To get a Grade C you need to be able to use the unitary method to solve ratio and proportion problems.

b You might buy a smaller jar if you only want a small amount of coffee (or if you do not have £4.45 to spend).

Practise... 4.3 Using ratios and proportion  

D

1 A builder makes mortar by mixing cement and sand in the ratio 1 : 5

- **a** How many buckets of sand does he need to mix with 3 buckets of cement?
- **b** How many buckets of cement does he need to mix with 10 buckets of sand?
- **c** How many buckets of cement and sand does he need to make 30 buckets of mortar?

2 The numbers x and y are in the ratio 3 : 4

- **a** If x is 12, what is y?
- **b** If y is 12, what is x?
- **c** If x is 1, what is y?
- **d** If y is 1, what is x?
- **e** If x and y add up to 35, what are x and y?

C

3 Ewan is paid £87.50 for 14 hours work.

- **a** How much should he be paid for 20 hours work?
- **b** He is paid £100. How many hours has he worked?
- **c** What assumption do you have to make to answer these questions?

4

- **a** Amy and Bianca buy some euro before they go to Paris for the weekend. Amy gets 300 euro for £250.
  How many euro does Bianca get for £275?
- **b** When they return, they go back to the bank to sell the euro they have left. Amy gets £40 for 50 euro.
  How much does Bianca get for 105 euro?

5 5 miles is approximately equal to 8 kilometres.

- **a** The distance from Southampton to Sheffield is 195 miles. How far is this in kilometres?
- **b** The distance from Barcelona to Madrid is 628 kilometres. How far is this in miles?

6 A box of chocolates contains milk chocolates, plain chocolates and white chocolates in the ratio 4 : 3 : 2

- **a** What fraction of the chocolates is:
  i milk **ii** plain **iii** white?
- **b** Show how you can check your answers to part **a**.

7 The ratio of men to women on a holiday cruise is 3 : 5

What percentage of the people on the cruise are women?

 **8** The scale on a map is 1 : 50 000

- **a** The distance between two landmarks on the map is 12 centimetres.
  Find the actual distance in kilometres.
- **b** Kathy says that the distance between these landmarks on a map with a scale of 1 : 25 000 will be 6 cm.
  Is she correct? Explain your answer.

  9

Road signs use ratios or percentages to give the gradients of hills.

Ratio of vertical distance : horizontal distance

Vertical distance as a % of the horizontal distance

Which of the hills described by these road signs is steeper? Explain your answer.

  10

When you enlarge a photograph, the ratio of the height to width must stay the same. If the ratio is different, the objects in the photograph will look stretched or squashed.

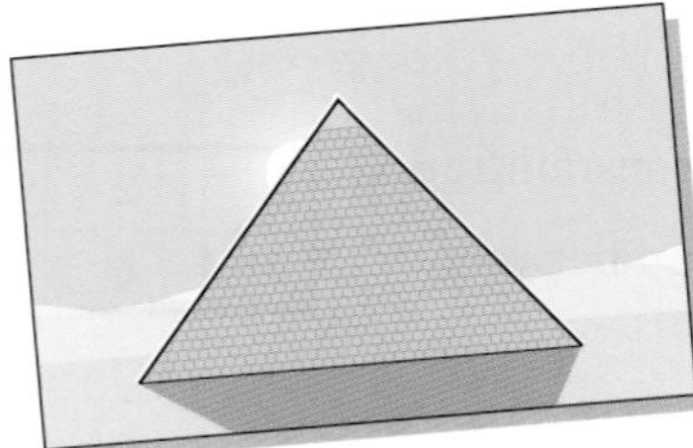

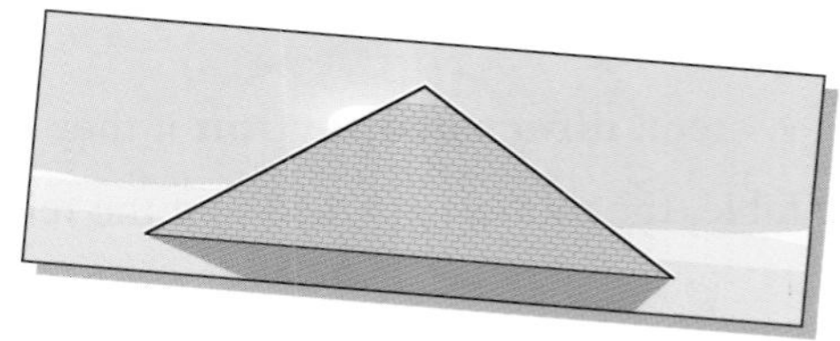

a Chloe has a photo of her favourite pop group that is 15 cm wide and 10 cm high. What is the ratio of width to height in its simplest form?

b Chloe wants to put an enlarged copy of the photo in a frame on her bedroom wall. The table gives the sizes of the frames she can buy.

Which of these frames is most suitable? Explain your answer.

Frame	Width (cm)	Height (cm)
A	25	20
B	30	25
C	40	30
D	45	30
E	50	40

  11

The table on the right gives the ages of the children who are booked into a pre-school crèche. The crèche has three rooms: one for children under two years old, one for two-year olds and one for children aged three years and over.

Age (years)	Number of children	
	morning	afternoon
0	2	1
1	4	2
2	6	9
3	5	8
4	2	1

The minimum adult : child ratios for crèches are given in the table below.

Age	Minimum adult : child ratio
Children under 2 years	1:3
Children aged 2 years	1:4
Children aged 3–7 years	1:8

How many staff does the crèche manager need:

a in the morning **b** in the afternoon?

  12

a Which size of shampoo bottle gives the best value for money?

You must show all your working and give a reason for your answer.

b Why might someone buy a different size?

Buy 1, get 1 free

20% extra free

10% off marked price

13 Students at a school can visit a theme park, a zoo or a safari park.
The table shows how many have chosen each place.

Choice	Number of students
Theme park	124
Zoo	76
Safari park	98

The school's policy is to have a maximum child : adult ratio of 8 : 1 on school visits.
There are 20 teachers available and some parents have offered to go on the visits if they are needed.

How many parent volunteers are needed?

Learn... 4.4 Direct and indirect proportion

Direct proportion

Two variables, x and y, are in **direct proportion** if they have a constant ratio.

x	2	4	6
y	6	12	18

For the values in the table, the ratio $x:y = 1:3$ and the relationship between x and y is $y = 3x$

If one of the quantities is multiplied or divided by a number, then so is the other.
For example, if x is doubled, so is y. If x is divided by 3, so is y.

In general, y **is directly proportional to** x and can be written as $y \propto x$. This means that $y = kx$ where k is a constant. The graph of y against x is a **straight line through the origin, (0, 0)**, with **gradient** k.

See also Unit 2 Chapter 6 Learn 6.2.
$y = kx + c$ represents a straight line with gradient k and intercept $(0, c)$ so $y = kx$ is a straight line with gradient k and intercept $(0, 0)$.

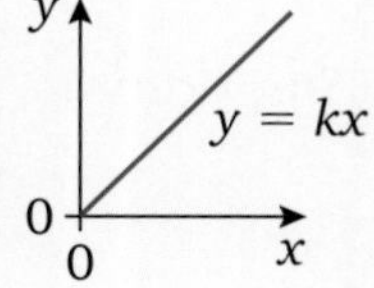

If $y = kx^2$, then the quantities x^2 and y have a constant ratio (but x and y change at different rates).
This gives a different type of direct variation.

In this case y is directly proportional to x^2. written as $y \propto x^2$

Similarly, if $y = kx^3$, then y is directly proportional to x^3 written as $y \propto x^3$

and if $y = k\sqrt{x}$, then y is directly proportional to $\sqrt{x}$. written as $y \propto \sqrt{x}$

k is called the **constant of proportionality.**

Indirect proportion

Two quantities, x and y, are in **indirect proportion** if $xy = k$ where k is a constant. As one quantity increases, the other decreases.

x	2	4	6
y	12	6	4

For the values in the table $xy = 24$
If x is doubled, then y is halved. If x is divided by 3, then y is multiplied by 3.

The equation $xy = k$ can be written as $y = \frac{k}{x}$ or $y = \frac{1}{x}$

So when y is indirectly proportional to x, $y \propto \frac{1}{x}$

There are also other types of indirect variation.

For example, if $y = \frac{k}{x^2}$, then y is indirectly proportional to x^2 written as $y \propto \frac{1}{x^2}$

and if $y = \frac{k}{\sqrt{x}}$, then y is indirectly proportional to $\sqrt{x}$. written as $y \propto \frac{1}{\sqrt{x}}$

Example: $y \propto x^2$ and when $x = 2$, $y = 36$

a Find the value of y when $x = 4$

b Find the value of x when $y = 4$

Solution: $y \propto x^2$ so $y = kx^2$

$x = 2$ when $y = 36$, so $36 = k \times 4$ and $k = 9$

the equation connecting x and y is $y = 9x^2$

a When $x = 4$, $y = 9 \times 16$

$y = 144$

b When $y = 4$, $4 = 9x^2$

$\frac{4}{9} = x^2$

$x = \pm\frac{2}{3}$

Example: y is directly proportional to $\sqrt{x}$.

Work out the missing value in the table.

x	400	
y	10	12

Solution: $y \propto \sqrt{x}$ and so $y = k\sqrt{x}$

When $x = 400$, $y = 10$, so $10 = k \times 20$ and $k = 0.5$

The equation connecting x and y is $y = 0.5\sqrt{x}$

When $y = 12$

$12 = 0.5 \times \sqrt{x}$

$\sqrt{x} = 24$

$x = 24^2$

$x = 576$

$\sqrt{x}$ means the positive square root of x.

Example: Which of these formulae indicate that the quantities x and y are inversely proportional to each other? (k is a constant.)

a $x = \frac{k}{y}$ **b** $xy = k$ **c** $x = yk$ **d** $y = \frac{x}{k}$ **e** $x + y = k$ **f** $xy = \frac{1}{k}$

Solution: Those formulae that can be written as xy = a constant are those in which x and y are inversely proportional to each other

a Yes **b** Yes **c** No **d** No **e** No **f** Yes, because $\frac{1}{k}$ is a constant.

Example: The length, l cm, and width, w cm, of a rectangle vary in such a way that its area is a constant 20 cm^2.

a Show that $l \propto \frac{1}{w}$ and write down the constant of proportionality.

b Draw a graph of l against w.

Solution: **a** The area of a rectangle is given by the product of its length and width.

In this case $lw = 20$ which can be rearranged to give $l = \frac{20}{w}$

So $l \propto \frac{1}{w}$ (l is inversely proportional to w) and the constant of proportionality is 20.

b Here is a table of some possible values for l and w.

w	1	2	4	5	10	20
l	20	10	5	4	2	1

The graph of $l = \frac{20}{w}$ is a curve (called a rectangular hyperbola).

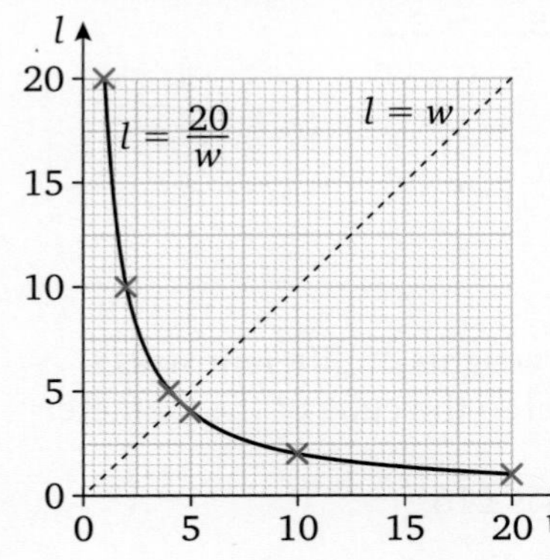

All inverse proportion graphs have this shape. The line $l = w$ is the line of symmetry.

The graph never touches the w- and l-axes but gets closer and closer to them as w gets larger and as l gets closer to zero.

The length and width of a rectangle cannot be negative. However, if w and l represent just numbers, they can be negative.

For example, in $l = \frac{20}{w}$, w could be -1 and l could be -20, or w could be -10 and l could be -2.

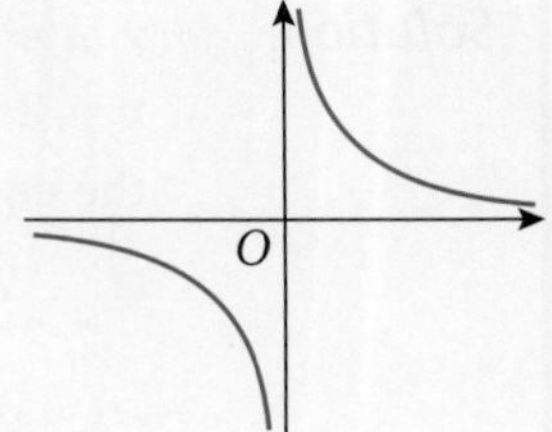

This means that there is another part of the graph of l against w, where both w and l are negative. This is a 180° rotation about the origin of the curve shown above.

The complete graph of an indirectly proportional relationship is of the form shown here.

AQA Examiner's tip

Make sure that you can recognise proportional relationships and their graphs.

Practise... 4.4 Direct and indirect proportion

k! D C B A A*

A

1 y is directly proportional to the square of x. When $x = 5$, $y = 100$

a Work out an equation connecting y and x.

b Draw a graph of y against x for $0 \leqslant x \leqslant 5$

2 T is inversely proportional to V. When $V = 60$, $T = 4$

a Obtain an equation connecting T and V.

b Work out the value of V when $T = 12$

3 A triangle has a fixed area of $15\,\text{cm}^2$.

a Express the formula connecting the base, b cm, and the perpendicular height, h cm, in the form $b = \frac{k}{h}$, and find the value of the constant of proportionality, k.

b Find three possible pairs of values for b and h.

c Sketch a graph of b against h.

4 You are given that $x \propto \frac{1}{\sqrt{y}}$ and when $y = 2.25$, $x = 3.2$

a Find an equation connecting x and y.

b Work out the value of x when $y = 0.25$

c Work out the value of y when $x = 1.2$

5 A glass company makes hemispherical paperweights of different sizes. The price of each paperweight is proportional to the cube of the diameter of the paperweight.
The price of a paperweight of diameter 8 cm is £16.

What is the price of a smaller paperweight of diameter 6 cm?

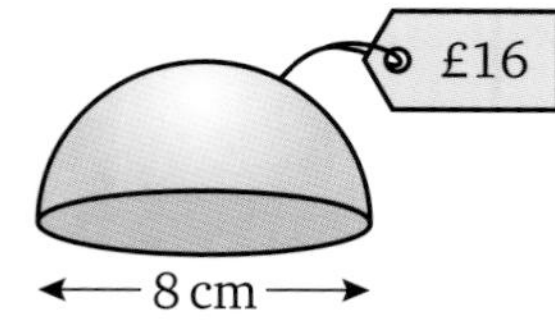

A

6 In each part, say which of the following relationships apply and find the constant of proportionality.

$y \propto x \qquad y \propto x^2 \qquad y \propto \frac{1}{x} \qquad y \propto \frac{1}{x^2}$

a Diameter of a circle $= x$ cm, circumference of the circle $= y$ cm

b Radius of a circle $= x$ cm, area of the circle $= y$ cm

c Price of bananas $= x$ pence per kilogram, the weight of bananas you can buy for £2 $= y$ kg

d Average speed of a car $= x$ mph, the time taken to travel 250 miles $= y$ hours

e Distance between two landmarks on a map with a scale of 1 : 50 000 $= x$ cm, actual distance between the landmarks $= y$ km

f Mass of each biscuit made at a factory $= x$ grams, number of biscuits the factory can make from 10 kilograms of biscuit dough $= y$

7 For each of these graphs, say whether it shows $y \propto x$ or $y \propto \frac{1}{x}$ or neither.

a

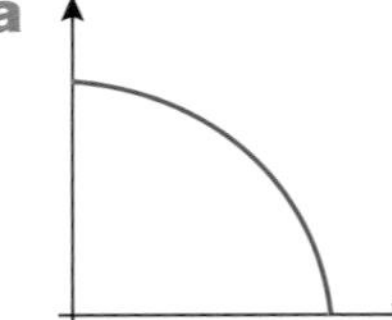

b

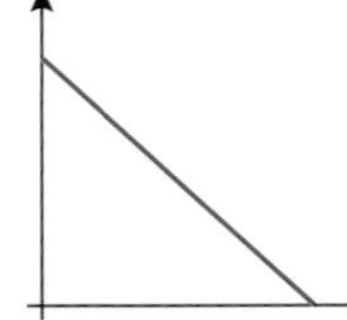

c

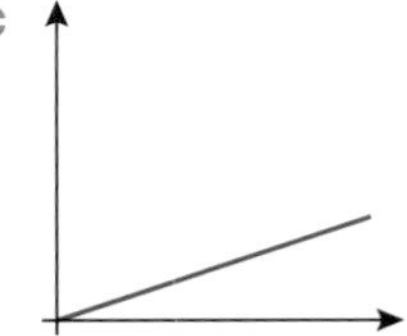

d

e

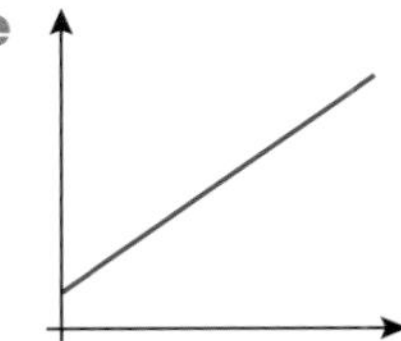

A*

8 You can use Body Mass Index (BMI) to work out whether your weight is underweight, healthy, overweight or obese. It is worked out from your weight (W kilograms) and height (h metres) using

$$\text{BMI} = \frac{W}{h^2}$$

a Sketch a graph of BMI against W (assuming a constant height h).

b Sketch a graph of BMI against h (assuming a constant weight W).

9 Peter and Carly work in a supermarket. Peter earns £8.75 an hour and Carly earns £10 per hour.

a Write down a formula to show:

i the relationship between Peter's pay and the hours worked

ii the relationship between Carly's pay and the hours worked.

b Sketch a graph to compare Peter's and Carly's earnings.

10 Calls on a mobile phone cost 25 pence per minute during peak periods and 15 pence per minute in off-peak periods.

a Draw a graph to show the cost of peak calls and the cost of off-peak calls for times up to 200 minutes.

b Find:

i the difference between the peak and off-peak cost of a call that lasts 45 minutes

ii the number of extra minutes you get for £15 during off-peak periods.

c A contract phone costs £15 per month. For this you get 60 minutes of 'free' calls and pay for all other calls at 20 pence per minute.

i Draw another line on your graph to show this.

ii Is the cost of calls on this mobile phone contract proportional to the length of the calls in minutes?
Give a reason for your answer.

11 z is inversely proportional to y^2 and y is directly proportional to $\sqrt{x}$.

Describe the relationship between z and x and sketch a graph of z against x.

4 Assess

D

1 A school buys 10 bottles of milk for drinks at a parents evening.
Each bottle holds enough for 25 drinks.
They make 235 drinks.

Calculate the percentage of milk used.

2 What percentage of this shape is blue?

5 cm
2 cm
2 cm 4 cm
Not drawn accurately

3 168 men and 210 women book a holiday cruise.

Write the ratio of the number of men to the number of women:

a in its simplest form

b in the form $1 : n$

D C

4 The table shows the amounts needed to make 24 fruit biscuits.

Ingredient	Amount for 24 biscuits (g)
Flour	300
Butter	150
Sugar	100
Fruit	120

Calculate the amount of each ingredient needed to make 36 fruit biscuits.

C

5 In the 1908 Olympic Games, Reggie Walker won the 100 metres in a time of 10.8 seconds.
A century later, in the 2008 Olympics, Usain Bolt won the 100 metres in 9.69 seconds.

Calculate the percentage decrease in the winning time.

6 It costs £259.80 for 20 m^2 of carpet.

How much does it cost for 36 m^2 of the same carpet?

7 Sun tan lotion is sold in two different sizes: small and large.

a Which bottle gives the best value for money?

b Give one reason why you might prefer to buy the other bottle.

B

8 A company has spent £25 000 on new equipment.
The value of the equipment falls by 18% in the first year and 12% in the second year after it is bought.

a How much is the equipment worth when it is two years old?

b The manager says that the value of the equipment has fallen by 30% in two years.

Is she correct? Explain your answer.

B

9 In a flu epidemic, the number of people who have flu is increasing at a rate of 10% per week.
It is estimated that 3200 people have flu this week.

a Predict, to the nearest hundred, the number of people who will have flu 8 weeks from now.

b Sketch a graph of the number of people who have flu against time.

c Describe this growth and give a reason why it cannot continue in the long term.

A

10 P and Q are both positive quantities. P is inversely proportional to Q.
When $Q = 12$, $P = 15$

a Express P in terms of Q.

b What is the value of P when $Q = 10$?

c What is the value of Q when $P = 10$?

d Sketch a graph of P against Q.

11 Here are sketches of three graphs.

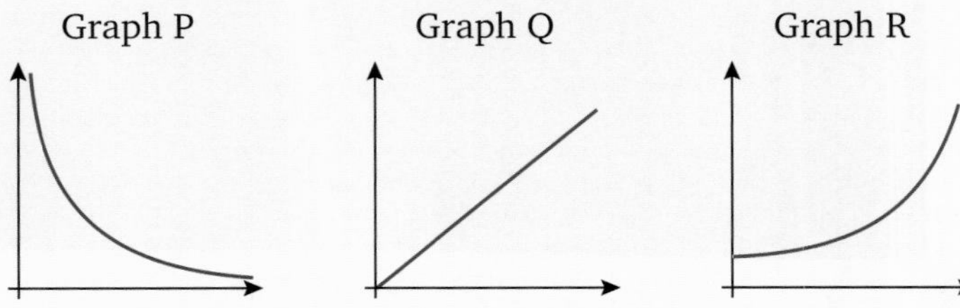

Copy and complete the following statements.

a $y = 20x$ matches Graph …

b $y = \frac{20}{x}$ matches Graph …

c $y = 20 \times 1.5^x$ matches Graph …

AQA Examination-style questions 

1 The size of the North Sea cod stock in 2008 was estimated at 250 000 tonnes.
Because of over-fishing it is decreasing at the rate of 11% per annum.
How many years will it be before the North Sea cod stock falls below the critical level of 70 000 tonnes?
You must show your working and justify your answer fully. *(4 marks)*

AQA 2008

2 75 scientists are trapped in the Antarctic.
They have enough food for 30 days on full rations.
After 16 days on full rations, a rescue party of 9 people arrives.
The rescue party brings enough food to increase existing supplies by 60%
The weather then gets worse and both the scientists and the rescue party are trapped.
They decide to go on half rations.
How many more days before the food runs out? *(4 marks)*

AQA 2008

5 Area and volume 1

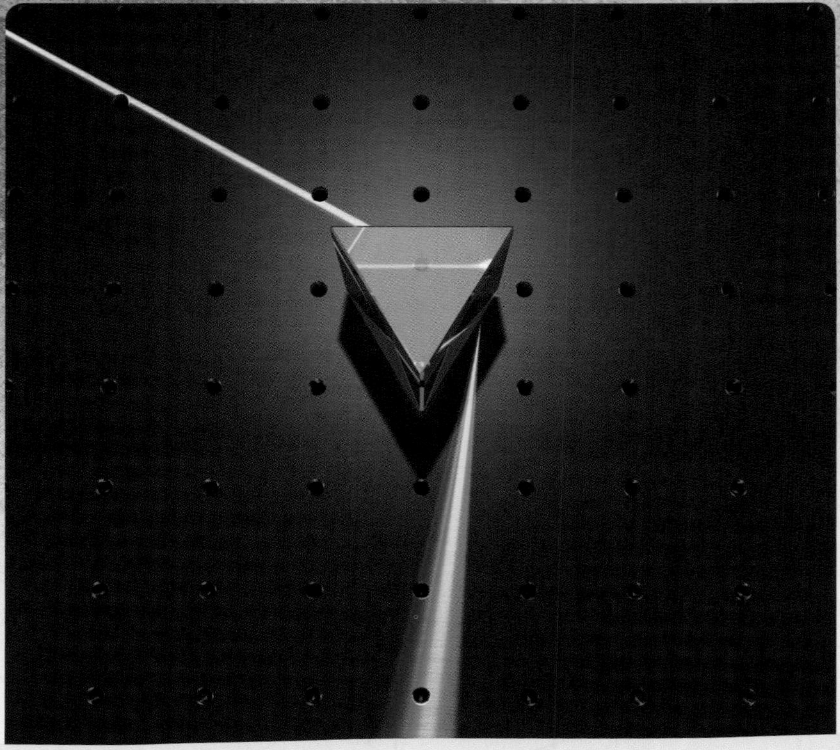

Objectives

Examiners would normally expect students who get these grades to be able to:

D

convert between square units such as changing 2.6 m^2 to cm^2

C

convert between cube units such as changing 3.7 m^3 to cm^3

find the volume of prisms including cylinders

find the surface area of simple prisms.

Did you know?

Optical prisms

In the study of light, a **prism** is a transparent optical element with flat, polished surfaces that can be used to refract light (break light up) into different colours.

In Isaac Newton's time, it was believed that white light was colourless, and that the prism itself produced the colour. Newton's experiments convinced him that all the colours already existed in the light, and that particles of light were fanned out because particles with different colours travelled with different speeds through the prism. This causes light of different colours to leave the prism at different angles, creating an effect similar to a rainbow. So a prism can be used to separate a beam of white light into its spectrum of colours.

The traditional geometrical shape is that of a triangular prism with a triangular base and rectangular sides, but there are other shapes that are also known as prisms.

Key terms

dimension
solid
cross-section
prism
volume
surface area
face
net

You should already know:

- ✔ how to find the area of rectangles, triangles, parallelograms and circles
- ✔ how to convert between metric units, such as centimetres to metres
- ✔ how to find the volume of a solid by counting cubes
- ✔ that area is measured in square units and volume in cube units
- ✔ how to draw simple nets.

Learn... 5.1 Volume of a prism

Shapes such as rectangles have two **dimensions**: length and width.
Shapes that have a third dimension such as thickness or height are called **solids**.
Solids that have the same **cross-section** all the way through the shape are called **prisms**.
Here are some prisms with their cross-sections shaded.

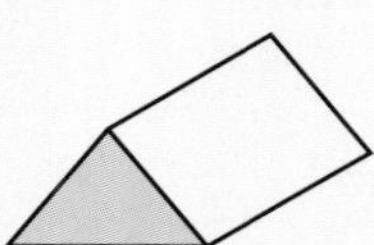
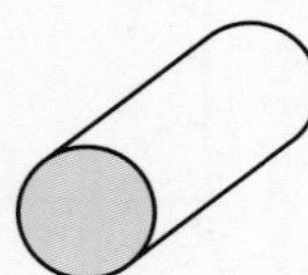
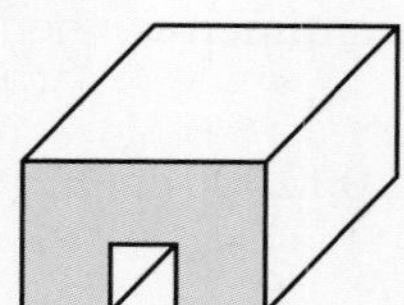
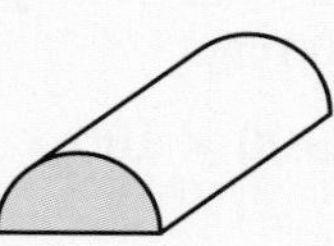

The formula for calculating the **volume** of a prism is:

volume = area of cross-section × length

Remember that the units should always be stated.

AQA *Examiner's tip*
This formula will be given on the exam paper.

Example: Work out the volume of each of these prisms.

Not drawn accurately

a

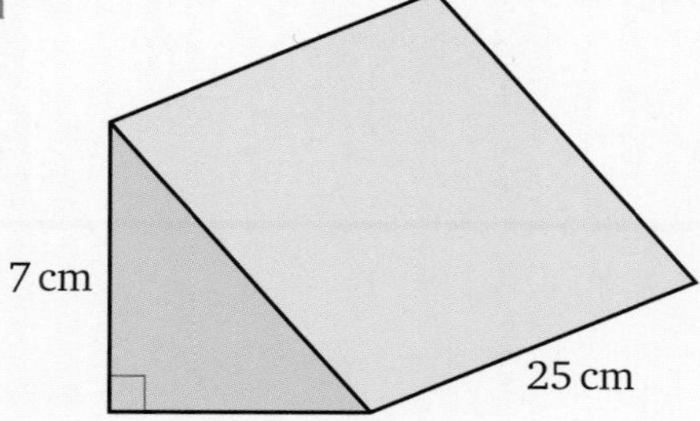

b

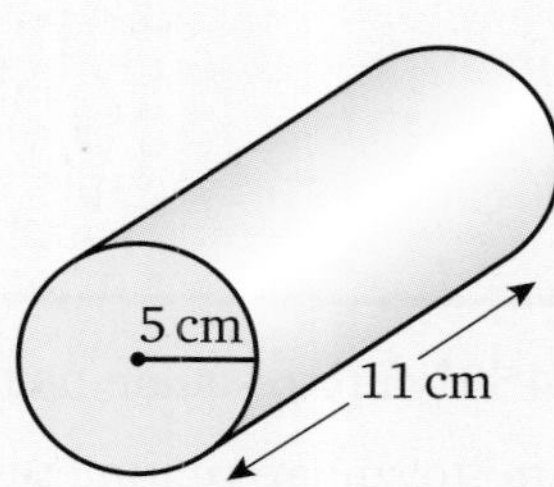

c

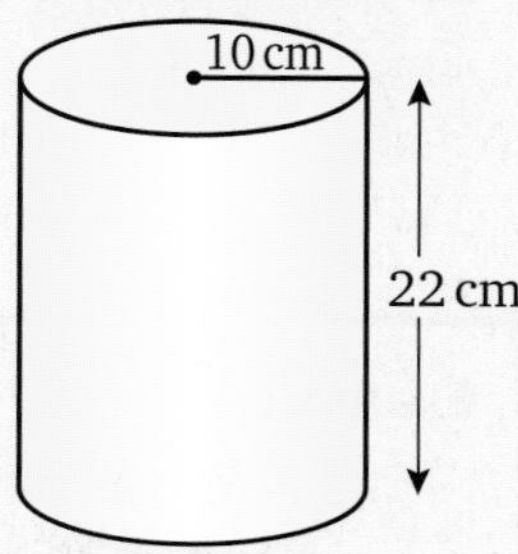

Solution:

a The cross-section of this prism is a right-angled triangle with base 6 cm and perpendicular height 7 cm.

Area of cross-section $= \frac{1}{2} \times \text{base} \times \text{height} = \frac{1}{2} \times 6 \times 7 = 21\text{ cm}^2$

Volume = area of cross-section × length

$= 21 \times 25 = 525\text{ cm}^3$

AQA *Examiner's tip*
Remember:
- area is measured in square units (such as cm^2)
- volume is measured in cube units (such as cm^3).

b The cross-section of this prism is a circle of radius 5 cm.

Area of a circle $= \pi r^2$ so area of cross-section $= \pi \times 5^2 = 25\pi\text{ cm}^2$

Volume = area of cross-section × height

$= 25 \times \pi \times 11 = 863.9\text{ cm}^3$ (to 1 d.p.)

c The cross-section of this prism is a circle of radius 10 cm.

Area of a circle $= \pi r^2$ so area of cross-section $= \pi \times 10^2\text{ cm}^2 = 100\pi\text{ cm}^2$

Volume = area of cross-section × height

$= 100 \times \pi \times 22$

$= 2200\pi\text{ cm}^3$

$= 6911.5\text{ cm}^3$ (to 1 d.p.)

Link
See Chapter 14 for the effects of enlargement on volume and area.

Notice that all the lengths in the prism (cylinder) in **c** are two times the corresponding lengths in the prism in **b**. The volume of the prism in **c**, however, is eight times the volume of the prism in **b**.)

Volume is always measured in cube units such as mm^3, cm^3, m^3.

Sometimes it is necessary to convert between these units.

You should know that there are 100 centimetres in 1 metre but how many cubic centimetres (cm^3) are there in 1 cubic metre (m^3)?

This cube has sides of 1 metre. Its volume is $1 \times 1 \times 1 = 1\,\text{m}^3$

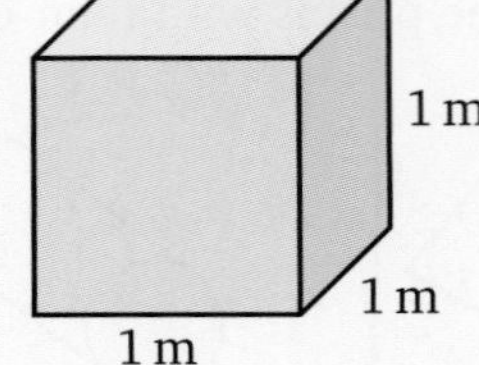

If the dimensions of the cube were given in centimetres, then each side would measure 100 cm.

The volume would be $100 \times 100 \times 100 = 1\,000\,000\,\text{cm}^3$

So $1\,\text{m}^3 = 1\,000\,000\,\text{cm}^3$

In the same way, the area of the base can be found in square metres or square centimetres.

The area of the base of this cube is $1 \times 1 = 1\,\text{m}^2$

In centimetres the area of the base is $100 \times 100 = 10\,000\,\text{cm}^2$

So $1\,\text{m}^2 = 10\,000\,\text{cm}^2$

So:

$1\,\text{m} = 100\,\text{cm}$

$1\,\text{m}^2 = 100\,\text{cm} \times 100\,\text{cm} = 10\,000\,\text{cm}^2$

$1\,\text{m}^3 = 100\,\text{cm} \times 100\,\text{cm} \times 100\,\text{cm} = 1\,000\,000\,\text{cm}^3$

Example:

a Convert 24 500 cm^2 to square metres.

b Convert 5 m^2 to square centimetres.

c Convert 7 250 000 cm^3 to cubic metres.

Solution:

a To convert cm^2 to m^2 divide by 10 000.
$24\,500 \div 10\,000 = 2.45\,\text{m}^2$

b To convert m^2 to cm^2 multiply by 10 000.
$5 \times 10\,000 = 50\,000\,\text{cm}^2$

c To convert cm^3 to m^3 divide by 1 000 000.
$7\,250\,000 \div 1\,000\,000 = 7.25\,\text{m}^3$

Practise... 5.1 Volume of a prism 

D

1 Find the volume of these solids.

a

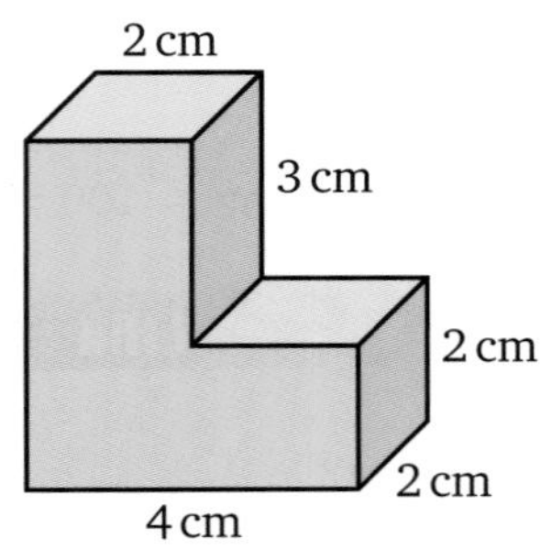

b

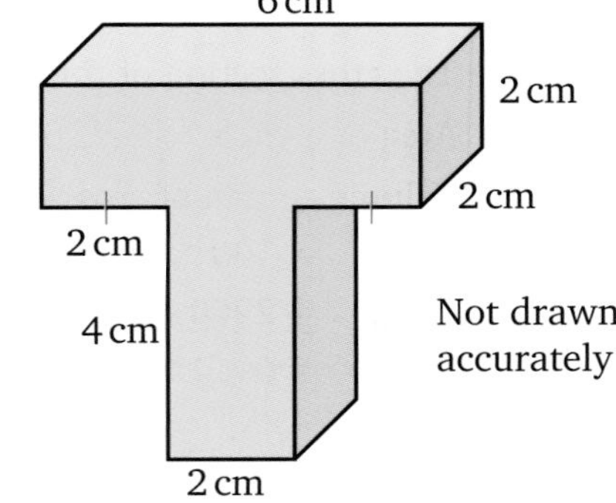

Hint

Divide each solid into cuboids.

D

2 **a** Convert the following areas to square centimetres.

i $4.6\,m^2$ **iii** $9\,m^2$ **v** $270\,mm^2$

ii $23\,m^2$ **iv** $0.5\,m^2$ **vi** $8000\,mm^2$

b Convert the following areas to square metres.

i $300\,000\,cm^2$ **iii** $57\,600\,cm^2$

ii $75\,000\,cm^2$ **iv** $8500\,cm^2$

3 This cuboid has a volume of $432\,m^3$.

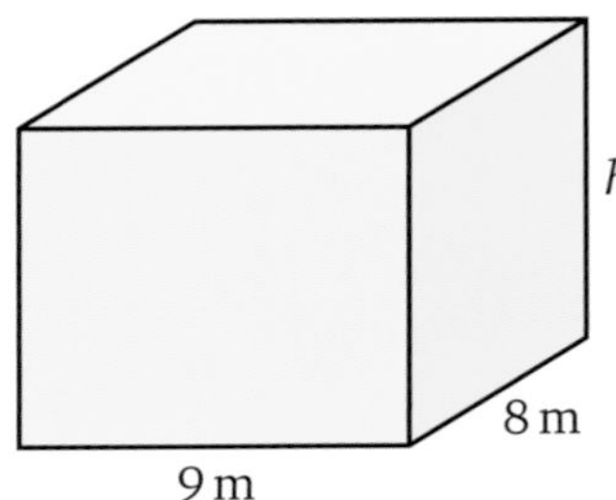

Not drawn accurately

Work out the height, h, of the cuboid.

C

4 The area of cross-section of each prism is given in these diagrams.

Work out the volume of each prism.
Remember to state the units of each answer.

a

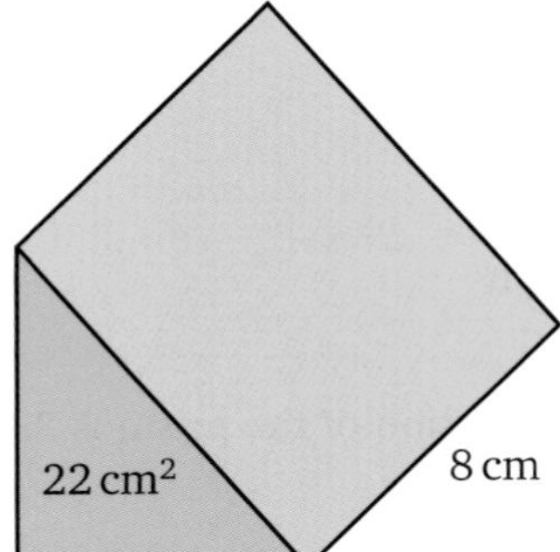

c

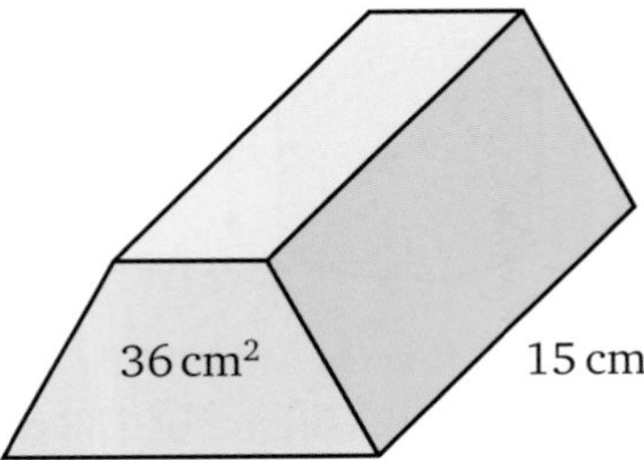

Not drawn accurately

b

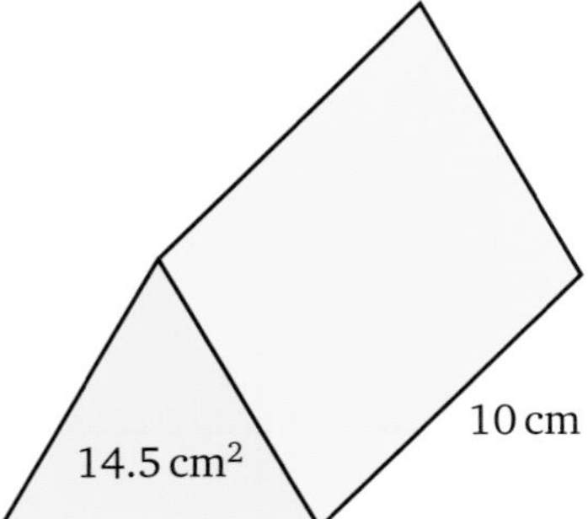

d

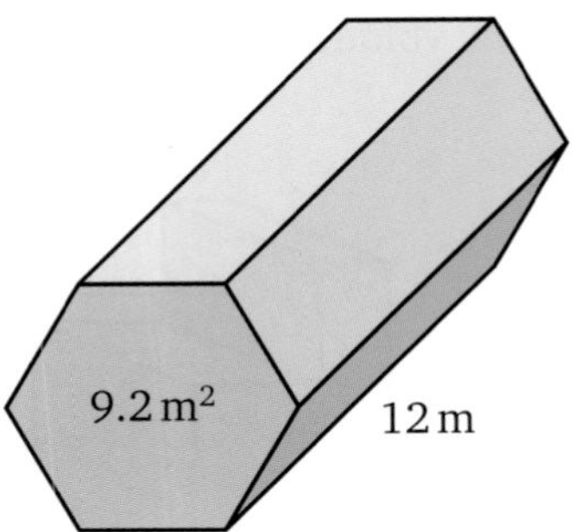

5 **a** Convert the following volumes to cubic centimetres.

i $8\,m^3$ **iii** $0.765\,m^3$ **v** $15\,600\,mm^3$

ii $3.2\,m^3$ **iv** $0.0568\,m^3$ **vi** $950\,mm^3$

b Convert the following volumes to cubic metres.

i $2\,360\,000\,cm^3$ **ii** $56\,000\,000\,cm^3$ **iii** $473\,100\,cm^3$

6 Work out the volume of each of these triangular prisms.

a

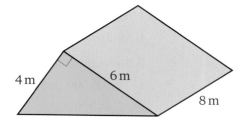

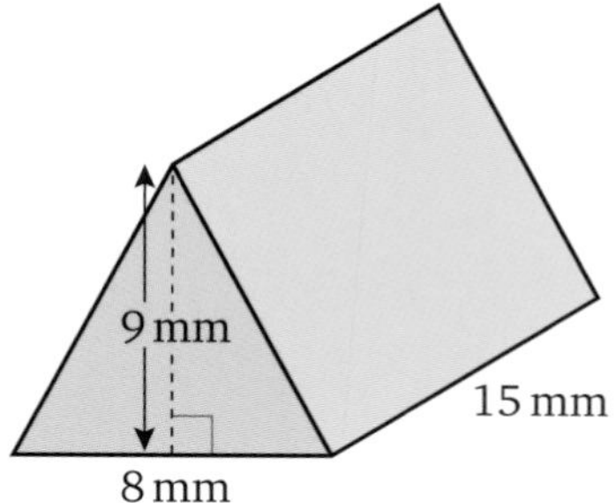

b

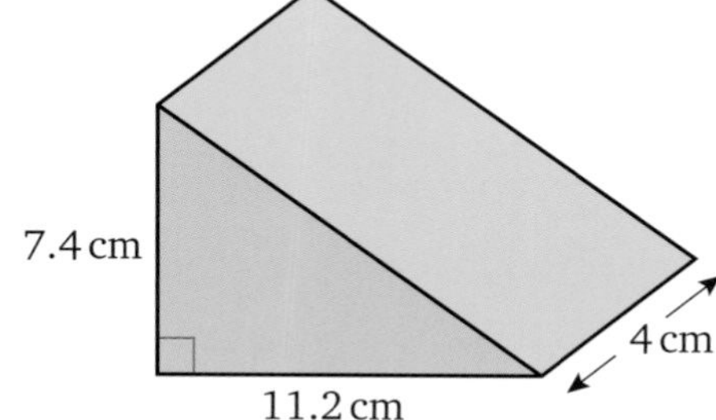

c

4 m
6 m
8 m

Not drawn accurately

7 Work out the volume of each cylinder.

a

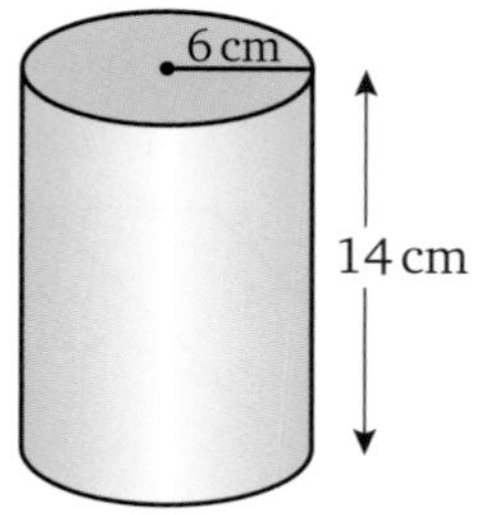

b

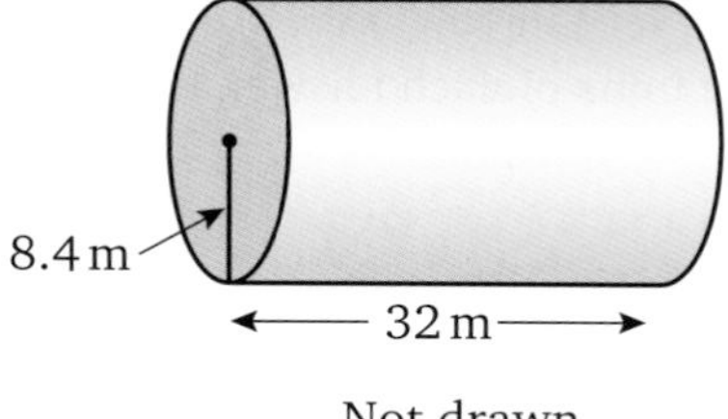

c

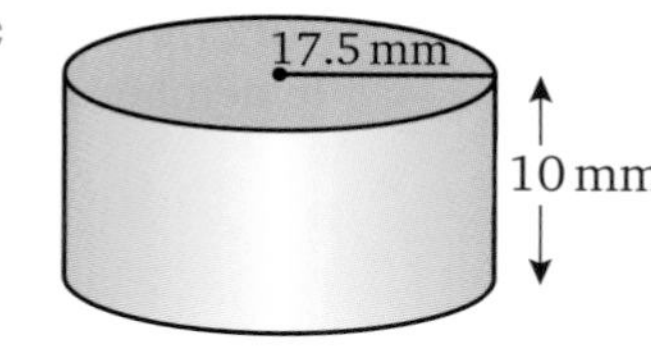

Not drawn accurately

8 A prism has a volume of $132\ \text{cm}^3$. The area of the cross-section of the prism is $33\ \text{cm}^2$.

Work out the height of the prism.

9 For each prism shown below, work out:

i the area of the cross-section

ii the volume.

Hint

Draw a sketch of the cross-section to help you.

a

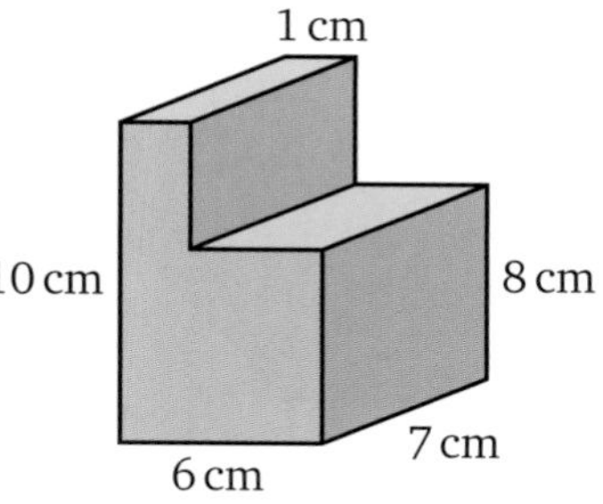

c

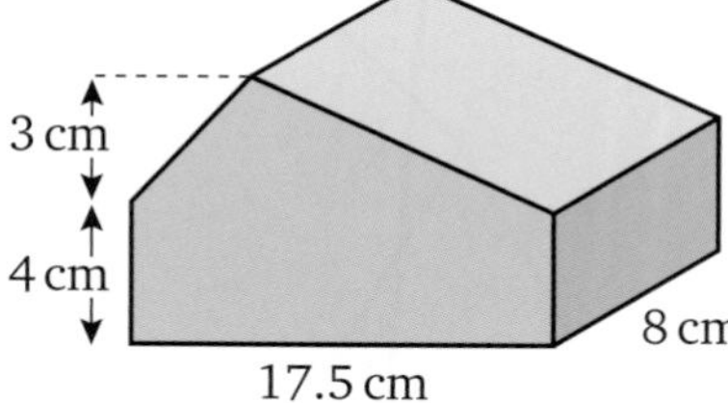

b

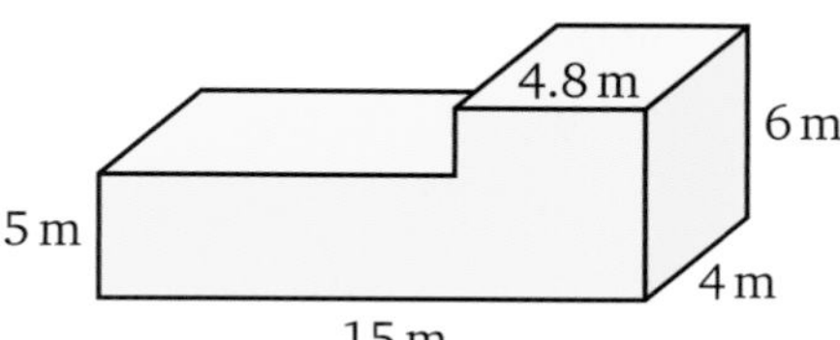

Not drawn accurately

C

10 A cylinder of height 3.2 m has a volume of 15.68 m^3.

Work out the area of the base of the cylinder.

Bump up your grade

To get a Grade C you need to know how to work out the volume of a solid from its dimensions. You also need to know how to do reverse calculations where you are given the volume and a dimension, and then work out the area of cross-section.

B

11 A 25 mm square hole is cut right through the centre of a cuboid as shown.

Find the volume of the remaining cuboid.

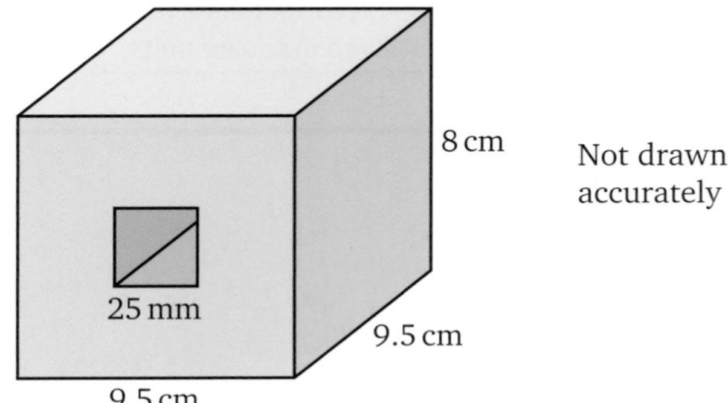

Not drawn accurately

12 The diagram shows a plastic pipe of internal radius 4 cm and length 60 cm.
The plastic has a thickness of 1 cm.

Calculate the volume of plastic in the pipe.

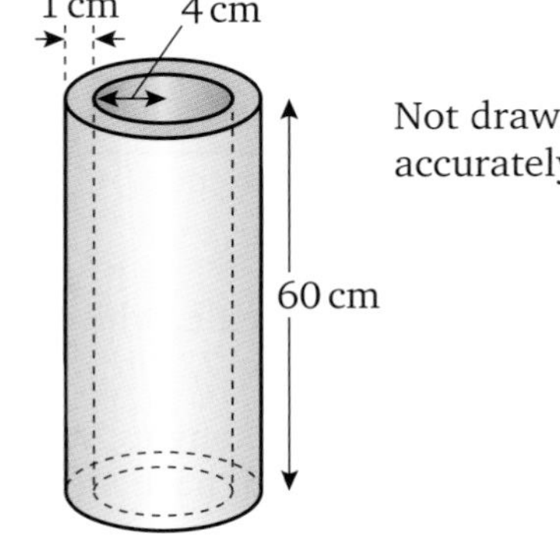

Not drawn accurately

13 The diagram shows a swimming pool.
The pool is filled with water at a rate of 350 litres per minute.

How long will it take to fill the pool?
Give your answer in hours and minutes. (1 m^3 = 1000 litres)

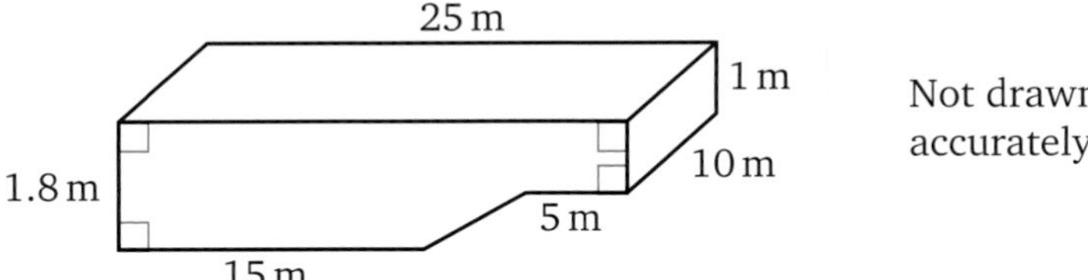

Not drawn accurately

14 At a pre-school playgroup, each of the 36 children are given a beaker of milk.
The beakers are cylinders of radius 3 cm and height 8 cm and are three quarters full.
Each milk carton contains 2.2 litres of milk.
Susie says that three cartons will be enough for all the children.

Is she correct?
Show your working.

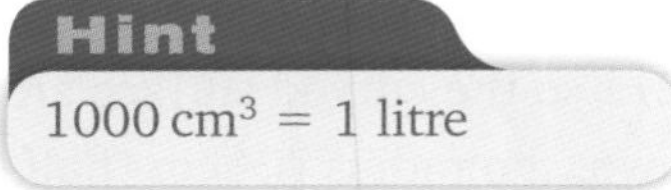

Hint

1000 cm^3 = 1 litre

15 The volume of a prism is 90 cm^3.

Find three different types of prism with this volume, giving the dimensions of each one.

Learn... 5.2 Surface area of a prism

The total **surface area** of a three-dimensional shape is the sum of the area of all the **faces** (sides) of the shape.

For example, a cube of side 3 cm is made up of six square faces, each measuring 3 cm by 3 cm.

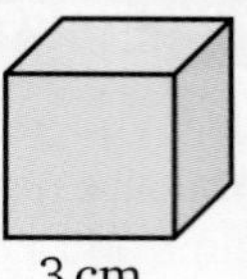

The area of each face is $3 \times 3 = 9\,\text{cm}^2$

The total surface area of all six faces is $6 \times 9 = 54\,\text{cm}^2$

It is often useful to draw the **net** of a solid to help you to see the individual areas.

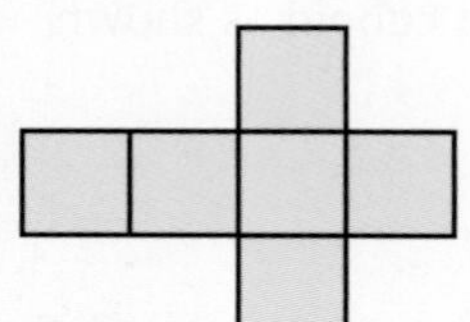

AQA Examiner's tip

Remember to give the correct units for your answer. Area is always measured in square units.

Example: Work out the surface area of this triangular prism.

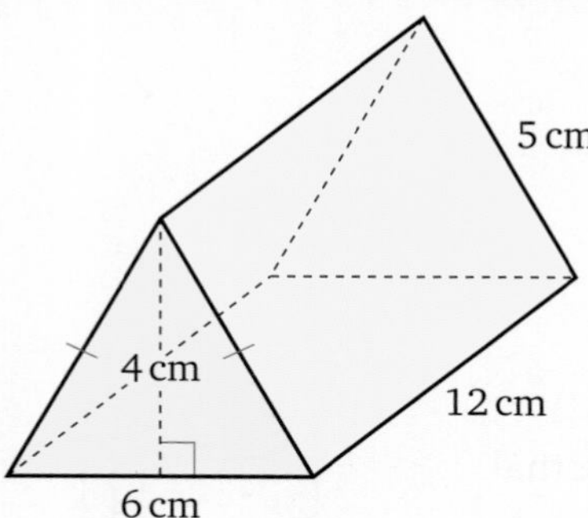

Not drawn accurately

Solution: Draw the net.

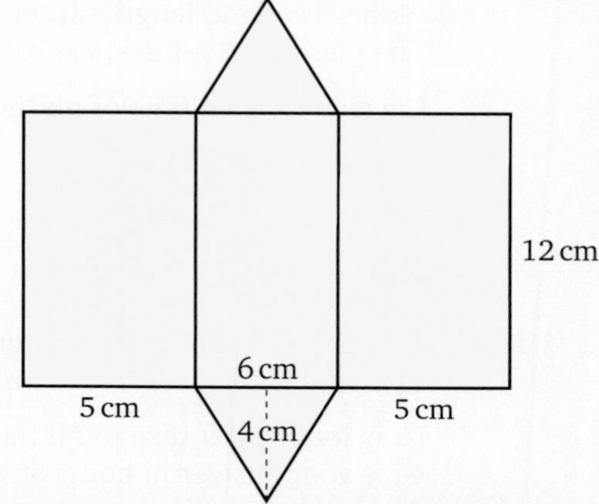

The middle rectangle, which is the base, has area $6 \times 12 = 72\,\text{cm}^2$

The two outer rectangles, which are the sloping faces, each have area $5 \times 12 = 60\,\text{cm}^2$

The triangles each have area $\frac{1}{2} \times 6 \times 4 = 12\,\text{cm}^2$

Total surface area $= 72 + (2 \times 60) + (2 \times 12)$
$= 216\,\text{cm}^2$

Example: Work out the surface area of this hollow cylinder.

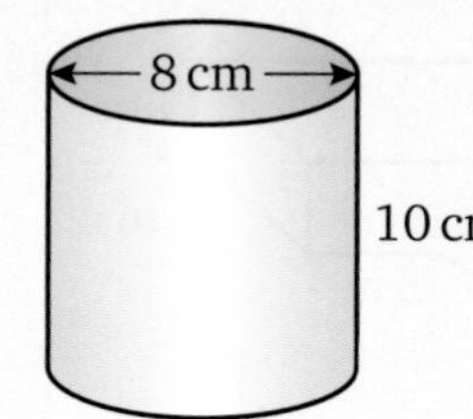

Not drawn accurately

Solution: A hollow cylinder has no base or top and is also called an open cylinder.

The curved surface of the cylinder forms a rectangle when the net is drawn.

h

$2\pi r$

The length of the rectangle is equal to the circumference of the circular end of the cylinder.

Circumference $= 2\pi r$ so the curved surface area $= 2\pi r \times h$

In the cylinder shown, the radius is 4 cm (half the diameter).

Area of the curved surface $= 2 \times \pi \times 4 \times 10 = 80\pi$ or $251.3\,\text{cm}^2$ (to 1 d.p.)

If a cylinder has a top or a base then the area of the circular top and/or base must be added to find the total surface area.

Practise... 5.2 Surface area of a prism

D C B A A*

The shapes in these exercises are not drawn accurately.

D

1 Here is a net of a cuboid.

Work out the surface area of the cuboid.

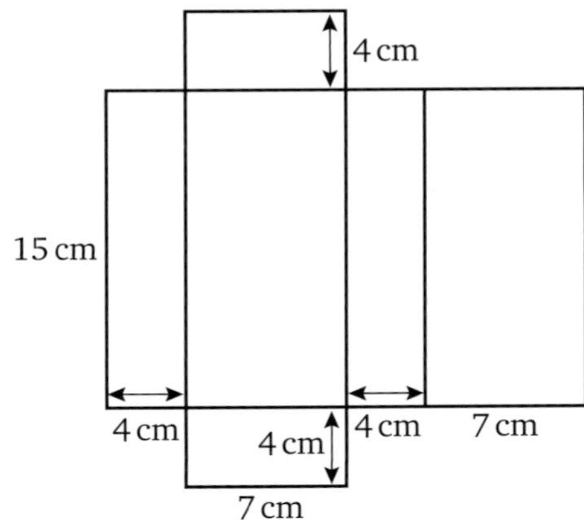

2 Here is a net of a triangular prism.

Work out the surface area.
Give your answer in:

a m^2 **b** cm^2

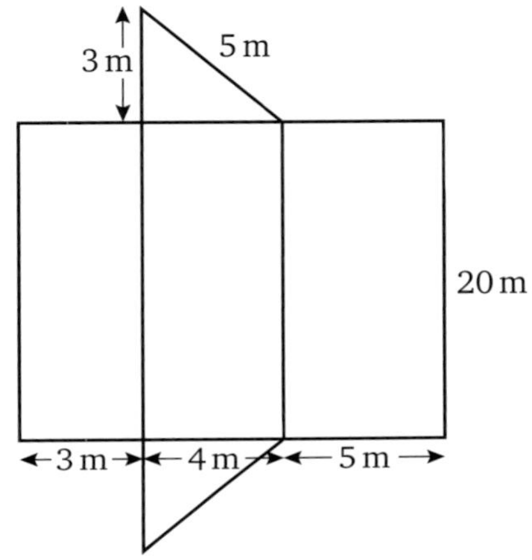

C

3 Calculate the total surface area of cubes with these side lengths.

a 7 cm **b** 10 cm **c** 5.4 cm

4 Calculate the total surface area of these cuboids.

a

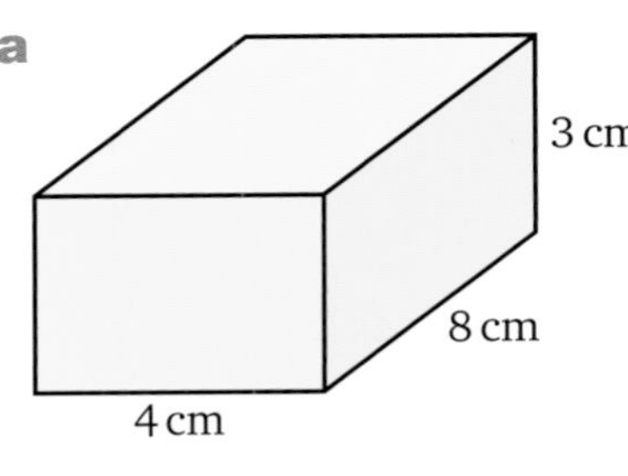

b

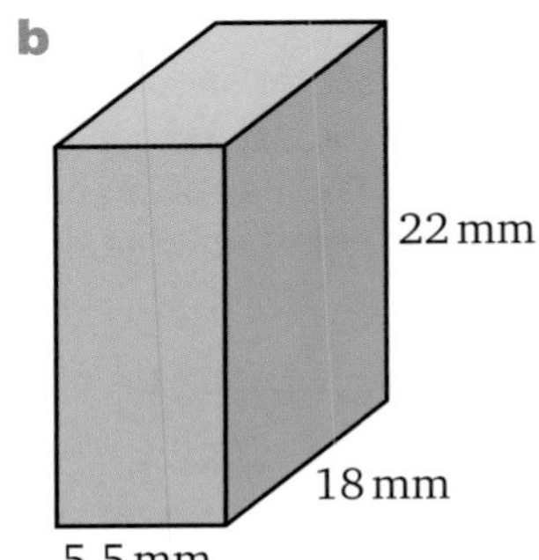

c

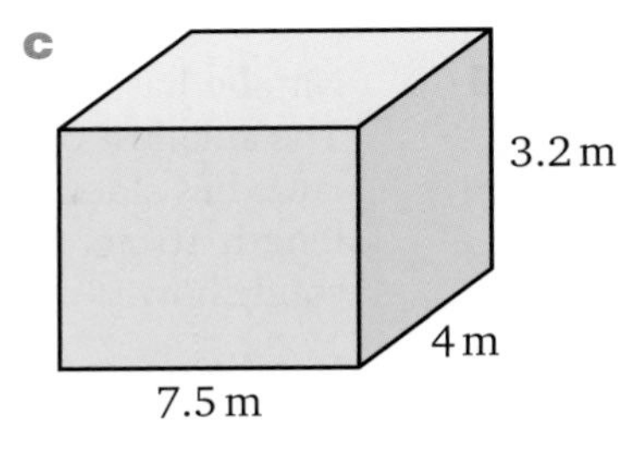

5 Work out the surface area of each triangular prism.

AQA Examiner's tip
Sketch a net to help you.

a

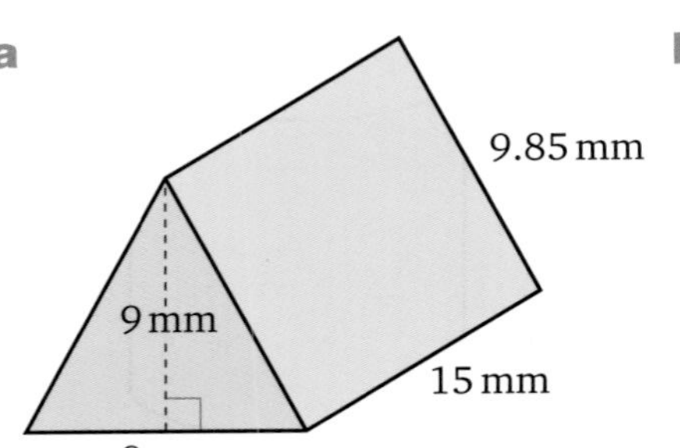

b

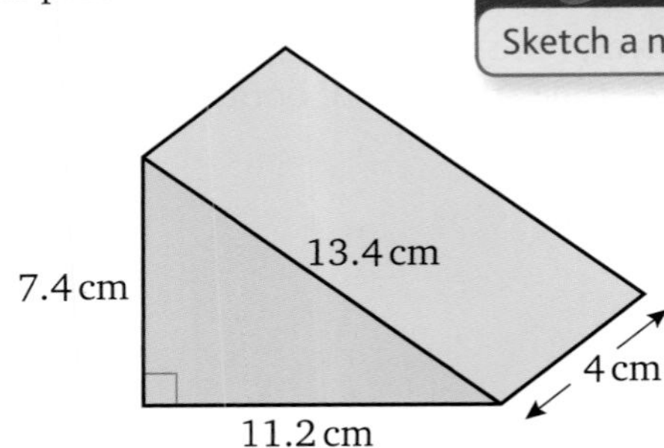

C

6 Calculate the curved surface area of each of these cylinders.

a
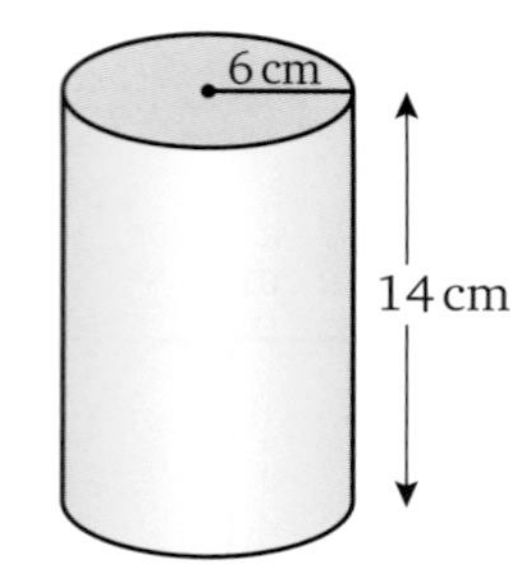

c
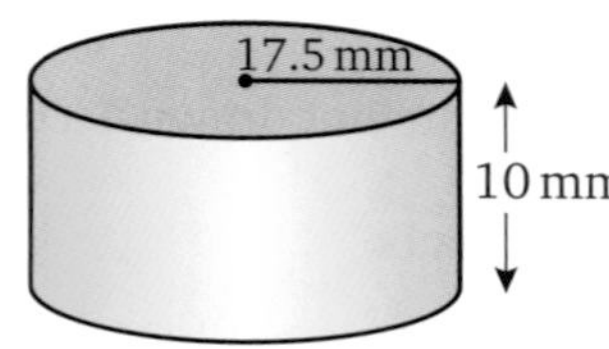

b
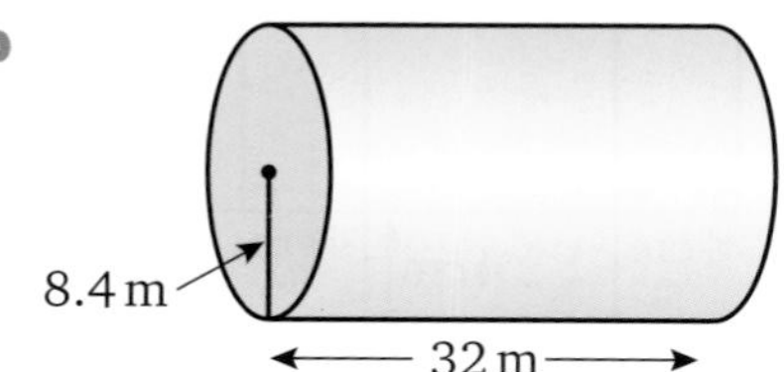

7 Here are two closed cylinders. (They have a top and a base.)
Work out the surface area of each cylinder.

a
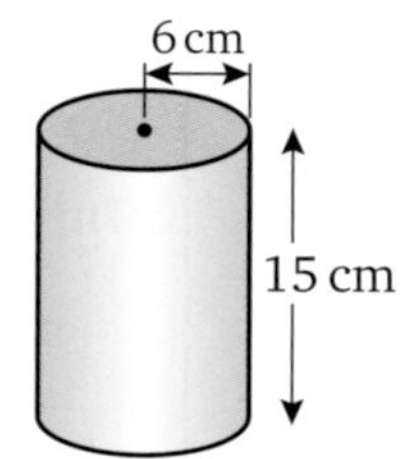

b
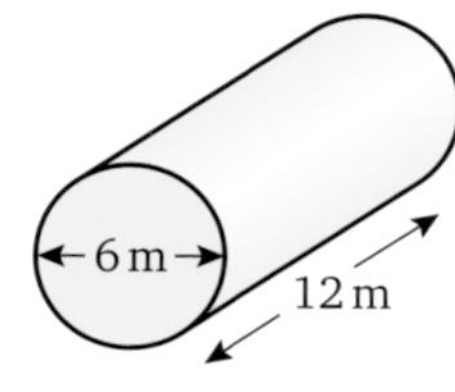

8 Andrew makes wooden jewellery boxes in the shape of cuboids of length 20 cm, width 15 cm and depth 10 cm. Each box has a lid.
The wood costs £18.75 for 1 square metre and on average he wastes 10% of each square metre.
He varnishes the outside of each box when he has made them.
A tin of varnish will cover an area of 1.5 m^2 and costs £7.99.

What is the cost of making each jewellery box?

9 Natalie has three wooden cubes that she is using in DT.
The smallest cube has sides of length 3 cm. The medium-sized cube has sides of length 6 cm. The largest cube has sides of length 10 cm. She sticks them together to make the solid shown.

Natalie wants to paint the solid red. The tin of red paint she uses will cover an area of 0.5 m^2.

Will she have enough paint? Show your working.

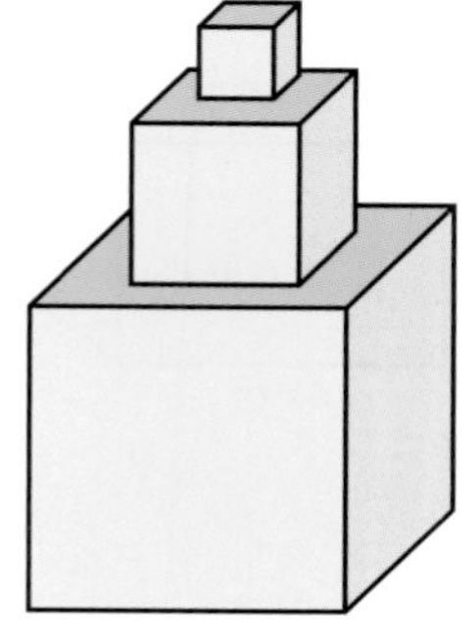

Not drawn accurately

10 The area of the curved surface of this cylinder is equal to three times the area of both ends added together.
Express h in terms of r.

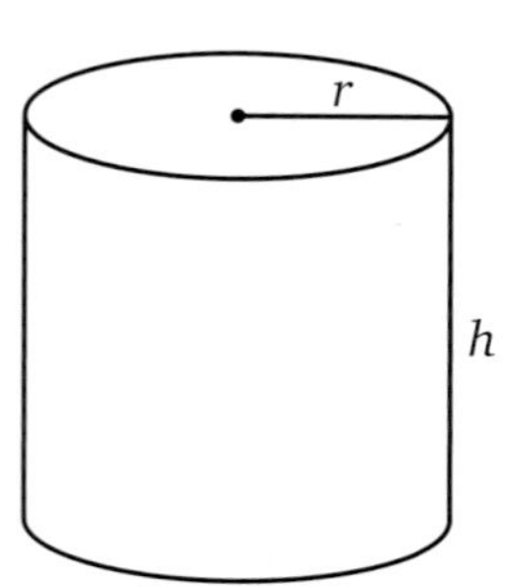

Not drawn accurately

5 Assess

C

1 A prism has a total surface area of $0.75\ m^2$.

Write this area in cm^2.

2 The table shows the measurements of some cuboids.

Calculate the total surface area of each cuboid. Remember to state the units of each answer.

Cuboid	Length	Width	Height
a	14 cm	6 cm	3 cm
b	45 mm	22 mm	10 mm
c	3.2 m	6 m	4 m
d	12.4 cm	15.5 cm	11 cm

3 Calculate the volume of this cuboid.

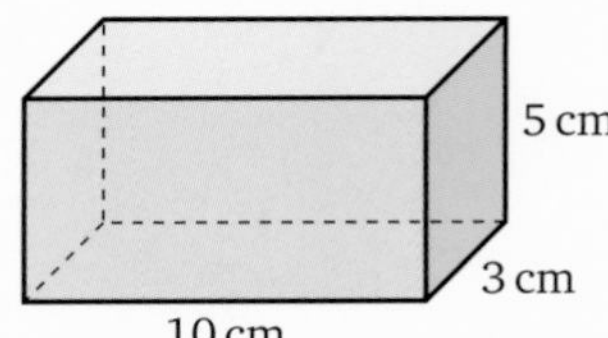

Not drawn accurately

Give your answer in: **a** cm^3 **b** mm^3

4 Calculate the volume of each of these prisms.

a

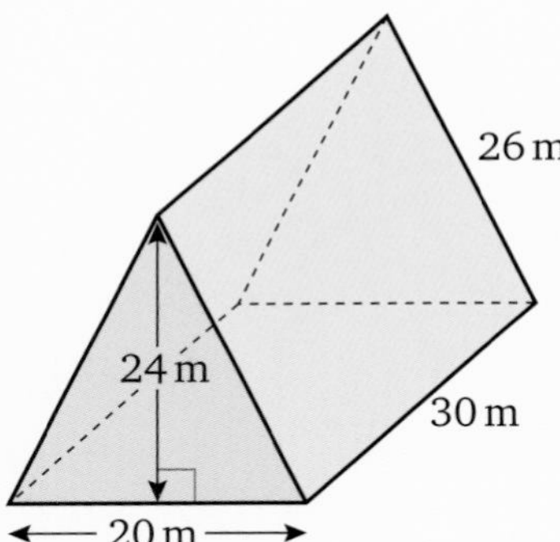

b

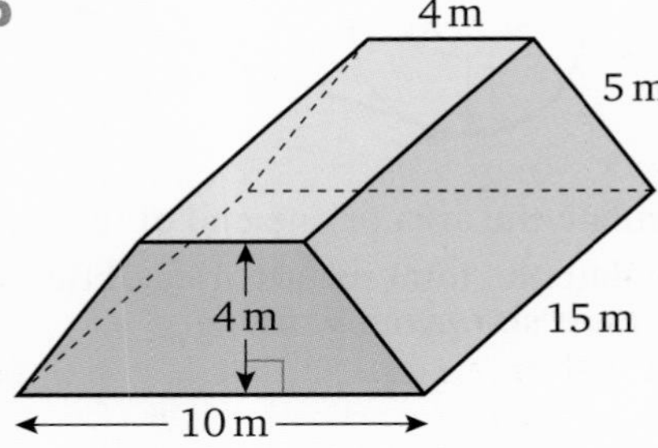

5 **a** Work out the total surface area of this triangular prism.

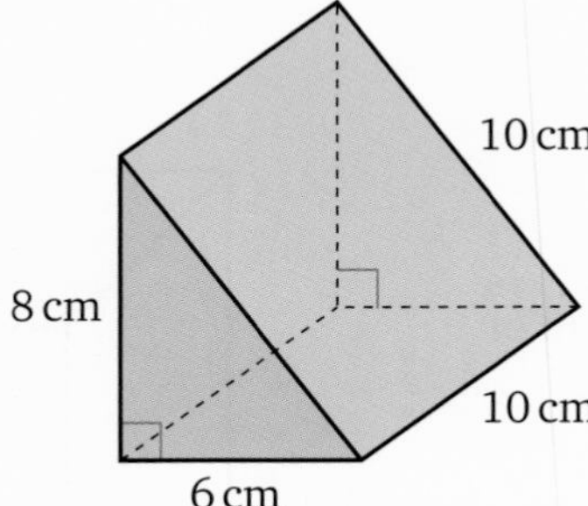

Not drawn accurately

b Another triangular prism has measurements double those of the triangular prism above.

What is its total surface area?

Copy and complete the statement 'The new surface area is _____ times the original surface area.'

6 A metal pole is in the shape of a cylinder. It has a radius of 1.5 m and a length of 17 m.

Work out the volume of metal used for the pole.

AQA Examination-style questions

1 The diagram shows a block of wood with uniform cross-section.
The cross-section is made of rectangles.
The block is 65 cm long.

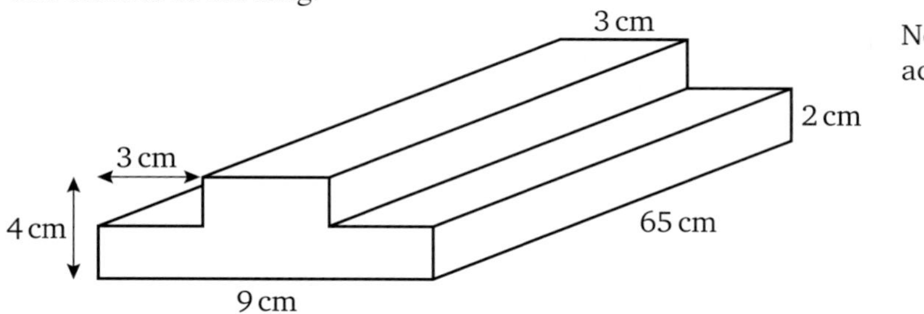

Calculate the volume of the block.
State the units of your answer. *(5 marks)*

AQA 2008

2 The diameter of a solid cylinder is 9.2 cm.
The cylinder is 25 cm long.

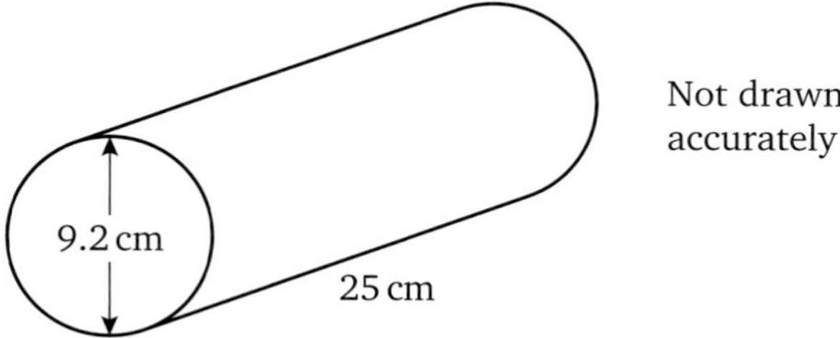

a Calculate the area of one end of the cylinder. *(2 marks)*

b Calculate the **total** surface area of the cylinder.
You **must** show your working. *(3 marks)*

AQA 2007

3 A cuboid is made from centimetre cubes.
The area of the base of the cuboid is 5 cm^2.
The volume of the cuboid is 10 cm^3.
Work out the surface area of the cuboid.
State the units of your answer. *(5 marks)*

AQA 2005

4 The diagram shows a cube.
The volume of the cube is 1000 cm^3.

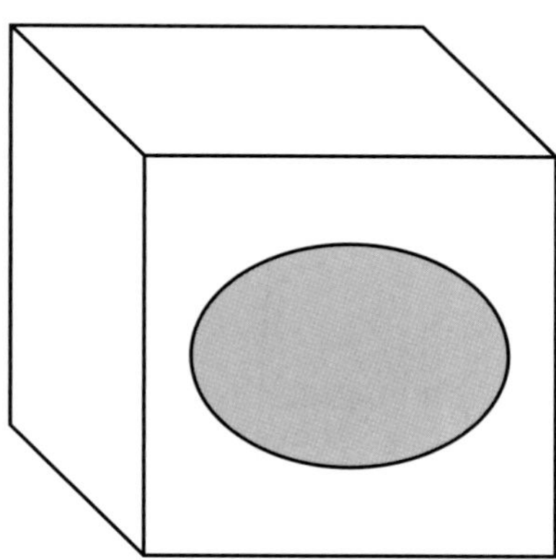

a A label covers half the area of the front of the cube.
Calculate the area of the label.
Show your working. *(3 marks)*

b The cube contains 200 cm^3 of water.
How much more water is needed for the cube to be three-quarters full?
Give your answers in litres. *(3 marks)*

AQA 2007

6 Equations and formulae

Objectives

Examiners would normally expect students who get these grades to be able to:

D

solve equations such as $3x - 4 = 5 + x$ or $2(5x + 1) = 28$

distinguish between an expression, an equation, an identity and a formula

substitute numbers into formulae such as $C = \frac{(A + 1)D}{9}$

derive complex expressions and formulae

C

solve equations such as $3x - 12 = 2(2x - 5)$, $\frac{2x}{3} - \frac{x}{4} = 5$ or $\frac{(7 - x)}{3} = 2$

B

solve equations such as $\frac{(2x - 1)}{6} + \frac{(x + 3)}{3} = \frac{5}{2}$

Did you know?

Falling down

The following formula shows how far a body will fall under gravity if air resistance is ignored:

$$h = \tfrac{1}{2}gt^2$$

h is the vertical distance travelled.

g is the acceleration of gravity on the Earth's surface.

t is the length of time the body falls.

Aristotle believed that if two objects were dropped, the heavier one would fall faster (and travel further in a given time). By dropping cannonballs from the leaning tower at Pisa, Galileo showed that all objects will travel the same distance at the same rate regardless of the mass of the body.

Fantastic formulae

Formulae exist that will make your brain spin!

Formulae can be used for finding anything, for example:

- how long it takes a coin to drop to the bottom of a well
- how long it will take a rocket to reach the moon
- how long the air will last in a fireman's breathing apparatus
- how fast a car was travelling when it crashed.

You should already know:

- ✔ how to collect like terms
- ✔ how to simplify expressions
- ✔ facts about angles in polygons
- ✔ facts about angles on straight and parallel lines
- ✔ how to expand brackets
- ✔ how to find areas of simple shapes
- ✔ about the order of operations (BIDMAS).

Key terms

equation
denominator
formula
substitution
identity

Learn... 6.1 Solving equations with x on both sides

Follow these steps to solve an **equation**:

- Collect all the terms that contain x on one side of the equation.
- Collect all the remaining terms on the other side of the equation.

Remember that positive and negative signs belong to the term following the sign.

In addition to these steps you also need to combine your equation-solving skills with knowledge from other areas of mathematics to solve problems.

Example: Solve $2x - 6 = 8 - 3x$

Solution: Collect terms in x on the left-hand side of the equation as $2x > -3x$
This produces a positive term x.

$2x - 6 = 8 - 3x$	The LHS has more x's as $2x$ is more than $-3x$.
$2x + 3x - 6 = 8 - 3x + 3x$	Add $3x$ to both sides.
$5x - 6 = 8$	
$5x - 6 + 6 = 8 + 6$	Add 6 to both sides.
$5x = 14$	
Answer: $x = 2.8$	Divide both sides by 5.

Example: $ABCD$ is a parallelogram.

Work out the value of p.

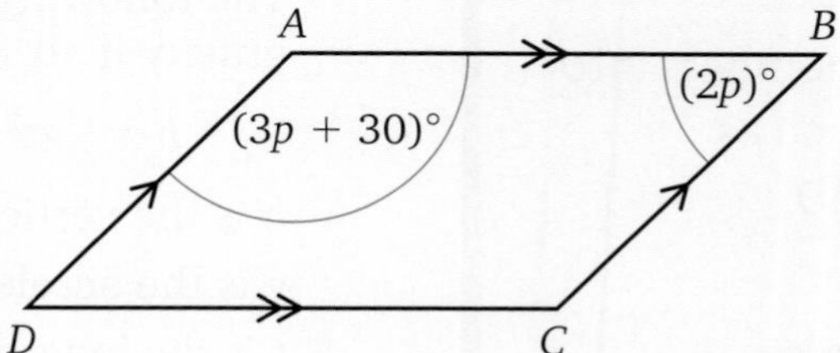

Solution: The angles in a quadrilateral add up to 360°.

Your knowledge of the angles in a parallelogram tells you that the opposite angles of a parallelogram are equal.

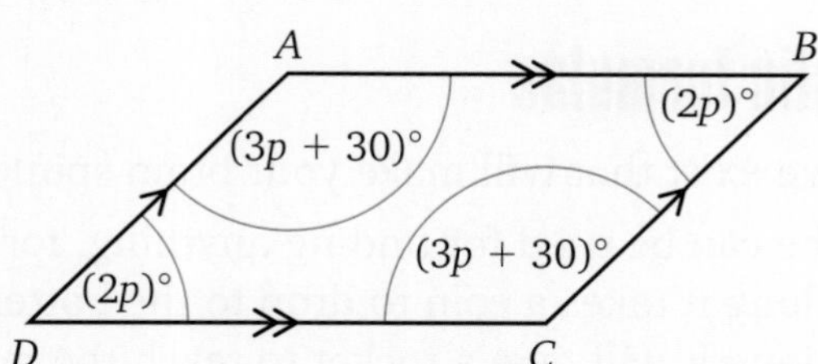

Write an equation and solve it to find p:

$3p + 30 + 2p + 2p + 3p + 30 = 360°$	
$10p + 60 = 360°$	Collect like terms.
$10p = 300°$	Subtract 60 from both sides.
$p = 30°$	Divide both sides by 10.

AQA Examiner's tip

Set your working out clearly, one step at a time. The examiner will give one mark for the first correct step, then marks for further correct steps when solving equations.

Practise... 6.1 Solving equations with x on both sides

1 Solve these equations.

a	$3x + 1 = x + 13$	**e**	$8t - \frac{1}{2} = 4t + \frac{1}{2}$	**i**	$5b + 16 = 8b + 10$
b	$5x + 8 = 2x$	**f**	$3p = 7p - 2$	**j**	$-7c - 3 = 30 - 4c$
c	$6y + 4 = -24 - y$	**g**	$2 + 2p = 4p - 1$	**k**	$10d - 0.6 = 0.9 - 5d$
d	$4z + 1.5 = 2z - 3$	**h**	$3 + 4k = 7k + 2$	**l**	$3 - 2n = 4 - 5n$

2 Jared solves the equation $9x + 7 = 9 - x$
His first step is $8x + 7 = 9$

What mistake has Jared made?

3 Rick solves the equation $3y - 4 = 6 + 2y$
He gets the answer $y = 2$

Can you find Rick's mistake?

4 AB and CD are straight lines.

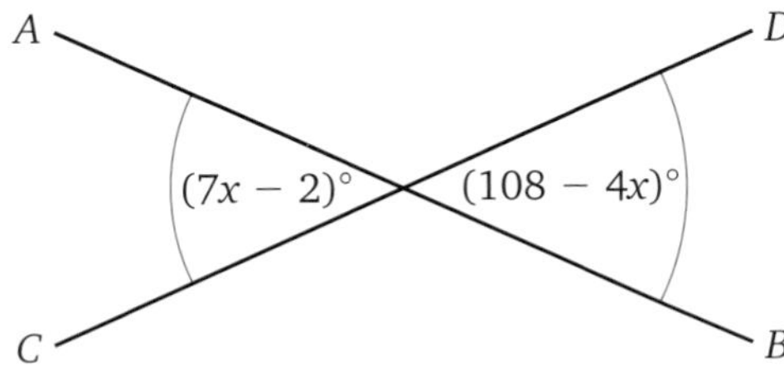

Work out the value of x.

5 The perimeters of the equilateral triangle and the rectangle below are equal.
All of the dimensions are in centimetres.

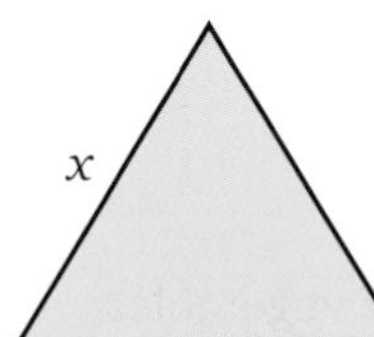

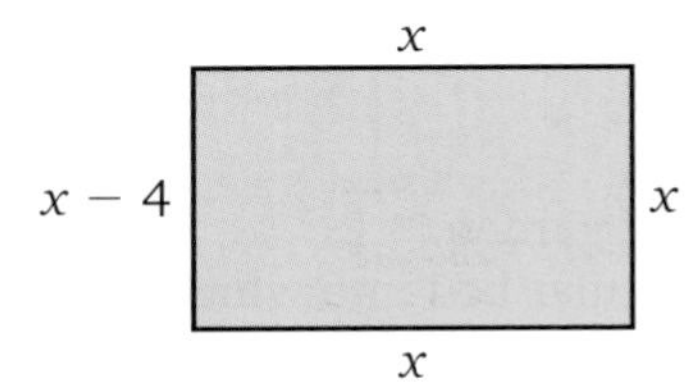

Not drawn accurately

a Write down an equation in x.

b Solve your equation to find the value of x.

> **AQA Examiner's tip**
> First write down your equation without simplifying it, as there will be a mark for this in your exam. If you try to do too much at once, you may make a slip and lose marks.

6 The perimeters of the regular hexagon and the equilateral triangle below are equal.
All of the dimensions are in centimetres.

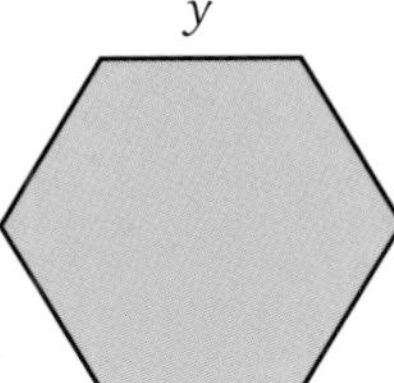

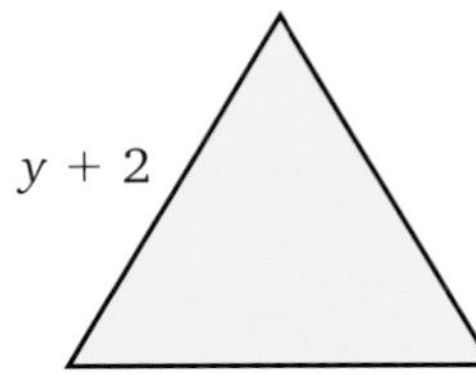

a Write down an equation in y.

b Solve your equation to find the value of y.

c Work out the perimeter of each shape.

> **AQA Examiner's tip**
> If algebra is used in the question, you must write down and solve an equation, which will then lead to the answer of the question asked.

D

7 The perimeters of each of these shapes are equal.
All of the dimensions are in centimetres.

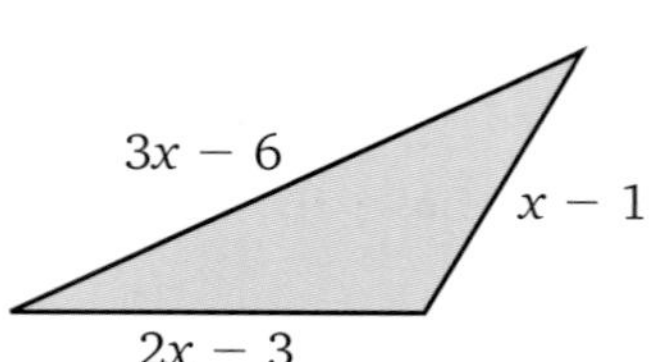

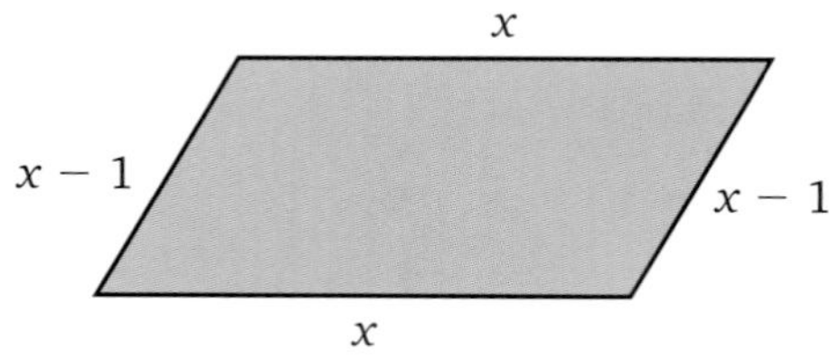

Not drawn accurately

a Write down an equation in x.

b Solve your equation to find the value of x.

c What is the actual perimeter of each of the shapes?

8

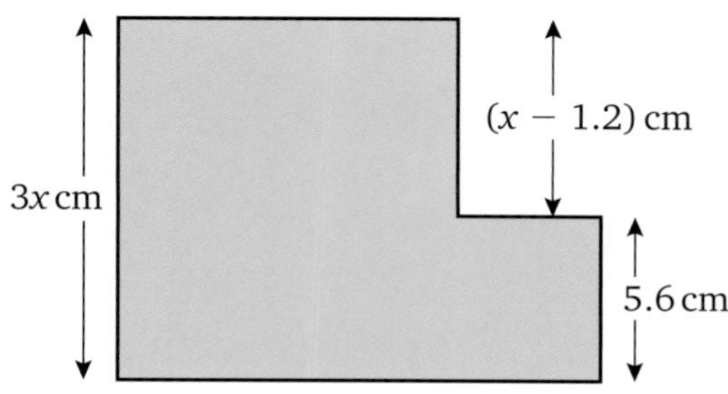

a Write down an equation in x for this shape, which is made up of rectangles.

b Solve your equation to find the value of x.

9 The diagram contains two parallel lines.

a Use what you know about the angles on parallel lines to write down an equation in x.

b Solve your equation to find the value of x.

c Redraw the diagram showing the numerical value of each of the three angles originally shown.

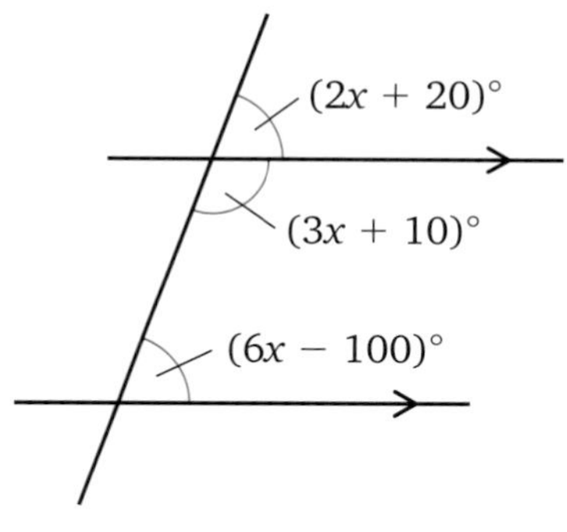

10 The diagram shows a rectangular garden.
The gardener has made a rectangular bed into which he wants to grow vegetables.

B and F are the midpoints of the sides AC and AE respectively.

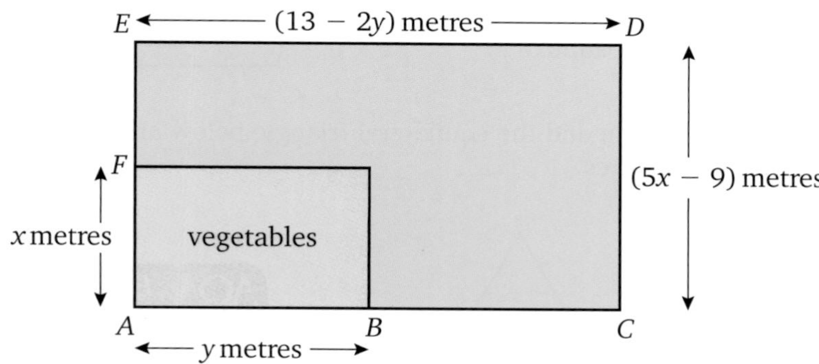

a Form two equations, one in x and one in y.

b Solve your equations to find the values of x and y.

c What are the dimensions of the vegetable patch?

d What is the area of the whole garden?

11 Jim is saving for his dad's birthday present. It costs £31. Twice as much money as he has now would be £5 more than he needs.

How much money does he have now?

12 Javed drives on minor roads for two hours. He then travels on a motorway for four hours, travelling three times as fast on average. He travels a total distance of 322 miles.

a Work out the average speed when travelling on the minor roads.

b Work out the average speed when travelling on the motorway.

13 Ian thinks of two consecutive numbers. The larger one added to twice the smaller one gives a total of 52.

Find the two numbers.

14 One class has three times as many students as another. 12 students are moved from the larger class to the smaller class, this makes the two classes the same size.

How many students were originally in each class?

15 Mabel thinks of a number. She multiplies it by 3 then adds 13.
Mac starts with the same number, he multiplies it by 7 then subtracts 23.
They both get the same answer.

Find the number they first thought of.

Learn... 6.2 Equations with brackets

To solve equations with brackets it is usually best to expand the brackets first.

When you expand a bracket, remember to multiply everything inside the bracket by the term in front of the bracket.

Take special care with signs. Remember that signs belong to the term following the sign.

Link

You can use this to help expand brackets, just as you did in Chapter 3.

Example: Solve the following equation.
$3(2x + 4) = 8 - 2(x - 5)$

Solution:

$3(2x + 4) = 8 - 2(x - 5)$	
$6x + 12 = 8 - 2x + 10$	Expand the brackets.
$6x + 12 = 18 - 2x$	Collect like terms.
$6x + 2x + 12 = 18 - 2x + 2x$	Add $2x$ to both sides
$8x + 12 = 18$	
$8x + 12 - 12 = 18 - 12$	Subtract 12 from both sides.
$8x = 6$	Divide on both sides by 8.
$x = \frac{6}{8} = \frac{3}{4}$	
$= 0.75$	

AQA Examiner's tip

Take extra care when you are expanding brackets with negative numbers. As in the example, remember that when you have -2 in front of the brackets, everything inside the brackets must be multiplied by -2.

Example: The area of this shape is 18 cm^2.

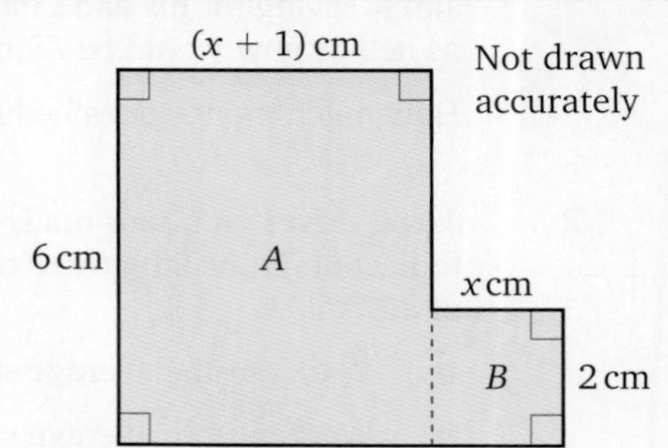

a Write down an equation for the area in terms of x.

b Solve your equation to find the value of x.

c Redraw your shape, replacing all measurements by numbers. This does not have to be to scale.

Solution:

a The area consists of two rectangles. Begin by writing down the area for each of them separately.

area of $A = 6(x + 1)$ area of $B = 2x$

total area = area of A + area of B

total area $= 6(x + 1) + 2x = 18\text{ cm}^2$

b

$$6(x + 1) + 2x = 18$$

$6x + 6 + 2x = 18$ Multiply out the brackets.

$8x + 6 = 18$ Simplify.

$8x + 6 - 6 = 18 - 6$ Subtract 6 from both sides.

$8x = 12$

$\frac{8x}{8} = \frac{12}{8}$ Divide both sides by 8.

$x = 1\frac{4}{8}$ or $x = 1\frac{1}{2}$

c

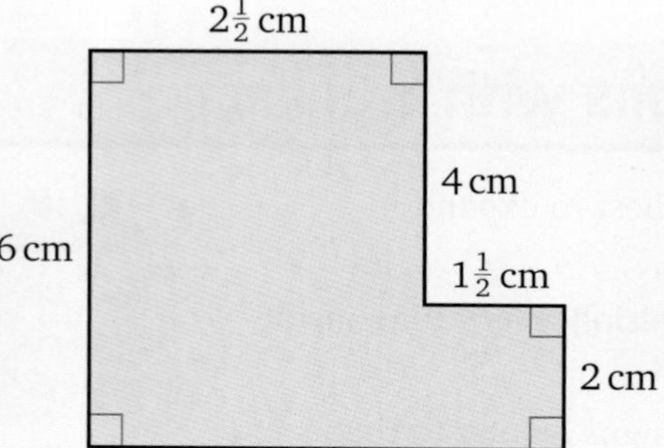

Bump up your grade

Multiply out the brackets to get the first mark. Then write down your working carefully, one step at a time to gain the rest of the marks. To get a Grade C you need to be able to solve this type of equation.

Sometimes it is easier to divide both sides of the equation by the term outside the brackets, instead of multiplying the brackets out. For example, to solve the equation $5(x + 1) = 20$, it is quicker to divide throughout by 5. This gives the simpler equation $x + 1 = 4$, for which the solution of $x = 3$ is easily found.

Practise... 6.2 Equations with brackets 

D C B A A*

D

1 Solve these equations.

a $3(y + 2) = 15$ **b** $5(x + 1) = 7$ **c** $4(1 + p) = 9p$ **d** $6x = 3(x + 4)$

2 The diagram shows a regular pentagon with perimeter 50 cm. Each side is $(x - 4)$ cm long.

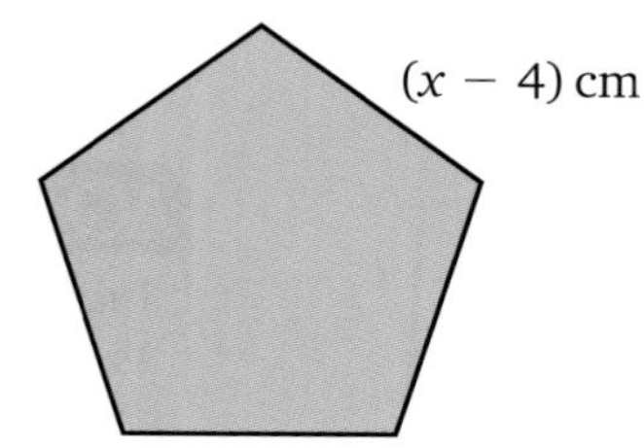

a How can you tell that x must be more than 4 without doing any working out?

b Copy and complete this equation for the perimeter of the shape.

5(   ) =

c Solve your equation to find the value of x.

d Find the length of each side.

C

3 Solve these equations.

a $4(2x + 3) = 3(x + 2)$

b $3(2 + x) = 8 - 2x$

c $3(1 + 2x) = 4(2 - x)$

d $2(3x + 1) = 4x + 8$

e $5(k - 2) = 3k + 4$

f $3(4 - a) = a - 12$

g $4(y + 3) = 7(8 - y)$

h $3x = 2(3x - 1)$

4 Ray is solving an equation.

Here are the first two lines of his working, with a piece of the paper torn out.

$2(2x - 10) + 3(2x - 4) = 17$

$4x - 20 + 6x -$ $\quad = 17$

$10x -$ $\quad = 17$

Copy and complete his work.

5 Solve the following equations.

a $2(3x + 1) + 3(2x - 1) = 13$

b $3(4x - 8) + 4(2x - 11) = 12$

c $3(x + 5) - 2(3x + 2) = 5$

d $6(5 - x) - 2(4x - 2) = 6$

e $3(5 - x) + 7(x + 8) = 4(5 + 2x)$

f $4(3 + 2x) - 3(5 - 2x) = 3 - 5(7 + 2x)$

6 What is the size of each angle in these shapes?
The diagrams are not drawn accurately.

a

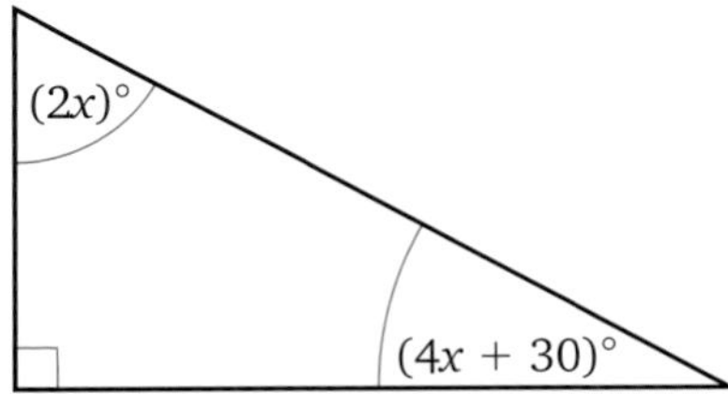

b

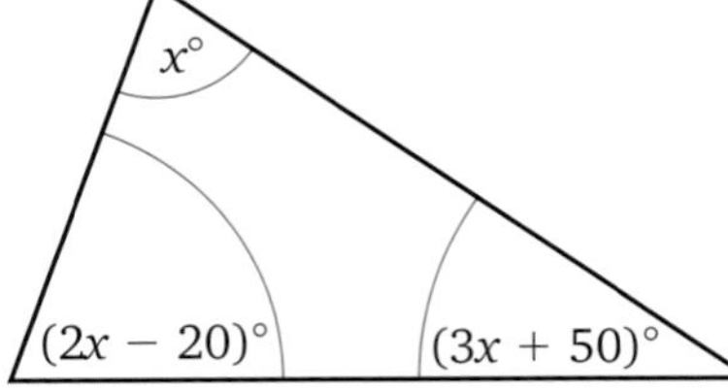

AQA Examiner's tip

Do not try to measure diagrams in an exam when told that they have not been drawn accurately.

7 The perimeter of this rectangle is 200 cm.

$(4x - 20)$ cm

$(x + 5)$ cm

Write down an equation in x and solve it to find the lengths of the sides.

C

8 This kite has a perimeter of 96 cm.

Write down an equation in x and solve it to find the lengths of the sides.

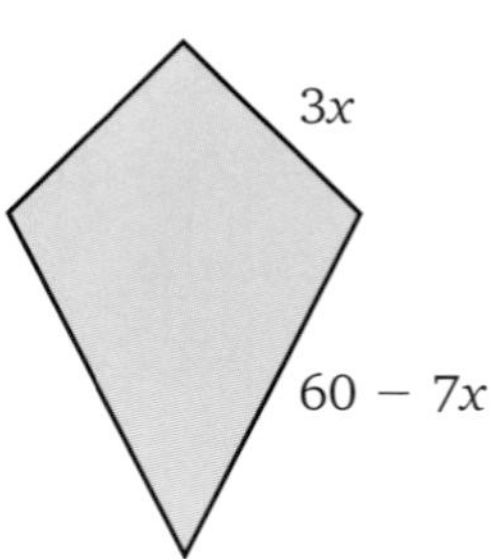

9 An interior and exterior angle of a regular polygon are shown.

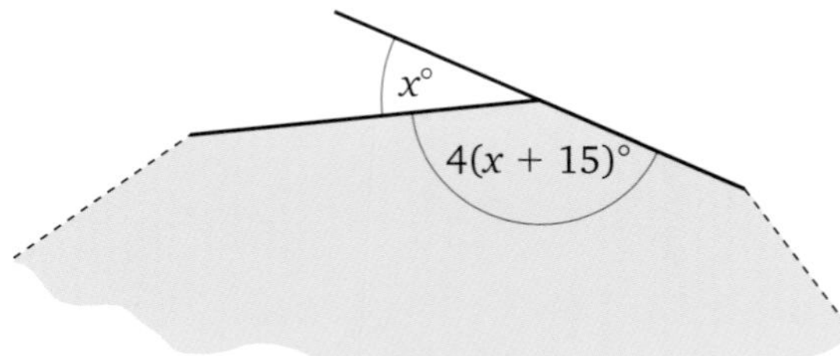

Work out the number of sides of the polygon.

10 I think of a number, double it, then add 3, and double the result. My answer is 26.

Write down an algebraic equation and solve it to find the number.

11 Phil thinks of a number. He multiplies it by 3 then adds 27. The answer is 4 times as big as when his number is doubled and 7 subtracted.

What is Phil's number?

B

12 The diagram shows a rectangular garden. Part of the lawn has been removed to make a rectangular pond.

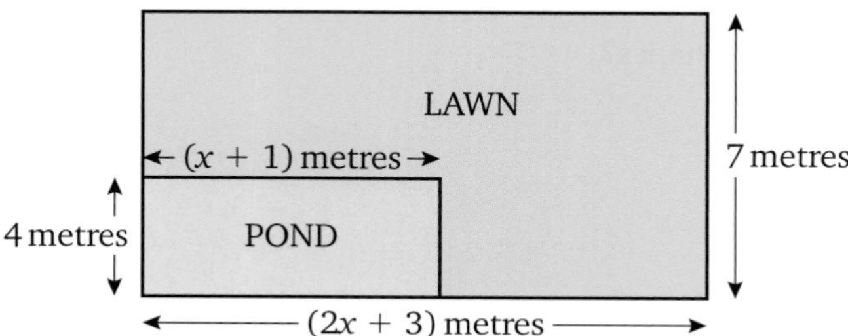

The area of the lawn is 35 m^2.

Work out the area of the pond.

13 Angle A is 20% larger than angle B.

$A = x + 64$, $B = 2x + 30$

Work out x.

14 The areas of the two shapes are in the following ratio:

area rectangle : area triangle = 3 : 2

Work out x.

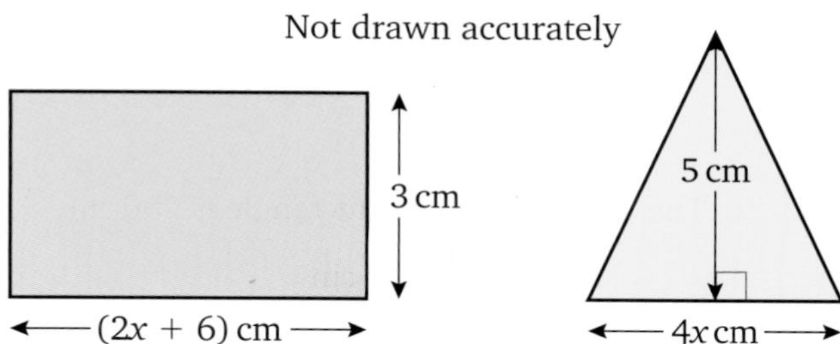

15 A gardener uses a wheelbarrow to move blocks of turf from one place to another. If he takes 7 at a time, he uses 10 fewer trips than when he takes 5 at a time.

How many journeys does he make if he takes 7 at a time?

16 Jack is 47 years of age. His son is 19. In how many years will Jack be twice the age of his son?

17 Andy and Trevor start an evening out with £45 and £49 respectively. They both spend equal amounts. At the end of the evening one has three times as much left as the other.

How much did they each spend?

18 Jayne travels 41 miles in 5 hours. Part of the time she walks at 4 mph and the rest of the time she cycles at 10 mph.

How long does Jayne cycle?

Learn... 6.3 Equations with fractions

To solve simple equations containing fractions, you need to multiply by the **denominator**. For example, if your equation has the single fraction $\frac{x}{2}$ then you will need to multiply by 2.

If your equation has more than one fraction, it is usually easier to multiply by all the denominators first. If your equation involves $\frac{x}{2}$ and $\frac{x}{3}$ then you will need to multiply by 2 and by 3. Or you could start by finding the lowest common multiple of all the denominators, in this case 6, and then multiply the whole equation by that number.

The following examples show you some methods of simplifying different types of equations with fractions.

Example: Solve the equation $\frac{x}{4} + 3 = 5$

Solution: This is an example of the simplest type of equation with a fraction.

As there is only one fraction in the equation, multiplying by 4 should not be the first step.

Start by collecting the number terms together.

$\frac{x}{4} + 3 - 3 = 5 - 3$ Subtract 3 from both sides.

$\frac{x}{4} = 2$ The fraction term is now on its own.

$\frac{x}{4} \times 4 = 2 \times 4$ Multiply both sides by 4 (the denominator).

$x = 8$

Example: Solve the equation $\frac{2x}{3} - \frac{x}{2} = 1$

Solution: This is an example where there is more than one fraction.

You need to multiply by both denominators, in this case, $3 \times 2 = 6$

Multiply **each** term by 6.

$$6 \times \left(\frac{2x}{3}\right) - 6 \times \left(\frac{x}{2}\right) = 6 \times 1$$

Remember to multiply all terms on both sides.

Taking the left-hand side one term at a time:

$$6 \times \left(\frac{2x}{3}\right) = \frac{12x}{3} = 4x$$

$$6 \times \frac{x}{2} = \frac{6x}{2} = 3x$$

and now the right-hand side:

$$6 \times 1 = 6$$

The equation now becomes:

$$4x - 3x = 6$$

$$x = 6$$

Example: **a** Solve the equation $\frac{3x + 4}{2} = \frac{3x + 1}{3}$

b Solve the equation $\frac{x + 5}{3} + \frac{x - 2}{5} = 5$

Solution: **a**

$$\frac{3x + 4}{2} = \frac{3x + 1}{3}$$

$$\frac{6(3x + 4)}{2} = \frac{6(3x + 1)}{3}$$ Multiply both sides by the common denominator 6 (= 2 × 3).

$$3(3x + 4) = 2(3x + 1)$$

$$9x + 12 = 6x + 2$$ Expand the brackets.

$$9x - 6x + 12 = 6x - 6x + 2$$ Subtract $6x$ from both sides.

$$3x + 12 = 2$$

$$3x + 12 - 12 = 2 - 12$$ Subtract 12 from both sides.

$$3x = -10$$

$$x = -\frac{10}{3}$$ Divide both sides by 3.

$$x = -3\tfrac{1}{3}$$

b

$$\frac{x + 5}{3} + \frac{x - 2}{5} = 5$$

$$\frac{15(x + 5)}{3} + \frac{15(x - 2)}{5} = 15 \times 5$$ Multiply both sides by 15 (= 3 × 5).

$$5(x + 5) + 3(x - 2) = 75$$

$$5x + 25 + 3x - 6 = 75$$ Expand the brackets.

$$8x + 19 = 75$$

$$8x + 19 - 19 = 75 - 19$$ Subtract 19 from both sides.

$$8x = 56$$

$$x = 7$$ Divide both sides by 8.

AQA Examiner's tip

Remember to multiply every term on both sides of the equation when you multiply by the denominators of fractions.

Practise... 6.3 Equations with fractions

D C B A A*

C

1 Solve these equations.

a $\frac{x}{2} + 3 = 5$

b $7 + \frac{x}{3} = 9$

c $8 - \frac{x}{4} = 3$

d $4 + \frac{2}{3}x = 8$

e $10 - \frac{3}{8}x = 4$

f $8.7 = \frac{5}{8}x + 1.2$

g $11.5 = \frac{3}{4}x - 3.5$

h $3.9 = 12.3 - \frac{7}{8}x$

2 Solve these equations.

a $\frac{x}{2} + 3 = 10$

b $\frac{x + 13}{4} = 8$

c $\frac{x + 8}{2} = x$

d $\frac{8x + 3}{4} = 3x + 2$

B

3 Solve each of these equations.

a $\frac{x + 1}{4} = \frac{x - 2}{3}$

b $\frac{2x + 3}{4} = \frac{1}{5}$

c $\frac{2x + 5}{3} = \frac{x}{4}$

d $\frac{4x + 2}{5} = \frac{5x - 3}{6}$

B

4 Write an equation to find Janet and John's starting numbers in the following two problems.

a Janet thinks of a number. She multiplies the number by 2 then adds 3.
She divides the result by 5.
Her answer is 3.

b John thinks of a number. He multiplies it by 3 then subtracts 4.
He divides the result by 2.
His answer is 7.

5 A piece of wood is $(3x + 25)$ cm long. One-fifth of the total length is cut off. There is 80 cm left.

Write an equation and solve it to find x.

6 Each of these shapes is made from a right-angled triangle and a rectangle. The areas of the two shapes are equal.

Find the value of x.

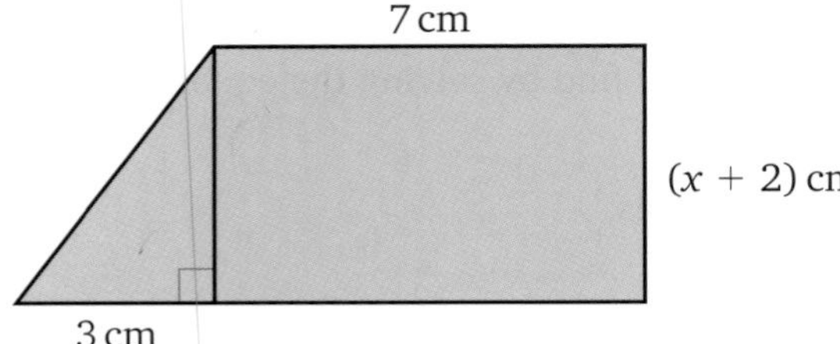

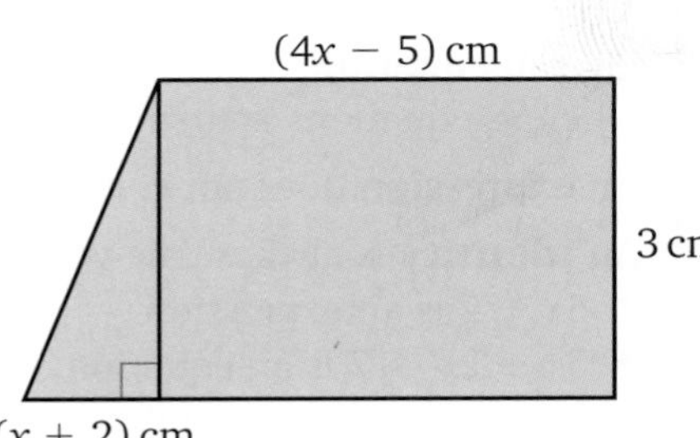

Not drawn accurately

7 A man is 28 years old when his son is born. How many years later will it be when the son's age will be half the age of the father?

8 I think of a number, subtract 7 from it, divide the result by 4 then add 13. My answer is 20.

What was the number I first thought of?

9 5 kg of tea priced at £12 per kg are mixed with a certain quantity of a different kind of tea priced at £8 per kg. The mixture obtained is worth £8.50 per kg.

How many kg of the cheaper tea were used?

10 Solve the following equation.

$$\frac{x-4}{3} - \frac{5-2x}{4} = \frac{2x+5}{12}$$

11 These two shapes have the same area.

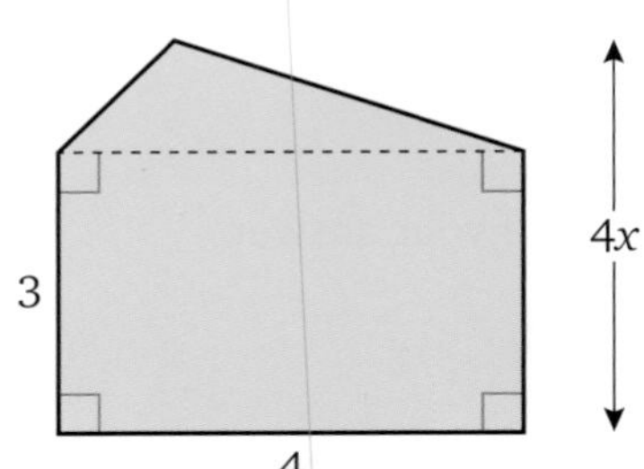

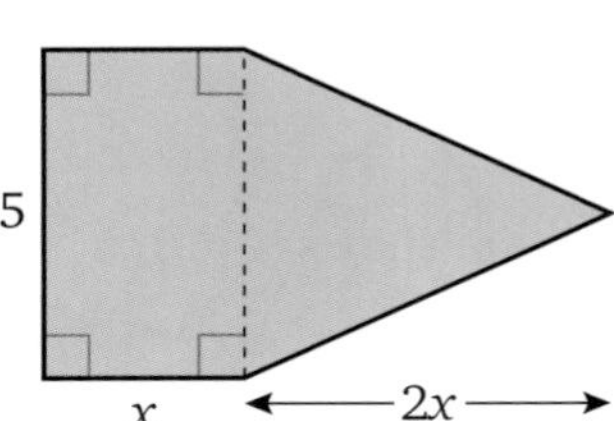

Find and solve an equation to find the value of x.

? **12** Two motorists make the same journey. One drives at an average of 28 mph, the other drives at an average of 24 mph. The first arrives 30 minutes before the second.

How far was their journey?

13 Mr D River takes $3\frac{3}{4}$ hours to travel 120 miles.
His average speed is 20 mph in towns and 35 mph elsewhere.
How many miles of his journey are in towns?

14 A boy took 25 minutes to walk to school and back home again. If he had been 4 minutes quicker on his way to school, he would have taken half the time he took to get back home again.
How long did he take to walk to school?

Learn... 6.4 Finding and using formulae

Formulae tell you how to work something out. For example, you can work out the perimeter of a rectangle by doubling the length and the width and adding them. This can be written using symbols as $P = 2L + 2W$

Substitution is where you replace letters with numbers. Substitution can be made in expressions and formulae.

You are expected to be able to use formulae from mathematics, science and other subjects.

Like an **equation**, a formula always includes an equals sign. A formula is true for a range of values whereas equations are true only for certain values which you find by solving the equation.

An **expression** does not contain an equals sign.

An **identity** is always true whatever the value of the symbols,

- $5x + 2$ is an expression
- $5x + 2 = 17$ is an equation: it is only true when $x = 3$
- $P = 5x + 2$ is a formula for finding P when you know the value of x
- $5y + 3 \equiv 3 + 5y$ is an identity $=$ is replaced by $\equiv$ to show it is always true.

You are expected to be able to identify whether an algebraic statement is a formula, an equation, an expression or an identity.

Example:

a Use the formula $A = \frac{1}{2}(a + b)h$ and the information given to find the area of this trapezium.

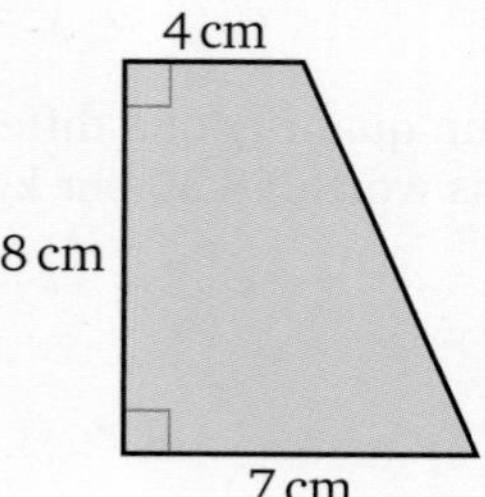

b The following diagram shows a trapezium. The area of the trapezium is 52 cm^2.
Use the formula to find the height h.

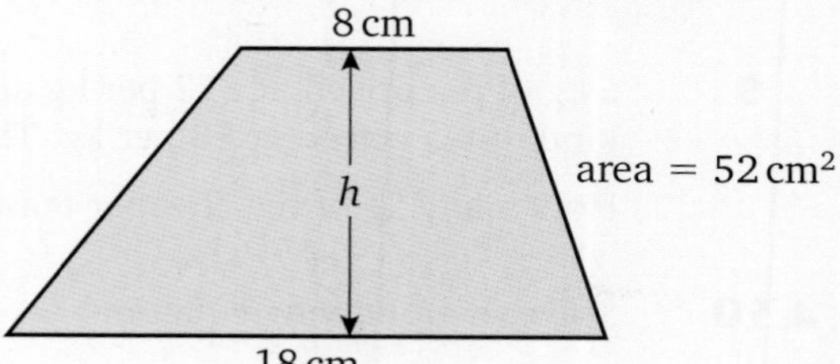

Solution:

a The diagram shows $h = 8$, $a = 4$, $b = 7$.

$A = \frac{1}{2}(a + b)h$ Copy the formula carefully.

$A = \frac{1}{2}(4 + 7)8$ Substitute.

$A = 4(4 + 7)$

$A = 4 \times 11$

$A = 44\text{ cm}^2$ Remember to state the units of your answer.

b The diagram shows $a = 8$, $b = 18$, $A = 52$

$A = \frac{1}{2}(a + b)h$ Copy the formula carefully.

$52 = \frac{1}{2}(8 + 18)h$ Substitute.

$52 = \frac{1}{2} \times 26h$

$52 = 13h$

$\frac{52}{13} = h$

$h = 4\text{ cm}$

AQA Examiner's tip

When you use a formula, the examiner is also testing whether you know how to substitute into a formula. So remember to show that you have substituted into the formula to gain this mark.

Practise... 6.4 Finding and using formulae

D

1 Dean is using the formula $H = \frac{1}{2}(50 - 0.1W)$

H is the height in centimetres of the bottom of his car above the ground.

W is the weight in kilograms he puts in the back of his car.

a One day Dean puts a weight of 90 kg in the back of his car.
Calculate the value of H.

b On a different day Dean has three people in the back of his car.
He knows they weigh 60 kg, 80 kg, and 85 kg.
What is the height of the bottom of his car above the ground?

c Dean knows that the bottom of the back of his car must be at least 10 cm above the ground for it to be safe to drive.
He has two people weighing 76 kg and 82 kg in the back of his car.
He has a lawnmower weighing 30 kg in his car boot.
How much extra weight could Dean put in his boot and it still be safe to drive?

2 Rachel is doing some research on the internet about how quickly young children learn words. She finds the formula $n = 60m - 900$. The age of the child is m months and n is the average number of words the child knows.

a Find n for a child aged 20 months.

b Can the formula be used for a child of 6 months? Explain your answer.

c How many words would a 2-year-old child know, according to the formula?

3 Mal is using the formula $s = ut + \frac{1}{2}at^2$ in science.

a Find s when $u = 12$, $t = 5$, $a = -9.8$

b Find u when $s = 20$, $t = 3.9$, $a = -9.8$

4 May is using the formula $A = \pi r^2$ in mathematics to find the area of circles with radius r cm. Use the formula to find:

a the area of a circle with radius 5 cm

b the radius of a circle with area 78 cm^2.

Give your answers to an appropriate degree of accuracy.

5 Young's rule is used to calculate a child's dose of medicine. The formula is:

$$C = \frac{An}{(n + 12)}$$

C = child's dose of medicine

n = age of child

A = adult's dose

a Use the rule to find the dose for a child aged 3 when the adult's dose is 15 ml.

b Use the rule to find the dose for a child aged 8 when the adult's dose is 30 ml.

c Find the adult's dose when the dose for a child aged 6 is 25 ml.

6 Pierre is using the formula $v^2 = u^2 + 2as$.

Find v when $u = 10$, $a = 9.8$, $s = 5$. Give your answers to one decimal place.

7 An isosceles triangle has angles as shown.

a Find a formula for x in term of y.

b Find a formula for y in terms of x.

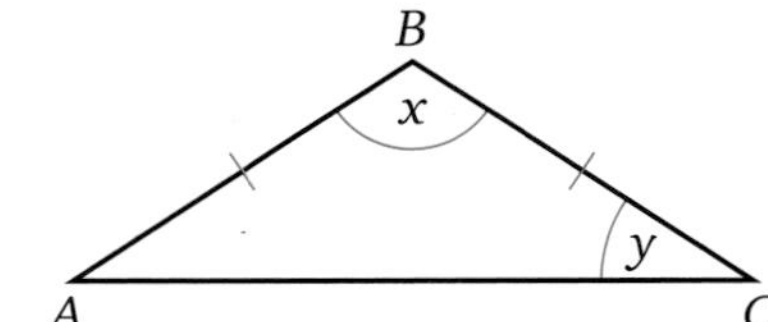

6 Assess

D

1 Solve the following equations.

a $3x - 4 = 5 + x$ **b** $2(5x + 1) = 28$

2 Use the formula $C = \frac{(A + 1)}{9}$ to find the value of C when $A = 10$.

3 The diagram shows a lawn with a corner removed for a pond.

All the measurements are given in metres.

Find formulae for

a P, the perimeter of the lawn

b A, the area of the lawn.

3
x − 2
x + 2
8

C

4 Solve the following equations.

a $3x - 12 = 2(2x - 5)$ **b** $\frac{2x}{3} - \frac{x}{4} = 5$ **c** $\frac{7 - x}{3} = 2$

5

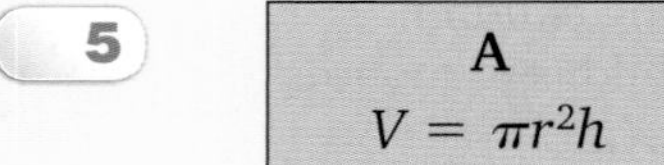

A	B	C
$V = \pi r^2 h$	$x + y \equiv y + x$	$2rh$

These are examples of three of the following:

expression, equation, formula, identity

Match **A**, **B** and **C** to their correct descriptions and write your own example to match the missing description.

AQA Examination-style questions

1 All areas in this question are in square centimetres.
Here is a rectangle of area R, a square of area S and a trapezium of area T.

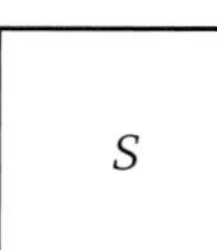

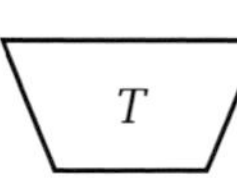

a The area of the shape below is given by $A = R + 2T$

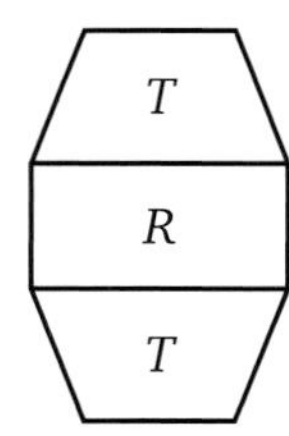

Find the value of A when $R = 7.5$ and $T = 6.3$ *(2 marks)*

b Here is a different shape.

(i) Write down an expression for the area of this shape.

(ii) Which of the following is correct?

$3R = S$ $2R = S$ $R = 2S$ $R = 3S$ *(1 mark)*

AQA 2007

7 Properties of polygons

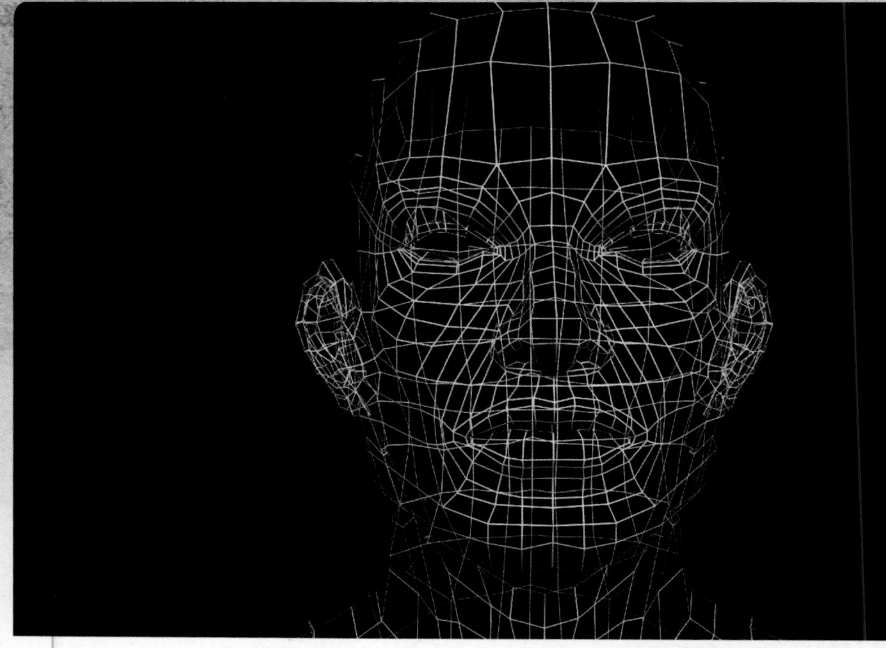

Objectives

Examiners would normally expect students who get these grades to be able to:

D

classify a quadrilateral using geometric properties

C

calculate exterior and interior angles of a regular polygon.

Did you know?

Polygons and video games

Objects in video games are made up of lots of polygons. Pictures are made up of a series of polygons such as triangles, squares, rectangles, parallelograms and rhombuses. The more polygons there are, the better the picture looks.

The polygons are all given coordinates. The computer changes and rotates the coordinates to match your position in the game. This gives the impression of movement.

For example, if you move away, the computer shrinks all the coordinates of the polygons. This makes the polygons appear smaller on the screen so they look further away.

Key terms

quadrilateral
polygon
diagonal
bisect
perpendicular
exterior angle
interior angle
triangle
pentagon
hexagon
regular
octagon
decagon
nonagon

You should already know:

- ✔ how to use properties of angles at a point, angles on a straight line, perpendicular lines, and opposite angles at a vertex
- ✔ the difference between between acute, obtuse, reflex and right angles
- ✔ how to use parallel lines, alternate angles and corresponding angles
- ✔ how to prove that the angle sum of a triangle is 180°
- ✔ how to prove that the exterior angle of a triangle is equal to the sum of the interior opposite angles
- ✔ how to use angle properties of equilateral, isosceles and right-angled triangles
- ✔ how to use angle properties of quadrilaterals.

Learn... 7.1 Properties of quadrilaterals

A **quadrilateral** is a **polygon** with four sides.

You need to know the names and properties of the following special quadrilaterals.

Square – a quadrilateral with four equal sides and four right angles.

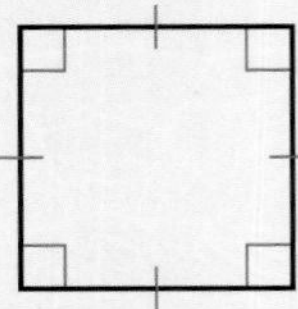

Rectangle – a quadrilateral with four right angles and opposite sides equal in length.

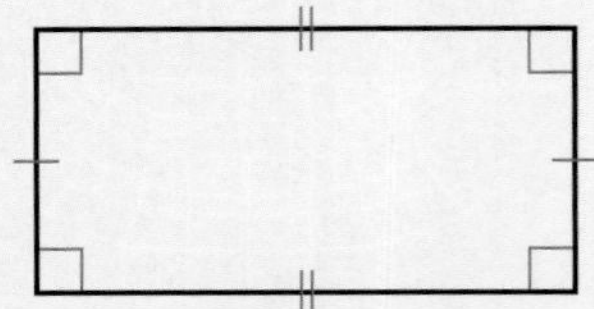

Kite – a quadrilateral with two pairs of equal adjacent sides.

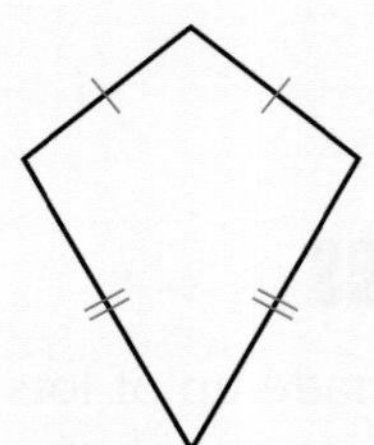

Trapezium – a quadrilateral with one pair of parallel sides.

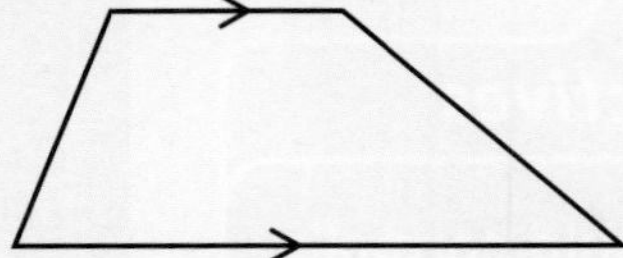

Parallelogram – a quadrilateral with opposite sides equal and parallel.

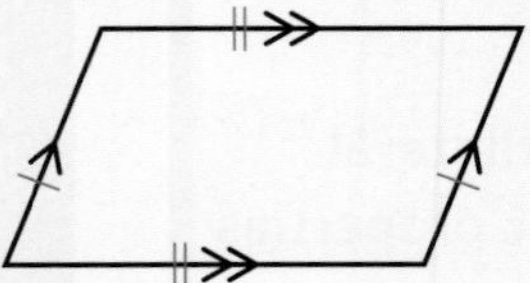

Rhombus – a quadrilateral with four equal sides and opposite sides parallel.

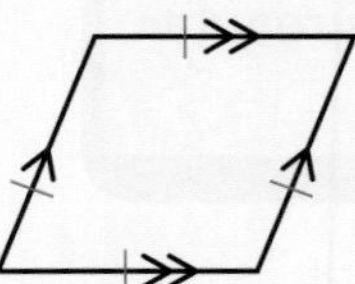

Isosceles trapezium – a trapezium where the non-parallel sides are equal in length.

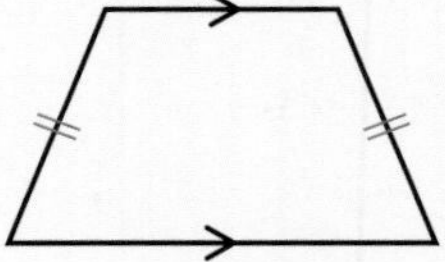

All quadrilaterals have four sides and four angles.
A quadrilateral can be split into two triangles.

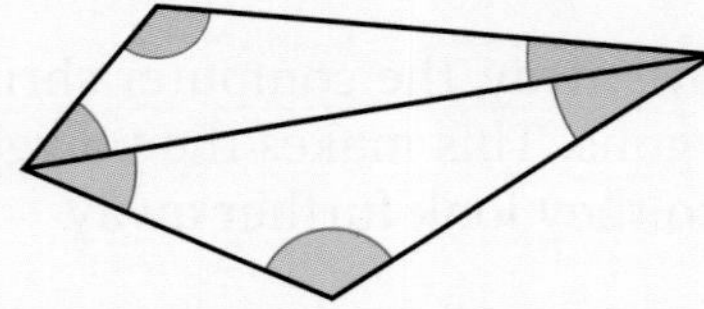

The angles in a triangle add up to 180°.
The quadrilateral is made up of two triangles.
The angles in a quadrilateral add up to $2 \times 180° = 360°$

The diagonals of a quadrilateral

A diagonal is a line joining one corner of a quadrilateral to another.

All quadrilaterals have two **diagonals**.

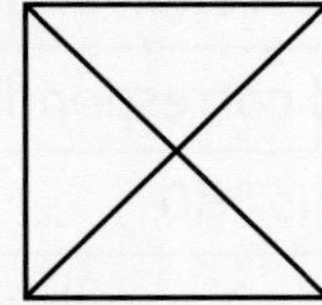

The square has two diagonals.

The diagonals are the same length.

The diagonals **bisect** one another.

The diagonals are **perpendicular**.

Bisect means they cut one another in half.

Perpendicular means at right angles.

Example: Calculate the angles marked with letters in the shape below.

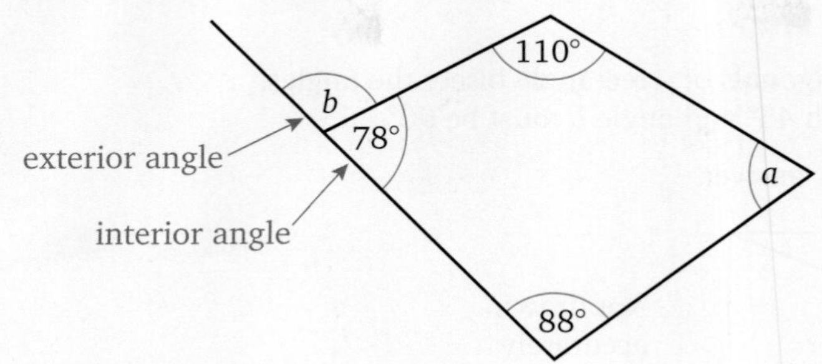

Not drawn accurately

Solution: The angles in the quadrilateral add up to 360°, so

$a = 360° - (78° + 88° + 110°)$
$\quad = 360° - 276°$
$\quad = 84°$

The **exterior** and **interior angles** add up to 180°, so

$b = 180° - 78°$
$\quad = 102°$

AQA Examiner's tip

Always make sure that your answer is reasonable. Angle b is an obtuse angle so the answer calculated here is reasonable.

Practise... 7.1 Properties of quadrilaterals  

D

1

square rectangle parallelogram rhombus kite trapezium

a Which of these quadrilaterals have all four sides equal?

b Which of these quadrilaterals have opposite sides that are parallel?

c Which of these quadrilaterals have adjacent sides that are equal?

d Which of these quadrilaterals has only one pair of parallel sides?

2 Barry measures the angles of a quadrilateral. He says that three of the angles are 82° and the other one is 124°.

Could he be right? Explain your answer.

3 Make an accurate drawing of each quadrilateral listed in the table, then draw the diagonals.
Use your diagrams to complete a copy of the table.

Shape	Are the diagonals equal? (Yes/No)	Do the diagonals bisect each other? (Yes/No/Sometimes)	Do the diagonals cross at right angles? (Yes/No)	Do the diagonals bisect the angles of the quadrilateral? (Yes/No/Sometimes)
Square				
Kite				
Parallelogram				
Trapezium				
Isosceles trapezium				
Rectangle				
Rhombus				

D

4 Rajesh says that he has drawn a quadrilateral. Its diagonals are equal.

What shapes might he have drawn? (Use the table from Question 3 to help you.)

D C

5 Michelle says that the diagonals of a rectangle bisect the angles.
So angles a and c are both 45° and angle b must be 90°.

Is she right? Explain your answer.

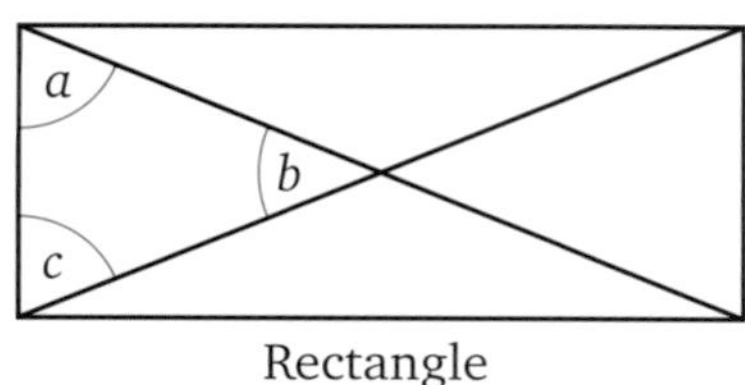

Not drawn accurately

Rectangle

6 The diagram shows a rhombus $ABCD$.
AC and BD are the diagonals.
Angle $ADB = 32°$

Calculate angle DAC.

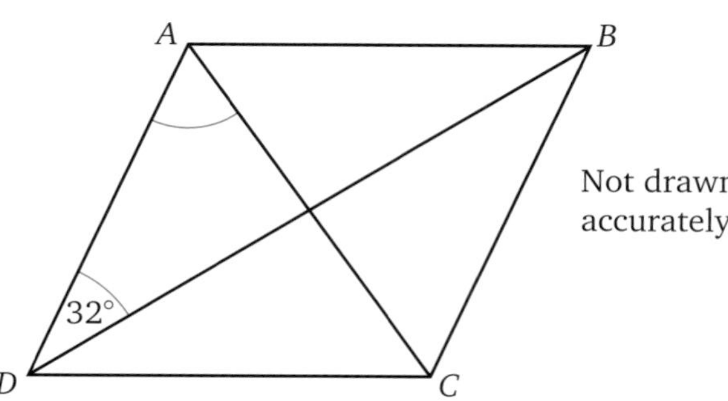

Not drawn accurately

C

7 Calculate the angles marked with letters in the diagrams. You will need to use parallel line facts and the properties of diagonals. Explain how you know the size of the angles.

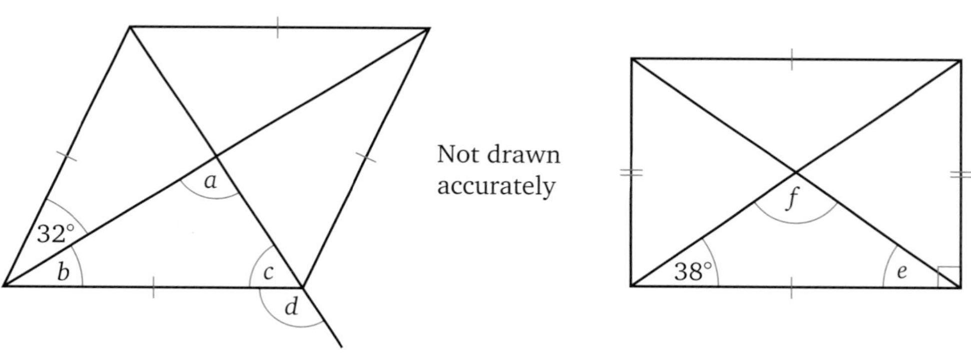

Not drawn accurately

8 In the diagram, EF is parallel to GH, and AB is parallel to CD.

IJ is perpendicular to AB, and IK is equal to JK.

Calculate the angles a to f, giving reasons for your answers.

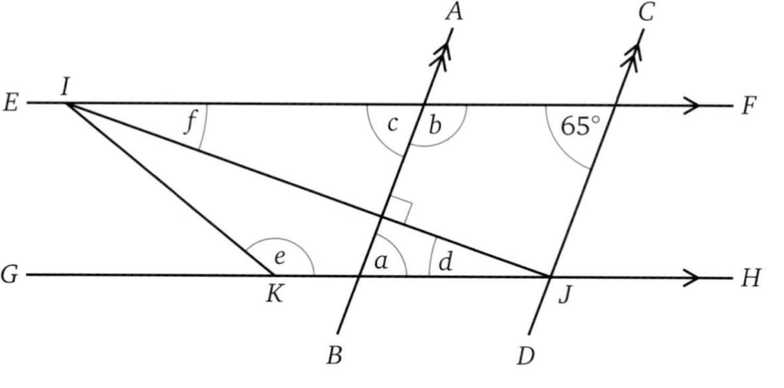

Not drawn accurately

9 Matt, Tess and Sam all draw kites with one angle of 76° and one angle of 60°. All three kites are different. Matt's kite has two obtuse angles. Sam's kite has a larger angle than the other two kites.

What are the angles of each kite? Draw diagrams to help.

10 *EDC* is a straight line and angle *DAB* = angle *ABC*

Work out the angle *ABC*.

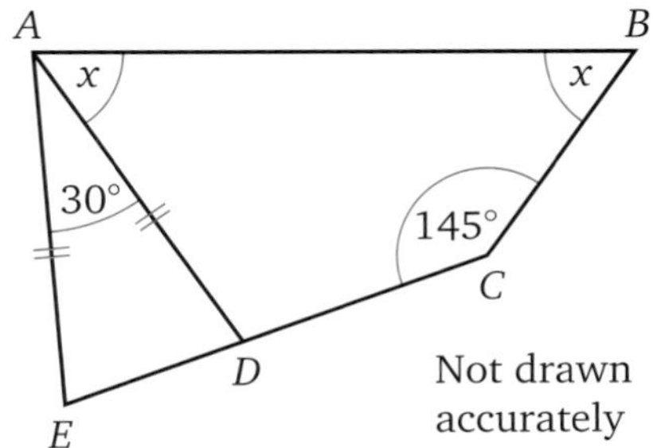

Learn... 7.2 Angle properties of polygons 

The interior angles of a **triangle** add up to 180°.

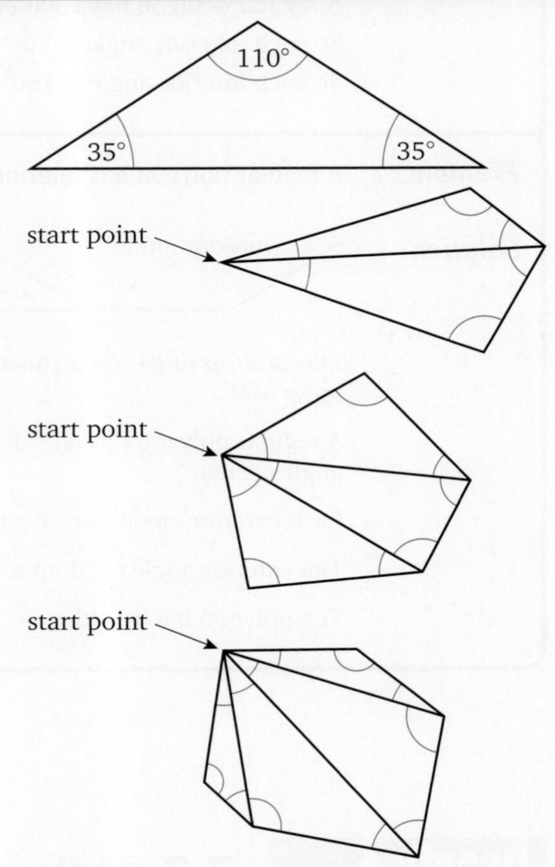

A quadrilateral has four sides and can be split into two triangles by drawing diagonals from a point.

The sum of the angles is 2 × 180° = 360°

A **pentagon** has five sides and can be split into three triangles by drawing diagonals from a point.

The sum of the angles is 3 × 180° = 540°

A **hexagon** has six sides and can be split into four triangles by drawing diagonals from a point.

The sum of the angles is 4 × 180° = 720°

In general, a polygon with n sides can be split into (n − 2) triangles.

The sum of the angles is (n − 2) × 180°

The interior angles of a polygon are the angles inside the polygon.

In a **convex** polygon all the internal angles are less than 180°

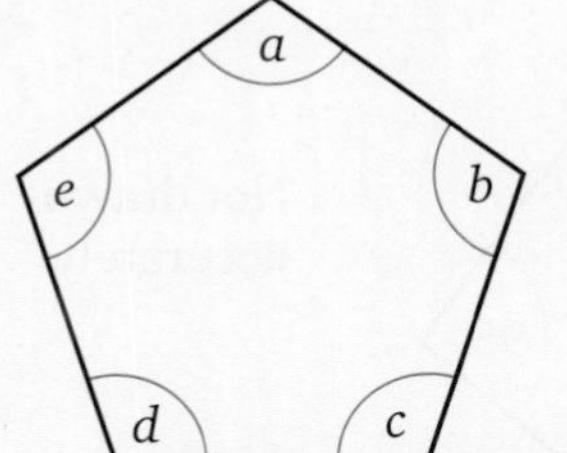

a, *b*, *c*, *d* and *e* are interior angles.

The exterior angles of a polygon are the angles between one side and the extension of the side.

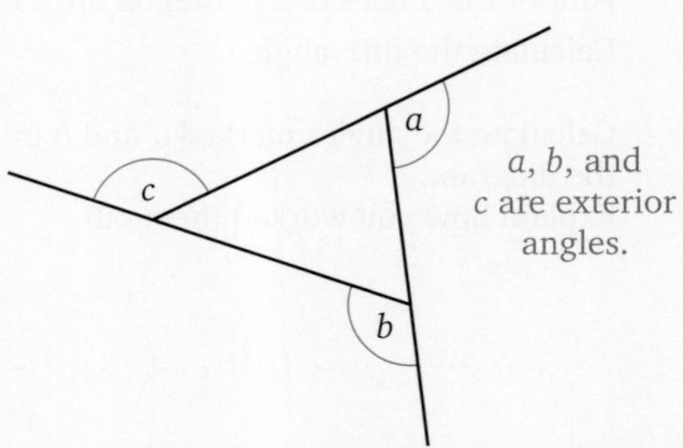

The exterior angles of a polygon add up to 360°.

Example: Find the interior angle of a **regular octagon**.

Solution: **Either:**

An octagon has eight sides.

So the sum of the interior angles is $(8 - 2) \times 180° = 1080°$

A regular octagon has all angles equal, so each interior angle is $1080° \div 8 = 135°$

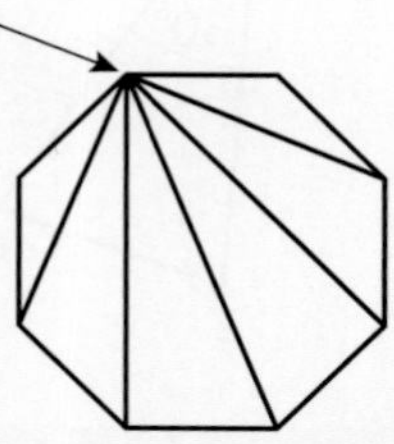

Or:

A regular octagon has eight equal exterior angles.

So each exterior angle is $360° \div 8 = 45°$

So each interior angle is $180° - 45° = 135°$

> **Bump up your grade**
>
> You need to know how to calculate interior and exterior angles of any polygon for a Grade C.

Example: A regular polygon has interior angles of 144°. How many sides does it have?

Solution:

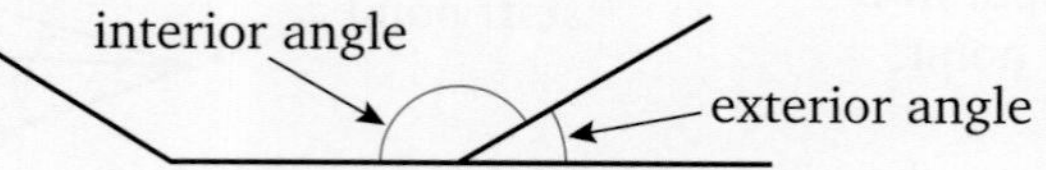

144°

The exterior angles of a convex polygon add up to 360°.

A regular polygon has all sides equal and all angles equal.

Each exterior angle must be $180° - 144° = 36°$

The exterior angles add up to 360°, so there must be $360° \div 36° = 10$ exterior angles

The polygon has 10 sides.

> **AQA Examiner's tip**
>
> Always draw a diagram to help answer the questions.
>
> You can then label the diagram to keep track of what you know.

Practise... 7.2 Angle properties of polygons  D C B A A*

C

1 Four of the angles of a pentagon are 110°, 130°, 102° and 97°.
Calculate the fifth angle.

2 Calculate the angles marked a and b in the diagram.
Explain how you worked them out.

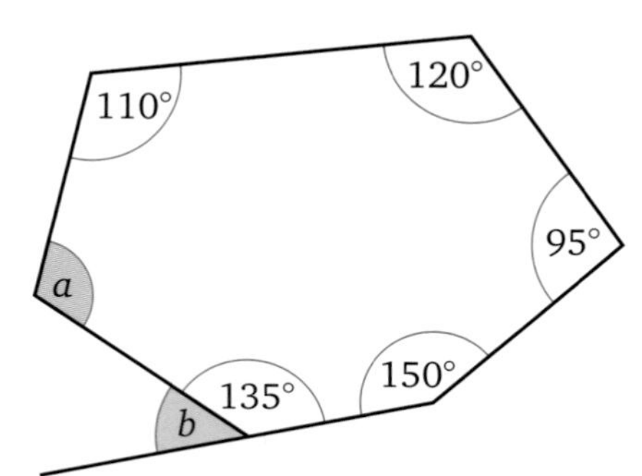

Not drawn accurately

C

3 A regular polygon has an exterior angle of 60°.

How many sides does it have?

4 Calculate the difference between the interior angle of a regular **decagon** (ten-sided shape) and the interior angle of a regular **nonagon** (nine-sided shape).

5 James divides a regular hexagon into six triangles as shown.

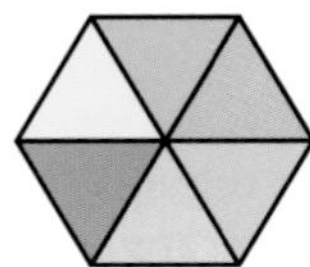

He says the angle sum of a regular hexagon is $6 \times 180°$.

Is he correct? Give a reason for your answer.

6 Lisa says that a regular octagon can be split into two trapeziums and a rectangle as shown.

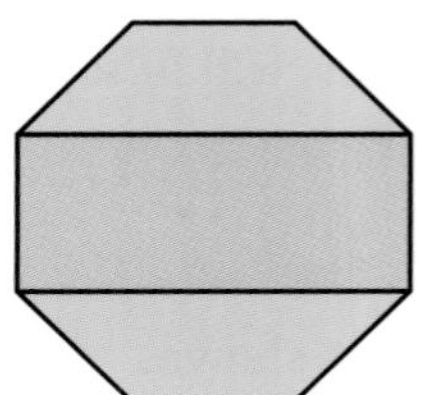

She says the angle sum of the octagon is $3 \times 360°$.

Show that Lisa is correct.

C B

7 The diagrams show how you draw an equilateral triangle and a regular pentagon inside a circle. You do this by dividing the angle at the centre equally.

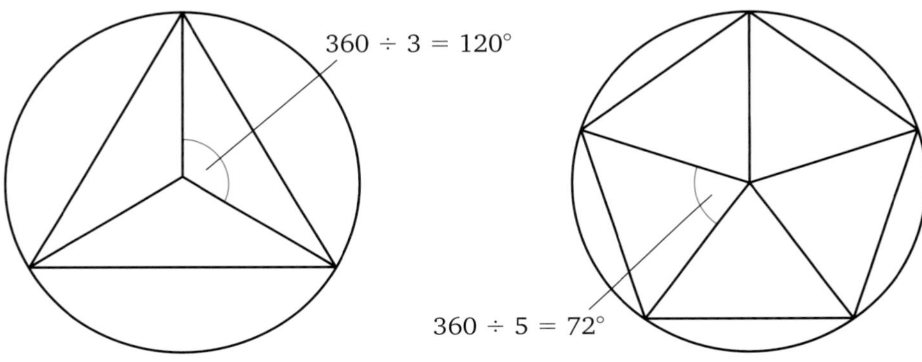

Use the same method to draw a regular hexagon and a regular nonagon (nine-sided shape) inside a circle.

8 The diagram shows a regular pentagon *ABCDE* and a regular hexagon *DEFGHI*.

Calculate:

a angle *EDC*

b angle *EDI*

c obtuse angle *CDI*

d angle *BAC*

e angle *CAE*

f angle *HIG*

g angle *DIG*.

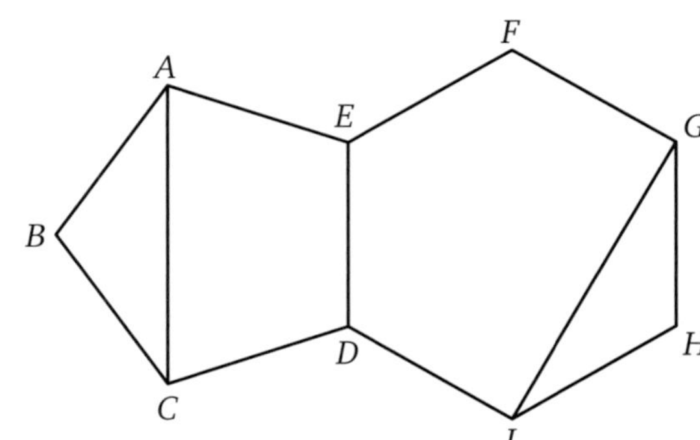

B

9 A badge is in the shape of a regular pentagon. The letter V is written on the badge.

What is the size of the angle marked x?

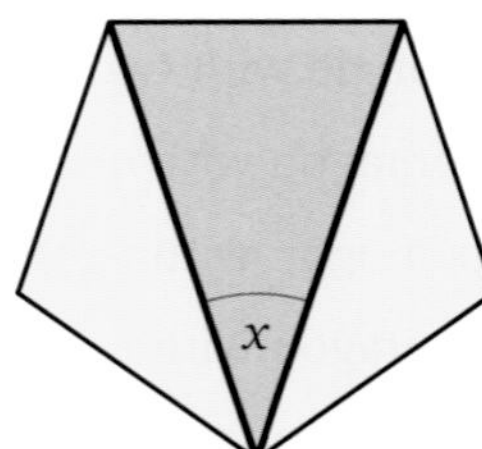

10 A convex polygon is one where all the interior angles are less than 180°.

Show that a convex polygon cannot have more than three acute angles.

Hint
Think about the sum of the exterior angles.

11 Show that if a convex polygon has more than six sides, then at least one of the sides has an obtuse angle at both ends.

12 Penny fits regular pentagons together in a circular arrangement.
Part of the arrangement is shown in the diagram.

Show that exactly ten pentagons will fit together in this way before meeting up.

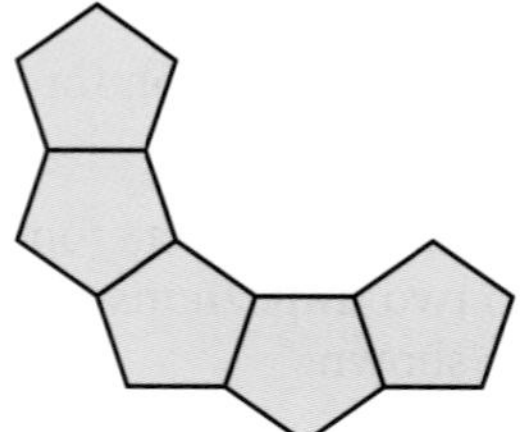

13 A company makes containers as shown.
The top is in the shape of a regular octagon.

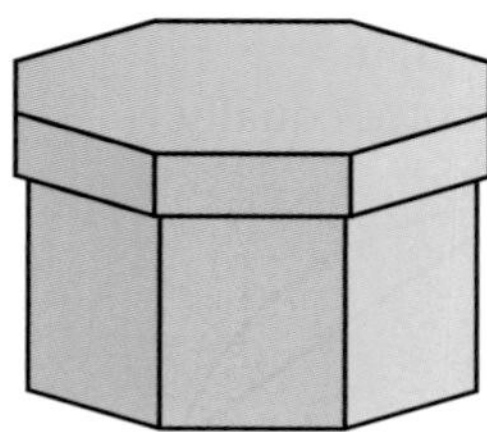

a What is the size of each interior angle?

b When the company packs them into a box, will they tessellate?
If not, what shape will be left between them?

Assess

D

1 Which of the following polygons are possible and which ones are not possible?
Make an accurate drawing of each one that is possible.

a a kite with a right angle

b a kite with two right angles

c a trapezium with two right angles

d a trapezium with only one right angle

e a triangle with a right angle

f a triangle with two right angles

g a pentagon with one right angle

h a pentagon with two right angles

i a pentagon with three right angles

j a pentagon with four right angles

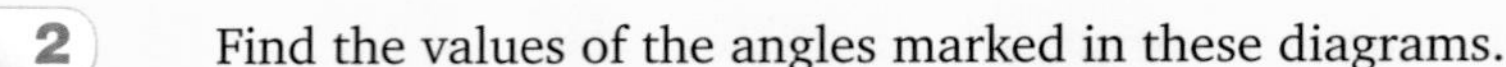

2 Find the values of the angles marked in these diagrams.

D C

a

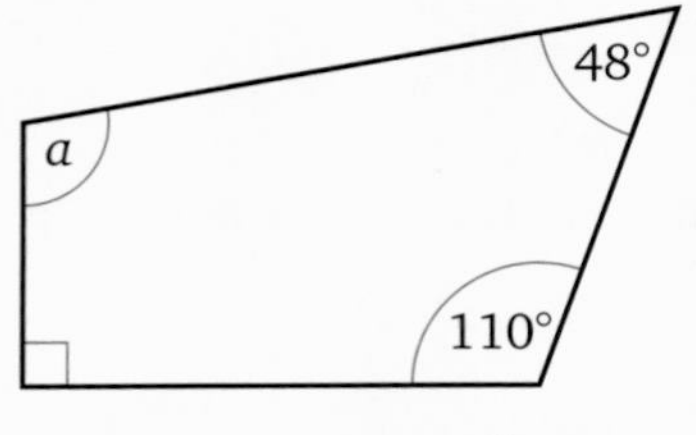

b

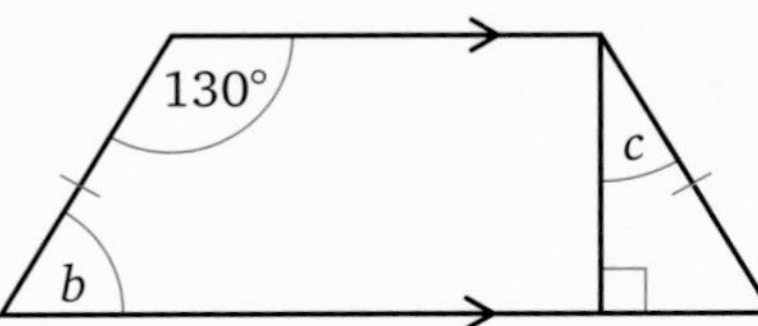

c

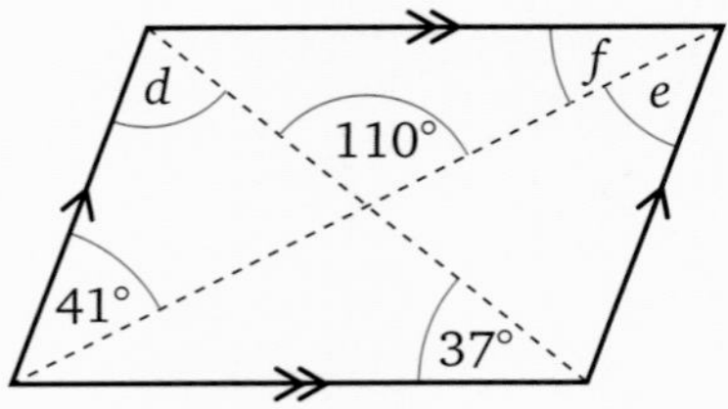

d

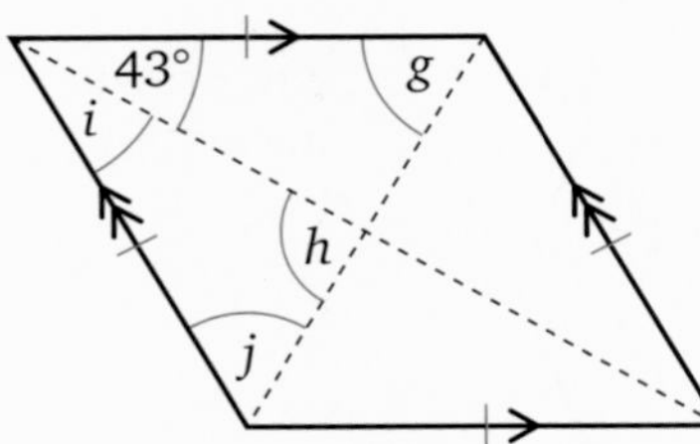

e

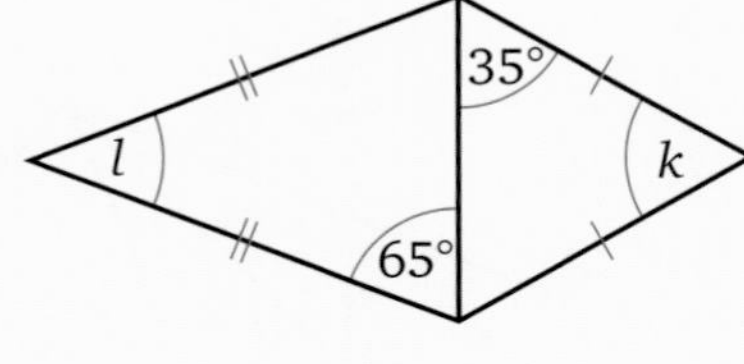

f

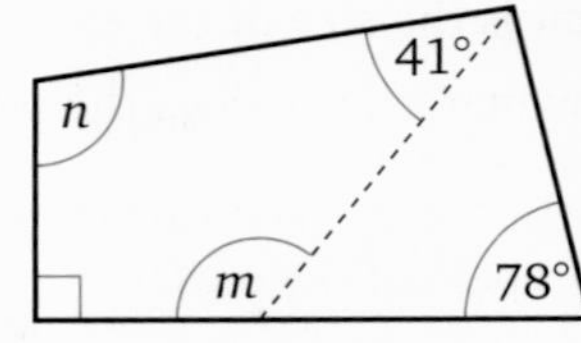

g

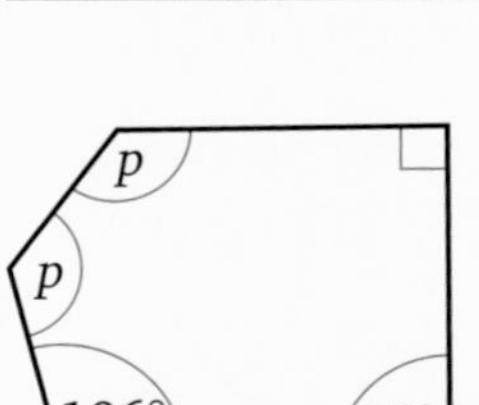

Not drawn accurately

h

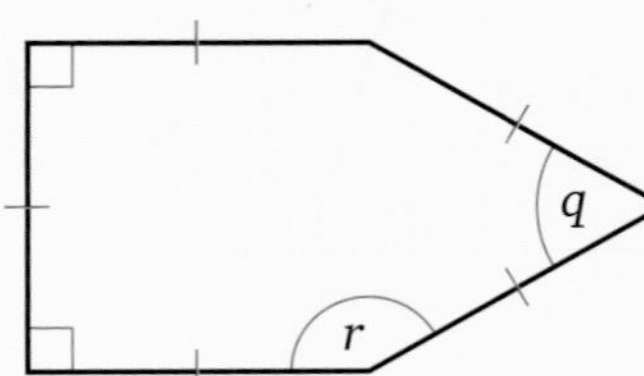

i

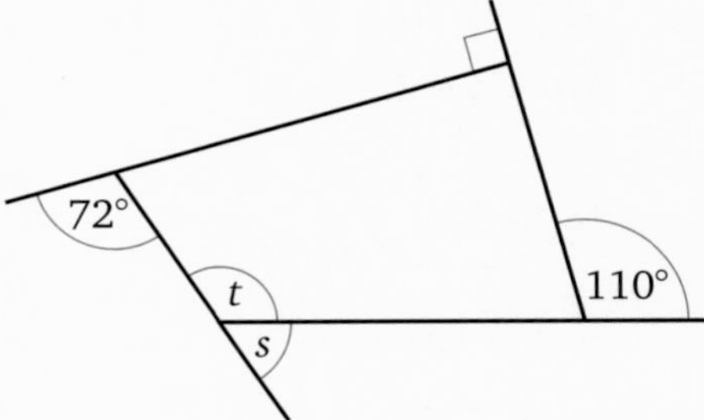

j

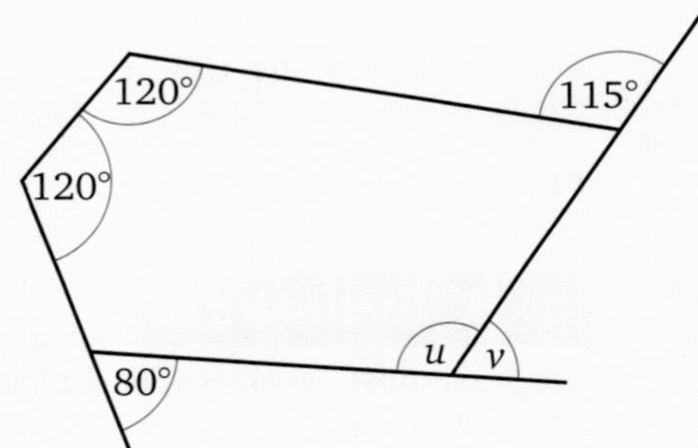

3 The only regular polygons that tessellate on their own are those whose interior angles divide exactly into 360°. Which ones are they?

C

4 Sophie says her regular polygon has an exterior angle of 40°.
Adam says that is not possible.

Who is correct? Give a reason for your answer.

5 The exterior angle of a regular polygon is 4°.

a How many sides does the polygon have?

b What is the size of each interior angle in the polygon?

c What is the sum of the interior angles of the polygon?

6 **a** A pentagon has angles of 110°, 155° and 75°.
The other two angles are equal. What size are they?

b A hexagon has three angles of 120°.
The remaining three angles are x, $2x$ and $3x$.
What is the size of the largest angle?

C

7 A regular polygon has an interior angle of 144°.

a What is the size of the exterior angles.

b How many sides does the polygon have?

8 A regular polygon has n sides where n is an odd number. Each interior angle is a whole numbers of degrees.

What is the largest possible value for n?

9 *ABCDE* is a regular pentagon.

DEG, *DCF* and *GABF* are straight lines.

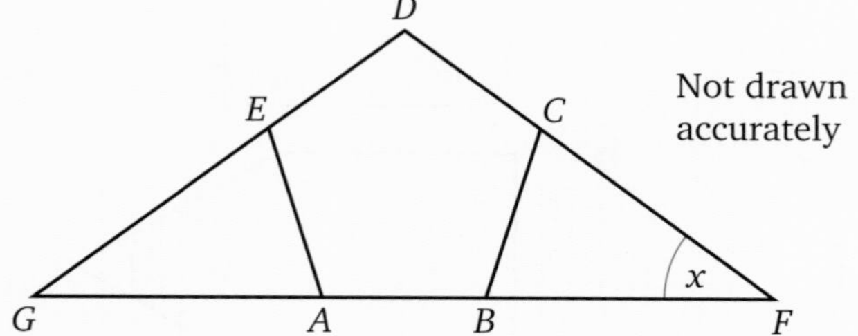

Work out the size of angle x.

B

10 *ABCDEFGH* is a regular octagon.

Work out the value of x.

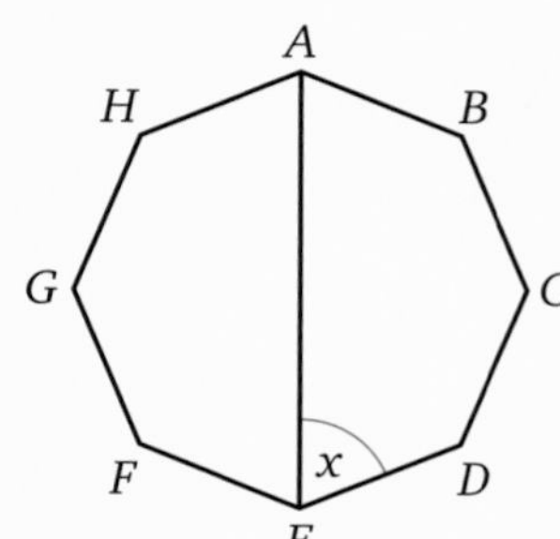

A

11 *ABCD* is a rhombus. *CDE* is an isosceles triangle. *BCE* is a straight line.

Prove that angle $BAD = 2x$

AQA Examiner's tip

A proof must use the general angle x. Do not use an example for x like 70°. Work round the diagram using geometrical facts to show angle $BAD = 2x$

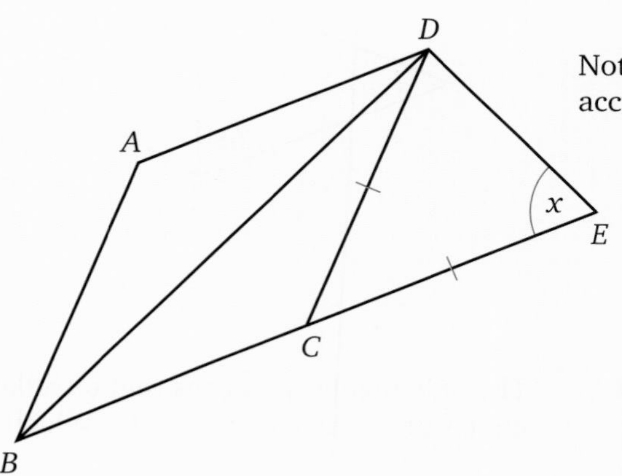

12 *ABCDEF* is a regular hexagon.

AFGH and *AJKB* are squares.

Prove that triangle *AHJ* is equilateral.

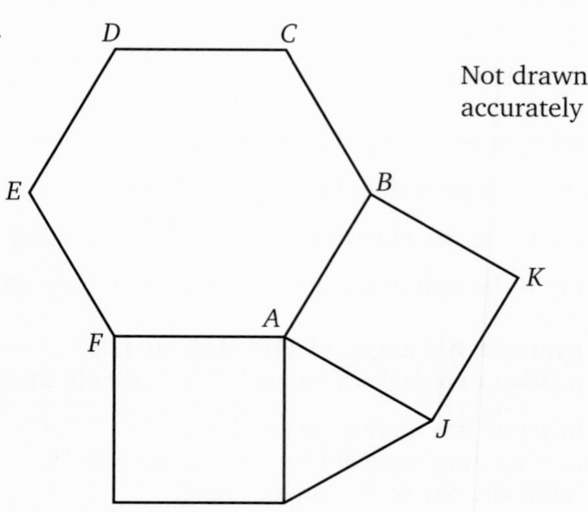

AQA Examination-style questions

1 The diagram shows a regular pentagon and a regular decagon joined at side XY.

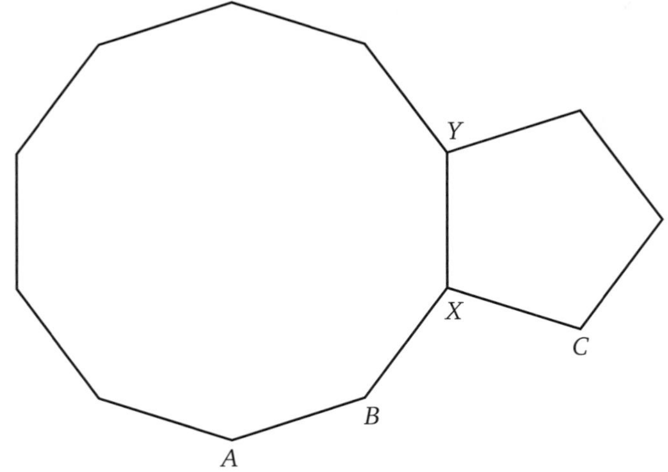

Not drawn accurately

Show that the points A, B and C lie on a straight line. *(5 marks)*

AQA 2009

8 3-D shapes, coordinates and graphs

Objectives

Examiners would normally expect students who get these grades to be able to:

D

draw the elevations of a solid on squared paper

make simple interpretations of real-life graphs

C

further interpret real-life graphs, for example find the average speed in km/h from a distance–time graph over time in minutes

B

use 3-D coordinates

discuss and interpret graphs modelling real situations.

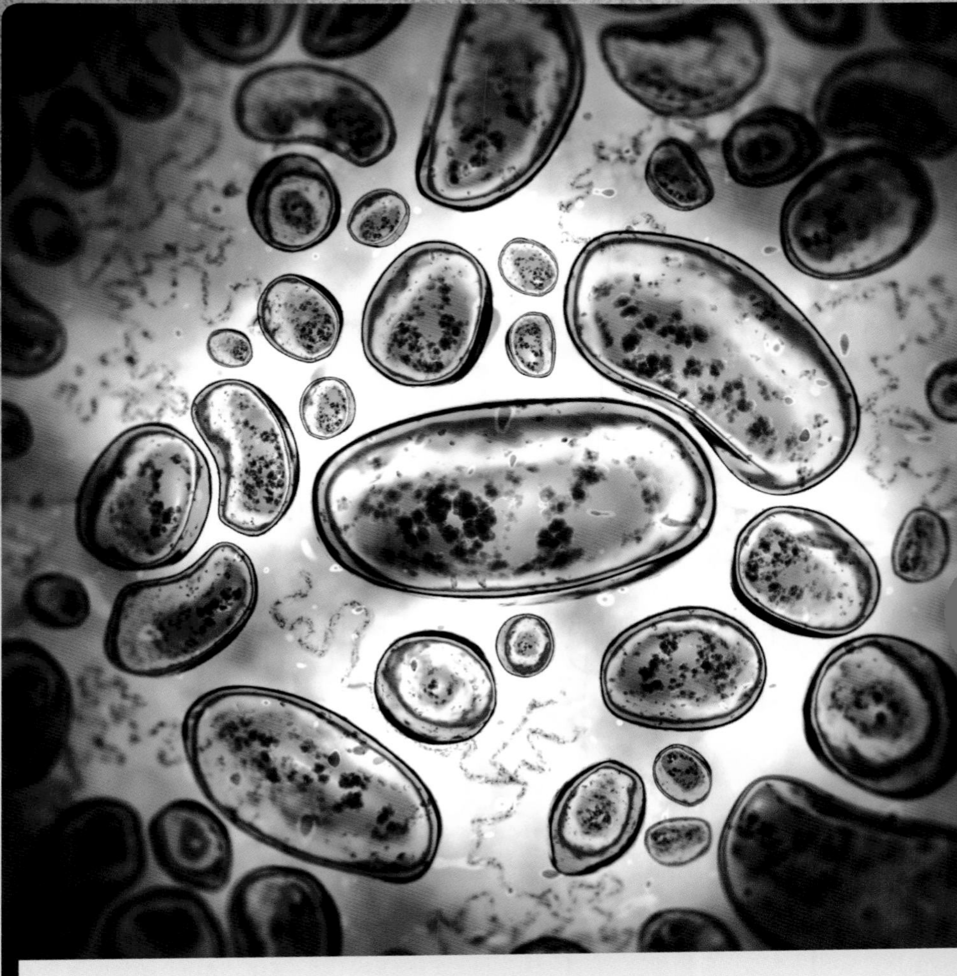

Did you know?

Modelling with graphs

Under laboratory conditions bacteria multiply rapidly, as rapidly as doubling their population every half hour. This type of growth is called exponential growth.

Graphs are used to plot experimental data. This can then be used for analysis to find relationships between variables.

Key terms

plan view
front elevation
side elevation
parabola

You should already know:

- ✔ what a plane is
- ✔ what a distance–time graph is
- ✔ coordinates in the x- and y-axes
- ✔ how to make nets of shapes.

Learn... 8.1 Plans and elevations

3-D objects may be viewed from different directions.

The view from above is called the **plan view**.

The view from the front is called the **front elevation**.

The view from the side is called the **side elevation**.

Sometimes the front elevation will look the same as the side elevation.

Lines that cannot be seen are drawn using dotted lines.

Drawing 2-D representations of 3-D shapes

There are two ways to represent cubes and cuboids in 2-D.

Method 1

This uses special paper, either **isometric paper** (with lines) or **isometric dotty paper** (which just has dots and is sometimes called triangular dotty paper).

Using isometric dotty paper:

The dots are arranged in triangles 1 cm apart.
You need to make sure that the paper is the correct way round in order for your diagrams to work.

You should be able to draw triangles like these.

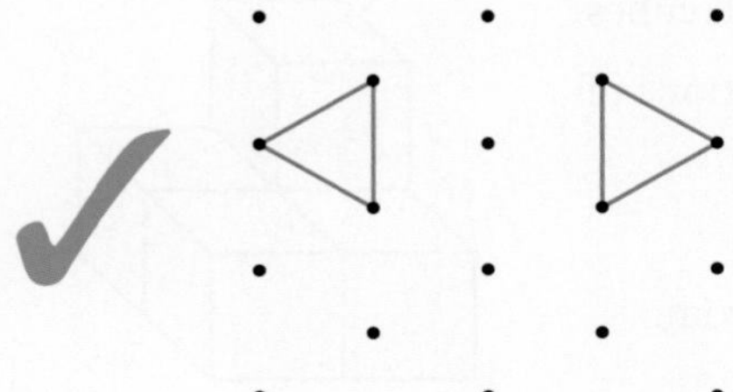

If you can only draw triangles like these then your paper is the wrong way round.

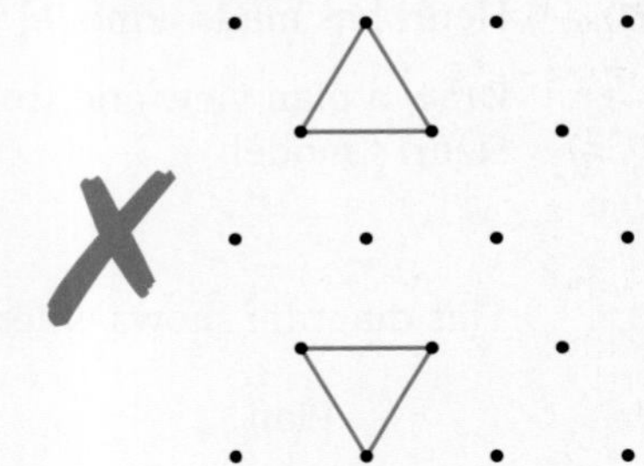

To draw a cube it is easiest to start with the top.

Join the dots to make a rhombus.

Then draw in the vertical edges.

Finally complete the edges on the base of the cube.

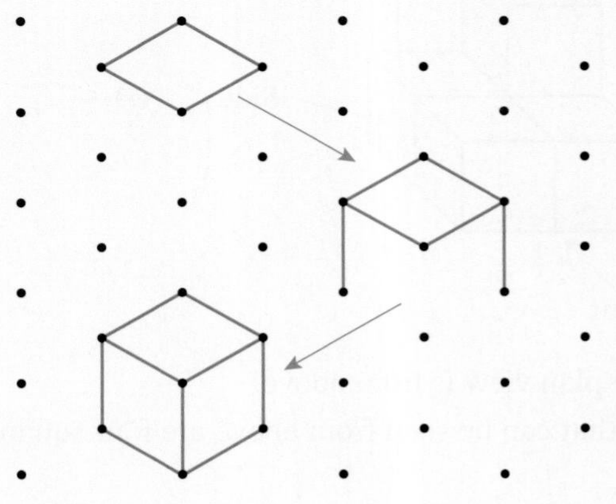

To draw shapes made of cuboids, start with the cube at the front, then work your way back.

Remember, you cannot see the front face of any cubes that are behind the cube at the front.

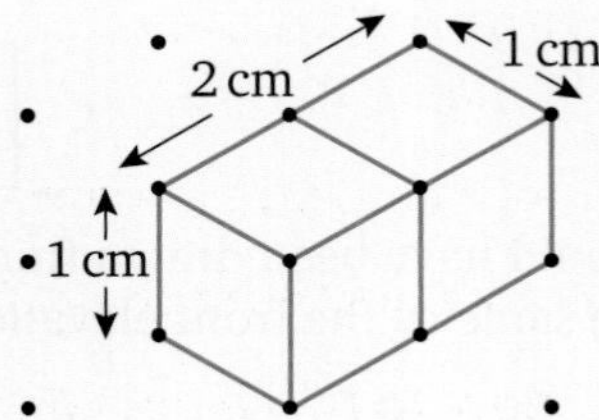

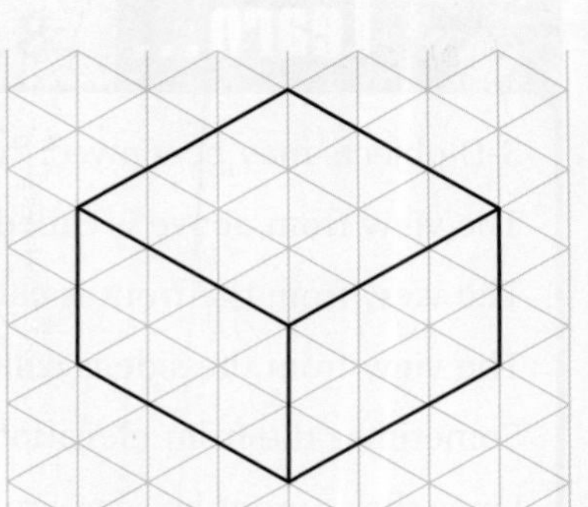

Using isometric paper.
This is just like the dotty paper, but the dots have all been joined up.

As with dotty paper, you need to make sure that it is the correct way round. You need to be able to draw the triangles the same way as for dotty paper. Another way to do this is to make sure that there are vertical lines on the page.

Method 2

This method can be used to sketch a cube when you don't have isometric paper or isometric dotty paper. Start with a square (this is the front).

Then draw three parallel lines going backwards like this (these are the edges going away from the front).
Any edges you cannot see need to be shown with dashed lines.

Add dashed edges to complete a square at the back and the fourth sloping line like this and you have sketched a cube.

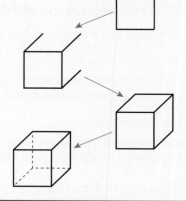

Example: Henri has made a model with five Multilink cubes.

Draw a plan view and front and side elevations of Henri's model.

Solution: This diagram shows where the views are from.

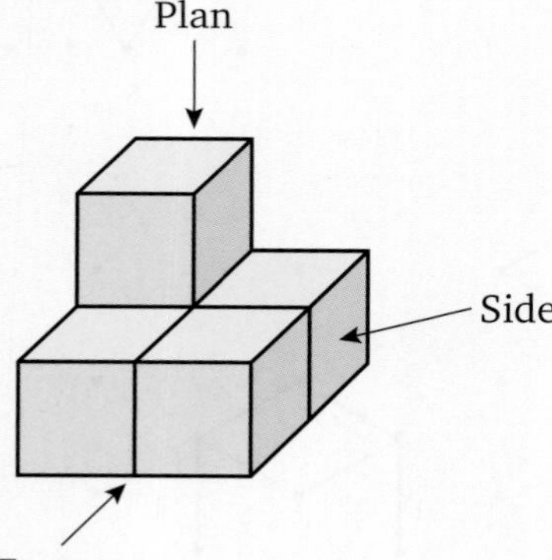

The plan view is from above:
All that can be seen from above are four squares.

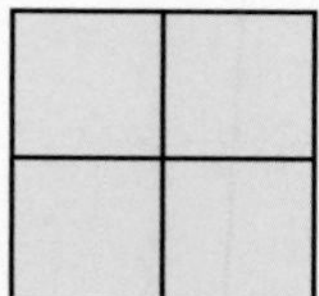

The front elevation is:

The side elevation is:

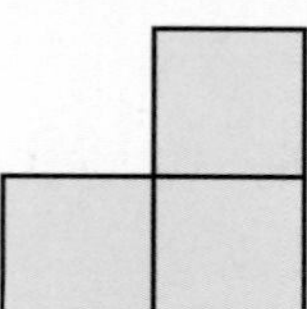

The side elevation could have been drawn from the 'other' side. If it had been, the side view would have been the same as the front elevation.

Example: Jez makes this model using five Multilink cubes.

Draw the plan view and the front and side elevations of Jez's model.

Solution: From above only three cubes can be seen.
The plan view is:

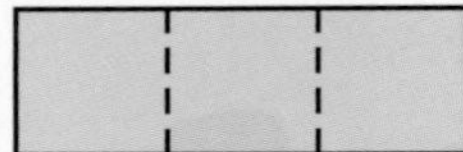

Note that the two dotted lines show where the gap is underneath.

The front elevation is:

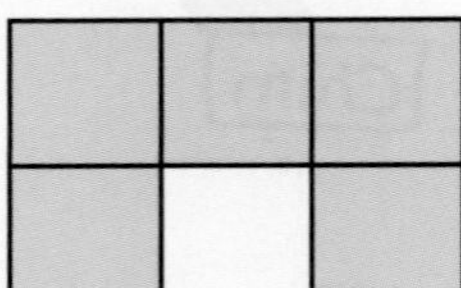

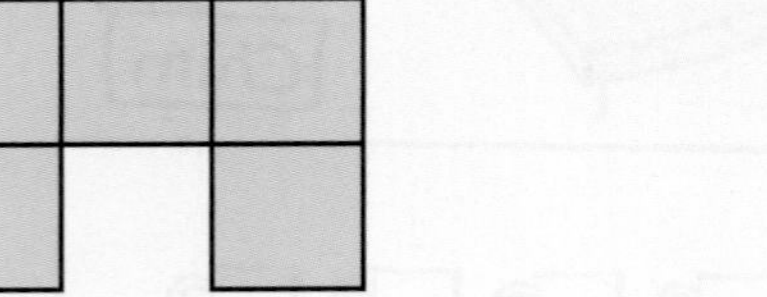

The side elevation is:

The dotted line shows the top of the gap underneath.

AQA Examiner's tip

When you use isometric paper, make sure you have the paper the right way round, otherwise you will not be able to draw 2-D representations of 3-D shapes on isometric or dotty paper.

Practise... 8.1 Plans and elevations

D

1 Each of the following shapes is made using multilink cubes.

For each shape, draw the plan view and the elevations from the directions labelled F (front) and S (side).

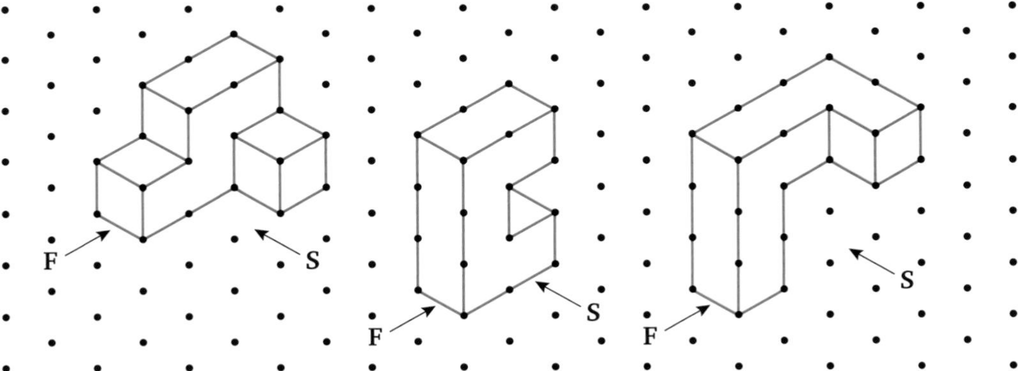

2 Here are the plan and elevations for an object made from multilink cubes.

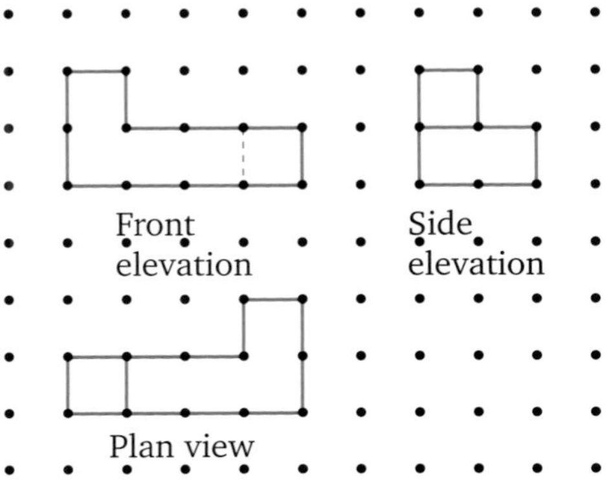

Draw the object using isometric dotty paper.

D

3 Four mugs of coffee are on a table as shown.

John sees the mugs as shown in the diagram.

a Sketch the mugs as Jane sees them.

b Sketch the view that Chim sees.

c Sketch the view that Charlie sees.

d Sketch the plan view.

4 The following diagrams show some pieces of furniture from a doll's house.

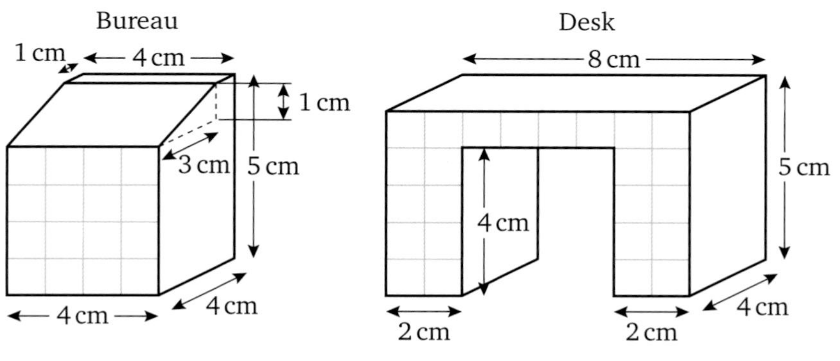

a Ian drew these plan views of the furniture.

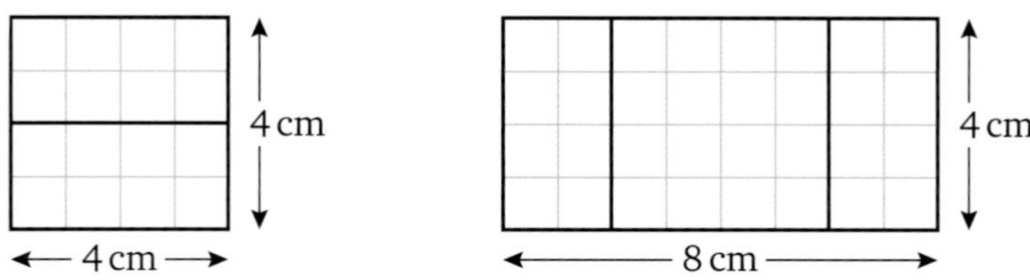

What mistakes has Ian made?

b Draw the plan views accurately.

D

5 Alan, Bridgette, Charlotte, and Dan are doing some work on plans and elevations in a mathematics lesson.

They have arranged two matchboxes on the table. They are sitting around the table as shown in the diagram.

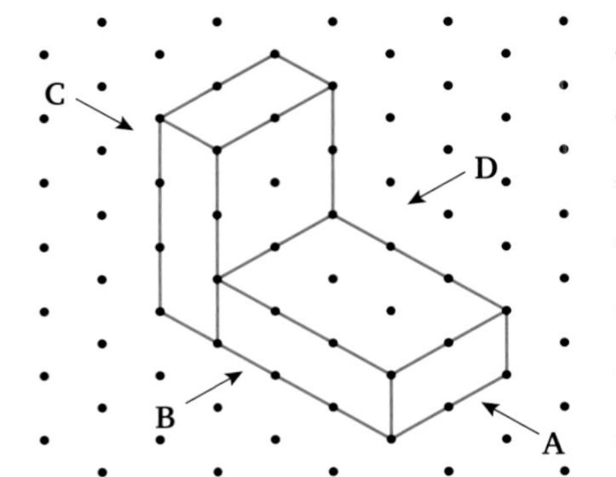

a They each draw the front elevation that they see. Which of the following does each draw?

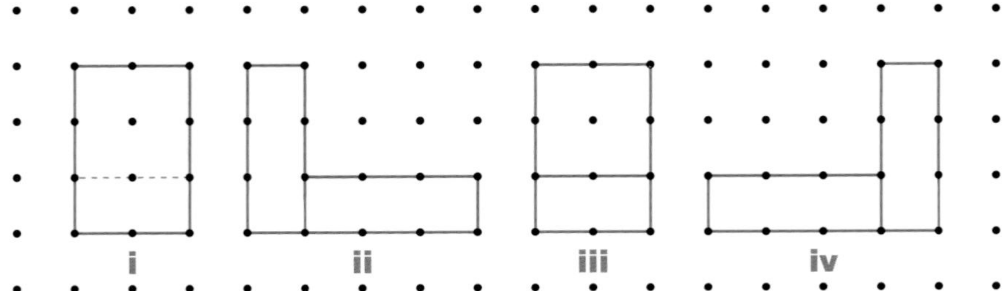

b They rearrange the matchboxes as shown.

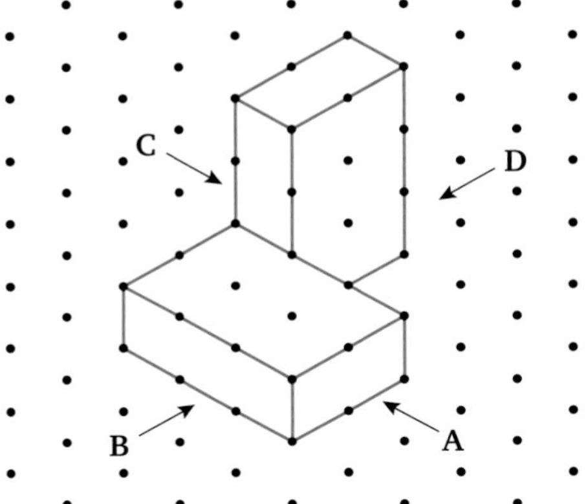

Draw the front elevation that each person sees. Label them clearly.

6 Jackie makes some models with multilink cubes. She draws a plan and elevations for each of her models.

Draw a representation of each model on triangular dotty paper.

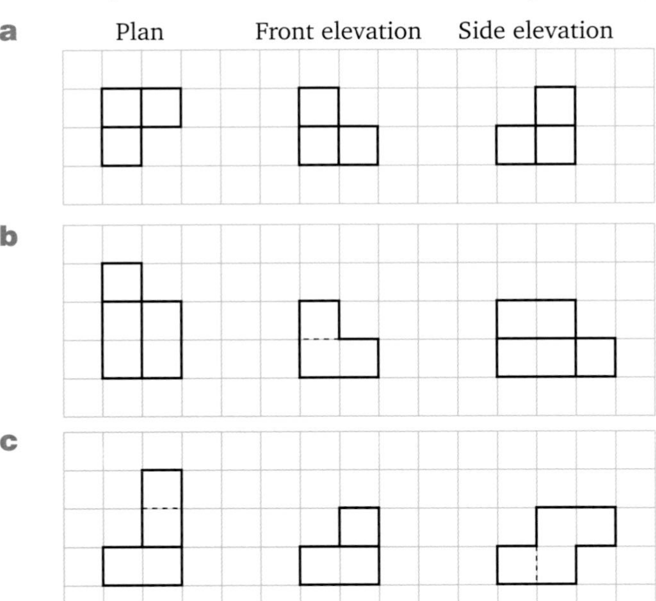

7 I have a shape. Both side elevations are a circle. The plan elevation is also a circle.

What shape have I got?

8 James is studying architecture and is doing a project on this building.

a Draw the side elevation.

b Draw the front elevation (don't forget the tree).

c Draw the plan view.

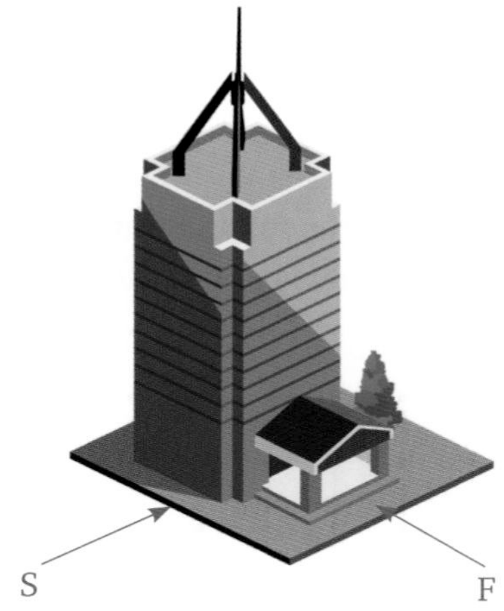

Learn... 8.2 Coordinates in 3-D

You have already learned about coordinates in two dimensions.

Consider a classroom which has a floor in the shape of a rectangle. You can use x- and y-coordinates to specify where something is on the floor of the classroom. You need a third coordinate to say how high something is above any point on the floor.

This third coordinate is called the z-coordinate.

The z-axis is perpendicular to both the x- and y-axes.

You write the coordinates in alphabetical order (x, y, z).

Example: In the diagram of the cuboid $OA = 2$ units, $AB = 5$ units and $AD = 3$ units. O is the origin.

Write down the coordinates of A, B, C and D.

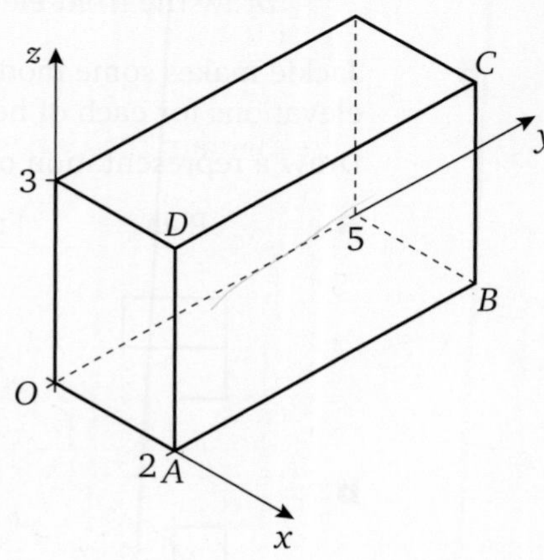

Solution: $A = (2, 0, 0)$
This is 2 units in the x-direction, 0 units in the y-direction and 0 units in the z-direction.

$B = (2, 5, 0)$
This is 2 units in the x-direction, 5 units in the y-direction and 0 units in the z-direction.

$C = (2, 5, 3)$
This is 2 units in the x-direction, 5 units in the y-direction and 3 units in the z-direction.

$D = (2, 0, 3)$
This is 2 units in the x-direction, 0 units in the y-direction and 3 units in the z-direction.

AQA Examiner's tip

Remember to get the coordinates in the correct order (x, y, z). (x, y, z) is in alphabetical order.

Practise... 8.2 Coordinates in 3-D 

D C B A A*

C

1 The diagram shows a model made from 1 cm cubes.
The point A has coordinates (1, 1, 0).

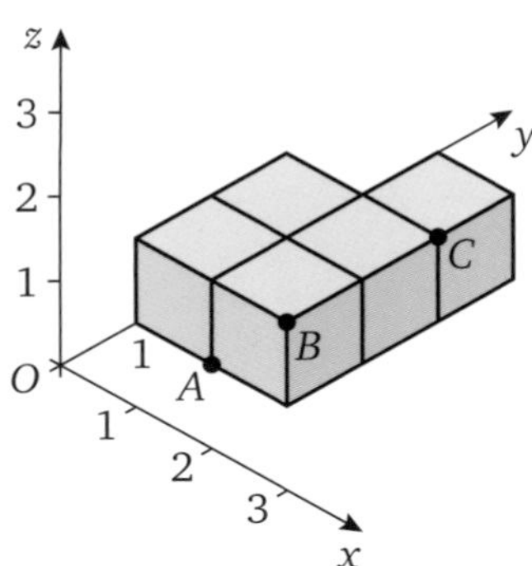

Write down the coordinates of points B and C.

2 The diagram shows a cube.

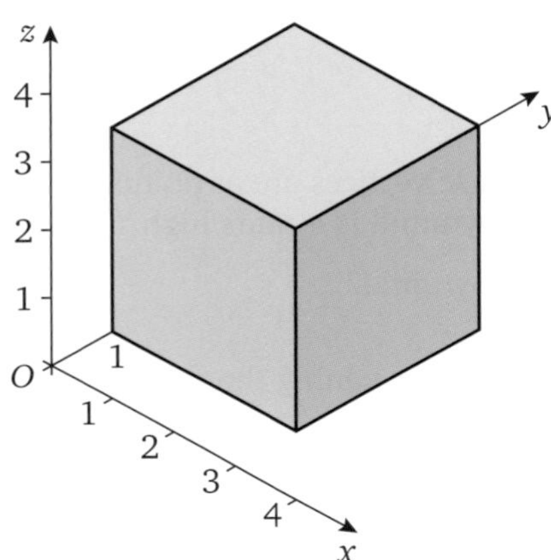

The coordinates of one vertex are (3, 1, 0). What are the coordinates of the other vertices?

B

3 The diagram shows a model made from 1 cm cubes.

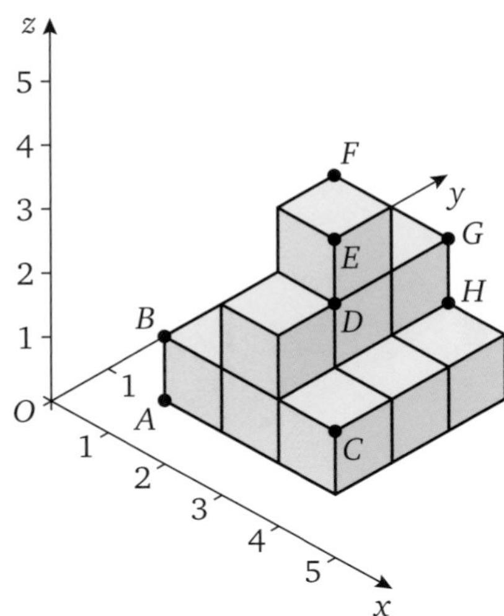

Write down the coordinates of the points labelled A to H in the diagram.

B

4 The diagram shows a cuboid.

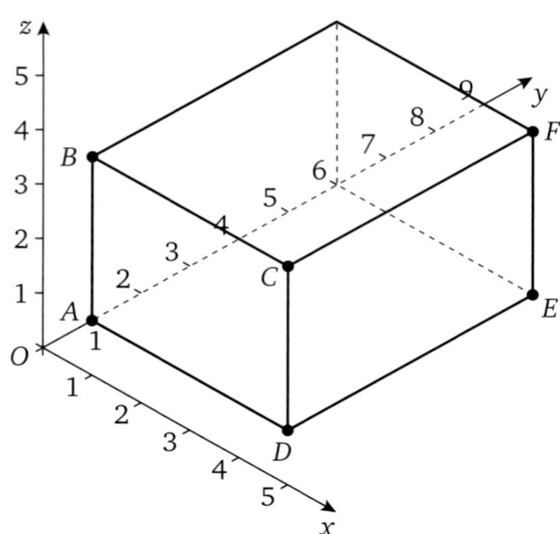

Write down the coordinates of the vertices labelled *A* to *F*.

5 A cube of side 3 cm has vertices at the points with coordinates (0, 0, 0), (0, 3, 0) and (3, 3, 0).

What are the coordinates of the other five vertices?

6 A square-based pyramid is placed so that three of the base vertices are at points with coordinates (0, 0, 0), (5, 5, 0) and (5, 0, 0). The pyramid is 4 units high.

What are the coordinates of the point at the top of the pyramid?

7 What are the coordinates of the midpoint of the line segment joining the points (2, 3, 4) and (8, 8, 10)?

Learn... 8.3 Graphs

Real-life graphs

Graphs are useful for tracking changes in a variable such as value, population size, height of a ball or temperature over time.

If you have ever been in a hospital you will have seen notes at the end of each bed. Included in these medical notes is a temperature chart. These record the temperature of patients over time.

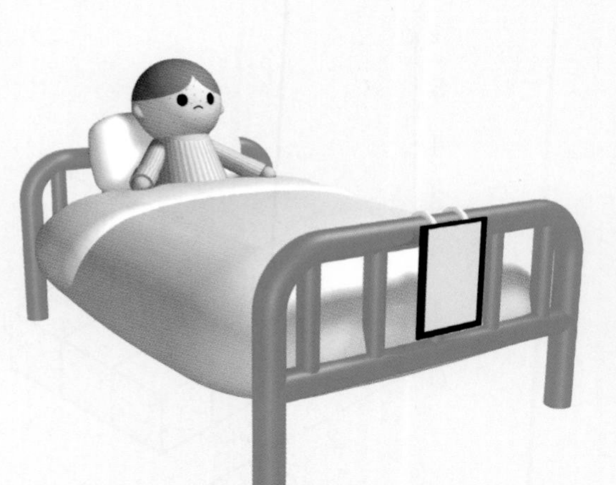

You have already met distance–time graphs. Distance–time graphs tell you about a journey of some kind. They are used to compare **speeds**.

Examples of real-life graphs

A. Graph showing temperature of a cup of coffee from the time it is made.

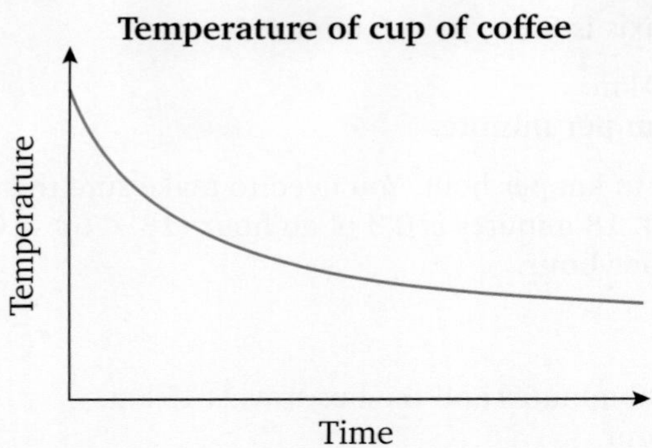

C. Graph showing population growth for a population.

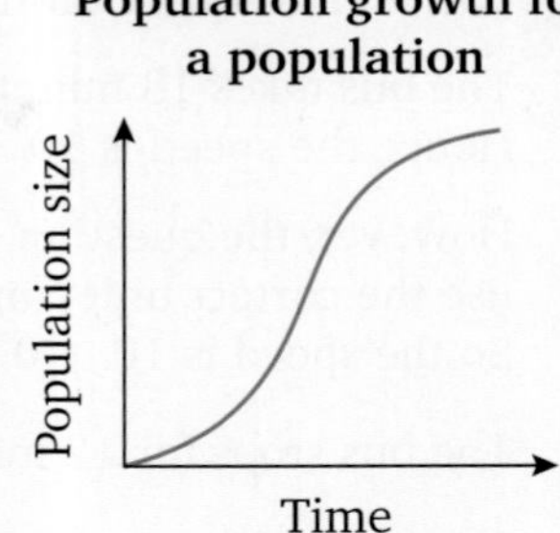

B. Graph showing height of a ball from the time it is thrown.

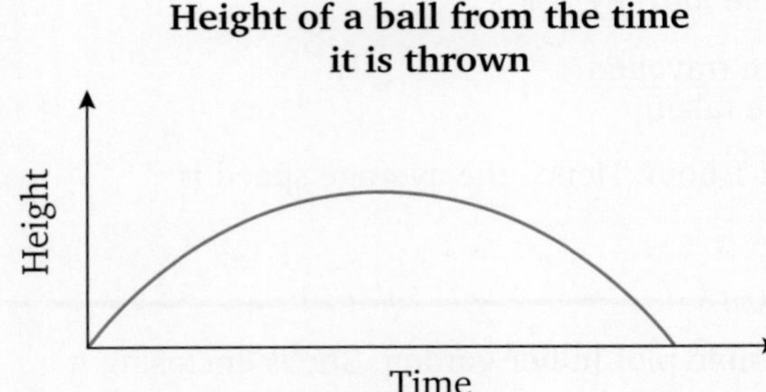

Graphs to help solve problems

Example: The distance–time graph shows the journey of the Year 10 football team in the school minibus to a nearby school for an after-school match.

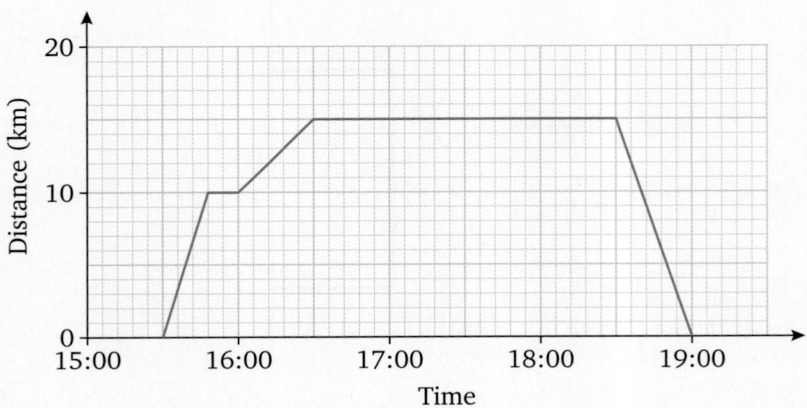

a What is the speed for the first 10 km of the journey to the football match (in km per hour)?

b On the way to the football match, the bus stops for fuel. How long did it stop for?

c What was the speed of the minibus on the way home (in km per hour)?

d At which part of the journey did the minibus travel the fastest?

e What was the average speed for the journey to the football match?

Solution:

a speed $= \dfrac{\text{distance travelled}}{\text{time taken}}$

The **gradient** (steepness) of the line is a measure of speed. The steeper the line the faster the speed. A horizontal line represents a speed of zero (i.e. stopped).

Each small square on the horizontal axis is $60 \div 10 = 6$ minutes.

The bus takes 18 minutes to travel 10 km.
Hence the speed is $10 \div 18 = 0.55$ **km per minute**.

However, the question asks for speed in km per hour. You need to make sure that you use the correct units for your problem. 18 minutes is 0.3 of an hour ($18 \div 60 = 0.3$). So the speed is $10 \div 0.3 = 33.3$ **km per hour**.

b The bus stops for 12 minutes for fuel.

c On the way home the journey takes 30 minutes and the bus travels 15 km.
The speed is $15 \div 0.5 = 30$ km per hour.

d The graph has the steepest gradient for the first part of the journey to the football match. Therefore, this was the greatest speed and the bus travelled the fastest in this part of the journey.

e To find the average speed for a whole journey, use:

$$\text{average speed} = \frac{\text{total distance travelled}}{\text{total time taken}}$$

The distance is 15 km and this took 1 hour. Hence the average speed is 15 km per hour.

Example:

A gardener is making a rectangular vegetable plot in her garden. She is enclosing it with a small hedge at one side of her garden. She has sufficient hedging plants for 10 metres of hedging.

The diagram shows the side of the garden shaded grey and the position of the hedge green. The side of the plot is labelled x. She wants to know how the area varies as x gets larger.

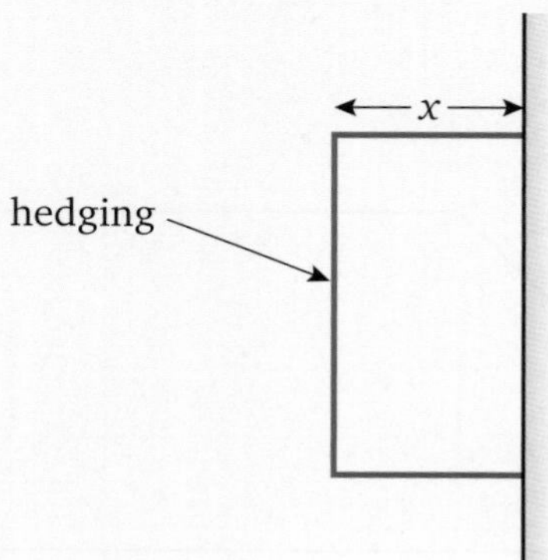

a Find a formula for the area y when you know the width x.

b Draw a table of values for different values of x.
Start with $x = 1$

c Draw a graph showing how the area changes.

d Find the largest possible area of the plot.

e Find the value of x when the area is $10\,\text{m}^2$.

Solution: **a** Length $= 10 - 2x$

Area $=$ length $\times$ width

$= (10 - 2x) \times x$

$y = x(10 - 2x)$

b

x	1	2	3	4	5
$(10 - 2x)$	8	6	4	2	0
y	8	12	12	8	0

c

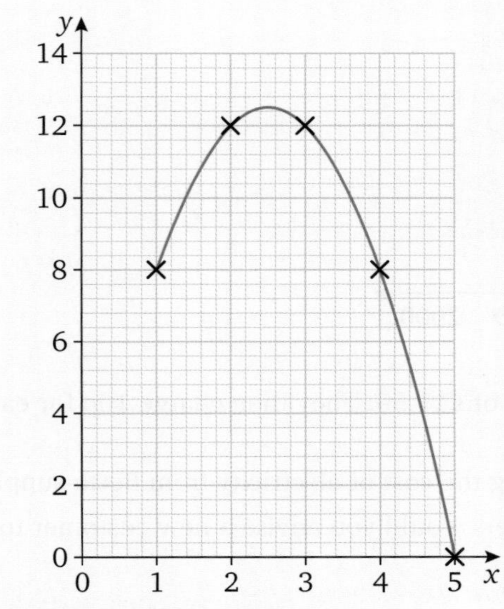

AQA Examiner's tip

Join the points with a smooth curve. Examiners do not give marks for using a ruler to join the points with straight lines.

d The largest possible area of the plot is when x is 2.5 (note the curve is a **parabola**, which is symmetrical) and you can read y is 12.5 m^2 from the graph.

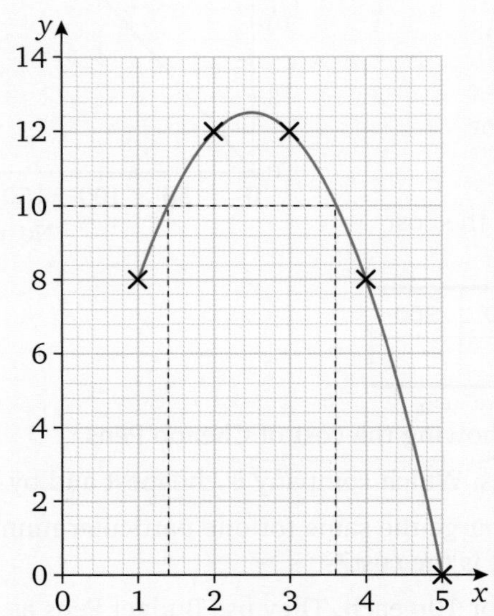

e Read across from 10 and down to the x-axis. $x = 1.38$ m and $x = 3.62$ m
When the width is 1.38 m or 3.62 m, the area is 10 m^2.

Practise... 8.3 Graphs

D

1 Ace-energy supply electricity. They have a standing charge of £8.50 each quarter. They then charge 10.5p per unit of electricity used.

a Copy and complete the table for the cost of electricity from Ace-energy.

Units used	0	500	1000	1500	2000
Cost (£)	8.50	61			

b Draw a graph showing this information using the scales shown on the axes below.

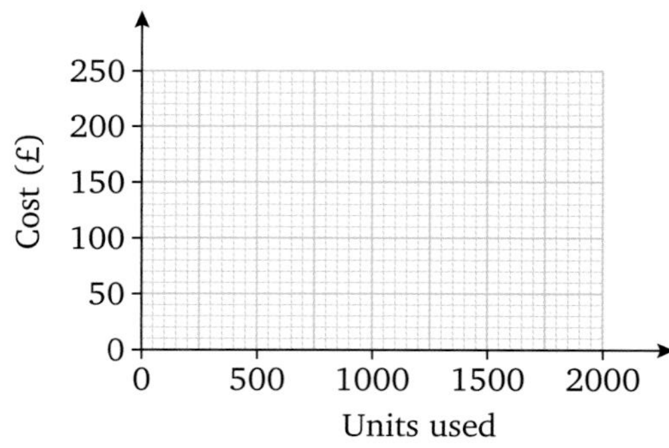

Betta-supplies have a standing charge of £11.50. They then charge 10p for each unit of electricity used.

c Add a line to your graph showing the cost of electricity from Betta-supplies.

d Which of these electricity suppliers would you advise a new customer to use? Explain your answer.

2 Budget Pens supply pens to schools. Their charges are shown in the graph. Large orders get a discount.

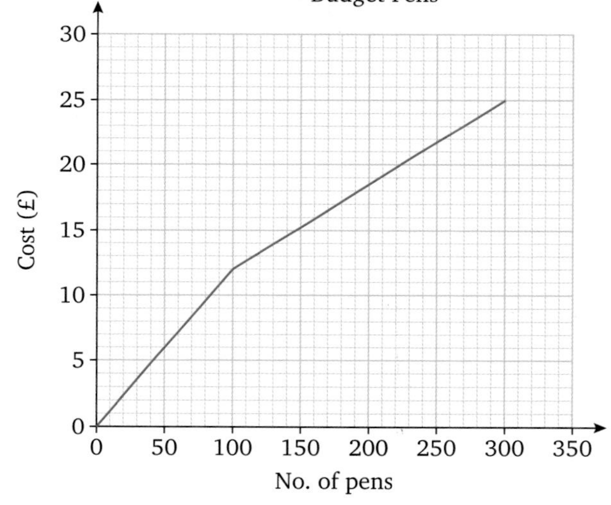

a What is the total cost of 80 pens from Budget Pens?

b A school buys 80 pens. What is the cost of each pen?

c What is the cost of 250 pens from Budget Pens?

Cheapo Pens are a rival company who also supply pens to schools. They charge 9p for each pen, no matter how big the order.

d Copy and complete the table for the cost of pens from Cheapo Pens.

Number of pens	100	200	300
Cost (£)	9		

e Copy the graph and add a line showing the cost of Cheapo Pens.

f A school wants to order 150 pens. Which company is cheapest and by how much?

g Budget Pens and Cheapo Pens charge the same for one particular number of pens. How many pens is this and what is the cost?

h School A sells pens to students for 10p each. They use Budget Pens as a supplier. How many pens do they need to sell before they start making a profit?

i School B buys 300 pens from Budget Pens.
How much do they need to sell them for if they are not to make a loss?
Give your answer to the nearest 1p. Will they make a profit? If so, how much?

C

3 The graph shows the distance of a train from Manchester in km.

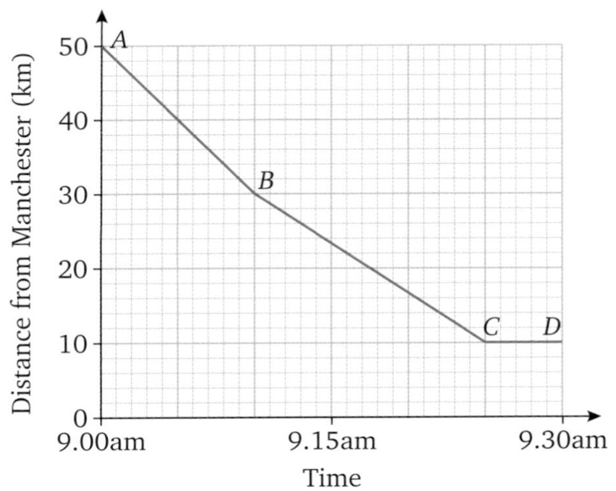

Bump up your grade

To get a Grade C you need to be able to find the speed in km/h when the time is given in minutes. You need to remember to change the minutes to hours before dividing. Remember that there are 60 minutes in one hour.

a Find the speed of the train at 9.05am.

b What is the speed of the train between the points marked B and C on the graph?

c Adam says the train is travelling at a constant speed between C and D.
Is Adam correct? Explain your answer.

d Bev says that the train is going downhill until it gets to C.
Is Bev correct? Explain your answer.

e Calculate the average speed of the train between the points marked A and C on the graph.

B

4 'Go-Monkey' is a company that makes playground resources for teenagers. One of their products is a rope bridge.

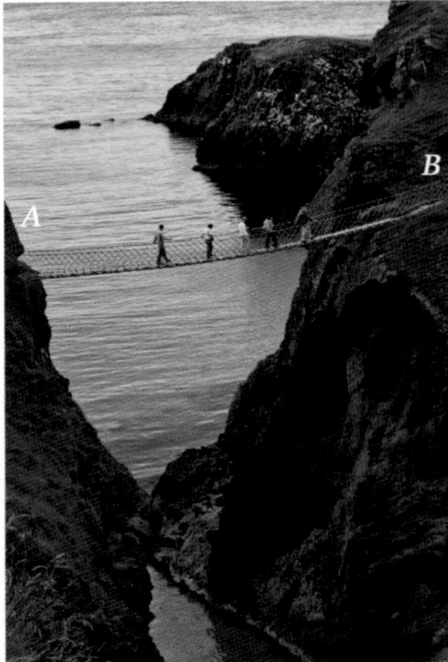

The equation for the curve of the rope is:

$$y = \frac{x(x - 20)}{100}$$

y is the vertical height above the end A and x is the horizontal distance from A.

a Complete the table of values.

x	0	5	10	15	20
y	0	−0.75			

b Copy the axes below, plot the points from the table and draw a smooth curve through them.

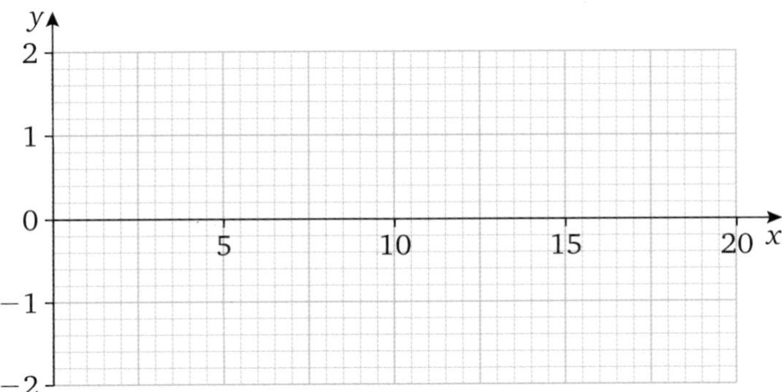

c Xavier says that the lowest point below the end is half way along and is 2 metres below the level of end A.
Is Xavier correct? Explain your answer.

B

5 Janet has a rectangular pond in her back garden.
It is 1 metre by 2 metres.
She wants to make a rectangular concrete border around the pond x metres wide. She wants to know how the area of concrete path changes as its width gets bigger.

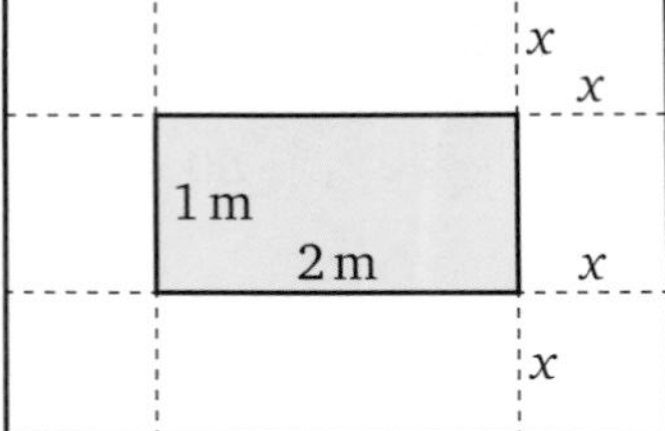

- **a** Find a formula for the area of concrete y when you know the width x.
- **b** Draw a table of values for values of x going from 0 to 5.
- **c** Draw a graph showing how the area changes as the width gets larger.
- **d** What is the area of concrete if the path is 1.5 metres wide?
- **e** What is the width of the path when the area of concrete is 25 m^2?

6 A caravan is bought new for £15 000. Each year the caravan loses 10% of its value at the start of the year.

- **a** Draw a graph to show the depreciation of the value of the caravan.
- **b** Use your graph to find:
  - **i** when the caravan is worth half the price it had when new
  - **ii** how much the caravan is worth after 10 years.

7 A rocket is launched vertically into the air. Its height, h metres, after t seconds is given by the formula $h = 96t - 16t^2$

- **a** Draw a graph to show the height of the rocket after t seconds.
- **b** How long does it take to reach a height of 144 metres? Explain your answer.

8 James is researching different bridges on the internet.
He finds the following on one of the websites he visits.

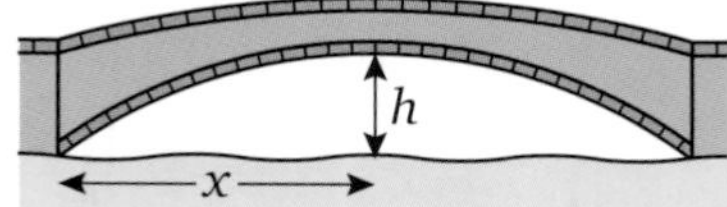

The height above the water of the arch of a bridge is given by the formula

$$h = \frac{5x}{2} - \frac{1}{20}x^2$$

where h metres is the height at a distance x metres from the bank.

What width of water is available to a yacht with mast height 25 m?

9 Rachel is doing some history homework on the internet. She is researching journeys using different methods of transport. She finds the following formula giving the cost £C of ship's voyage at V km/h is given by the formula:

$$C = \frac{9900}{V} + \frac{V^2}{2}$$

- **a** Plot the C–V curve for values of V from 10 to 30.
- **b** What is the most economical speed for the voyage, and also the total cost at that speed?

? 10 A polygon with n sides has $\frac{n(n-3)}{2}$ diagonals.

How many sides does a polygon with 35 diagonals have?

? 11 A farmer has 70 metres of fencing with which to enclose three sides of a rectangular sheep pen. The 4th side is a wall.

If the area of the pen is 600 m^2, find the length of the shorter sides.

Assess

D

1 Two matchboxes are arranged as shown.

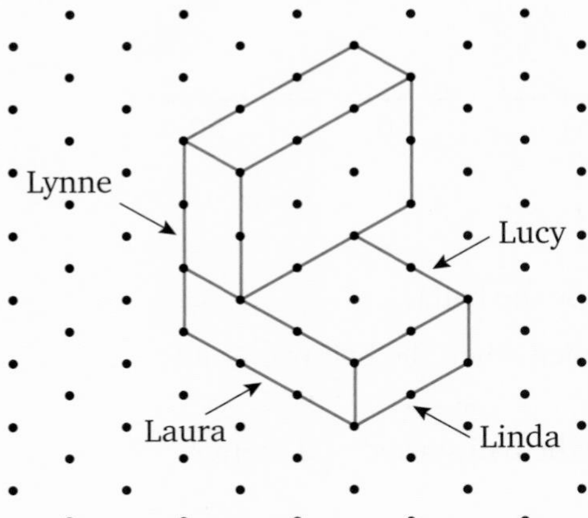

a Draw four diagrams showing the views that Linda, Laura, Lynne and Lucy see.

b Draw a plan view of the arrangement.

2 Rashid makes this object from four Multilink cubes.

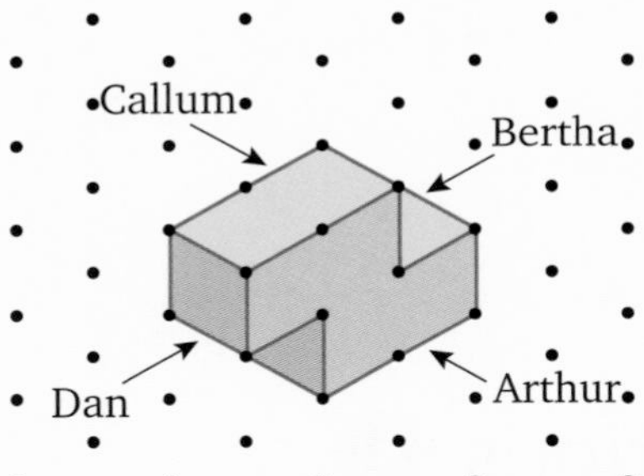

Draw the plan view of the object and the view each of his friends sees. Label your diagrams carefully.

C

3 The diagram shows a pile of 1 cm cubes.

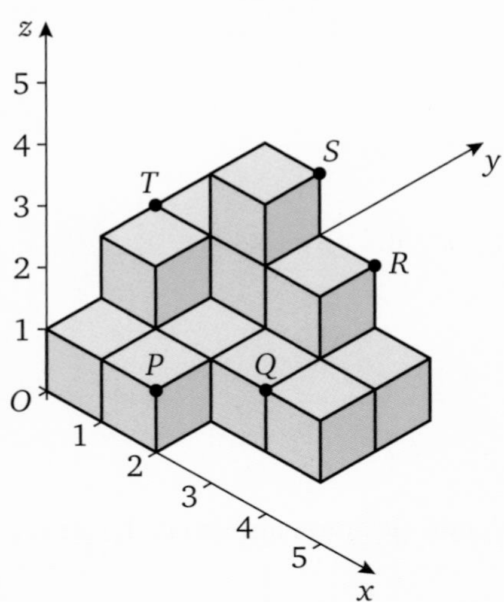

Write down the coordinates of the points P, Q, R, S and T.

C

4 The graph shows the path of a cricket ball through the air after it is thrown.

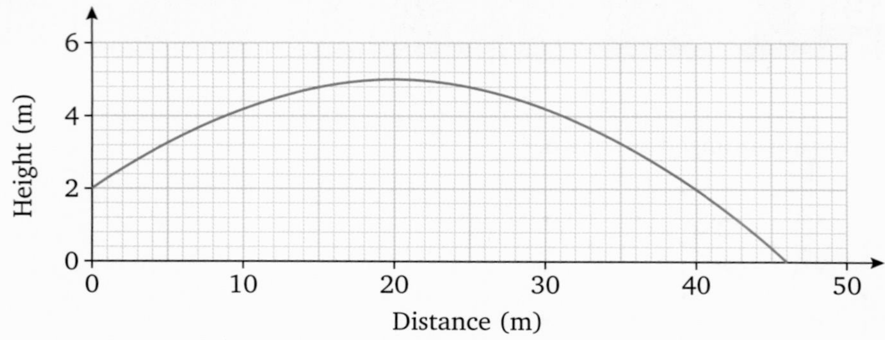

a What was the total distance thrown?

b What was the maximum height reached by the ball?

c What horizontal distance had been travelled when the ball was at its highest?

d Give a possible reason why the height of the ball started at 2 metres.

B

5 The area of a rectangle is $y\,\text{m}^2$. The length of the rectangle is 1 m greater than its width.

a Find a formula for the area, y, in terms of x. Use x for the width of the rectangle.

b Use values of x going from 0 to 5, draw a graph showing what happens to the area as the width changes.

c What is the width when the area is $10\,\text{m}^2$?

d What is the area when the length is 4.5 metres?

e Find the length when the area is $4\,\text{m}^2$.

AQA Examination-style questions 

1 The diagram shows a solid made from two cuboids.
The large cuboid is 5 cm by 4 cm by 3 cm.
The small cuboid is 3 cm by 1 cm by 1 cm.

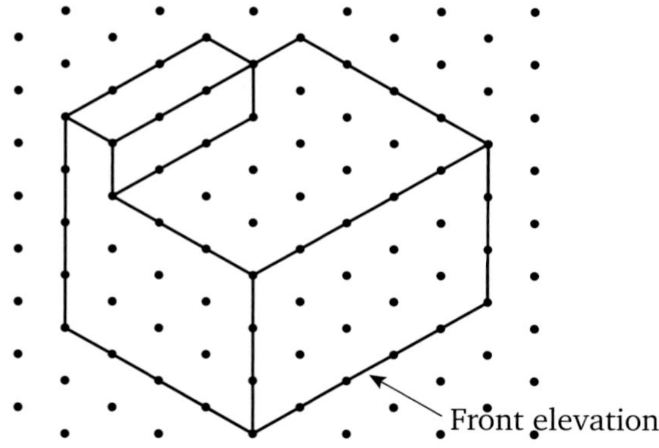

Copy the grids on the opposite page. Draw the plan view, side elevation and front elevation.

Plan view

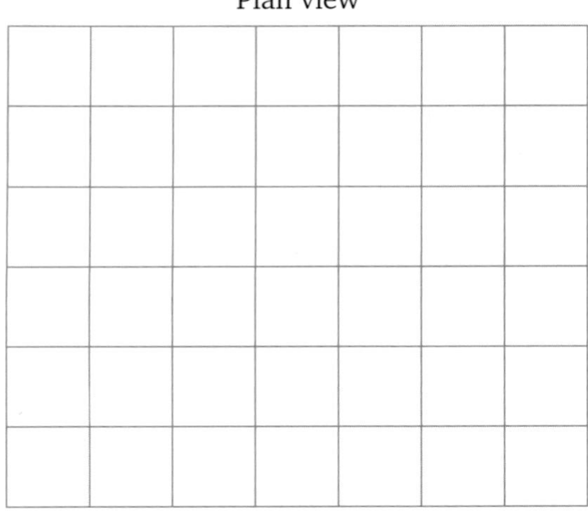

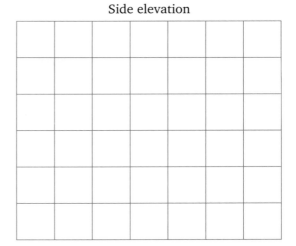

Side elevation

Front elevation

(3 marks)

AQA 2008

2 You may use graph paper to help you solve this question.

At 9am a man leaves point P and walks along a road at a steady speed of 6 kilometres per hour.

At 12 noon a cyclist leaves P, on the same road in the same direction, at a steady speed of 20 kilometres per hour.

After travelling for an hour, the cyclist gets a puncture, which delays her for 30 minutes.

She then continues at 20 kilometres per hour until she overtakes the walker.

a At what time did the cyclist overtake the walker? *(3 marks)*

b A motorist leaves P, travelling at a steady speed of 50 kilometres per hour.

The motorist overtakes the walker at the same time as the cyclist.

At what time did the motorist leave P? *(2 marks)*

AQA 2009

Consolidation 1

So far you have covered the following topics:

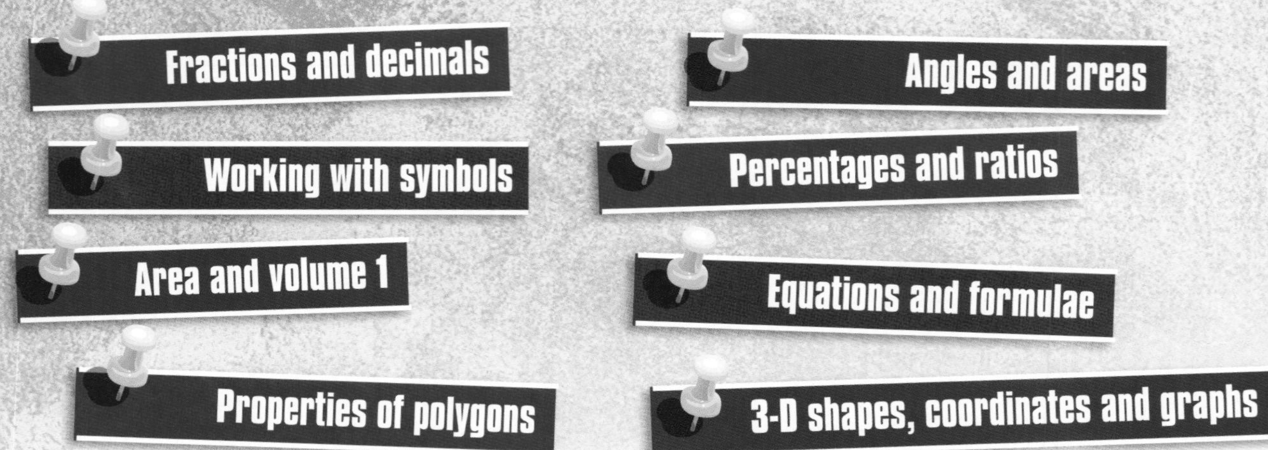

All these topics will be tested in this chapter and you will find a mixture of problem-solving and functional questions. You won't always be told which bit of maths to use or what type a question is, so you will have to decide on the best method, just like in your exam.

Example: A, B and C are three towns.

The bearing of A from B is 020°.
The bearing of A from C is 310°.
A and C are the same distance from B.

a Work out the angles marked x and y. *(3 marks)*

b Work out the bearing of C from B. *(2 marks)*

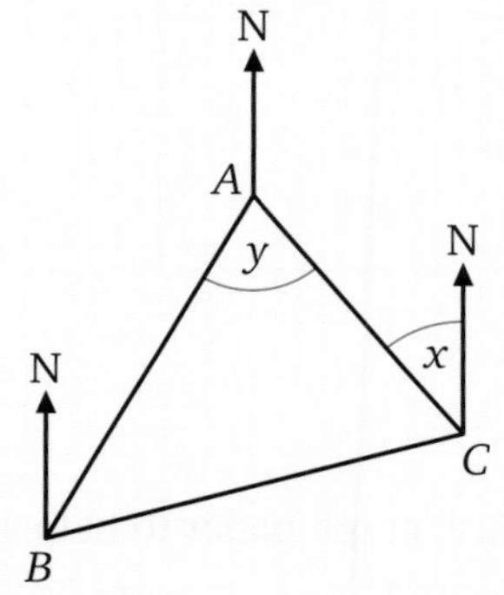

Not drawn accurately

Solution: **a** Firstly, mark on the diagram the information given in the question.
To do this, mark the 20° and 310° angles on the diagram.
Also, since A and C are the same distance from B, $AB = AC$
Mark the sides AB and AC with small lines to show they are equal.

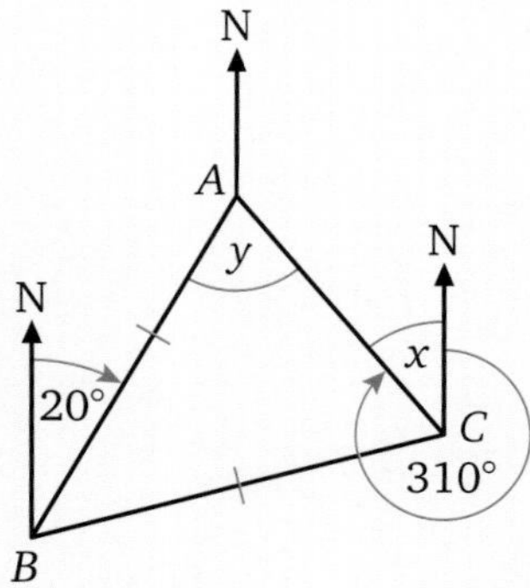

AQA Examiner's tip

Remember the bearing of A from B means you are **starting** from B, so the bearing is an angle **at** B.

The angle x must be $360° - 310° = 50°$
So angle x is 50°.

You can now use interior angles to work out y.

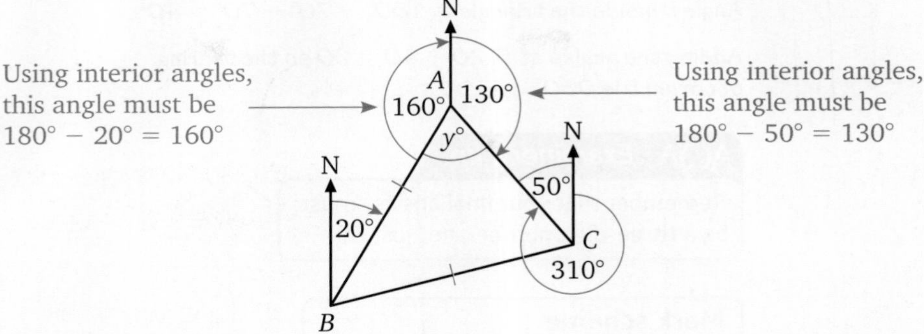

The three angles at A must add up to 360°.

So $y = 360° - 160° - 130° = 70°$

So angle y is 70°.

So the answer for part **a** is:

$x = 50°$ and $y = 70°$

Mark scheme

- 1 mark for the angle x.
- 1 mark for the angles of 20° drawn at B and 50° drawn at A.
- 1 mark for the angle y.

b Think about which angle gives the bearing of C from B.

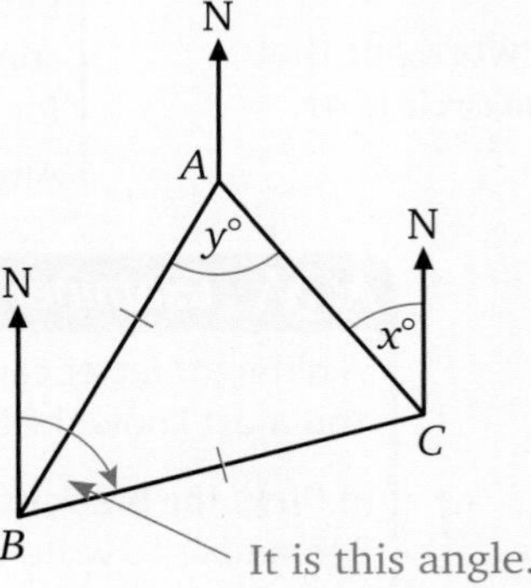

So you need to work out angle ABC inside the triangle.

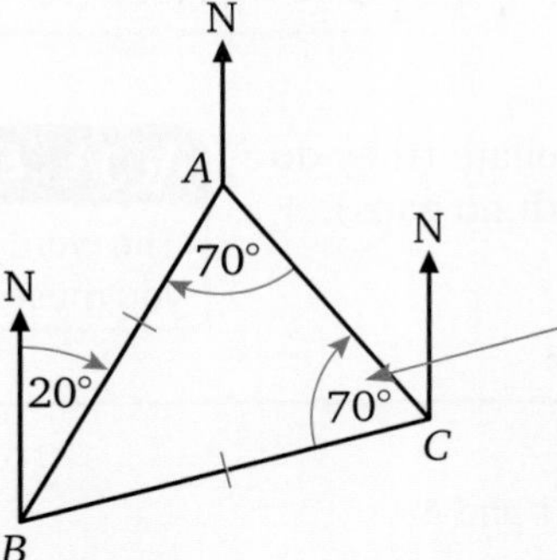

The triangle must be isosceles because we have already said that $AB = AC$. This angle must be the same as the angle at A.

The angle at C must be 70°.

Now you can work out the angle at B.

Angle B inside the triangle $= 180° - 70° - 70° = 40°$

Adding the angles at B, $20 + 40 = 60$ so the bearing of C from B is 060°.

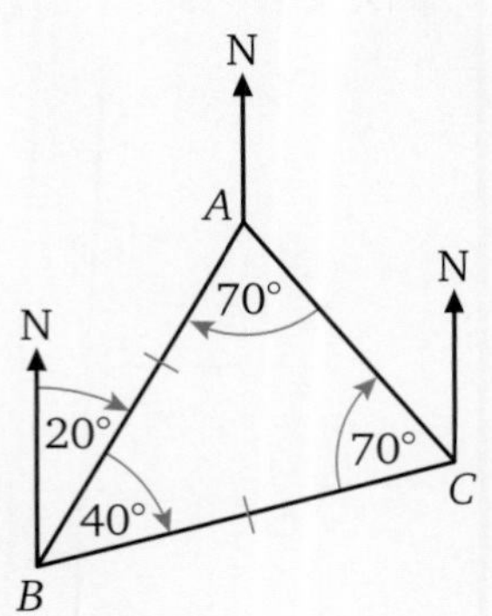

AQA Examiner's tip

Remember that your final answer must be a **three**-digit number, not just 60.

Mark scheme

- 1 mark for working out 40° at B.
- 1 mark for the final answer.

Example: The diagram shows one large circle, area L, and two small circles, each area S. They touch as shown. C is the centre of the large circle and is on the circumference of one of the small circles.

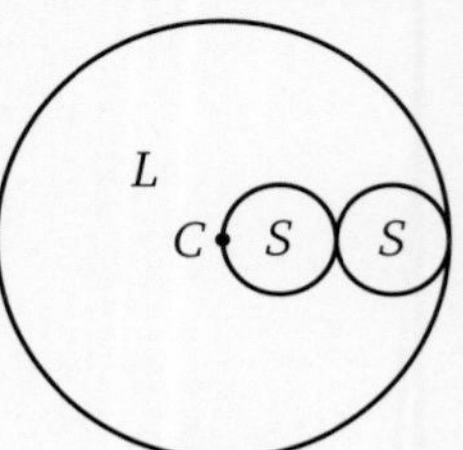

a Show that $L = 16S$ *(3 marks)*

b 16 of the small circles **cannot** be placed inside the large circle without any of them overlapping.

Give a reason why this is true. *(1 mark)*

AQA Examiner's tip

There are no numbers to work with. Because you are asked to prove something, you should not use numbers for the radii of the circles.

Always make it clear what you are doing.

Solution: **a** Let the radius of one small circle be r.

Use the diagram to work out that the radius of the large circle is $4r$.

$S = \pi \times \text{radius}^2$

$= \pi r^2$

$L = \pi \times \text{radius}^2$

$= \pi \times (4r)^2$

$= \pi \times 16r^2$

$= 16\pi r^2$

So, $L = 16S$

AQA Examiner's tip

A different letter can be used, e.g. x.

You must know the formula: area of circle $= \pi \times \text{radius}^2$

In $(4r)^2$, the brackets are necessary.

Remember to write expressions in their simplest form.

In some questions on circles, you need to use a calculator for the value of π. In this question the algebraic letter π can be used without needing to substitute a numerical value for it.

b Circles do not tessellate (they do not fit together with no gaps).

AQA Examiner's tip

The word tessellate does not have to be used but you must make your reason clear to the examiner.

Mark scheme

- 1 mark for using r and $4r$.
- 1 mark for obtaining $\pi \times (4r)^2$ or $\pi \times 16r^2$ or $16\pi r^2$.
- 1 mark for obtaining πr^2 and $16\pi r^2$.
- 1 mark for giving a correct reason.

D

Consolidation

1 Nisha left home at 9.00am to visit a friend.
This is how long her journey took.

Walk to the station	12 minutes
Wait for the train	21 minutes
Journey on the train	1 hour 15 minutes
Walk to her friend's house	12 minutes

a What fraction of the time was spent on the train?

b On the way back, the time for the whole journey took 15% less.
How long was the journey back?
Give your answer in hours and minutes.

2 The boat *Clarabel* breaks down and sends out a distress call.
The call is heard by two other boats, *Aramis* and *Bellamy*.
They set out to intercept *Clarabel*.
Aramis sets out on a bearing of 055°.
Bellamy sets out on a bearing of 260°.

Copy the diagram below and mark the position of *Clarabel*.

N

• *Bellamy*

Aramis •

3 Alice is four years younger than Ben. She is nine years older than Carly.
The total age of all three people is 79.

How old is Alice?

4 A rectangle has a length that is twice the width.
Four of these rectangles make this shape.

Work out the perimeter of the shape.

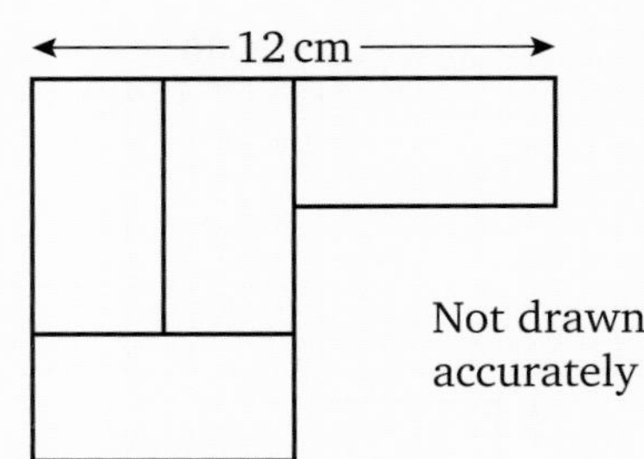

5 The diagram shows the cross-section of a roof.
The two sides of the roof are perpendicular.
One side of the roof slopes at 35° to the horizontal.

The support is made from a rectangular piece of wood.
To make it fit, the wood is cut along the dotted lines.

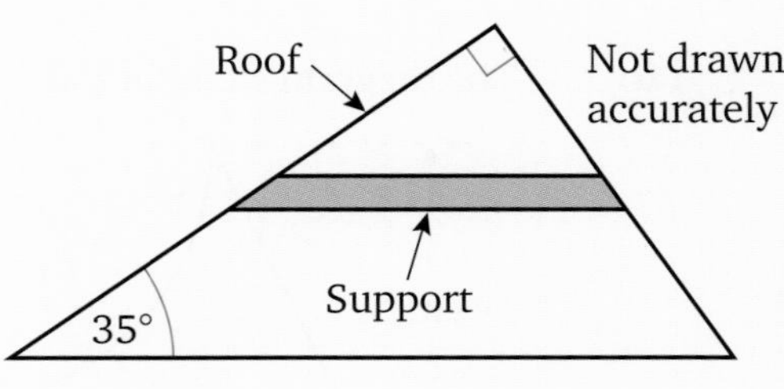

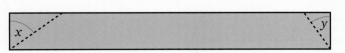

Work out the values of x and y.

D

6 To calculate a person's BMI this formula is used.

$$\text{BMI} = \frac{\text{weight in kilograms}}{(\text{height in metres})^2}$$

This table shows how people are classified using their BMI.

BMI	Classification
BMI < 20	Underweight
20 ⩽ BMI < 25	Healthy
25 ⩽ BMI < 30	Overweight
30 < BMI	Obese

a Ollie weighs 60 kilograms and is 1.8 metres tall.
Which classification is Ollie?

b Hannah is 1.54 metres tall. She weighs 68 kilograms.
Work out the least amount of weight that she must lose to be healthy.

D C

7 An activity centre organises climbing trips for the public.

a Each member of staff at the centre must not take more than 8 members of the public in their group.
A party of 76 people book a trip.
12 members of staff are available to take the party.

i Are there enough members of staff to take all the people?
You **must** show all your working.

ii What is the ratio of staff to people for this trip?
Give your answer in the form $1 : n$

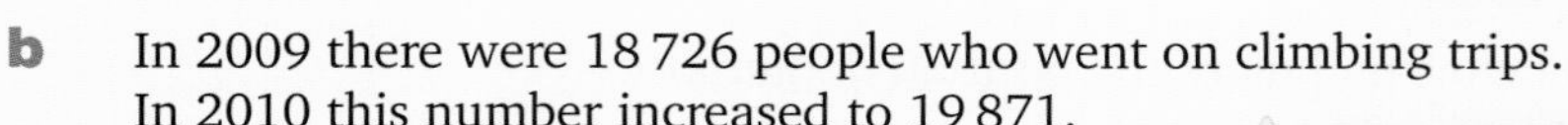

b In 2009 there were 18 726 people who went on climbing trips.
In 2010 this number increased to 19 871.
Work out the percentage increase.
Give your answer to one decimal place.

Bump up your grade

To get a Grade C you should be able to work out a percentage increase or decrease.

C

8 Three triangles fit together to make a parallelogram.

The base of the parallelogram is 10 cm long.
The area of triangle A is 28 cm².
The area of triangle B is 40 cm².

What is the area of triangle C?

A
B
C
10 cm
Not drawn accurately

9 Find an expression in terms of x for the area of this shape.
You must simplify your expression.

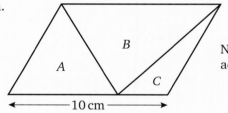

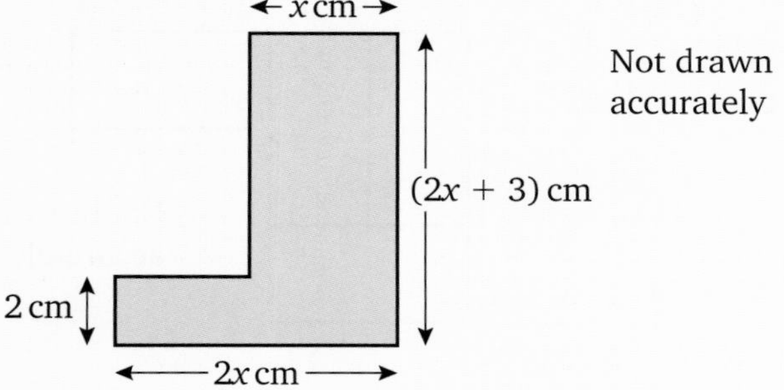

10 The diagram shows a shape made by removing a semicircle from a trapezium.

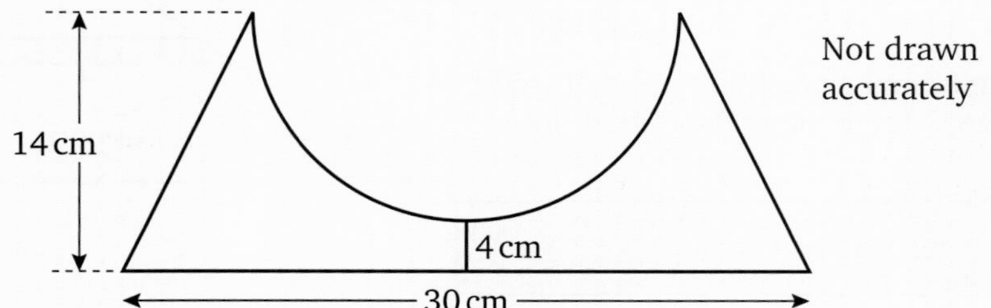

Work out the area of the shape.

To remind yourself how to find the area of a trapezium, look back at Chapter 2.

C

11 AB and CD are parallel lines.

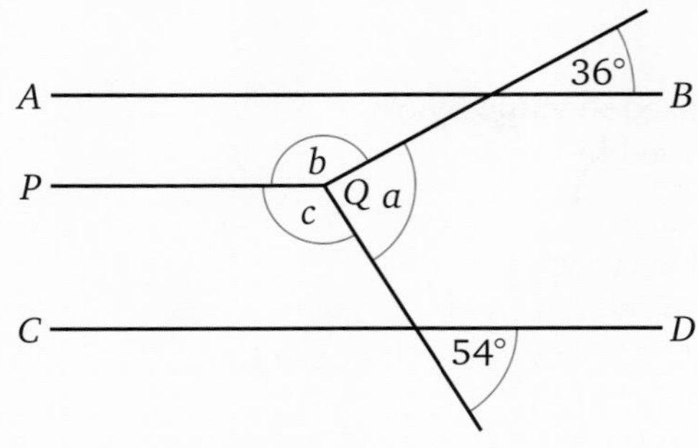

Not drawn accurately

a Show that angle $a = 90°$

b The ratio of angle b to angle c is $8:7$

Show that PQ is parallel to AB and CD.

12 A room contains a number of boys and girls. The ratio of boys to girls is $3:2$
When another 15 girls enter the room, the ratio of boys to girls changes to $2:3$

How many boys are there in the room?

13 **a** Work out the reciprocal of 8.

Give your answer as a decimal.

b Work out the reciprocal of 2.5

Give your answer as a fraction in its simplest form.

c Write down an expression for the reciprocal of $\frac{x}{2y}$

14 The points P, Q and R are on a straight line.

$PQ:QR$ is $3:1$

Work out the coordinates of point Q.

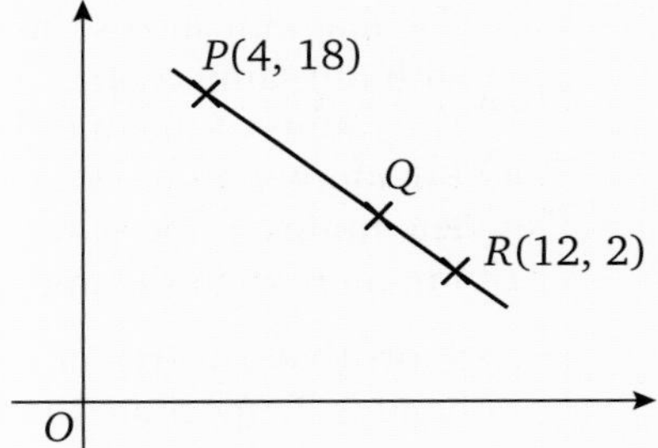

15 A 10 pence coin is approximately a cylinder of radius 12 mm and thickness 2 mm.

A vertical tower is made by stacking up £2 worth of 10 pence coins.

Calculate the volume of the tower.
Give your answer to the nearest cubic centimetre.

Not drawn accurately

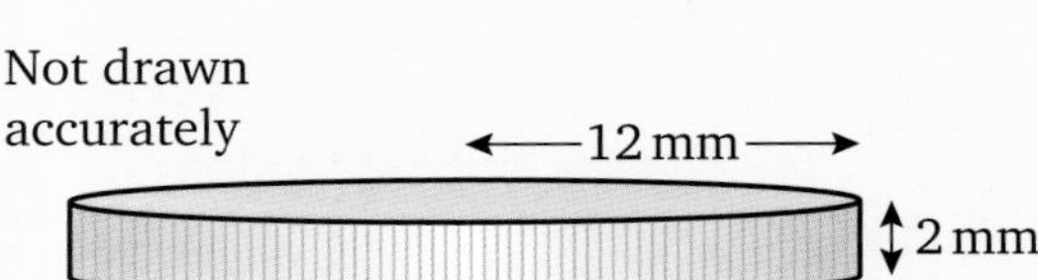

16 An interior angle of a regular polygon is 135°.
Another regular polygon has two sides fewer than this.

Work out one of the interior angles of this polygon.

B

17 Solve:

a $\frac{2x-1}{4} = 5 - x$ **b** $\frac{3x+2}{2} + \frac{2x-5}{3} = 1$

18 This tin of soup has height 14 cm and diameter 9.5 cm.
The label for the tin only covers the curved surface.
The label needs a 1 cm overlap to glue it down.

Calculate the area of the label in cm^2.

B

19 Expand and simplify $(c + 5)(c - 3)$.

20 A house was valued at £164 500 in 2005.
In each of the next two years the house increased in value by 6%
In the third year the value of the house decreased by 7.5%

Work out the value of the house after three years.
Give your answer to an appropriate degree of accuracy.

21 **a** Expand and simplify $(2x + 3)(x + 1)$.

b A rectangle has an area of 253 cm^2.
The lengths of the sides are integers greater than 1.
Using part **a**, or otherwise, work out the lengths of the sides of the rectangle.

A

22 In this cuboid, length : width : height = 4 : 1 : 1
The total surface area of the cuboid is 162 cm^2.

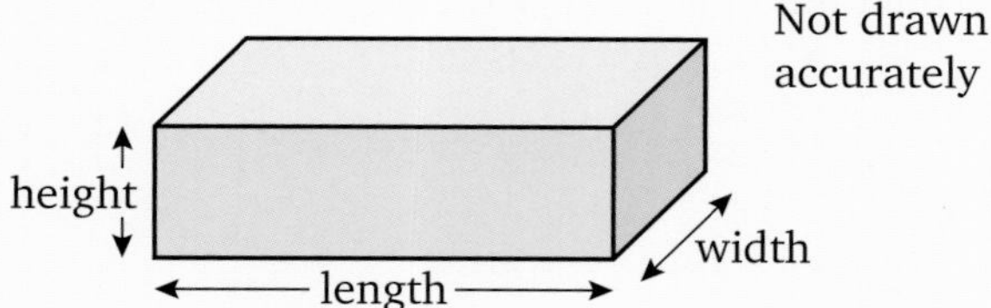

Not drawn accurately

Work out the volume of the cuboid.

23 **a** Simplify $(r + 2)^2 - (r + 1)^2$.

b The diagram shows three circles with the same centre O.
The radius of the middle circle is 1 m more than the inner circle.
The radius of the outer circle is 1 m more than the middle circle.

Show that the area of the outer band is 2π m^2 more than the area of the middle band.

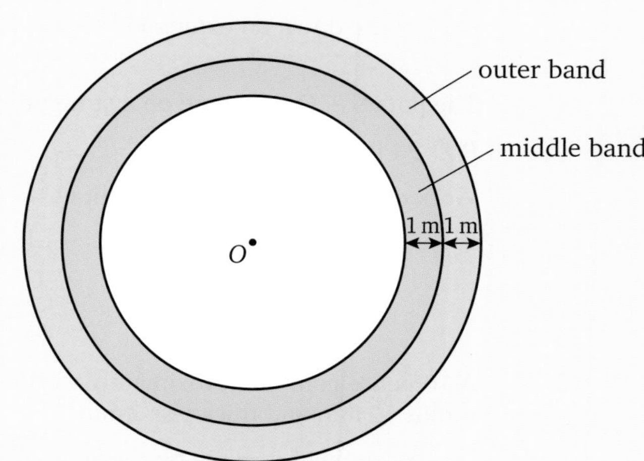

Not drawn accurately

24 A ball is dropped from a height, h metres.
It takes t seconds to reach the ground.
The height, h, is directly proportional to the square of the time, t.
It takes 10 seconds for the ball to reach the ground when dropped from a height of 490 metres.

If a ball takes 6 seconds to reach the ground, what height was it dropped from?

A*

25 Here is a flow chart.

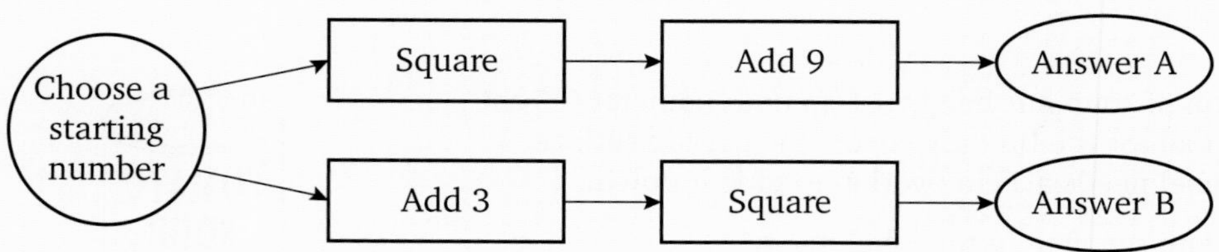

Samir says 'The difference between A and B = 6 × the starting number'

Prove that Samir is correct.

AQA Examination-style questions

1 The graph shows a journey.

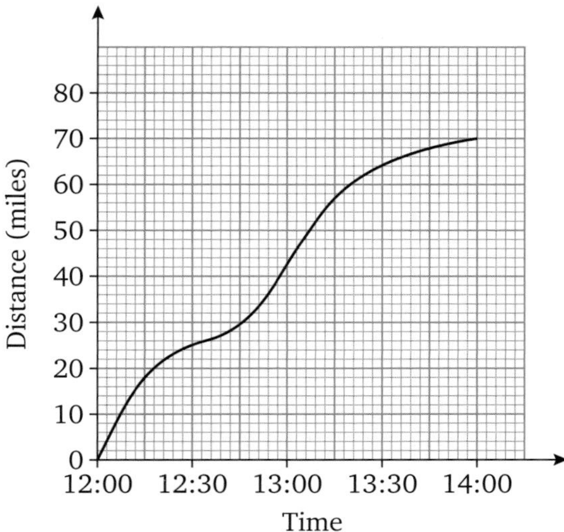

Work out the average speed of the journey.
State the units of your answer. *(4 marks)*
AQA 2008

2 The diagram shows the plan view of a landfill site on a centimetre grid.

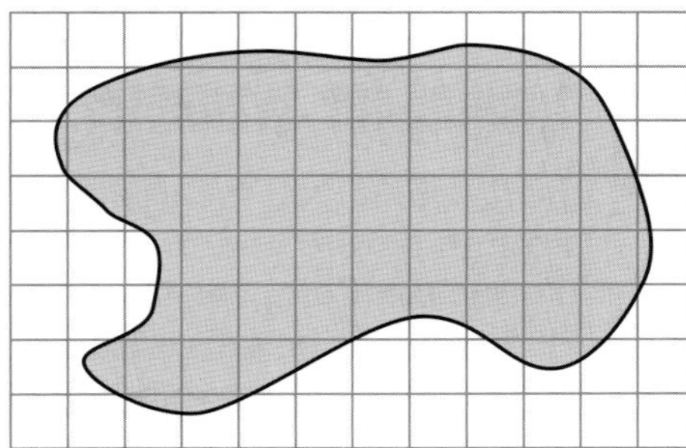

A landscape gardener is going to cover the site with turf (grass).
The table shows the cost of turf for different areas in (m^2).

Area of turf (m^2)	Cost per square metre
40 – 59	£2.83
60 – 130	£2.33
131 – 240	£2.03
241 – 480	£1.78
481 – 640	£1.53
641 – 960	£1.40
961 – 1440	£1.23

On the diagram one square metre represents $4\,m^2$.
The landscape gardener must buy enough turf to cover the landfill site.
Work out how much he has to pay.
You **must** show your working. *(3 marks)*
AQA 2009

9 Reflections, rotations and translations

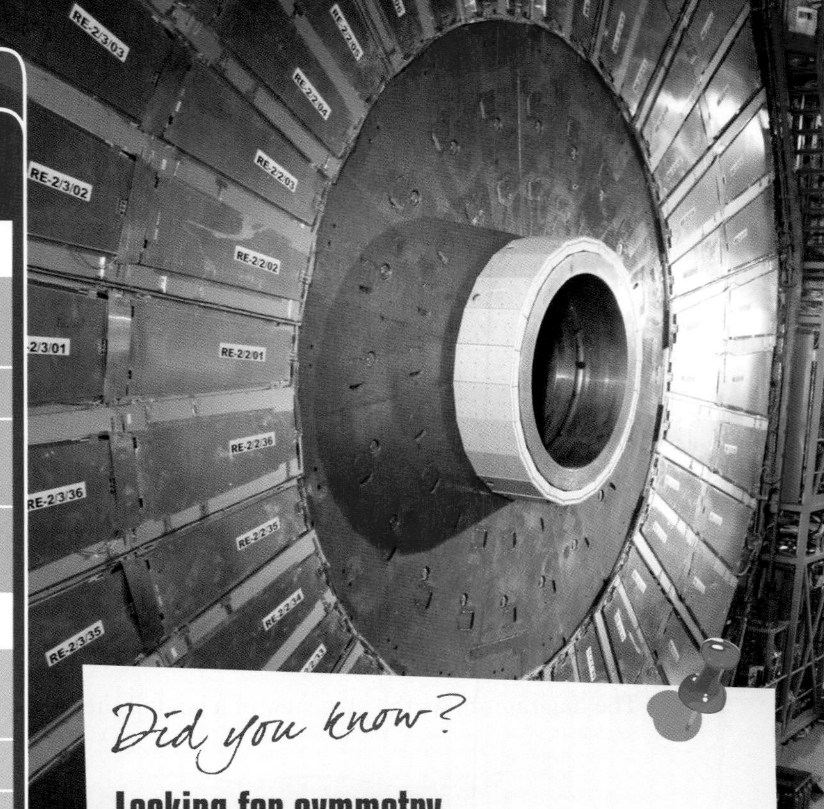

Objectives

Examiners would normally expect students who get these grades to be able to:

D

reflect shapes in lines parallel to the axes, such as $x = 2$ and $y = -1$

rotate shapes about the origin

describe fully reflections in a line and rotations about the origin

translate a shape using a description such as 4 units right and 3 units down

C

reflect shapes in lines such as $y = x$ and $y = -x$

rotate shapes about any point

describe fully reflections in any line parallel to the axes, $y = x$ or $y = -x$, and rotations about any point

find the centre of a rotation and describe it fully

transform shapes by a combination of translation, rotation and reflection

B

translate a shape by a vector such as $\begin{pmatrix} 4 \\ -3 \end{pmatrix}$

use congruence to show that translations, rotations and reflections preserve length and angle, so that any figure is congruent to its image under any of these transformations.

Did you know?

Looking for symmetry

Everywhere you look you can see symmetry: both in nature and in man-made constructions. This is the Compact Muon Solenoid, one of the detectors in the Large Hadron Collider. This is the world's largest and highest-energy particle accelerator. It is a symmetrical construction being used to advance science. It was built by scientists and engineers from over 100 countries. After initial problems, the collider has been in successful operation since November 2009.

You should already know:

- ✔ how to plot positive and negative coordinates
- ✔ equations of lines, such as $x = 3$, $y = -2$, $y = x$ and $y = -x$
- ✔ the names of 2-D and 3-D shapes
- ✔ how to draw all the lines of reflection on a 2-D shape and reflect shapes in the axes of a graph
- ✔ how to identify the order of rotational symmetry of a 2-D shape.

Key terms

reflection
image
coordinate
vertex, vertices
line of symmetry
mirror line
rotation
centre of rotation
perpendicular bisector
translation
congruent
vector
transformation

Learn... 9.1 Reflection

This shape, P, has been **reflected** in the line $x = 2$

The image of an object is labelled with a dash symbol: $'$.

The **image** of P is P'.

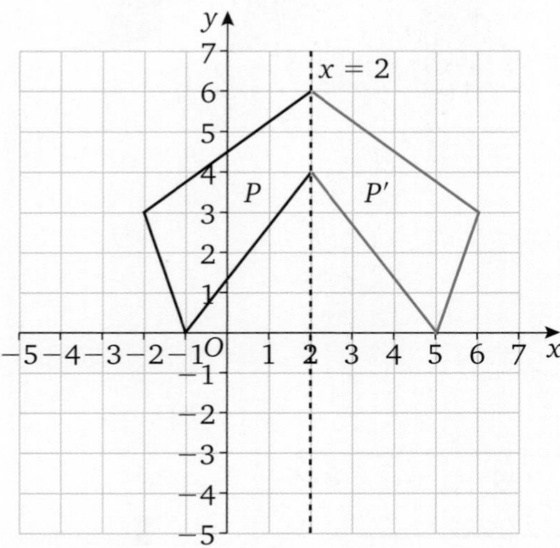

This shape, K, has been reflected in the line $y = x$

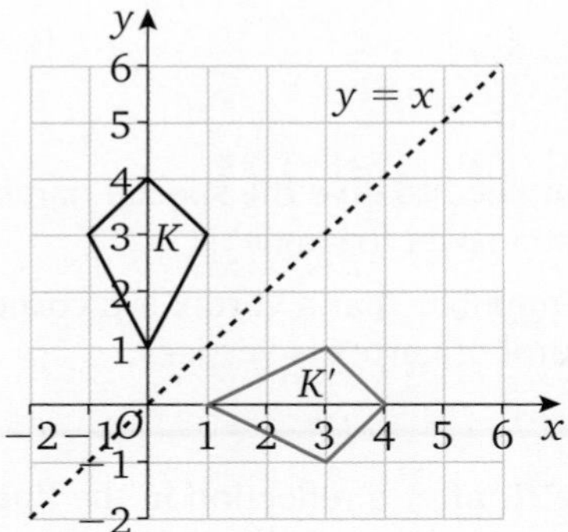

Coordinates of the **vertices** of K	Corresponding coordinates of the vertices of K'
(0, 1)	(1, 0)
(−1, 3)	(3, −1)
(0, 4)	(4, 0)
(1, 3)	(3, 1)

Notice the change in the coordinates of the vertices as K is mapped onto K'. The x-coordinate becomes the y-coordinate and the y-coordinate becomes the x-coordinate.

You can use tracing paper to check you have reflected a shape correctly. This method is particularly useful when the **line of symmetry** is diagonal because these are more difficult to visualise.

- Trace the shape.
- Draw in the line of symmetry.
- Fold the paper along the line of symmetry.
- You will then see clearly where the image should be.

Example:

a Draw a pair of x- and y-axes from -8 to 8.

b Draw a polygon, K, by plotting and joining these points: $(-2, 1)$, $(2, 1)$, $(0, 4)$

c What is the name of this polygon?

d Reflect the polygon in the line $y = -2$. Label your reflected shape K'.

e Write down the coordinates of the vertices of the image, K'.

Solution: **a**, **b** and **d**

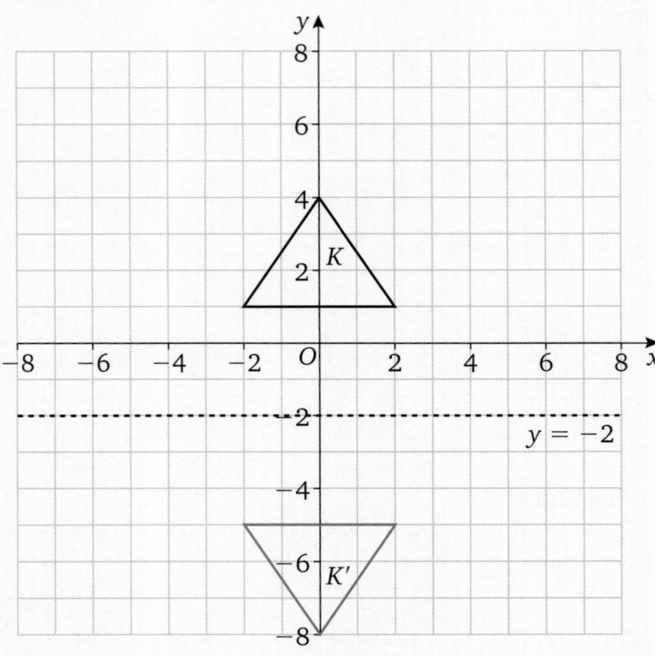

c The polygon is an isosceles triangle.

You need to give the special name and not just answer 'triangle'.

e The coordinates are: $(-2, -5)$, $(2, -5)$, $(0, -8)$

Remember that a vertex is a corner, and the plural of vertex is vertices.

Example: Find the coordinates of the image of the triangle, T, after a reflection in the line $y = -x$
The coordinates of T are: $(-1, 3)$, $(0, 2)$, $(2, 3)$

Solution: The coordinates of T' are: $(-3, 1)$, $(-2, 0)$, $(-3, -2)$

You can check your answer by sketching the triangle and its image under this **transformation**.

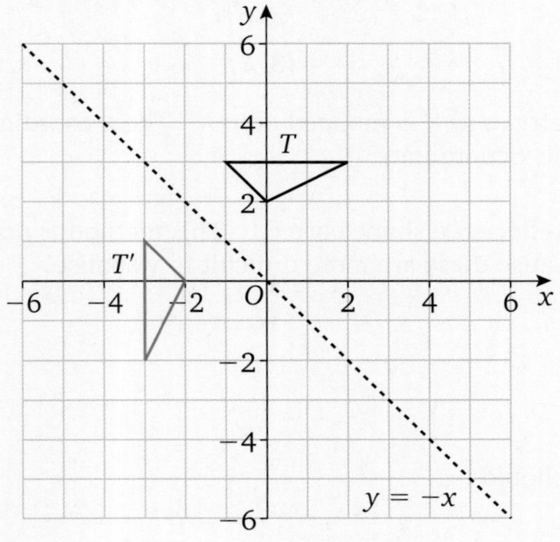

Practise... 9.1 Reflection

D

1 The diagram shows the triangle T.

a Write down the coordinates of the vertices of T.

b Use the diagram to find the coordinates of the vertices of the image of T after a reflection in each of the following lines.

i $x = 1$

ii $x = 2$

iii $y = -1$

iv $y = -2$

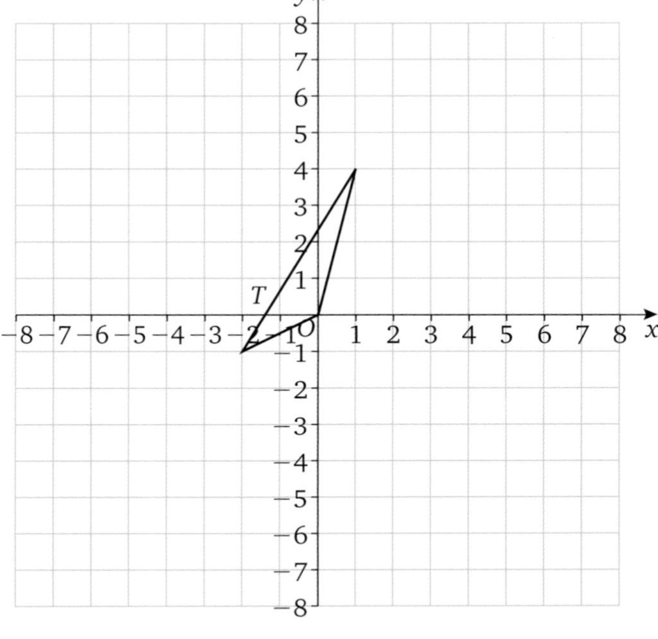

Hint

Remember that you can use tracing paper to help you with reflections.

2 Draw the following triangles on separate axes along with their image, a reflection in the given line.

a A (3, 1), (0, 1), (0, −2) $y = -1$

b B (−1, 0.5), (−1, 2), (1, 3) $y = -2$

c C (0, 0), (3, 0), (2, 2) $y = 1$

d D (0, 3.5), (−2, −2), (2, 2) $x = 1$

e E (−2, − 2), (−2, 2), (0, 0) $x = 2$

f F (0, −1.5), (3, 1), (1, 4) $x = -1$

3 For each diagram, find the equation of the line of reflection.

a

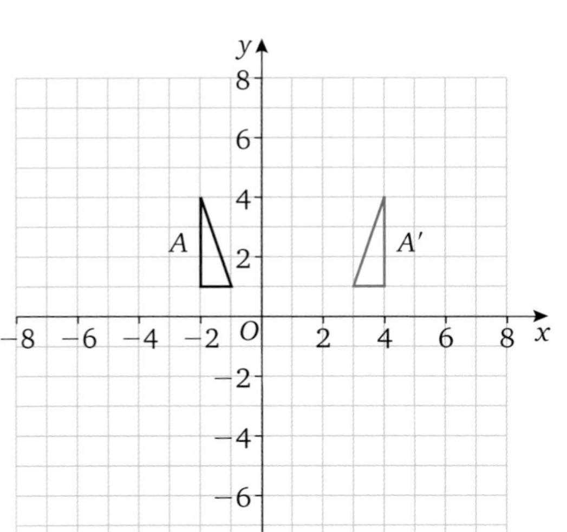

b

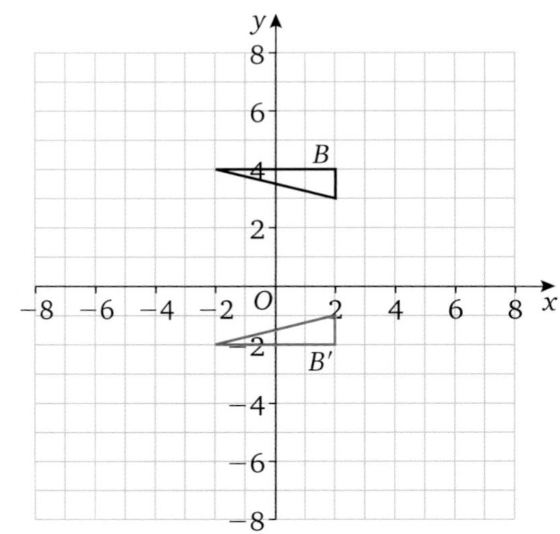

D

c

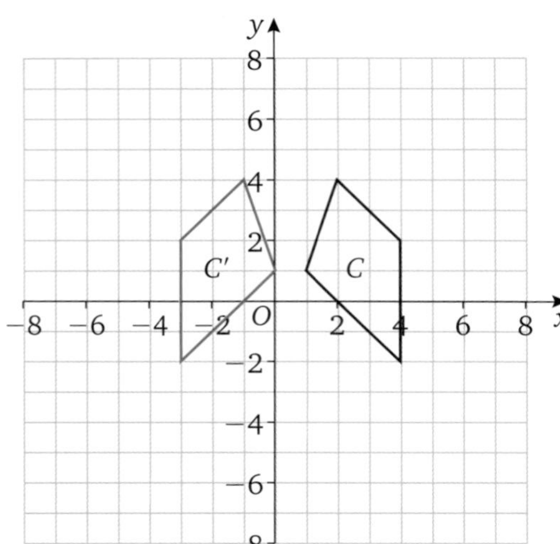

f

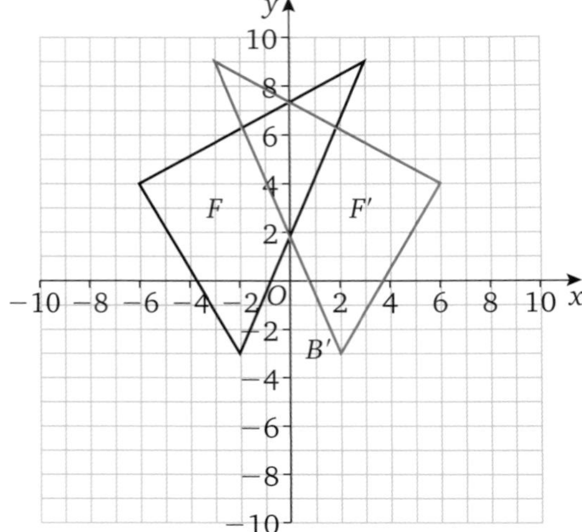

d

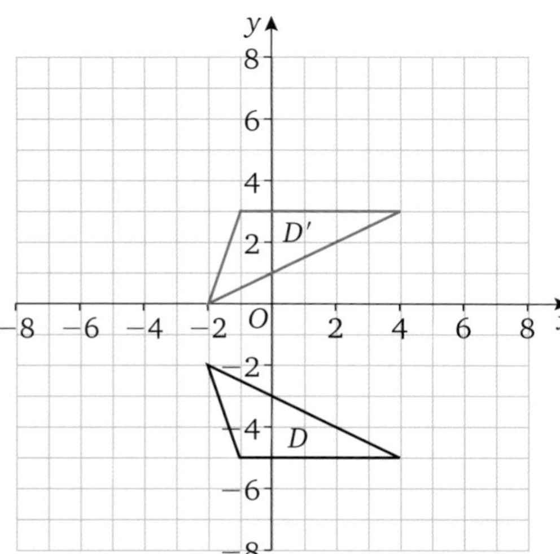

g

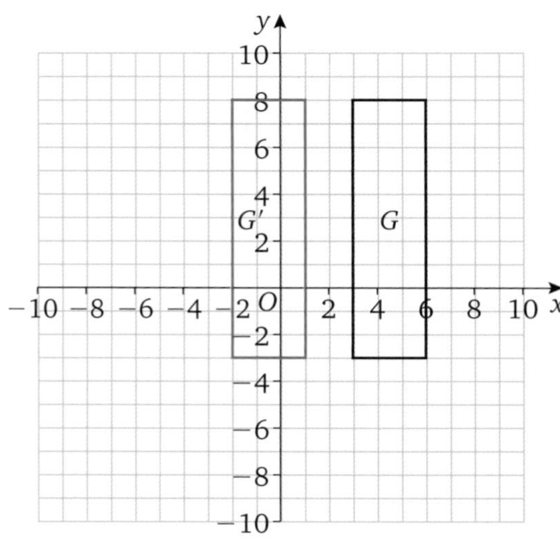

e

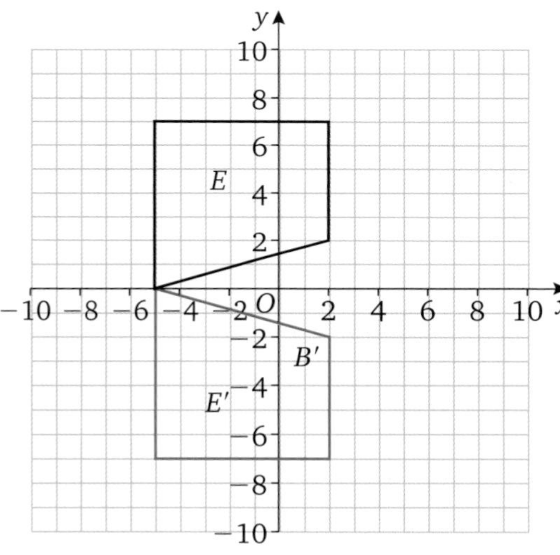

h

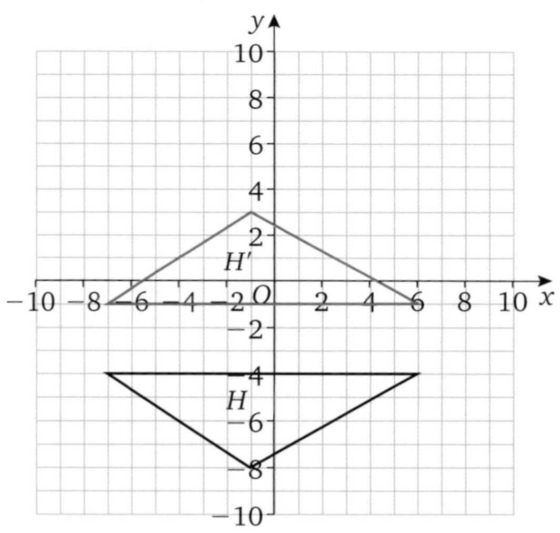

C

4 On isometric paper, draw a 3-D solid that has:

a no planes of symmetry **b** one plane of symmetry **c** two planes of symmetry.

5 **a** Draw a pair of x- and y-axes from -6 to 6.

b Draw a polygon, R, by plotting and joining these points:
$(1, -1)$, $(3, -1)$, $(3, -2)$, $(0, -3)$

c What is the mathematical name of the polygon?

C

d Reflect the polygon in the line $y = x$ and write down the coordinates of the corners of the image, R'.

e Reflect the polygon in the line $y = -x$ and write down the coordinates of the corners of the image, R''.

f Write down what you notice.

Hint

Two dashes are used after two transformations.

6 **a** Draw a polygon, P, by plotting and joining these points: (0, 0), (4, 0), (1, 3), (2, 3)

b What is the mathematical name of the polygon, P?

c Find the coordinates of P', the image of P after a reflection in the line $y = x$

d Draw P' on the same pair of axes as P.

Bump up your grade

To get a Grade C you need to know how to reflect in the lines $y = x$ and $y = -x$ as well as horizontal and vertical lines.

7 The diagram shows a pentagon.

a Write down the coordinates of the vertices of the pentagon.

b The pentagon is reflected in the line $y = -x$. What are the coordinates of the vertices of the image of G?

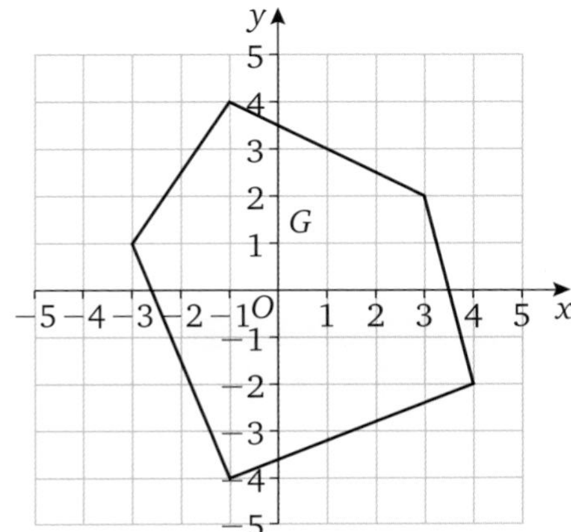

8 The diagram shows the line AB.

a What are the coordinates of the points A and B?

b Copy the diagram and draw the line BC, the image of AB after a reflection in the line $y = x$

c On the same diagram, draw the reflection of the lines AB and BC following a reflection in the line $y = -x$

d Describe the finished shape.

e Jason says that it doesn't matter which line you start with, you will always end up with this shape after a reflection in the line $y = x$ followed by a reflection in the line $y = -x$

Give an example that shows that Jason is wrong.

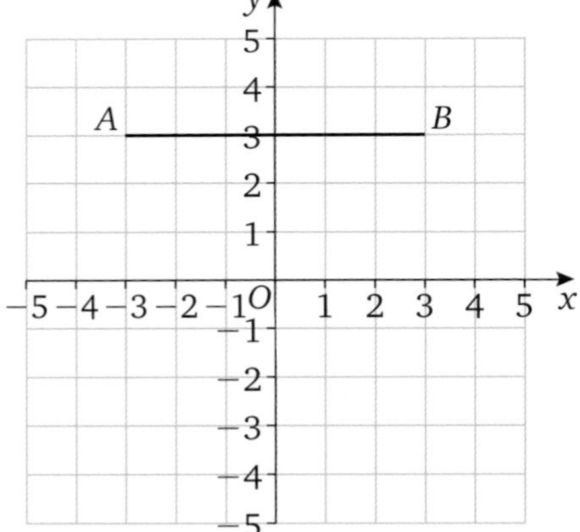

9 **a** Write down the coordinates of the vertices of these triangles after a reflection in the line $y = x$

i (0, 2), (3, 4), (1, 5)

ii (−2, −1), (−3, 4), (−2, 4)

iii (−1.5, 2.5), (0.5, 3.5), (−1, 1)

b Write down the coordinates of the vertices of these triangles after a reflection in the line $y = -x$

i (0, 2), (3, 4), (1, 5)

ii (−2, −1), (−3, 4), (−2, 4)

iii (−1.5, 2.5), (0.5, 3.5), (−1, 1)

10 This diagram shows a shape A and its reflection A'.

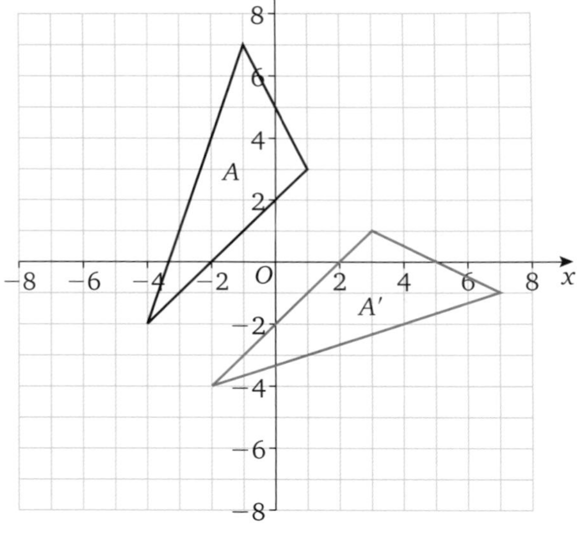

a Describe the reflection that maps A onto A'.

b Write down the coordinates of the vertices of both triangles.

c Find a rule that connects the coordinates after the reflection.

d A polygon has these coordinates:
(2, 3), (−1, −2), (0, −4), (2, −2)
An image of this polygon is created by reflecting it in the same line of reflection as in part **a**.

Use your rule from part **c** to work out the coordinates of the vertices of this image.

Try to work out the coordinates by following your rule and not by drawing.

e Can you find a rule for each of these lines of symmetry without drawing them?

i $x = 0$ **ii** $y = 3$ **iii** $x = -5$

Learn... 9.2 Rotation

Shapes can be **rotated** on grids or axes.

- The amount the shape **rotates** is given as an angle or fraction of a complete turn, for example, 270° or $\frac{3}{4}$ turn.
- The **direction of rotation** is given as clockwise or anticlockwise.
- The **centre of rotation** is the fixed point around which the object is rotated. It is given using coordinates.

The shape is called the object and the rotation is called the image.

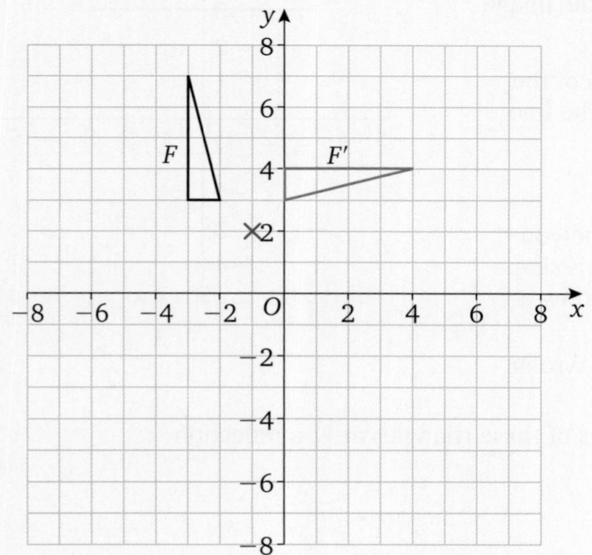

AQA Examiner's tip

In your exam you can ask for tracing paper to find the centre of rotation.

When describing a rotation you must give:

- angle or turn
- direction
- centre of rotation.

This shape has been rotated clockwise through 90°. The centre of rotation is the point (−1, 2).

If the object is labelled F then the image is usually labelled F'.

Angles of rotation can be any angle.

This shape has been rotated clockwise through 212°.

If you know or can work out the centre of rotation, you can use a protractor to measure the angle of rotation.

To describe a rotation fully, you must give the centre of rotation, the angle of rotation and the direction of rotation, either clockwise or anticlockwise.

To find the centre of rotation when you have the object and its image requires a number of constructions.

1 Draw a line joining corresponding vertices (blue line).

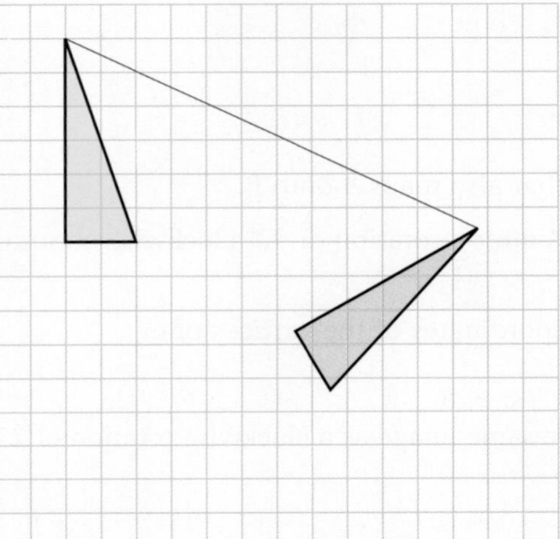

2 Draw the perpendicular bisector of this line (red line).

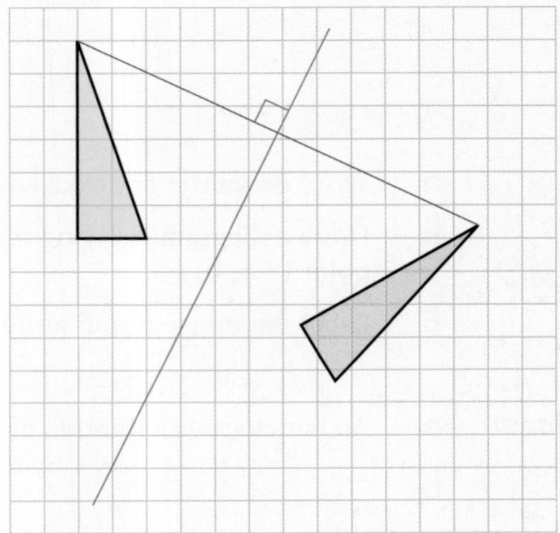

The **perpendicular bisector** is the line at right angles that cuts the line in half.

See Chapter 15, Learn 15.2 to find out how to construct a perpendicular bisector.

3 Repeat this for each vertex.

The point where the three perpendicular bisectors cross is the centre of rotation.

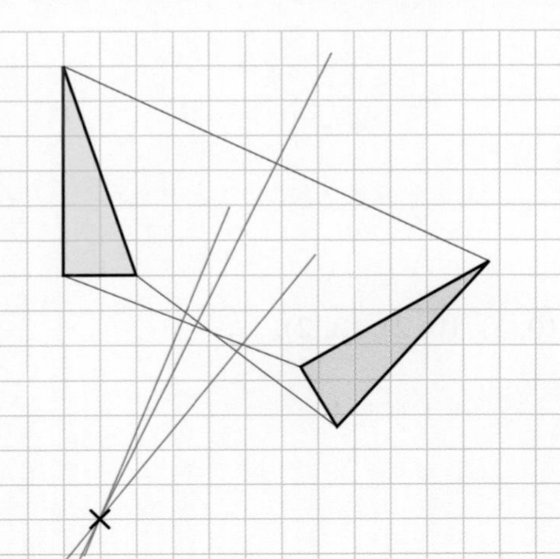

AQA Examiner's tip

Make sure that you use the correct terms when describing rotations. Use 'rotation' not 'turn'.

Example: In this diagram the right-angled triangle, A, has been rotated clockwise by 270° about the origin.

The image of A following this rotation is B.

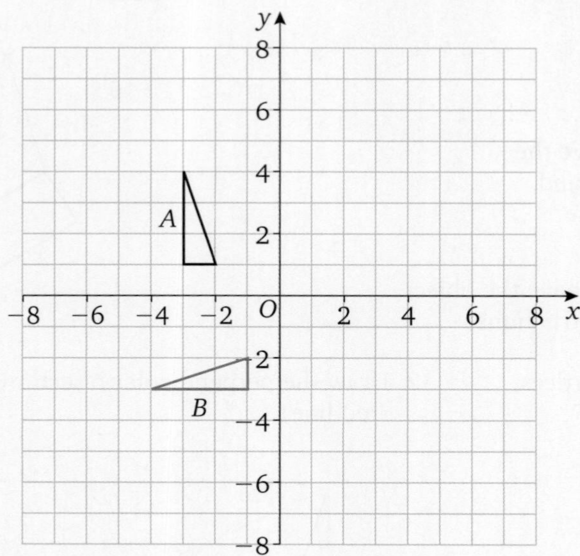

a Write down the anticlockwise rotation that also maps A onto B.

b Draw a diagram showing the image of A after a rotation of 90° clockwise about the point $(-1, 0)$.

c Label the image C and write down the coordinates of the vertices of C.

Solution: **a** An anticlockwise rotation of 90° gives the same image as a clockwise rotation of 270°.

b

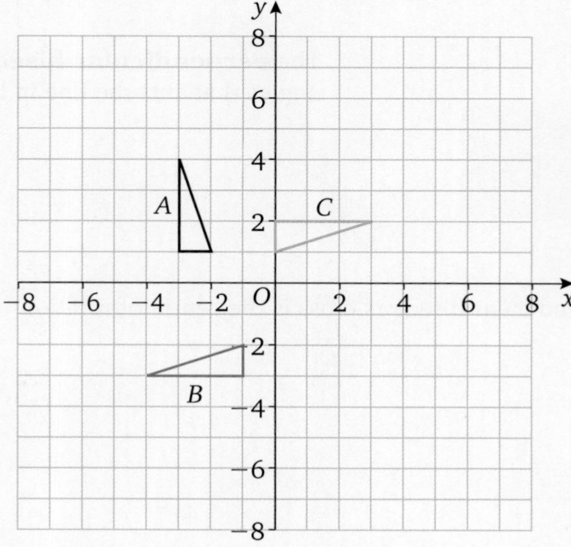

c The coordinates of the vertices of C are: $(0, 1)$, $(0, 2)$, $(3, 2)$.

Practise... 9.2 Rotation 

D C B A A*

Hint

You can use tracing paper to help you answer these questions.

1 **a** Draw a pair of x- and y-axes from -8 to 8 for each question part.

Copy each object and rotate it by 180°.
Use the origin (0, 0) as the centre of rotation.
Label each image R'–W'.

A rotation through 180° clockwise has the same outcome as a rotation through 180° anticlockwise. This means that the direction does not have to be given in the answer.

D

i

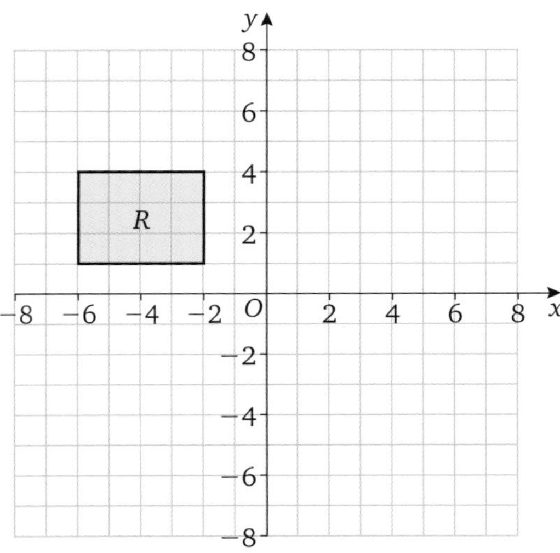

iv

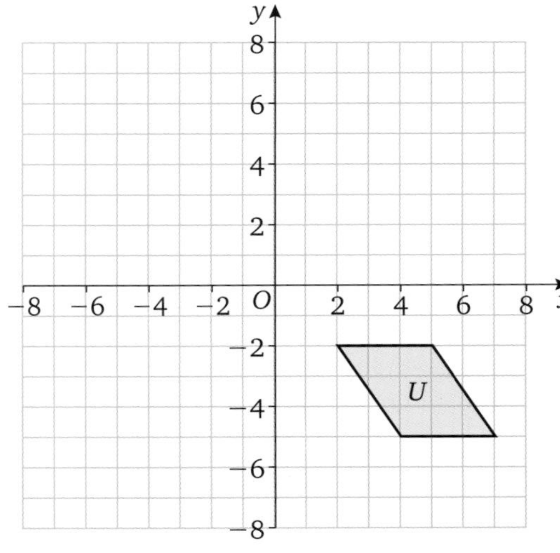

ii

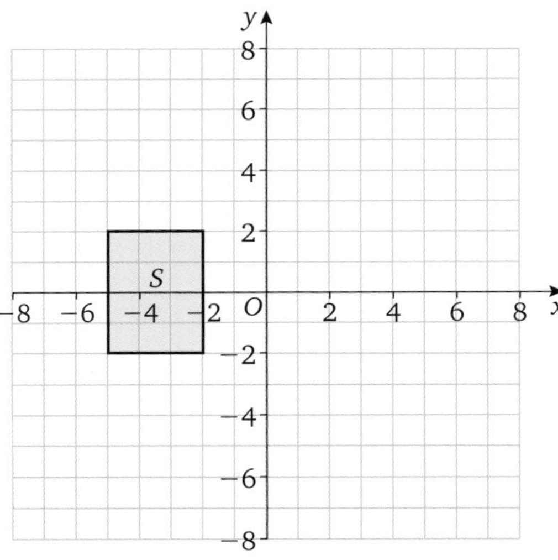

v

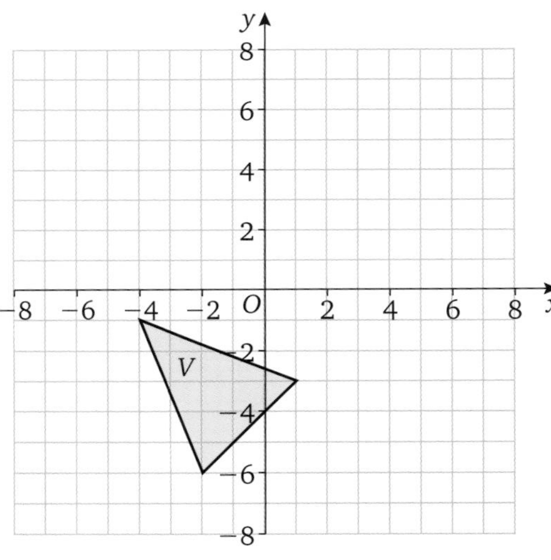

iii

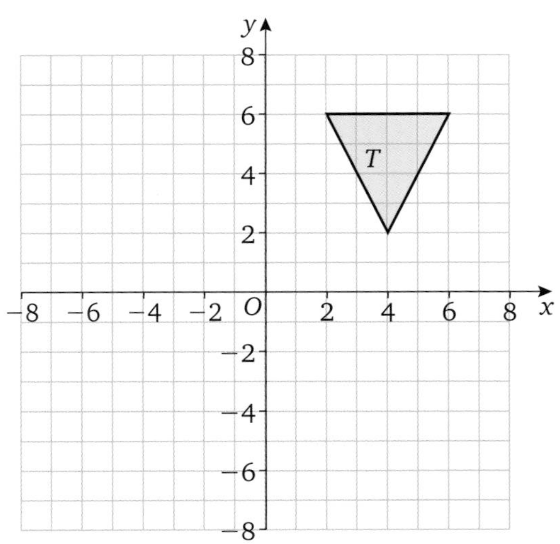

vi

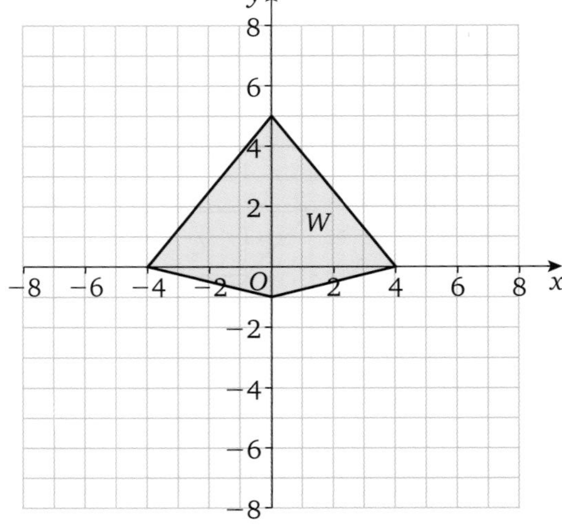

D

b Draw a pair of x- and y-axes from -8 to 8 for each question part.

Copy each object and rotate it by 90° clockwise.
Use the point (1, 2) as the centre of rotation.
Label each image A'–F'.

i

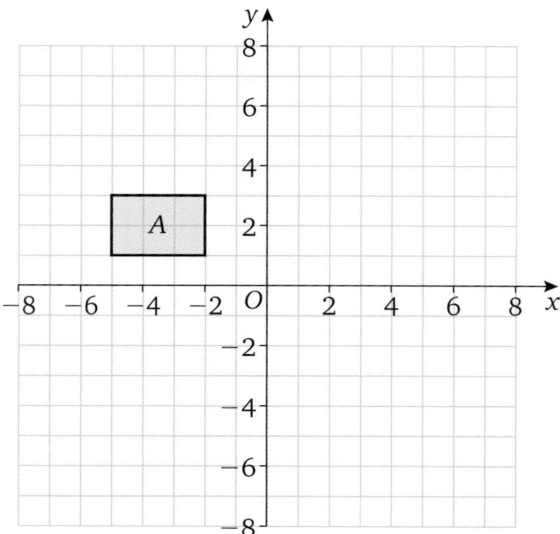

iv

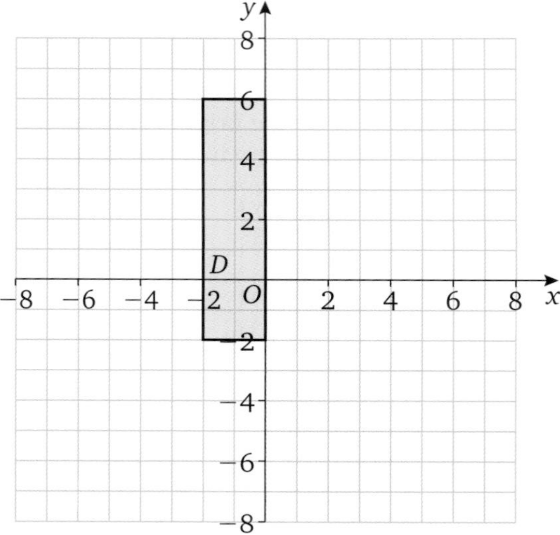

ii

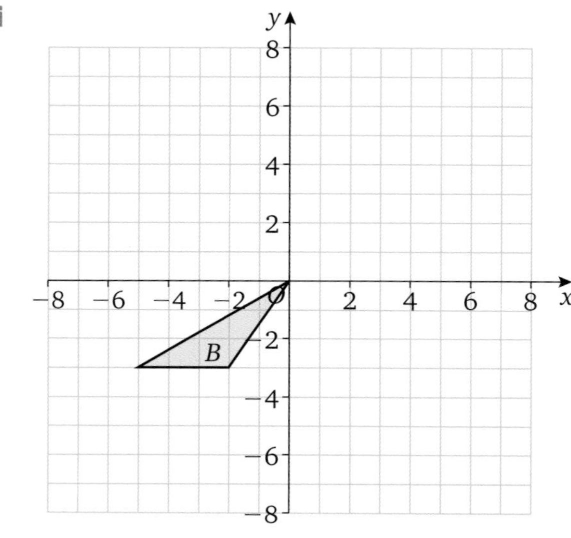

v

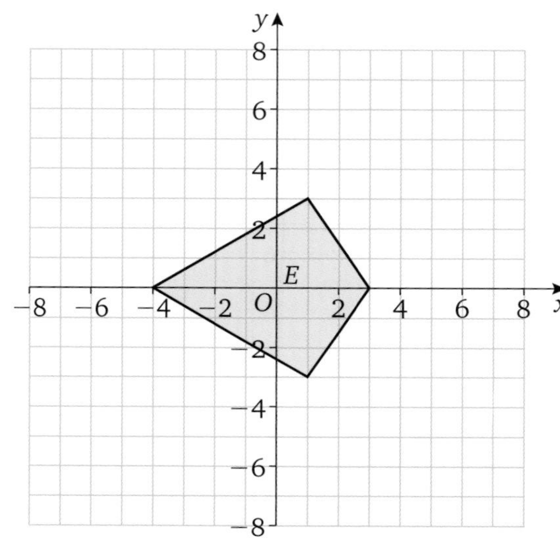

iii

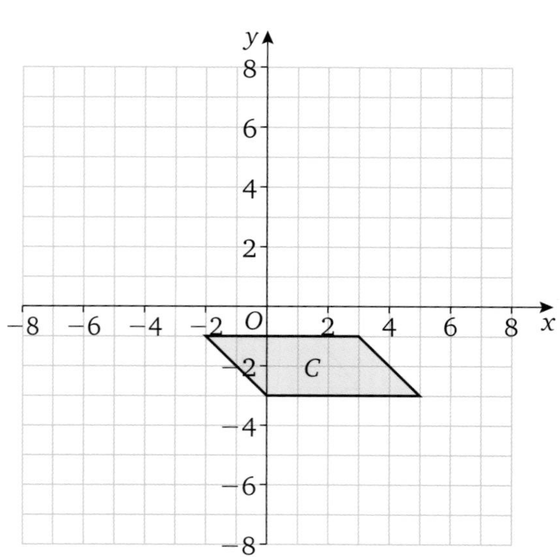

vi

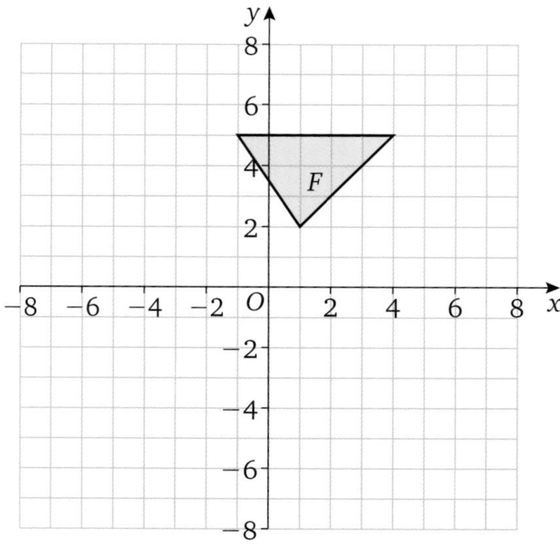

D

c Draw a pair of x- and y-axes from -8 to 8 for each question part.
Copy each object and rotate it by 270° clockwise.
Use the point $(-1, -1)$ as the centre of rotation.
Label each image G'–L'.

i

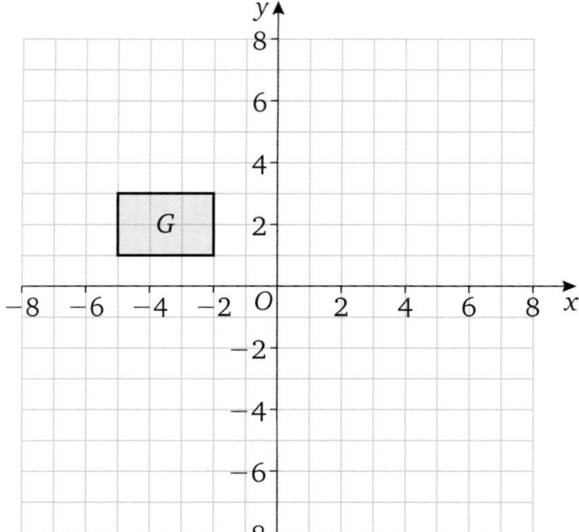

iv

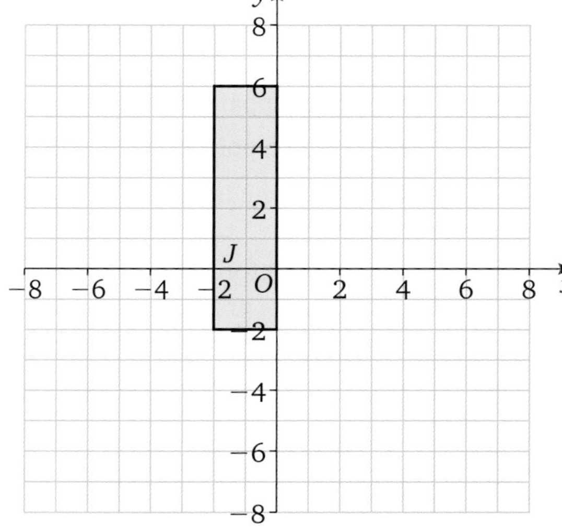

ii

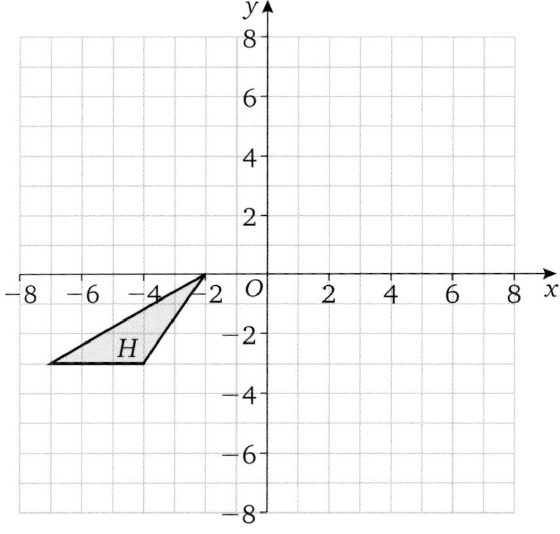

v

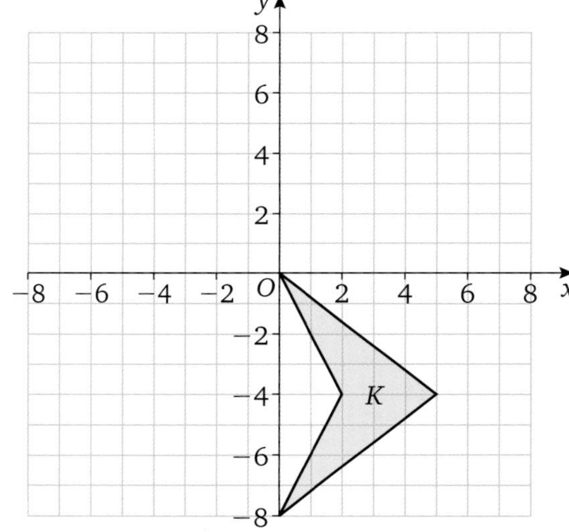

iii

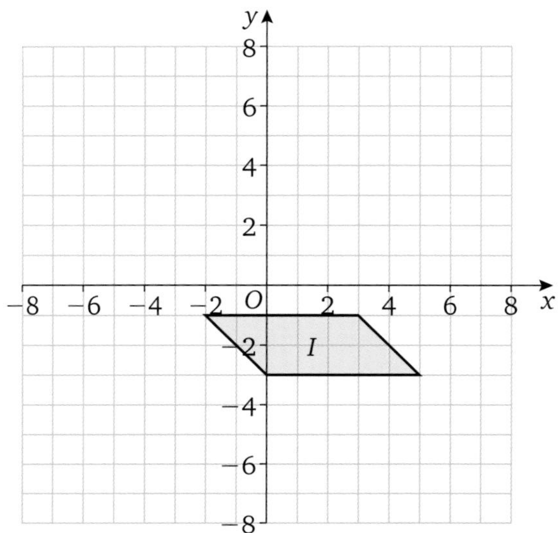

vi

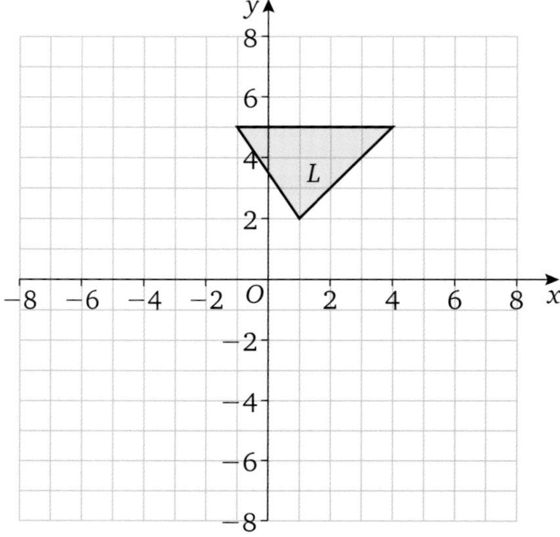

Bump up your grade

You need to be able to rotate objects around any point and not just the origin to get a Grade C.

C

2 a Draw a pair of x- and y-axes from -7 to 7.

b Draw a triangle T by plotting and joining these points: (0, 0), (-2, -1), (-1, -2)

c Draw the image V by rotating T by 90° clockwise about (0, 0).

d Draw the image W by rotating T by 180° about (0, 0).

e Draw the image X by rotating T by 270° clockwise about (0, 0).

3 a Draw a pair of x- and y-axes from -7 to 7.

b Draw a polygon Q by plotting and joining these points: (4, 1), (6, 1), (4, -1), (3, -1)

c Draw the image Q' by rotating Q by 90° clockwise about (2, 3).

4 Use a protractor to find the angle of rotation for each shape.

a

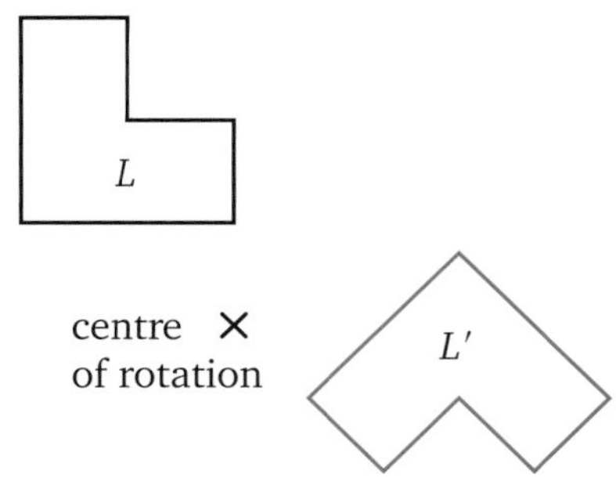

c

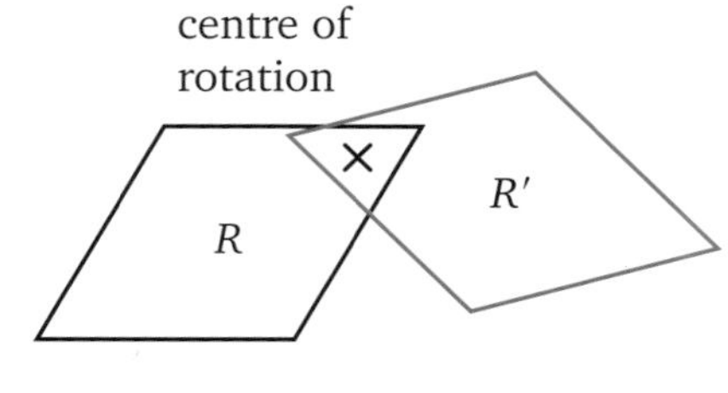

b

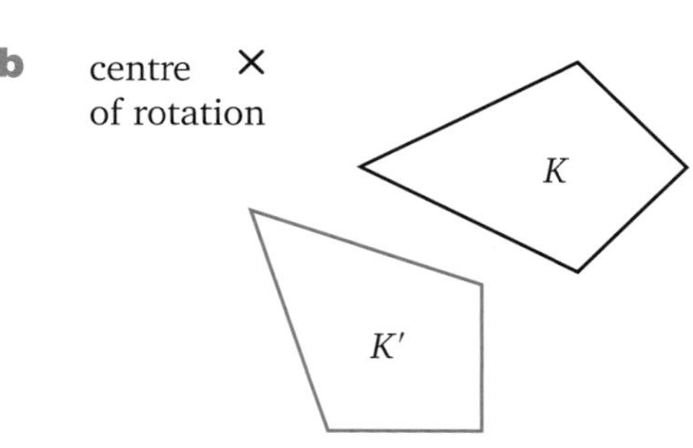

d

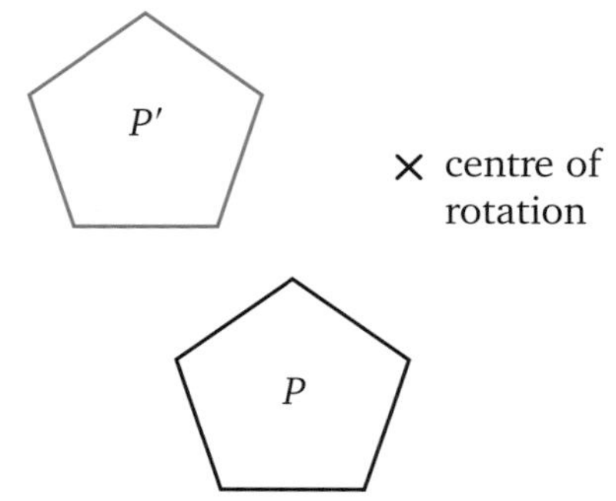

5 For each of the following mappings, give the angle of rotation, the direction of rotation and the centre of rotation.

a A onto B

b A onto C

c A onto D

d D onto B

e B onto F

f A onto E

g C onto G

h H onto A

i B onto E

j D onto C

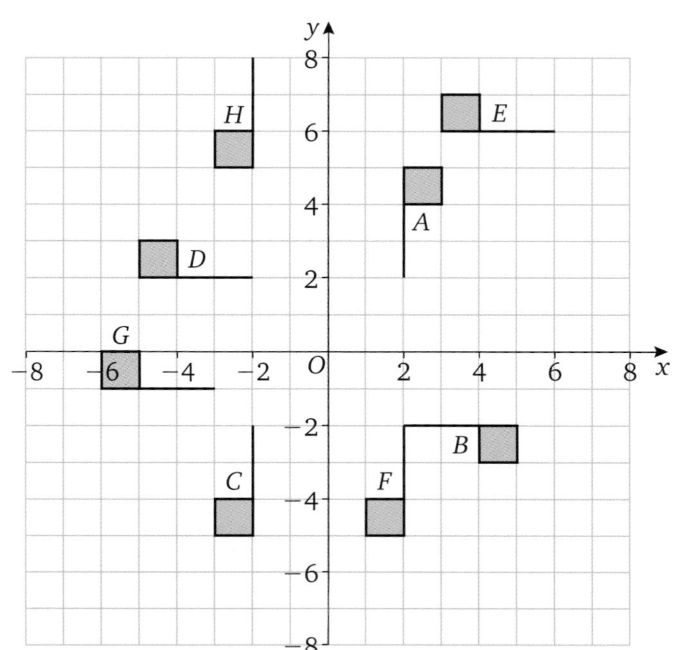

C

6 The pink triangle has been rotated clockwise through 90° to give its image, the blue triangle.

Work out which of the points A, B, C or D is the centre of rotation.
You can use tracing paper to help you.

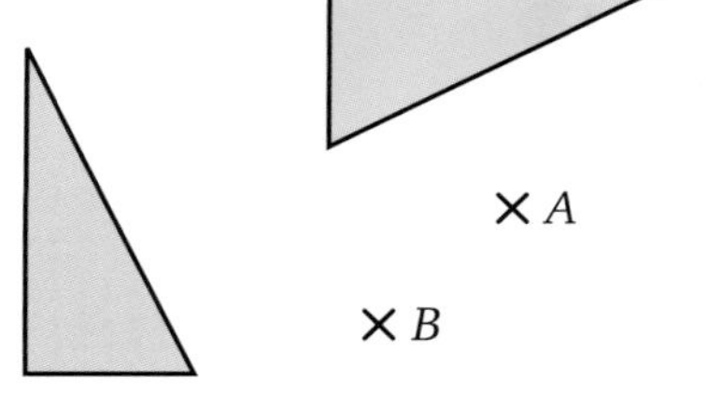

7 This diagram shows two different images of a triangle, A, after two rotations by the same angle.

Image B followed the first rotation clockwise about the point (1, 1).

Image C followed the other rotation of A anticlockwise about the point (−1, 1).

Find the position of triangle A.

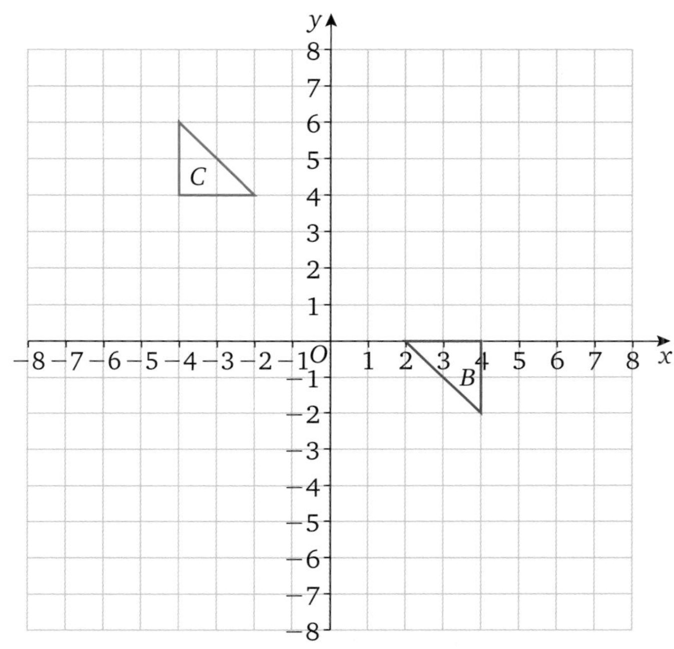

Hint

Use tracing paper to try out both centres of rotation until you find a position for triangle A that works for both images.

8 The yellow triangle has been rotated clockwise through 120° to give its image, the green triangle.

Trace the diagram. Then find the centre of rotation by constructing appropriate perpendicular bisectors.

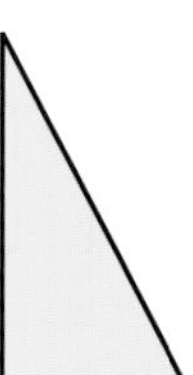

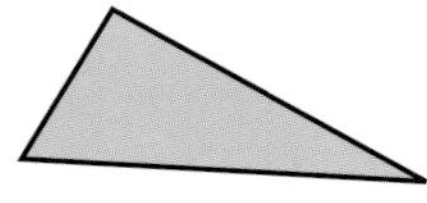

9 Rotational symmetry is often used in the design of company logos.
The arrows on this logo make it look as if it has rotational symmetry order 2.

a Explain why this logo does not have rotational symmetry order 2.

b This logo uses arrows too.

i What is the order of rotational symmetry of this logo?

ii What is the angle of rotation of the logo?

c **i** Design your own company logo with rotational symmetry, order 5

ii What is the angle of rotation of your logo?

Learn... 9.3 Translation

This shape has been **translated**.

Every point moves the same distance in the same direction.

The object and the image are **congruent**.
Two shapes are congruent if the lengths and angles stay the same when they are translated, rotated or reflected.

The distance and direction can be written as a **vector**.

Shape A has been mapped onto shape B by a translation of 2 to the right and 3 units up.

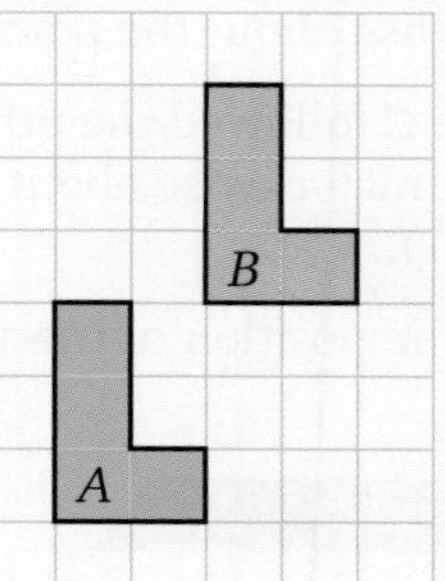

Written as a vector, this is $\begin{pmatrix} 2 \\ 3 \end{pmatrix}$ ← 2 units to the right, ← 3 units up

The vector that maps shape C onto shape D is $\begin{pmatrix} 0 \\ -4 \end{pmatrix}$

The top number is the horizontal move.

- If the number is positive the shape moves to the right.
- If the number is negative the shape moves to the left.

The lower number is the vertical move.

- If the number is positive the shape moves up.
- If the number is negative the shape moves down.

Vectors can be used on any grid with or without a pair of axes.

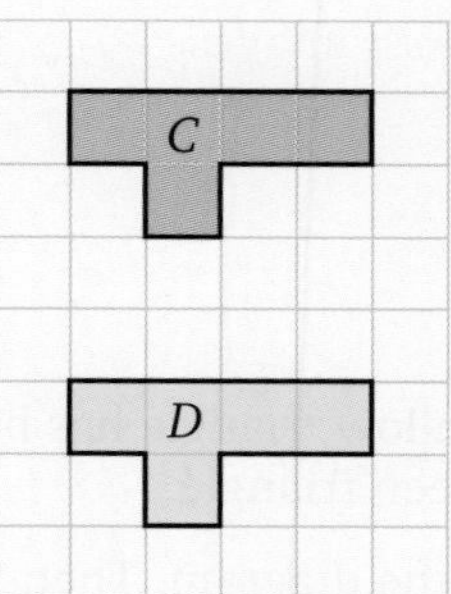

Example: Give the vector that translates shape A onto shape B.

a

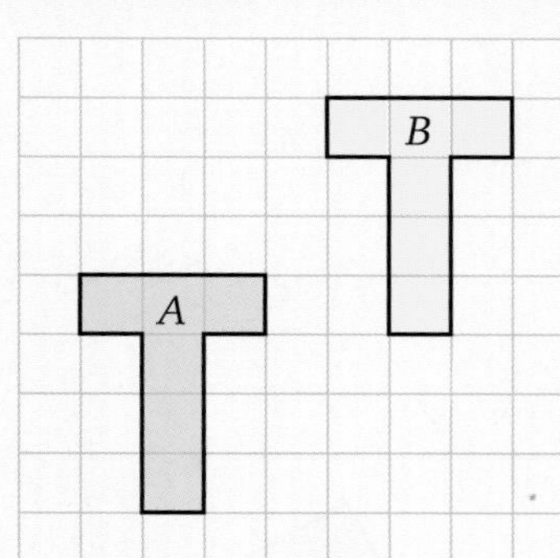

b

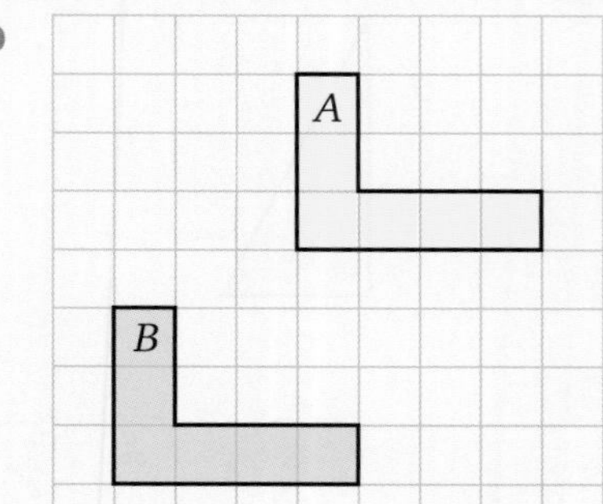

Solution:

a A has moved 4 to the right and 3 up so the vector translation is $\begin{pmatrix} 4 \\ 3 \end{pmatrix}$

b A has moved 3 to the left and 4 down so the vector translation is $\begin{pmatrix} -3 \\ -4 \end{pmatrix}$

Example: Copy the diagram, drawing triangle T with the coordinates (2, 4), (4, 4) and (4, 1). Then answer the questions below.

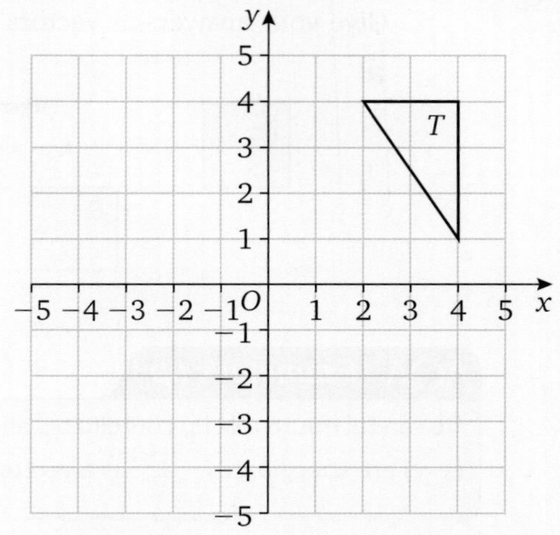

a Draw the image of T after the vector translation $\begin{pmatrix}-5\\-3\end{pmatrix}$. Label the image R.

b Write down the vector translation that maps R back onto T.

c What is the relationship between the coordinates of the vertices of T and the coordinates of R?

Solution: **a**

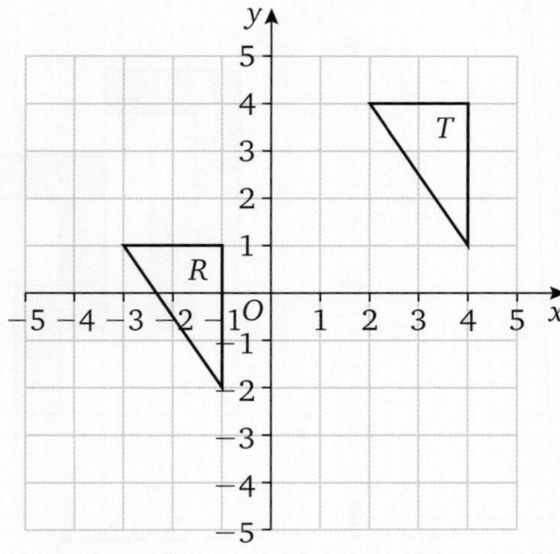

b The vector that maps R back onto T is $\begin{pmatrix}5\\3\end{pmatrix}$ 

c The x-coordinates of R are all less than the x-coordinates of T.
The y-coordinates of R are all less than the y-coordinates of T.

Practise... 9.3 Translation

1 Using squared paper, copy these shapes and translate each one by the given amount.

a 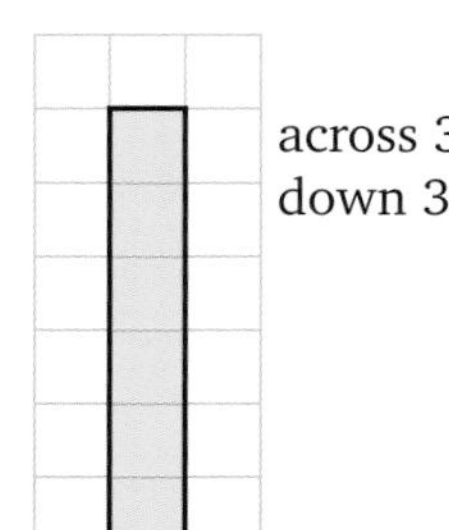 across 3 down 3

c 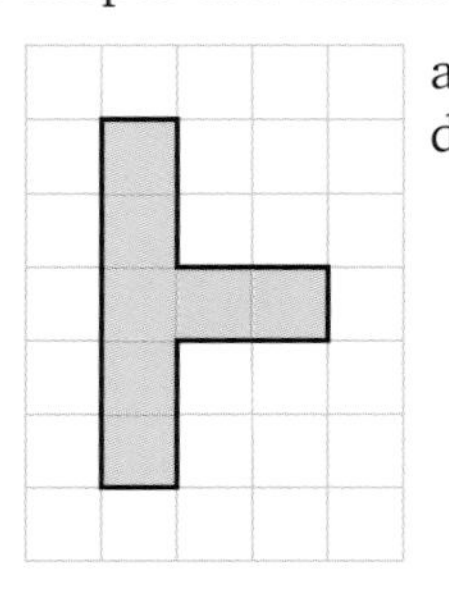 across 4 down 1

e 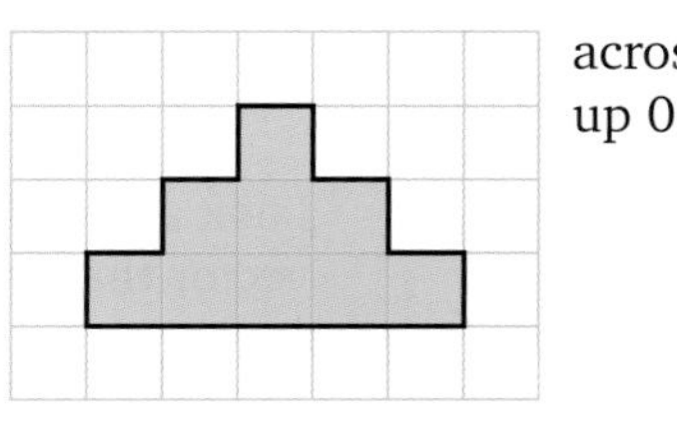 across 5 up 0

b 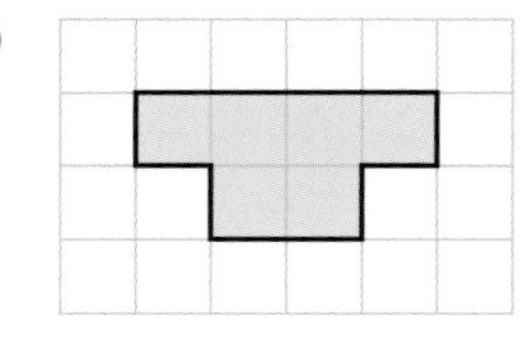 across 2 up 4

d 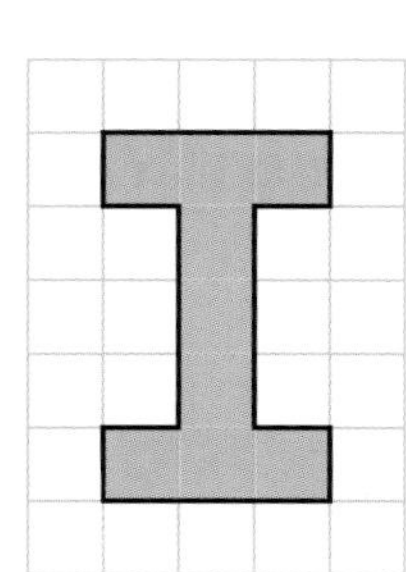 across 2 up 5

f 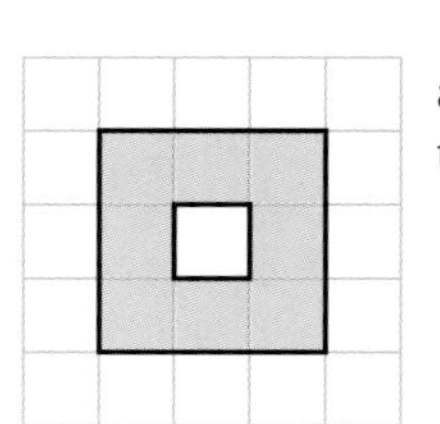 across 0 up 5

D

C

2 Describe the translation that maps shape A onto shape B in each diagram. Give your answers as vectors.

a

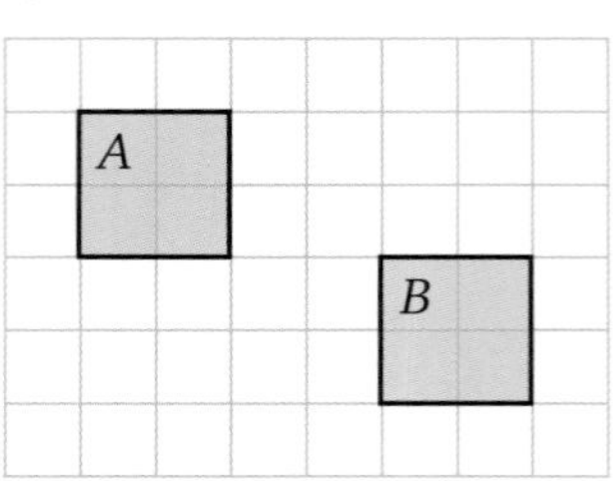

d

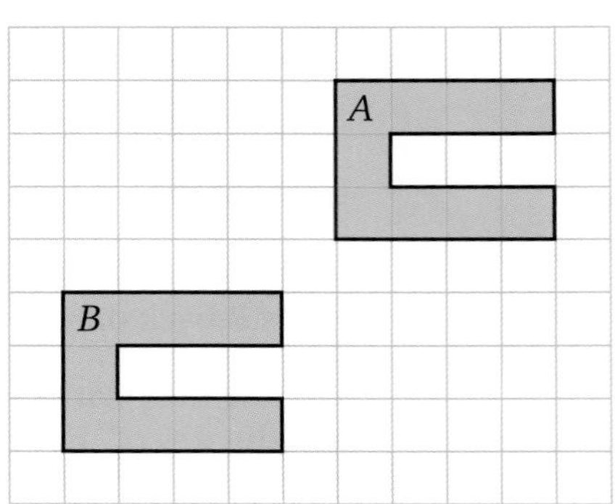

AQA Examiner's tip

Be careful not to mix up coordinates and vectors. (x, y) are coordinates. $\begin{pmatrix} x \\ y \end{pmatrix}$ is a vector.

b

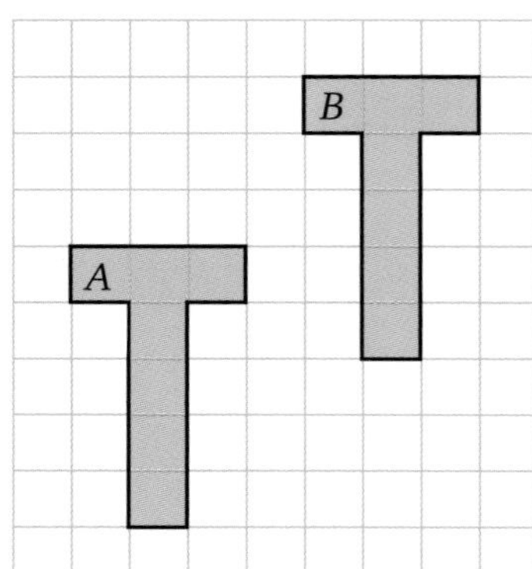

e

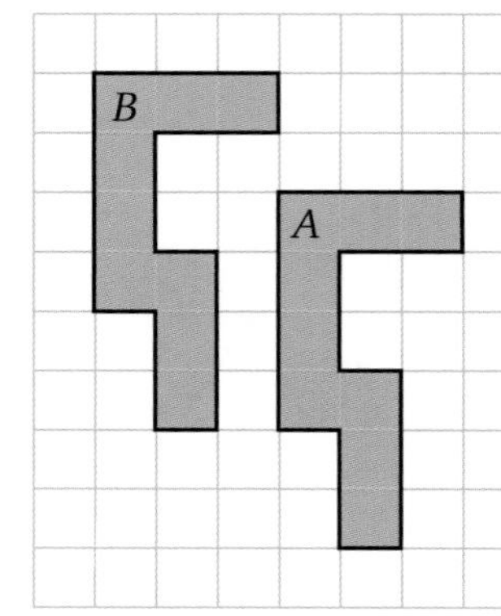

c

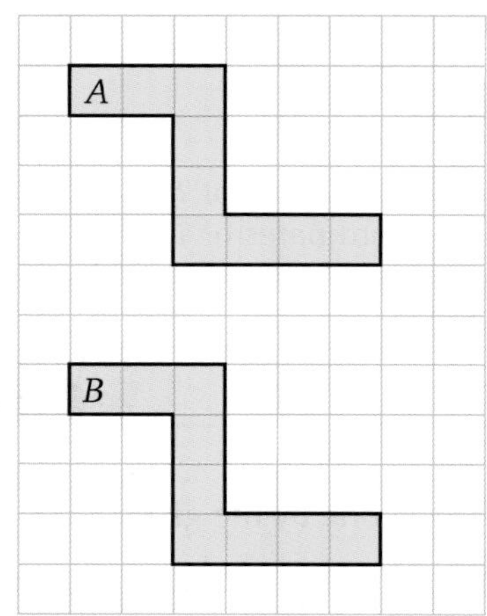

f

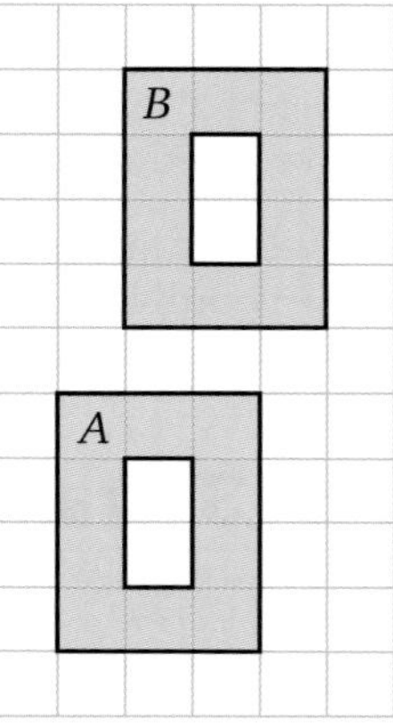

3 **a** Look at the diagram and write down the vector that translates:

i A onto B

ii A onto C

iii B onto C

iv C onto B.

b What is the relationship between the vector in part **a iii** and the vector in part **a iv**?

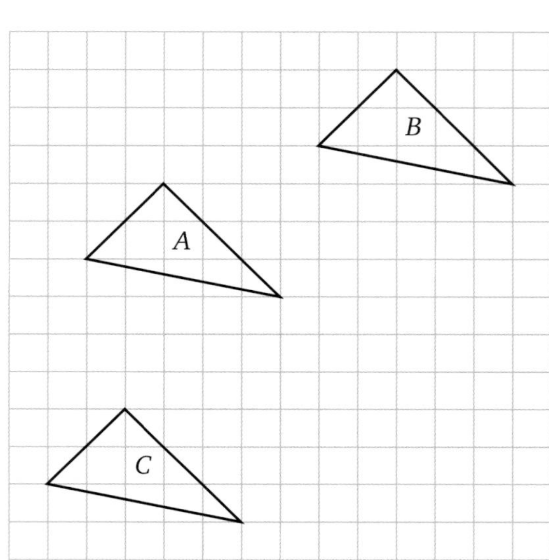

4 Look at the diagram to write down the vector that maps:

a A onto B

b C onto F

c E onto D

d B onto E.

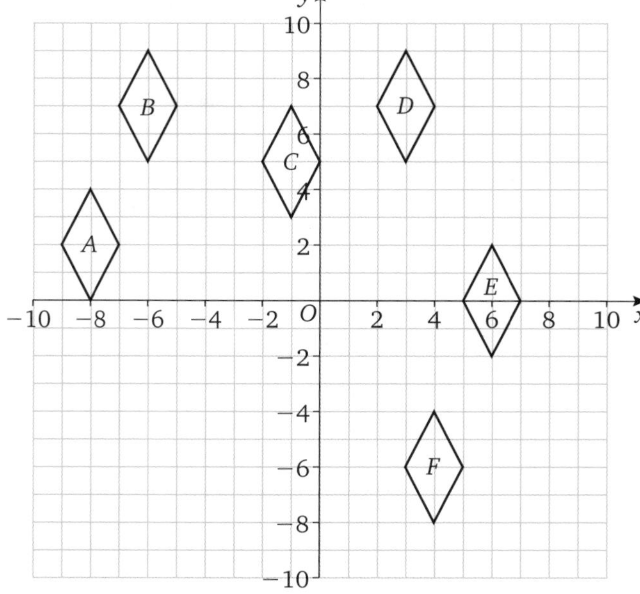

5 **a** **i** Draw a pair of x- and y-axes from -5 to 5.

ii Plot and join the points $(-4, -5)$, $(-1, -5)$, $(-4, 2)$. Label this shape T.

b Translate T using the vector translation $\begin{pmatrix}4\\3\end{pmatrix}$ to give the image U.

c Translate U using the vector translation $\begin{pmatrix}-5\\0\end{pmatrix}$ to give the image V.

d Translate V using the vector translation $\begin{pmatrix}6\\-1\end{pmatrix}$ to give the image W.

e Describe fully the single translation that maps T directly onto W.

6 **a** Draw a pair of x- and y-axes from -5 to 5.

b Begin at the origin as a starting point. After each of the following translations, put a cross.

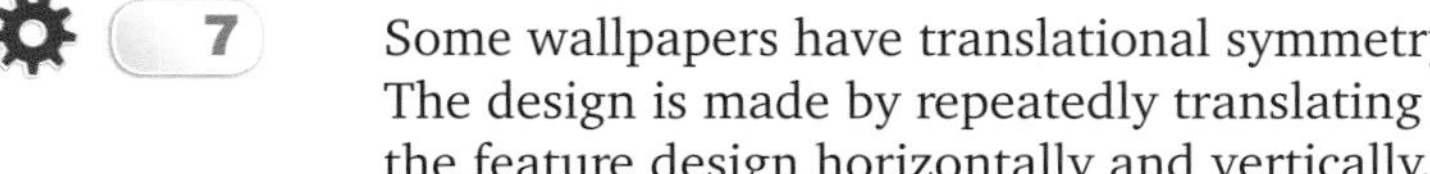

$\begin{pmatrix}2\\1\end{pmatrix}$ $\begin{pmatrix}1\\1\end{pmatrix}$ $\begin{pmatrix}1\\0\end{pmatrix}$ $\begin{pmatrix}1\\-1\end{pmatrix}$ $\begin{pmatrix}0\\-2\end{pmatrix}$ $\begin{pmatrix}-5\\-4\end{pmatrix}$

c Join the crosses you have drawn in the order in which you drew them.

d Reflect the shape in the y-axis.

7 Some wallpapers have translational symmetry. The design is made by repeatedly translating the feature design horizontally and vertically.

Emma is designing a wallpaper pattern on squared paper. She translates her design by the vector $\begin{pmatrix}4\\3\end{pmatrix}$

She then repeats the pattern in a different colour.

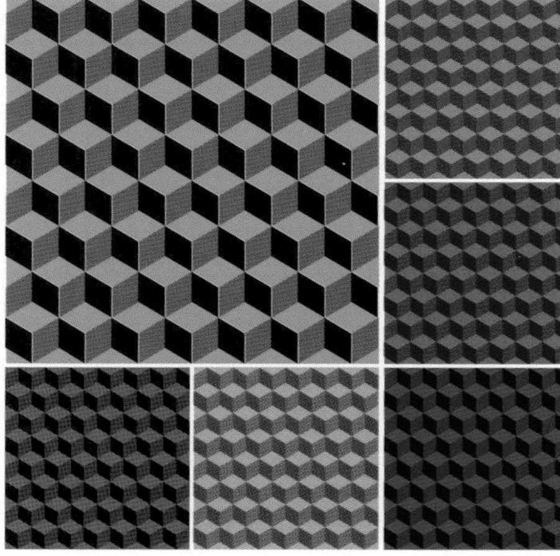

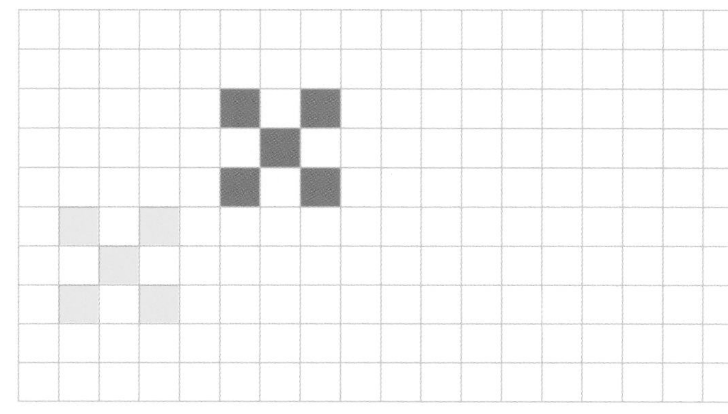

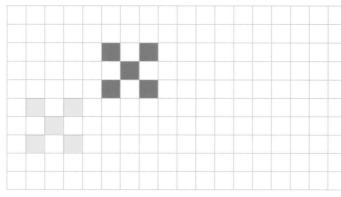

a On squared or isometric paper, create a simple design.

b Choose a translation vector. Repeat your pattern using your chosen translation vector to create your own wallpaper design.

8 Sam and Holly designed this board for a vector game.
A **vector route** starts at X and lands on each of the other squares before returning to X.

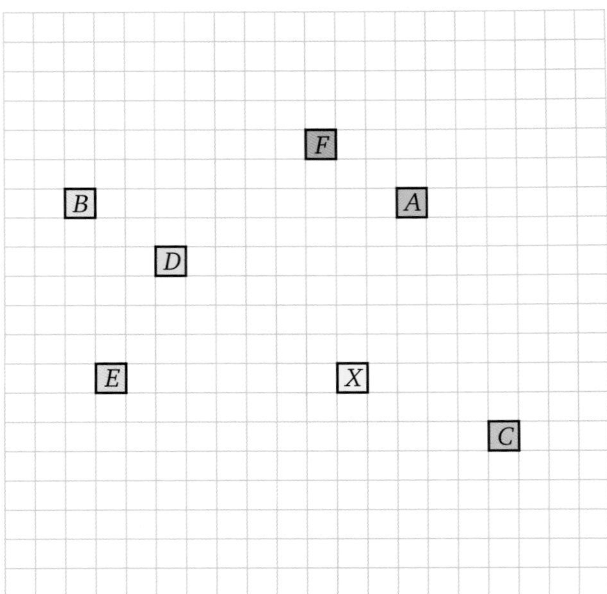

a Holly wrote down this vector route.

$\begin{pmatrix} 5 \\ -2 \end{pmatrix}$ $\begin{pmatrix} -13 \\ 2 \end{pmatrix}$ $\begin{pmatrix} -1 \\ 6 \end{pmatrix}$ $\begin{pmatrix} 3 \\ -2 \end{pmatrix}$

$\begin{pmatrix} 5 \\ 4 \end{pmatrix}$ $\begin{pmatrix} 3 \\ -2 \end{pmatrix}$ $\begin{pmatrix} -2 \\ -6 \end{pmatrix}$

Write down the order she landed on the squares.

b Sam chose to visit the triangles in this order: X F D C B A E X

Write down Sam's vector route.

c Create a vector route of your own and test it on a friend.

Learn... 9.4 Transformation and congruence

Reflections, rotations and translations can be combined into a single transformation.

A reflection in the y-axis, $x = 0$, followed by a reflection in the x-axis, $y = 0$, has the same effect as a rotation about the origin.

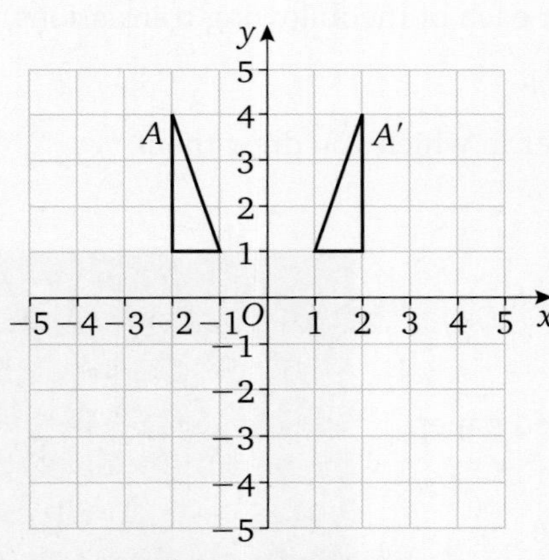

reflection in y-axis

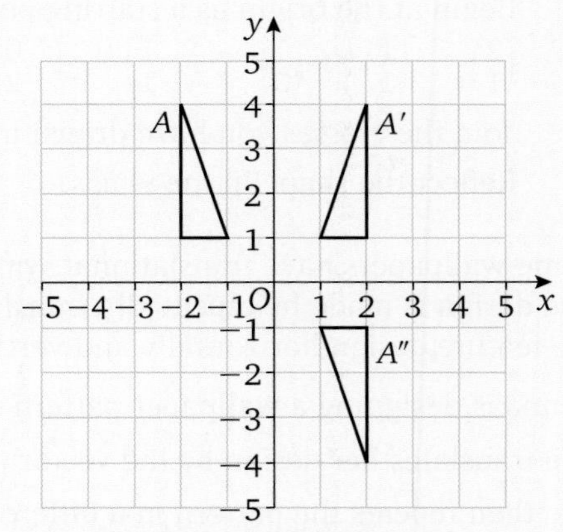

reflection in x-axis

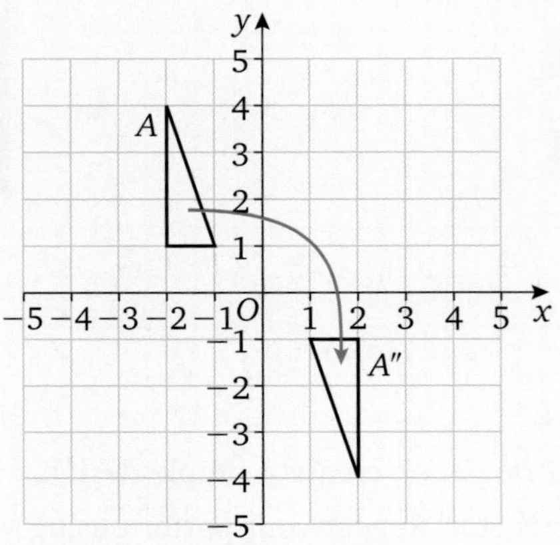

180° rotation about the origin

In the next diagram, the triangle A is mapped onto B by a 90° rotation clockwise about the origin.

The triangle B is mapped onto the triangle C by a reflection in the line $y = -1$

This can be described in one translation. Triangle A maps onto C by a vector translation $\begin{pmatrix} 8 \\ -10 \end{pmatrix}$

Rotation 90° clockwise about (0, 0)

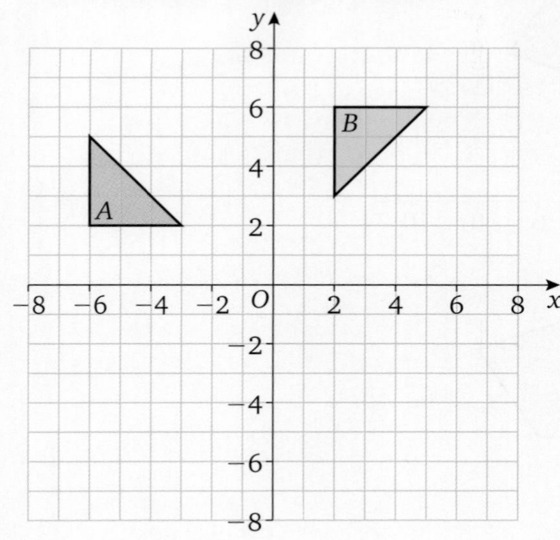

then reflection in the line $y = -1$

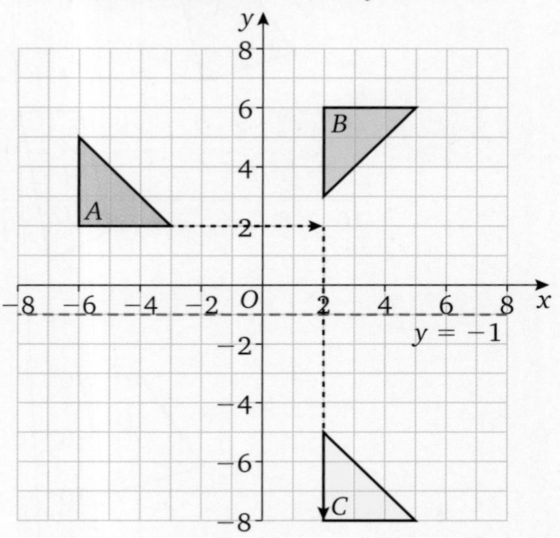

When a shape is transformed by a reflection, rotation, translation or combination of these, the shape and its image are said to be congruent. The size of the angles and lengths of the sides remain the same.

Example: In this diagram, A has been mapped onto B by a rotation clockwise by 90° about the origin.

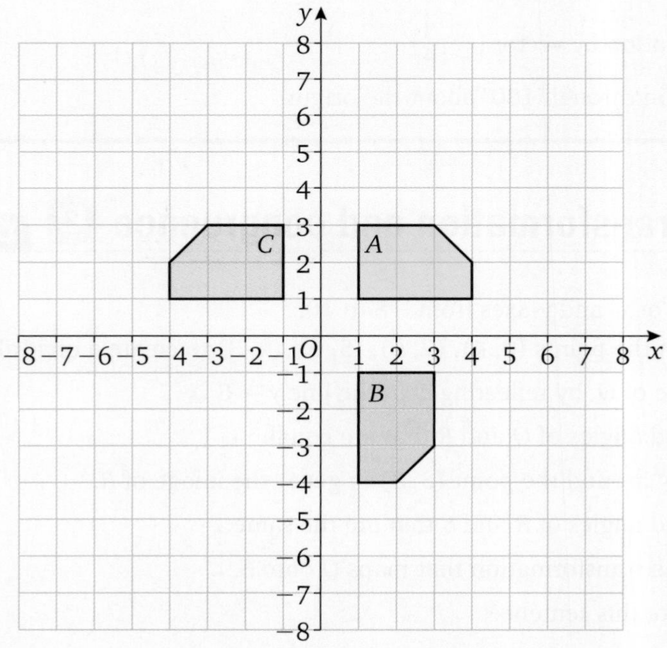

a Describe the transformation that maps B onto C.

b Describe the single transformation that maps A directly onto C.

Solution: **a** B has been reflected in the line $y = x$

b The single transformation that maps A onto C is a reflection in the y-axis ($x = 0$).

Example: The diagram shows the shape A and its image after a number of transformations.

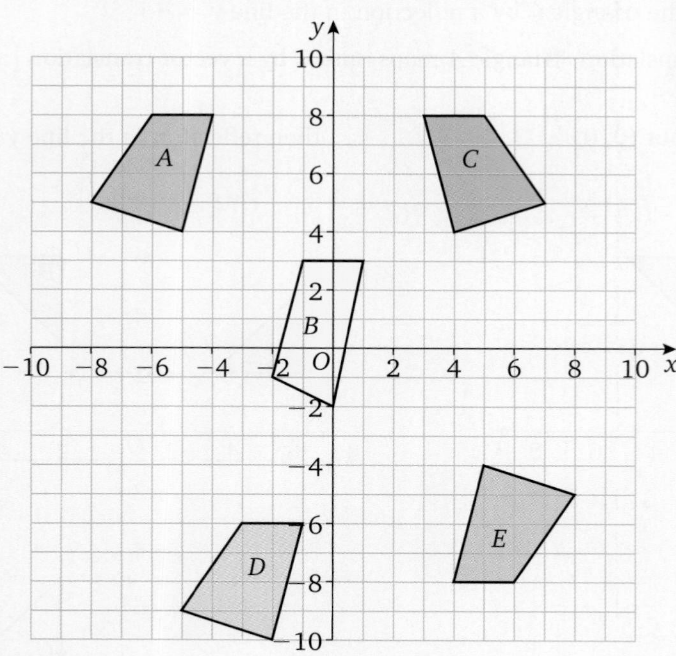

a Which of shapes B, C, D and E are congruent to A?

b Describe each single transformation that maps A onto each image.

Solution: **a** C, D and E are congruent to A.

B is not congruent because the length of each of its sides is not equal to the length of the corresponding side of A.

b C is a reflection in the y-axis.

D is a translation by vector $\begin{pmatrix} 3 \\ -14 \end{pmatrix}$

E is a rotation through 180° about the origin.

Practise... 9.4 Transformation and congruence

D C B A A*

B

1 **a** **i** Draw a pair of x- and y-axes from −8 to 10.

ii Plot and join the points (0, 4), (2, 6), (5, 6), (4, 3) to create a quadrilateral Q.

b Draw R, the image of Q, by reflecting Q in the line $y = 3$

c Mark the sides and angles of Q and R that are equal.

d Now rotate R 180° around the point (5, 3) to get S, the image of R.

e Mark the sides and angles of R and S that are the same.

f Describe the single transformation that maps Q onto S.

g Copy and complete this sentence.

Q, R and S are ____________.

2 **a** **i** Draw a pair of x- and y-axes from −5 to 5.

ii Draw the quadrilateral M by plotting and joining the points (−3, 2), (−3, −1), (−2, −1), (−1, 2).

b Draw the reflection of M in the line $y = -1$. Label the image M'.

c Now rotate M' 180° around the point (0, −1) and label the image M''.

d Describe fully the single transformation that maps M onto M''.

C

3 This diagram shows the triangle, T.

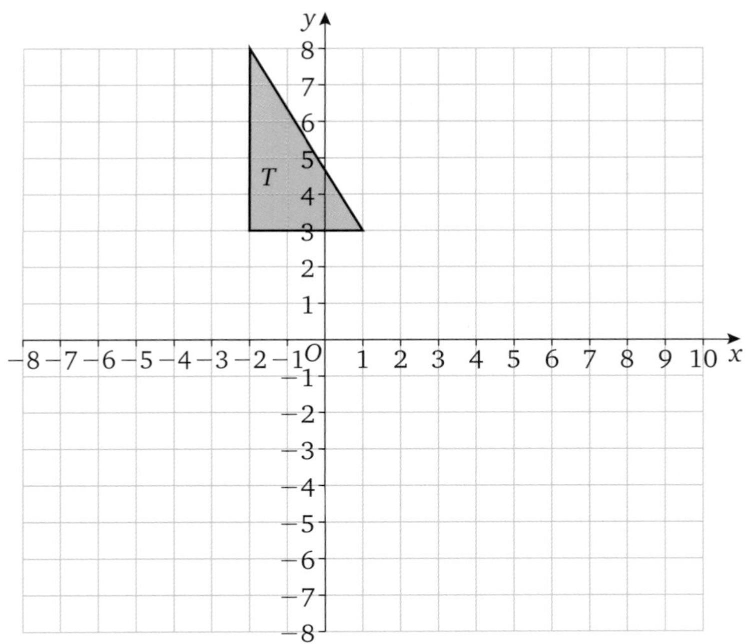

a Copy the diagram. Draw the image of T after a vector translation of $\begin{pmatrix}-4\\-9\end{pmatrix}$. Label this image U.

b Reflect U in the line $x = 2$ and label the image V.

c Reflect V in the line $y = -4.5$ and label the image W.

d Describe fully the single transformation that maps T onto W.

4 **a** Draw the triangle A with vertices (1, 4), (0, 0) and (0, 4) on x- and y-axes labelled -6 to 6.

b Rotate A by 180° around the point $(-1, -1)$. Label the image B.

c Translate B by the vector $\begin{pmatrix}6\\0\end{pmatrix}$. Label the image C.

d Reflect C in the line $y = -1$. Label the image D.

e Describe the single transformation that maps A onto D.

5 In this diagram, the kite K has been transformed using a number of different transformations.

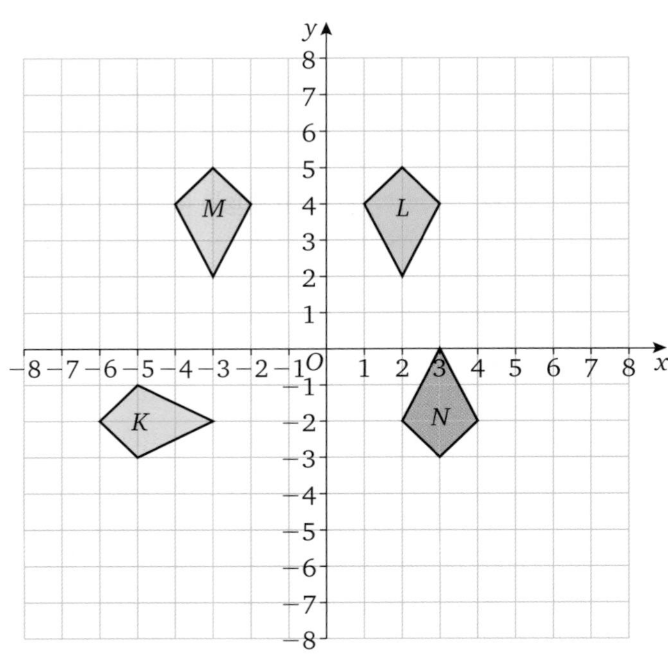

a Describe the transformation that maps K onto L.

b Describe the transformation that maps L onto M.

c Describe the transformation that maps M onto N.

d N can be mapped onto K with a clockwise rotation of 90°.

Find the coordinates of the centre of rotation.

B

6 Jake says that B is the image of A after a translation by the vector $\begin{pmatrix} 4 \\ -2 \end{pmatrix}$

Monique says Jake is wrong as A and B are not congruent.

Who is right, Jake or Monique?

Give an explanation for your answer.

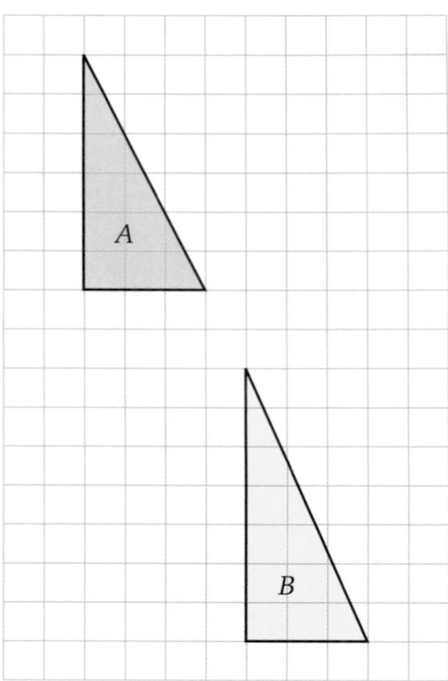

7 **a** Which of these shapes are congruent to P? Give a reason for each answer.

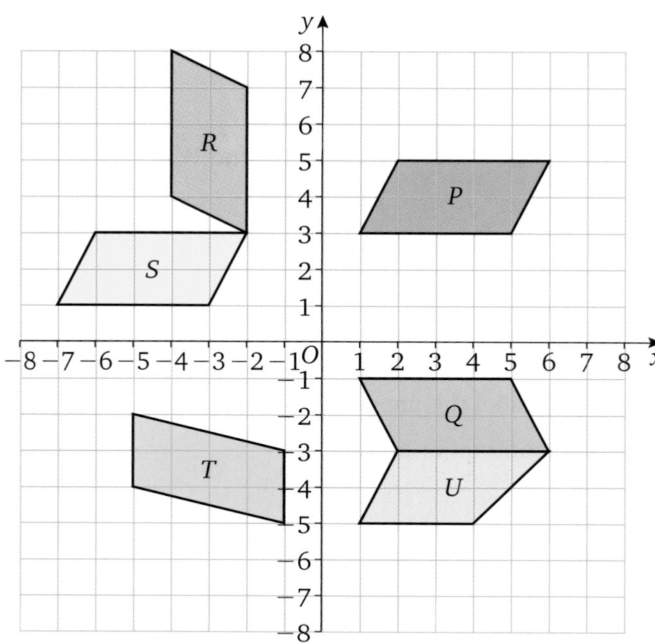

b Describe each single transformation that maps P onto each congruent image.

8 Josef is trying to answer these three questions.

- Does a reflection in the line $x = 2$, followed by a reflection in the line $y = 2$ have the same effect as a rotation around the point (2, 2)?
- Is this pattern true for negative values of x and y?
- Is this pattern true for all values of x and y?

a Draw some shapes and reflect and rotate them to help you answer all three questions.

b Draw at least three examples of each case to include in your answer.

9 Assess

D

1 **a** Copy the diagram.

b Rotate the shape 90° clockwise about the origin.

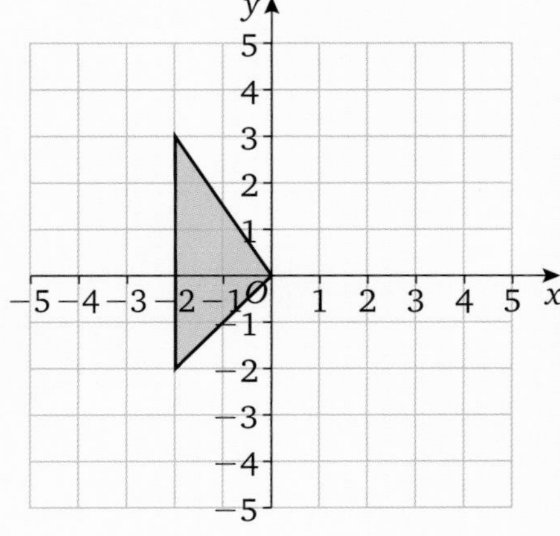

2 **a** Write down the coordinates of the vertices of the triangle A.

b A is rotated 90° clockwise about the origin.

Write down the coordinates of the vertices of the triangle after this rotation.

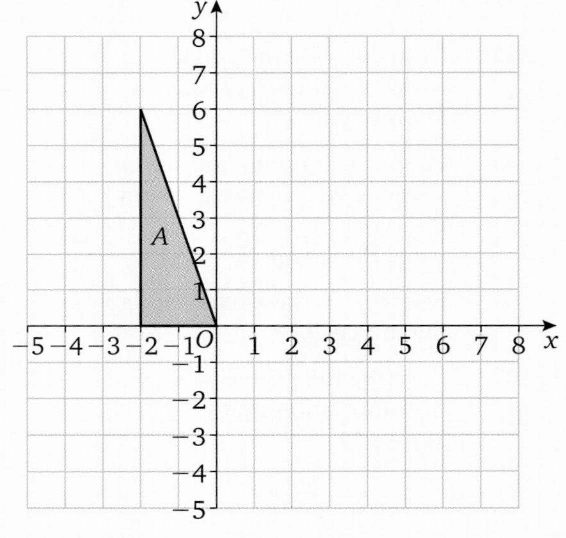

D C

3 The coordinates of the triangle A are: $(-3, 1)$, $(-5, 3)$, $(-6, 1)$

a On suitable axes, draw the triangle A and it's image A' after a reflection in the line $x = -2$

b On the same axes, draw A'', the image of A after a rotation through 90° anticlockwise about the point $(-2, 1)$.

4 Write down the equation of the line of reflectional symmetry in this diagram.

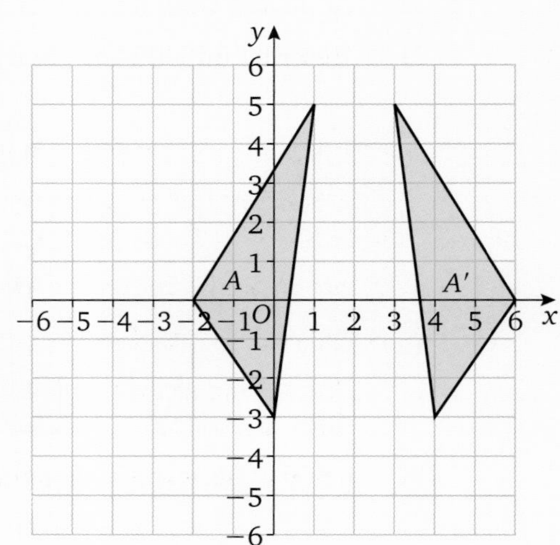

C

5 **a** Copy the diagram.

b Write down the coordinates of the vertices of the triangle.

c Reflect the triangle in the line $x = -y$

d Write down the coordinates of the vertices of the new triangle.

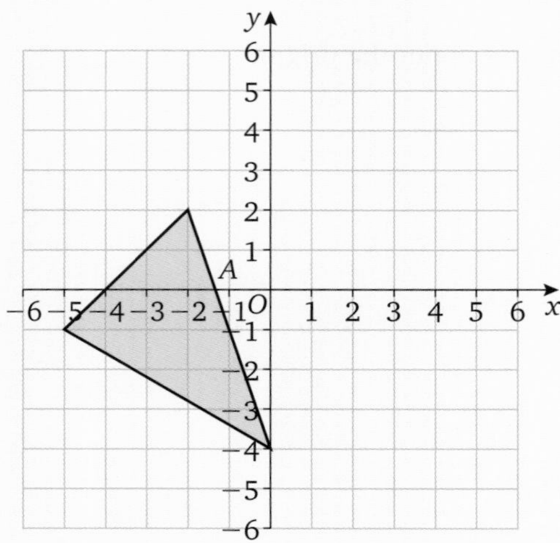

6 **a** Describe fully the single transformation that maps M onto N.

b Copy the diagram and rotate shape N 90° clockwise about the point $(1, -1)$. Label the image P.

c Reflect P in the x-axis and label the image Q.

d Describe fully the single transformation that maps M onto Q.

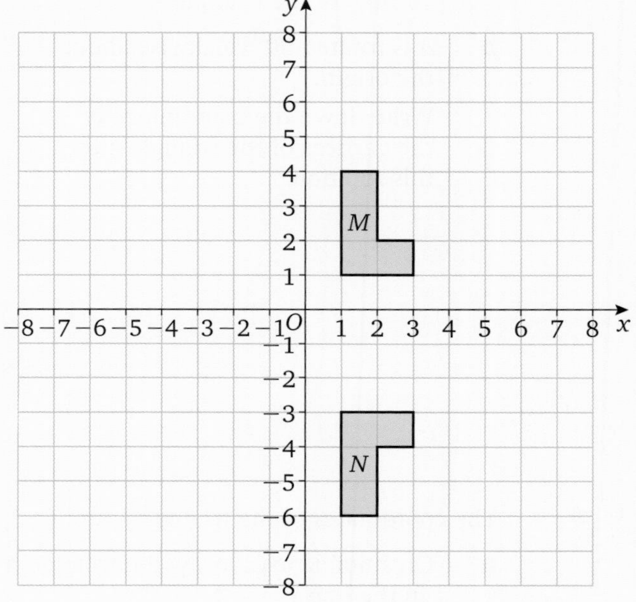

B

7 The shapes X and Y are congruent.

a Copy and complete the diagram of X and Y.

b Describe fully the **single** transformation that maps X onto Y.

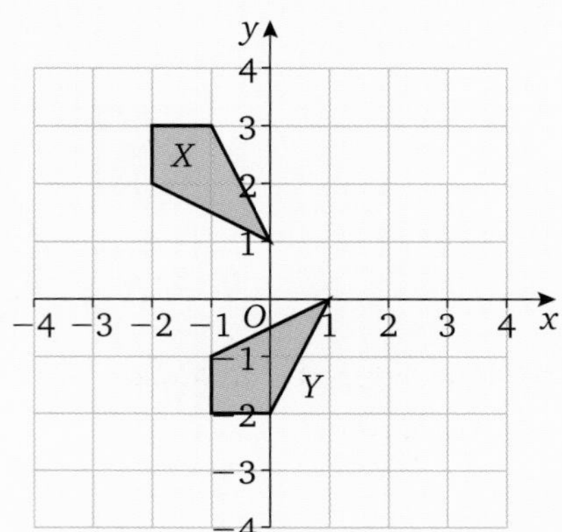

B

8 **a** Which of the shapes B, C and D are congruent to A?

b Which of the shapes B, C and D are a transformation of A? Give an explanation for your answer.

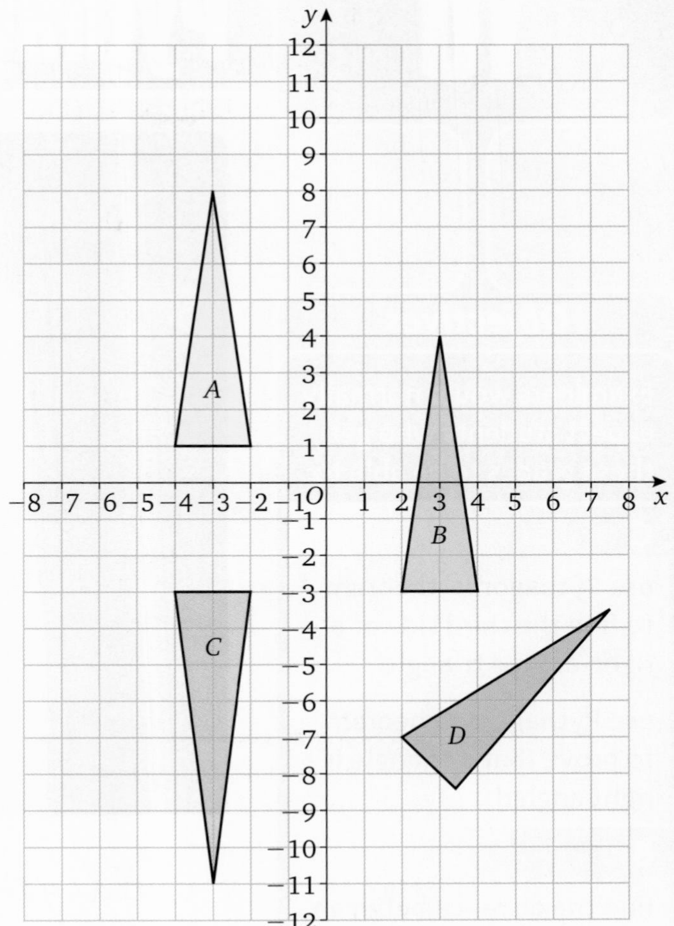

AQA Examination-style questions

1

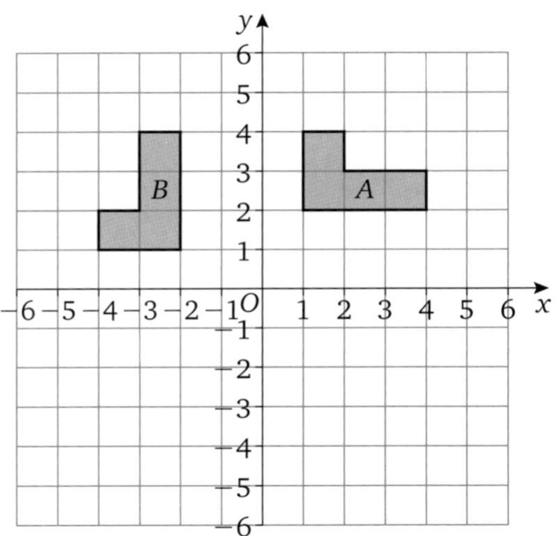

a Describe the single transformation that takes shape A to shape B. *(3 marks)*

b Copy the diagram. Reflect shape B in the line $y = -1$ *(2 marks)*

AQA 2008

10 Pythagoras' theorem

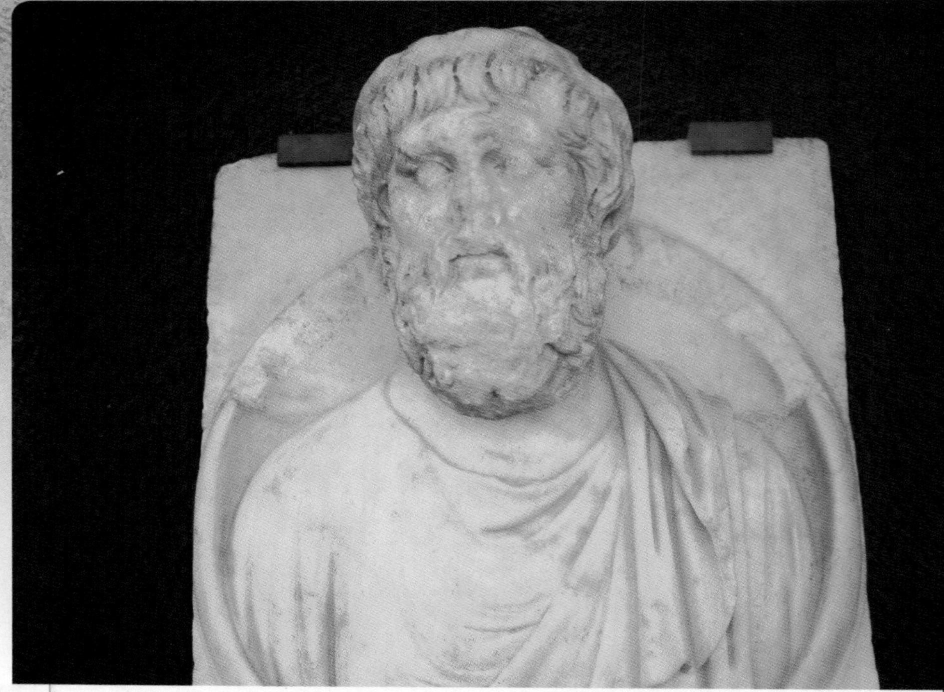

Objectives

Examiners would normally expect students who get these grades to be able to:

C

use Pythagoras' theorem to find the third side of a right-angled triangle

use Pythagoras' theorem to prove that a triangle is right angled

B

find the distance between two points from their coordinates

A/A*

use Pythagoras' theorem in 3-D problems.

Did you know?

Pythagoras

Pythagoras lived in the 5th century BCE, and was one of the first Greek mathematical thinkers.

Pythagoras is known to students of mathematics because of the theorem that bears his name:

'The square on the hypotenuse is equal to the sum of the squares on the other two sides.'

The Egyptians knew that a triangle with sides of length 3, 4 and 5 has a 90° angle. They used a rope with 12 evenly spaced knots like this one, to make right angles.

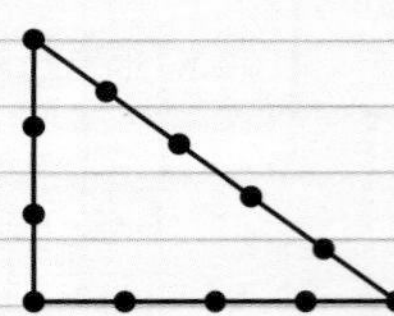

However, they did not extend the idea to triangles with other dimensions.

Other people such as the Chinese and the Sumerians also already knew that it was generally true, and used it in their measurements. However, it was Pythagoras who is said to have proved that it is always true.

Key terms

hypotenuse
Pythagoras' theorem

You should already know:

- ✔ how to find squares and square roots using a calculator
- ✔ the properties of triangles and quadrilaterals.

Learn... 10.1 Pythagoras' theorem

In any right-angled triangle the longest side is always opposite the right angle.

This side is called the **hypotenuse**.

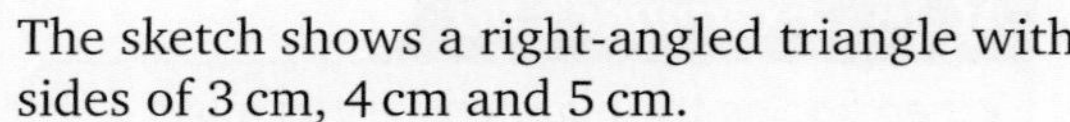

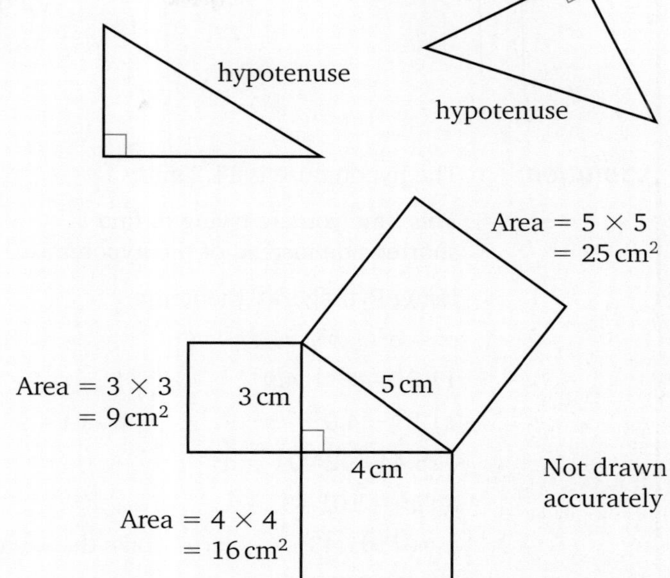

The sketch shows a right-angled triangle with sides of 3 cm, 4 cm and 5 cm.

Squares have been drawn on each side of the triangle and the area of each square is shown.

The area of the large square is equal to the sum of the areas of the two smaller squares:

$25 = 9 + 16$

This can be written as

$5^2 = 3^2 + 4^2$

This equation can be generalised for the sides of any right-angled triangle and is known as **Pythagoras' theorem**.

In general, in a right-angled triangle $c^2 = a^2 + b^2$

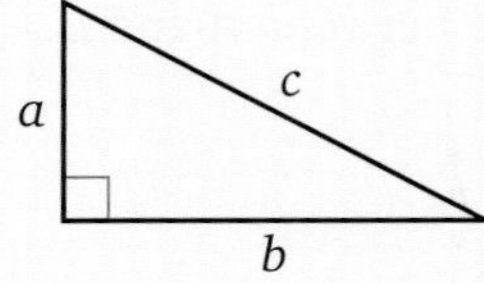

AQA Examiner's tip

You will not be asked to prove Pythagoras' theorem in the exam. One example of a proof is given here to show how it can be done.

Proof of Pythagoras' theorem

There are many proofs, both geometric and algebraic. Here is one algebraic proof using area.

A square of side $(x + y)$ is divided up as shown.

Each of the four right-angled triangles has area $\frac{1}{2}xy$.

The small square has area z^2.

Area of large square $= (x + y)^2 = (x + y)\,(x + y) = x^2 + 2xy + y^2$

This is also equal to the sum of the areas of the triangles and the small square.

So $\quad x^2 + 2xy + y^2 = 4 \times \frac{1}{2}xy + z^2$

$\quad x^2 + 2xy + y^2 = 2xy + z^2$ Take away $2xy$ from both sides.

Therefore $\mathbf{x^2 + y^2 = z^2}$

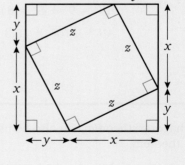

Example: Calculate the length of the hypotenuse (labelled c) of this triangle.

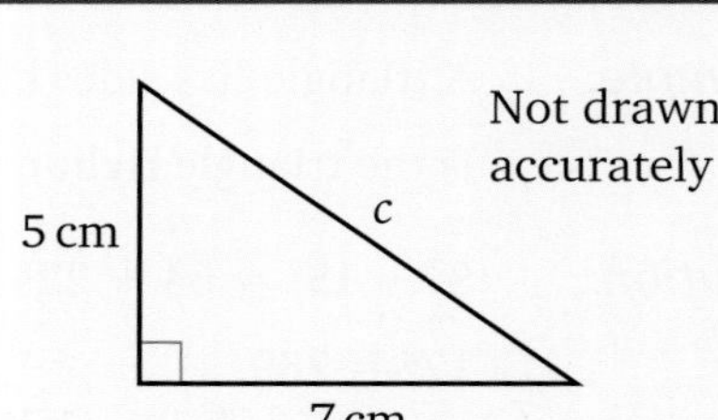

Solution: Using Pythagoras' theorem:

$c^2 = a^2 + b^2$

$c^2 = 5^2 + 7^2 = 25 + 49 = 74$

$c = \sqrt{74} = 8.6$ cm (to 1 d.p.) Take the square root of each side.

Example: Work out the length of side a.

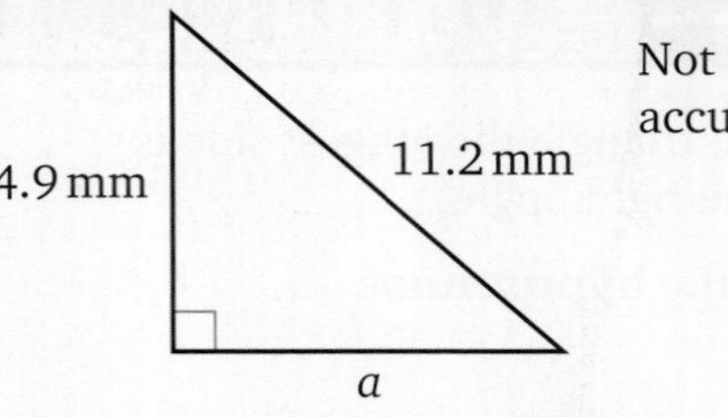

Solution: The hypotenuse is 11.2 mm.

This time you are trying to find a shorter side instead of the hypotenuse.

Using Pythagoras' theorem:

$c^2 = a^2 + b^2$

$11.2^2 = a^2 + 4.9^2$

$11.2^2 - 4.9^2 = a^2$ Subtract 4.9^2 from both sides.

$125.44 - 24.01 = a^2$

So $a^2 = 101.43$

$a = \sqrt{101.43}$ Take the square root of both sides.

$a = 10.0712462...$ mm

$a = 10.1$ mm (to 1 d.p.) Note the answer has been rounded to one decimal place as that is the same degree of accuracy as the question.

AQA Examiner's tip

When finding the hypotenuse you must **add** the squares of the other two sides. When finding a shorter side you must **subtract** the square of the known short side from the square of the hypotenuse.

Example: A is the point (2, 8) and B is the point (7, −1). Find the distance between A and B.

Solution: Draw a sketch to show the positions of A and B.

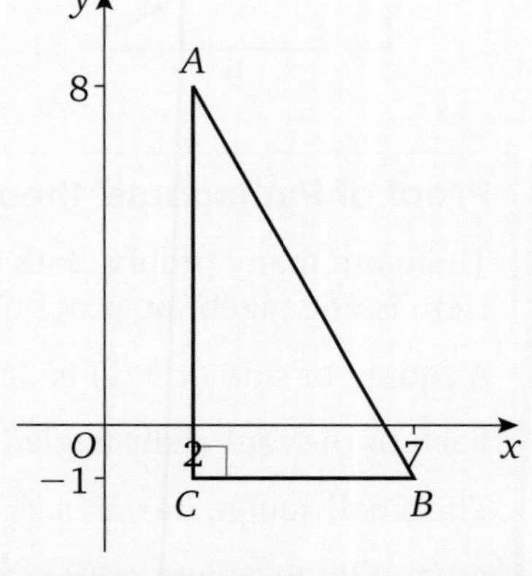

The line AB is the hypotenuse of the right-angled triangle ACB so Pythagoras' theorem can be used.

The length of AC is $8 - -1 = 9$ units

The length of CB is $7 - 2 = 5$ units

$AB^2 = 9^2 + 5^2 = 81 + 25 = 106$

$AB = \sqrt{106} = 10.3$ units (to 1 d.p.)

Pythagoras' theorem can also be used to test whether a triangle is right angled by showing that the sides fit the theorem.

To test for a right angle, first square the longest side.

Then add the squares of the two shorter sides.

If your results are equal, the triangle contains a right angle.

Example: A triangle has sides of 8 cm, 15 cm and 17 cm.

Is the triangle right-angled?

Solution: $8^2 + 15^2 = 64 + 225 = 289$

$17^2 = 289$

So $17^2 = 8^2 + 15^2$ and the sides of the triangle obey Pythagoras' theorem. Therefore, the triangle must be right-angled.

Right-angled triangles can be formed from other shapes. Some examples are shown below. Wherever a right-angled triangle is formed you can use Pythagoras' theorem to find the length of the third side.

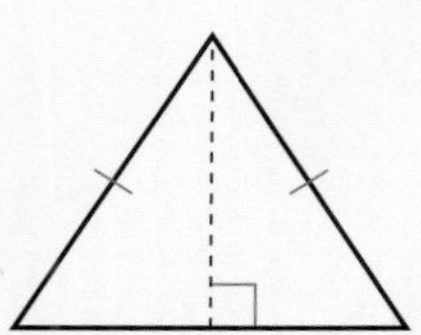
Isosceles triangle

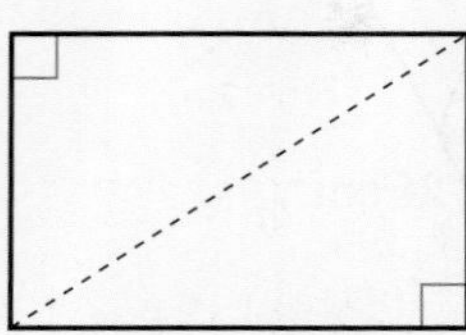
Rectangle

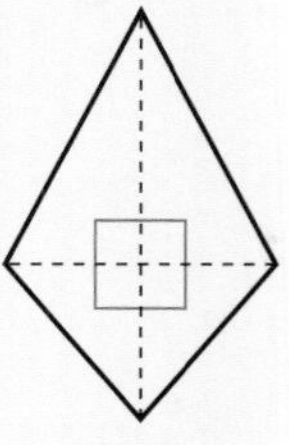
Kite

Bump up your grade

To get a Grade C make sure you understand Pythagoras' theorem and know how to apply it to solve problems.

Practise... 10.1 Pythagoras' theorem 

C

1 Find the length of the hypotenuse in each triangle.

a

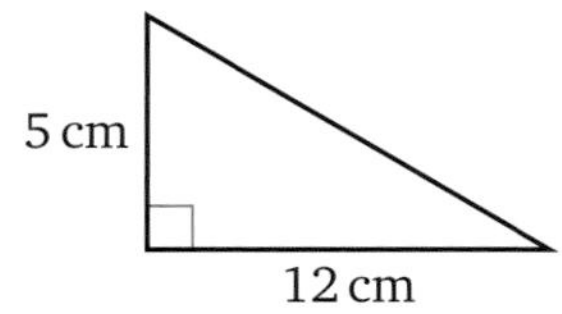

c

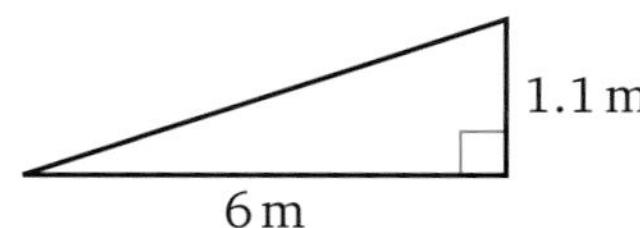

Not drawn accurately

b

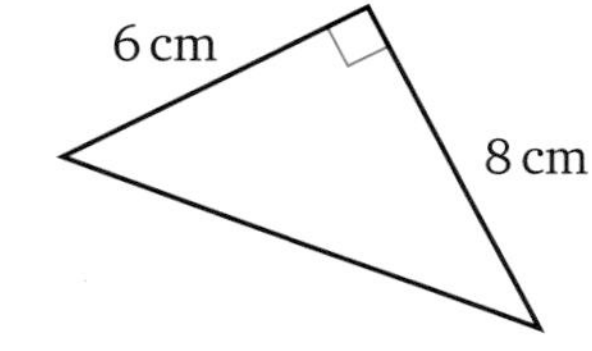

d

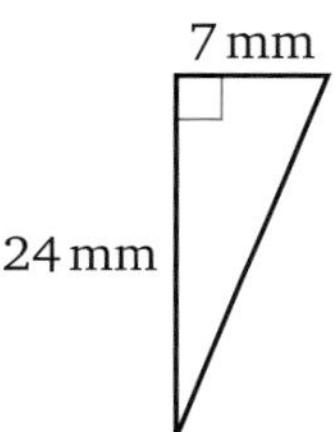

2 Find the length of the diagonal in each of these rectangles.
Give each answer correct to one decimal place.

a

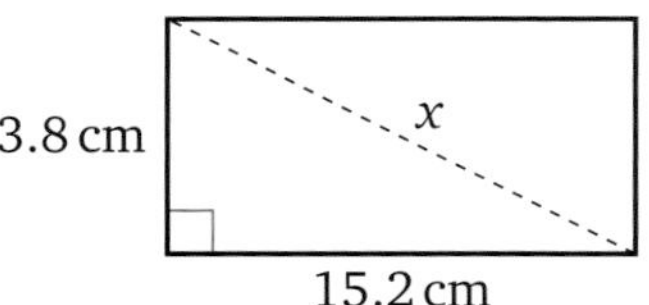

b

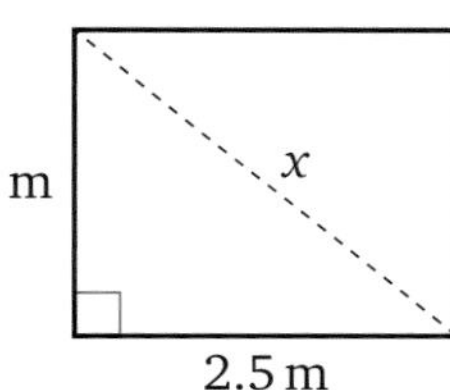

c

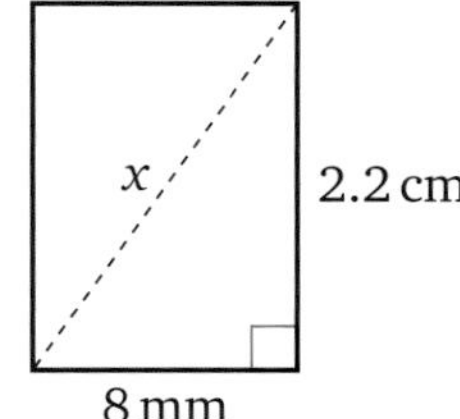

Not drawn accurately

3 Find the length of the missing side, marked x, in each of these triangles.
Give each answer correct to one decimal place.

a

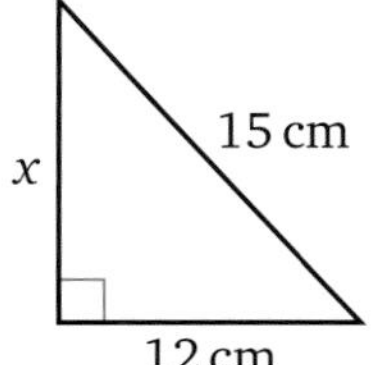

Not drawn accurately

b

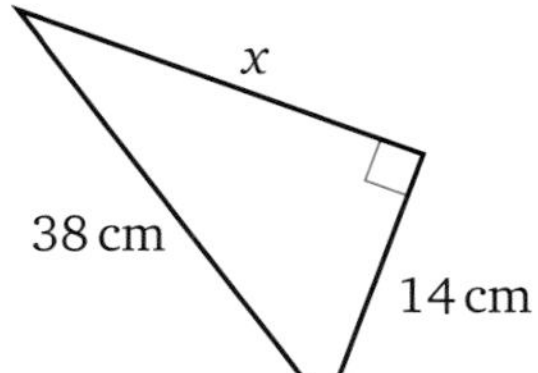

c

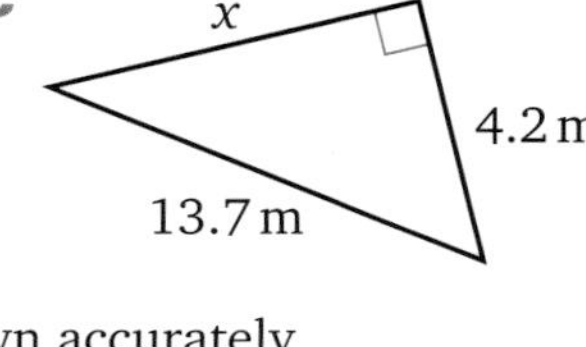

Not drawn accurately

4 Check whether each of these triangles is right angled. Show your working.

a Side lengths: 13 cm, 84 cm and 85 cm

b Side lengths: 3.5 m, 5.8 m and 4.2 m

c Side lengths: 12 mm, 34 mm and 46 mm

d Side lengths: 26 cm, 24 cm and 10 cm

5 Find the length of the side marked x in each of these triangles.
Give each answer to one decimal place.

a

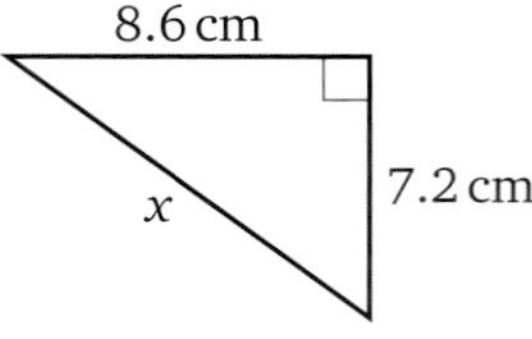

c

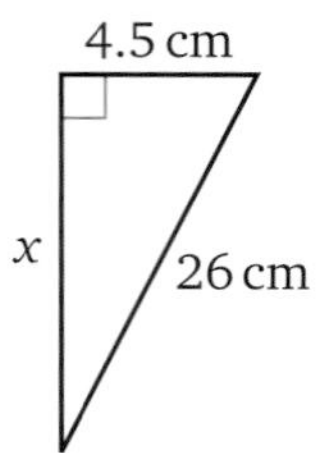

b

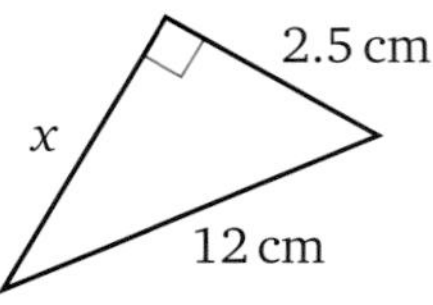

d

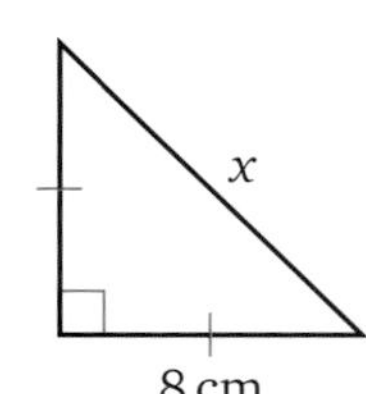

Not drawn accurately

6 Sarah and Ravi are working out the missing side in this triangle.

Sarah works out $7^2 + 22^2$ and says that $x = 23.1$ cm (to 1 d.p.).

Ravi works out $22^2 - 7^2$ and says that $x = 435$ cm.

Both of these answers are incorrect.

Explain each person's mistake and work out the correct value of x.

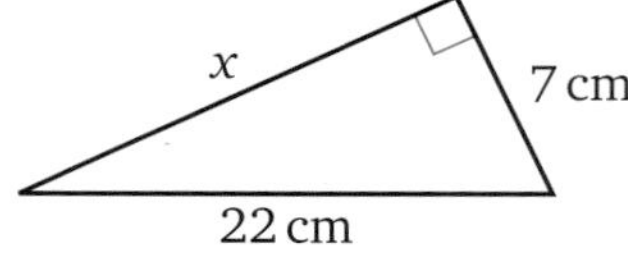

Not drawn accurately

7 An isosceles triangle has two sides of length 7.5 cm and one side of length 12 cm.

Calculate the height of the triangle.

Give your answer to one decimal place.

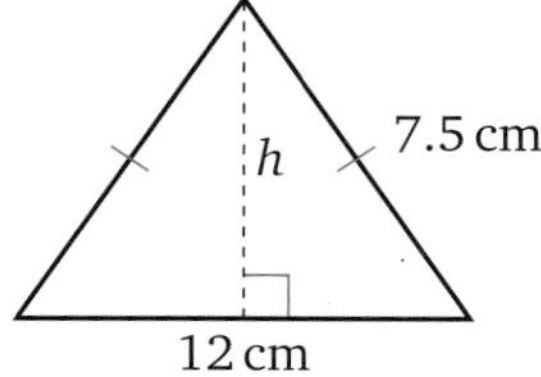

Not drawn accurately

8 Calculate the length of the chord AB of the circle.

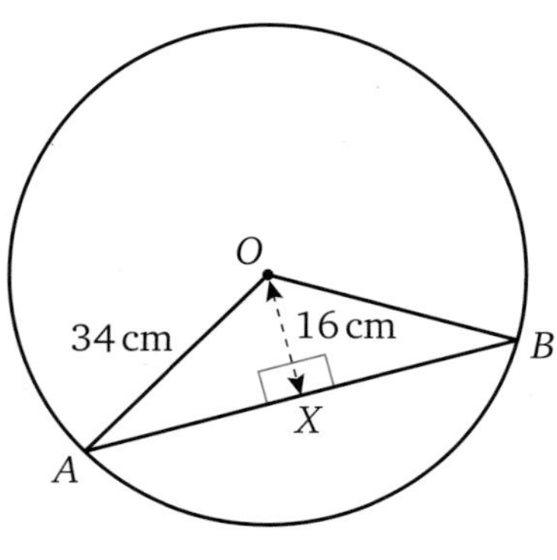

Not drawn accurately

9 Find the length of the side marked x in each diagram.

a

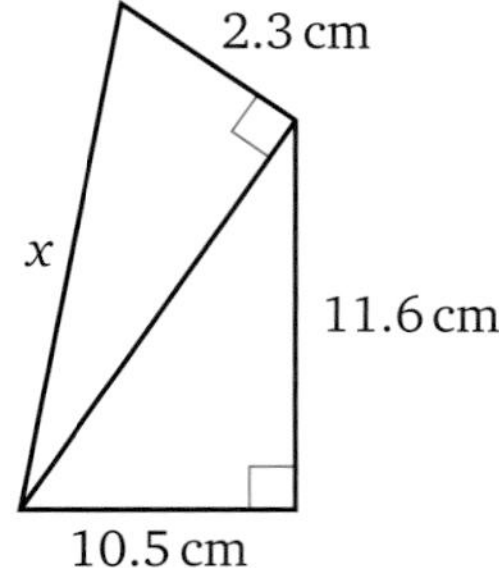

b

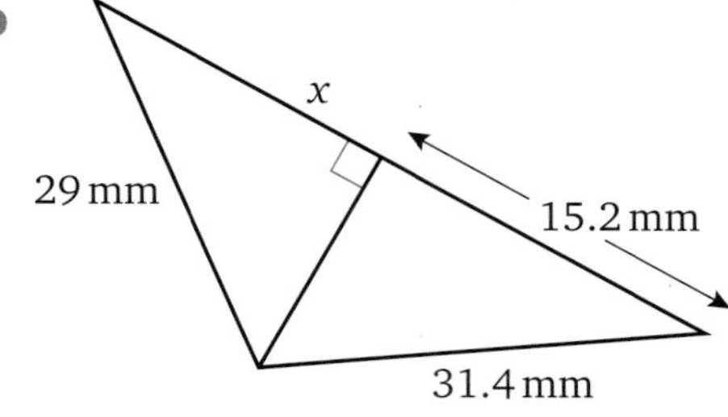

Not drawn accurately

10 **a** A is the point (2, 3) and B is the point (5, 7) as shown in the diagram.

Use Pythagoras' theorem to find the distance between points A and B.

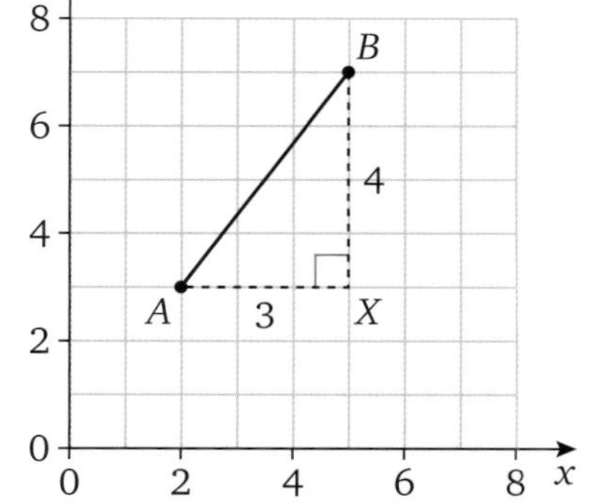

b Sketch a diagram and use Pythagoras' theorem to find the distance between each of the following sets of points.

- **i** (3, 5) and (7, 8)
- **ii** (0, 4) and (5, 10)
- **iii** (−2, 3) and (1, 7)

11 Leon walks 1.5 km due north from his house.

He then turns and walks 2 km due east.

How far is he now from his house?

12 A field is in the shape of a rectangle of length 45 metres and width 22 metres. A pipe runs from one corner of the field to the opposite corner.

How long is the pipe?

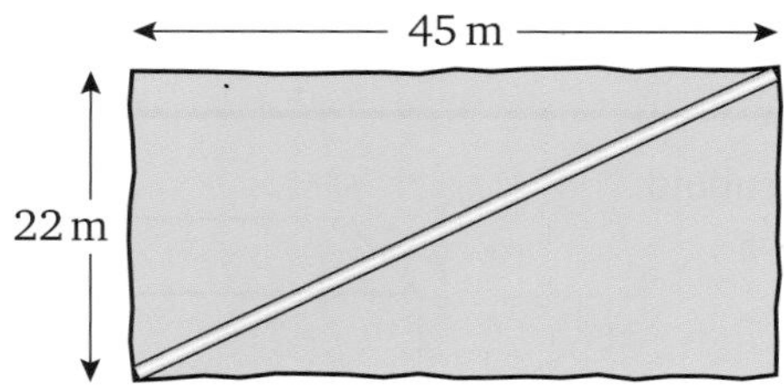

Not drawn accurately

13 A ladder is 6.5 metres long.

The safety instructions say that for a ladder of this length:

- the maximum safe distance of the foot of the ladder from the wall is 1.7 metres
- the minimum safe distance of the foot of the ladder from the wall is 1.5 metres.

What is the maximum vertical height that the ladder can safely reach? Give your answer to the nearest centimetre.

14 Cath has designed a pendant, in the shape of a kite, for a necklace.

Work out the length, x, of the green line.

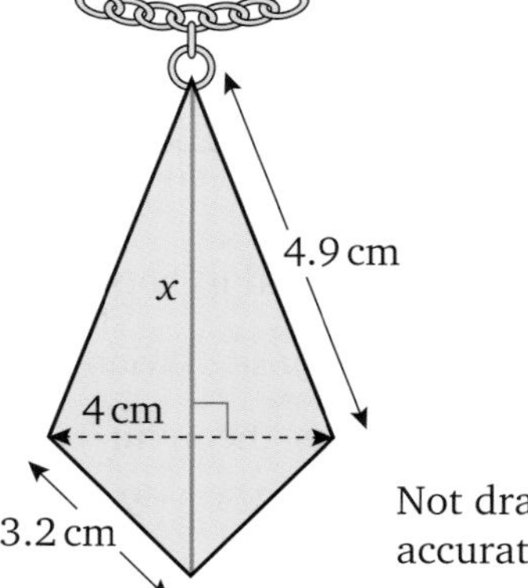

Not drawn accurately

15 Pythagorean triples are sets of three integers that fit Pythagoras' theorem. For example:

3, 4, 5　　5, 12, 13　　7, 24, 25

a Find other Pythagorean triples.

b What patterns can you see in the numbers?

c See whether you can find a rule to generate other triples.

16 **a** Calculate the length x.

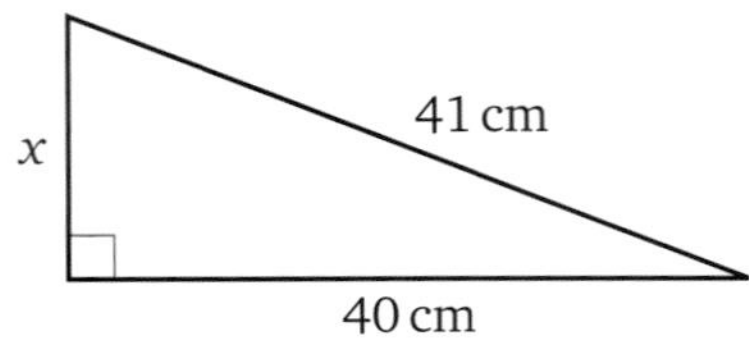

Not drawn accurately

b m and n are integers.

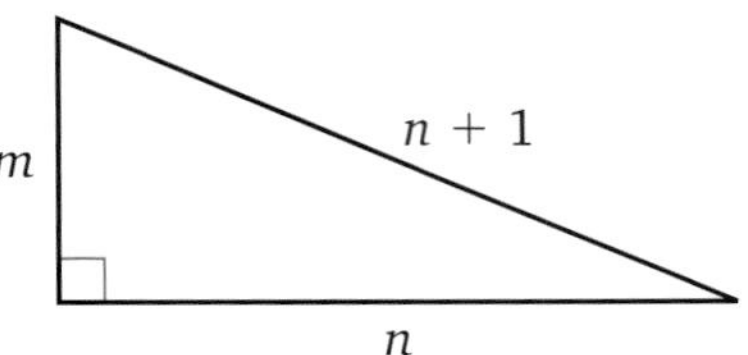

Prove that m is an odd number.

Learn... 10.2 Pythagoras' theorem in three dimensions

When working in three dimensions, it is sometimes necessary to find the length of a line joining two points. Often this length will form the side of a right-angled triangle and so Pythagoras' theorem can be used. The triangles outlined in red are right-angled triangles.

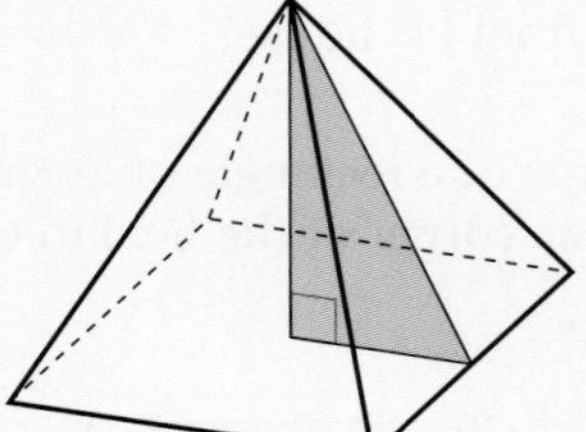

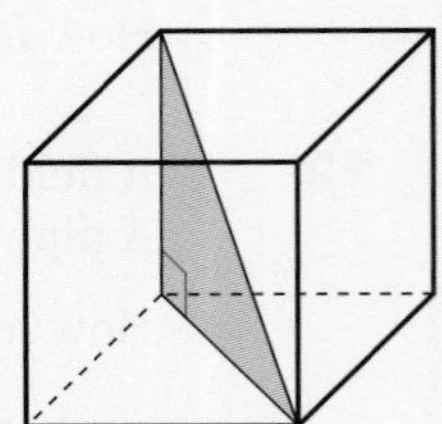

Example: Find the length of the diagonal AG in this cuboid. Give your answer to three significant figures.

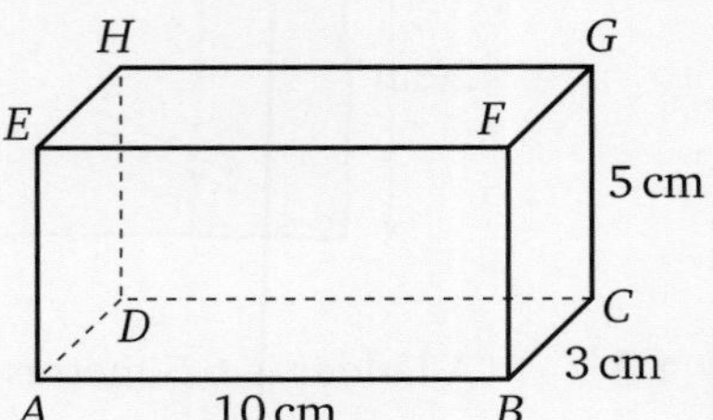

Not drawn accurately

Solution: Joining AG and AC gives the right-angled triangle ACG.

Make a sketch of this triangle so that you can see whether you are finding the length of the hypotenuse or a shorter side.

AQA Examiner's tip

Always draw out the right-angled triangle you are using.

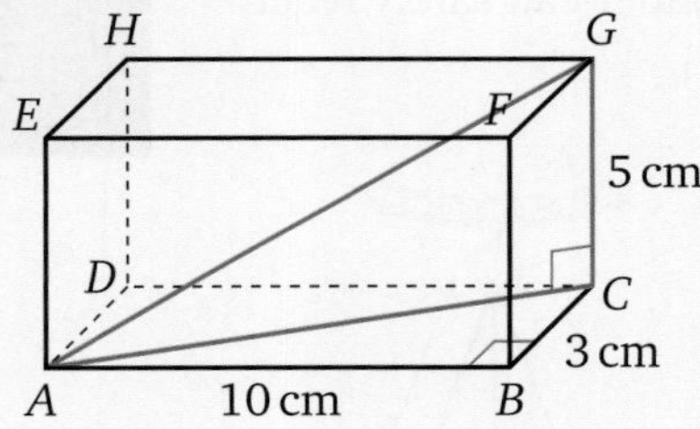

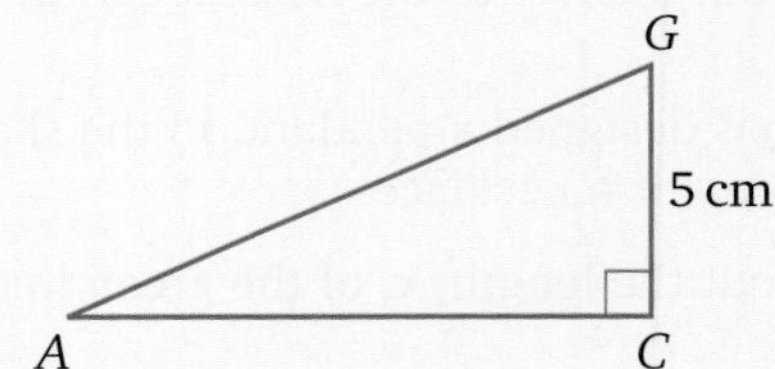

Not drawn accurately

To find AG you need to know the length AC.

AC is the hypotenuse of triangle ABC.

Using Pythagoras' theorem:

$AC^2 = 10^2 + 3^2 = 100 + 9 = 109$

$AC = \sqrt{109}$

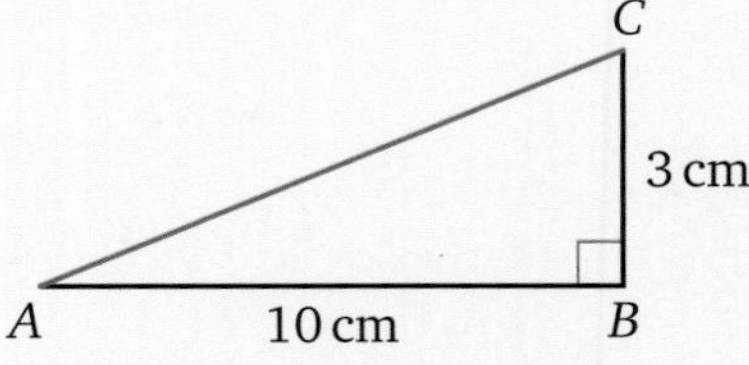

Now using triangle ACG:

$AG^2 = AC^2 + CG^2$

$= (\sqrt{109})^2 + 5^2$

$= 109 + 25 = 134$

$AG = \sqrt{134} = 11.5758369...$

$AG = 11.6$ cm (to 3 s.f.)

AQA Examiner's tip

Leave AC as $\sqrt{109}$ as you will be squaring it again in the next part of the solution.

Example: The square-based pyramid shown in the diagram has four triangular sides.

Each triangle is isosceles, with sides 13 cm, 13 cm and 10 cm.

Calculate the vertical height of the pyramid. Give your answer to three significant figures.

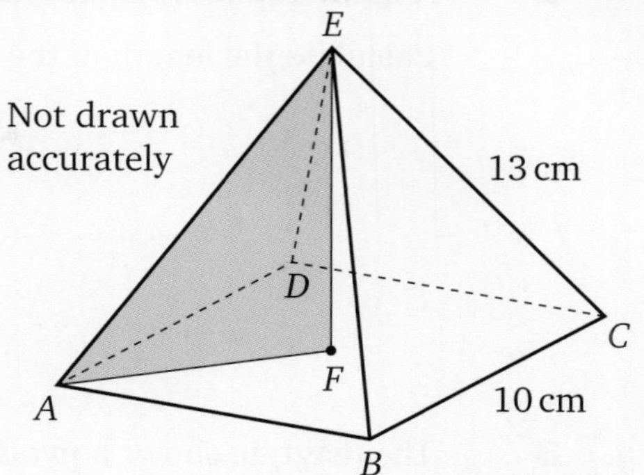

Solution: The vertical height is EF.

Triangle AEF is a right-angled triangle.

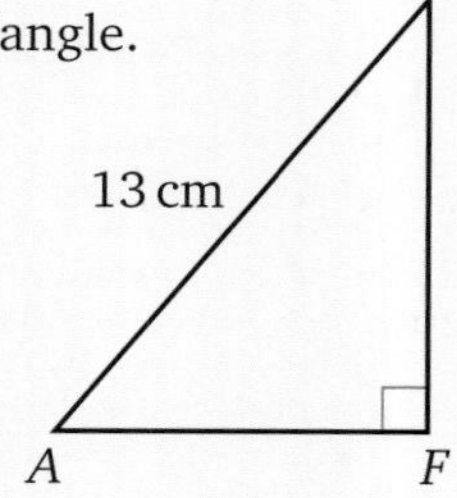

To find EF you first need to find the length of AF.

AF is $\frac{1}{2}AC$, where AC is the diagonal of the square base.
So ABC is a right-angled triangle.

Using Pythagoras' theorem,

$AC^2 = 10^2 + 10^2$

$AC^2 = 200$

$AC = \sqrt{200} = 10\sqrt{2}$

So $AF = \frac{1}{2}AC$

$= 5\sqrt{2}$

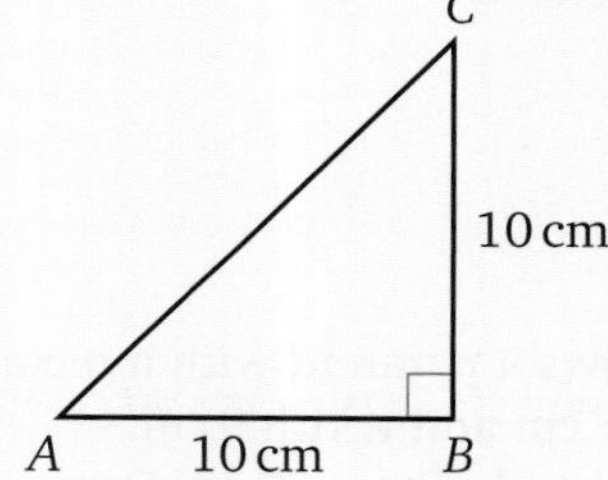

AQA Examiner's tip

You can leave the value of AF in surd form to avoid rounding until the end of the calculation.

Now using Pythagoras' theorem in the first triangle AFE:

AE is the hypotenuse.

$EF^2 = AE^2 - AF^2$. As you want to find EF (a shorter side), you must subtract.

$EF^2 = 13^2 - (5\sqrt{2})^2$

$= 119$

$EF = \sqrt{119} = 10.9$ cm

Practise... 10.2 Pythagoras' theorem in three dimensions

D C B A A*

The shapes in these exercises are not drawn accurately.

1 The diagram shows a cuboid in which $AB = 11$ cm, $BC = 7$ cm and $CG = 4$ cm.

Calculate the length of:

a AC **b** BG **c** CE

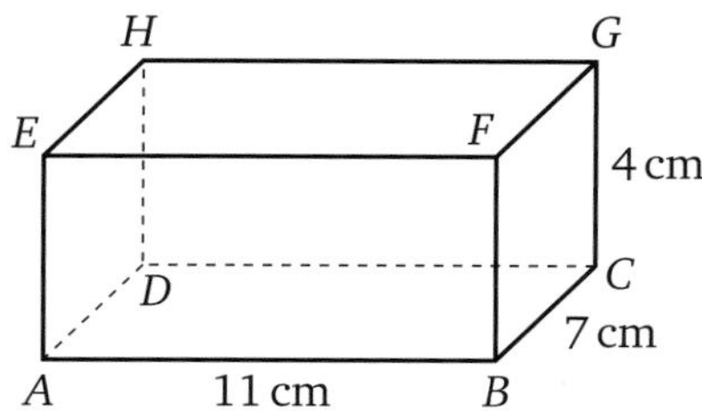

A

A

2 A metal rod just fits into this box as shown.

Calculate the length of the rod.

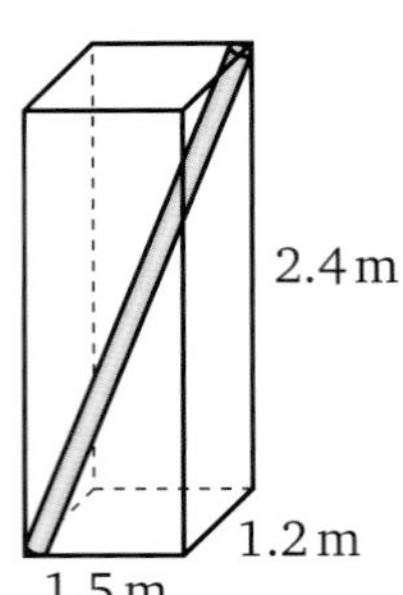

3 The diagram shows a pyramid with a square base of side 15 cm. The vertical height of the pyramid is 10 cm.

Calculate:

a the length of *BD*

b the length of *YB*.

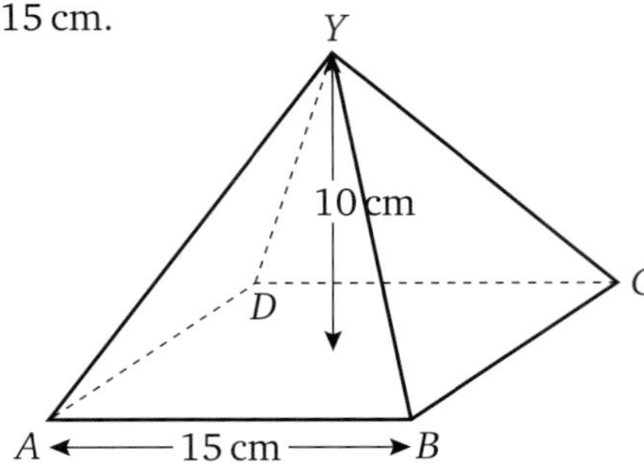

4 The diagram shows a cube of side 30 cm. *X* is the midpoint of *CG*.

Calculate the length of:

a *DX*

b *AX*

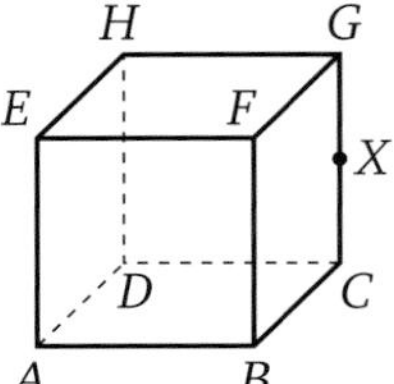

5 The diagram shows a pyramid with a rectangular base of length 15 cm and width 9 cm. Edges *YA*, *YB*, *YC* and *YD* are 20 cm long.

Work out the vertical height of the pyramid.

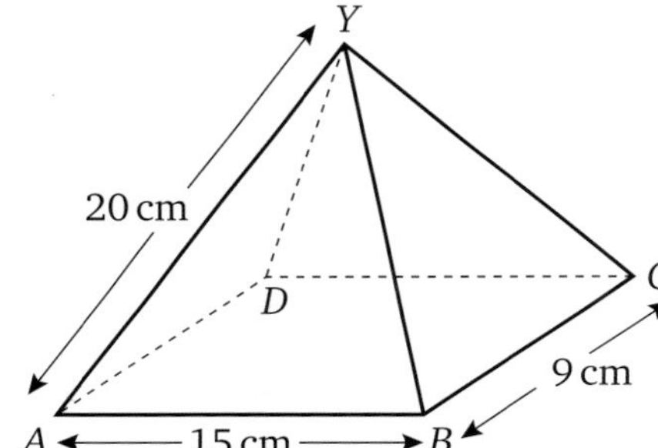

6 The diagram shows a cube of side 8 cm. *X*, *Y* and *Z* are the midpoints of *AE*, *BC* and *GH* respectively.

Calculate the area of triangle *XYZ*.

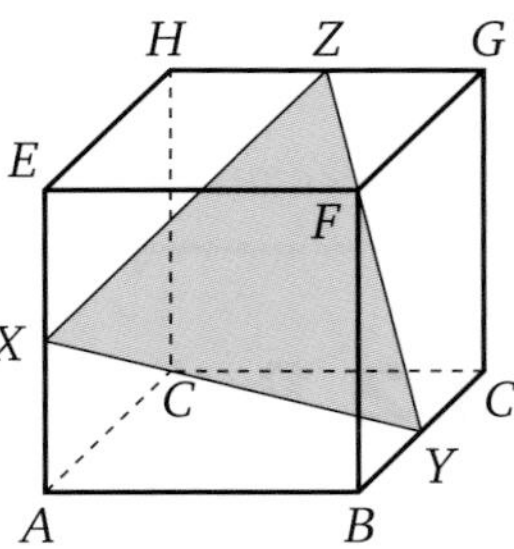

7 The diagram shows a skateboard ramp in a park. The ramp is a triangular prism. Jodie uses her skateboard to travel down the slope in a straight line from *X* to *B*. Kev uses his skateboard to travel down the slope in a straight line parallel to *XA*.

How much further does Jodie travel?

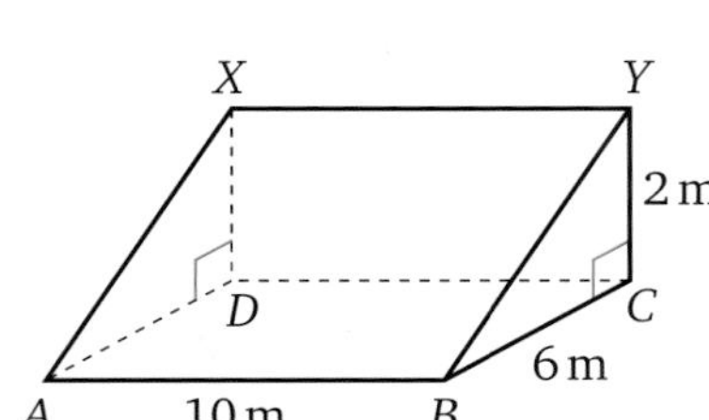

 8 The great Pyramid of Giza in Egypt is built with a square base of side approximately 230 metres.
The slanting edges are approximately 214 m in length.

Calculate an estimate of the height of the pyramid.

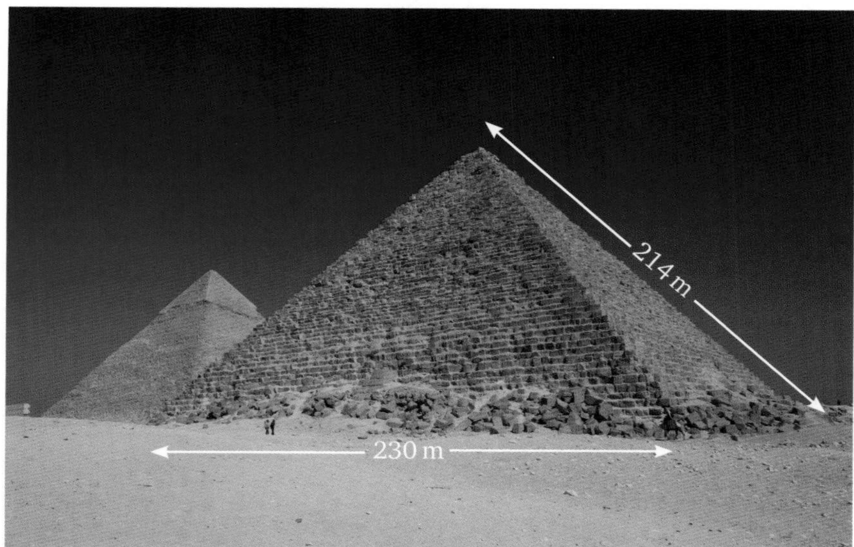

10 Assess

1 Find the length of side x in each triangle.

a
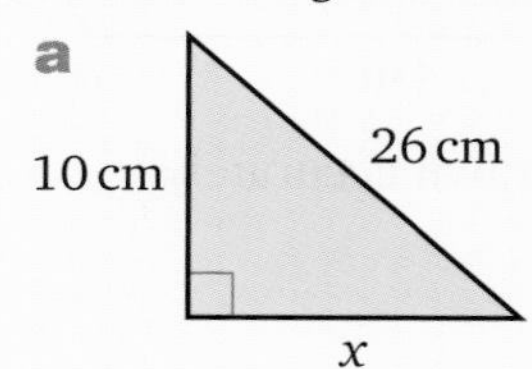

b
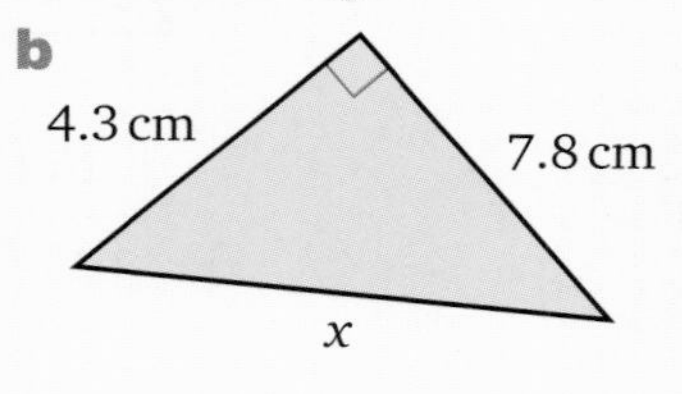

c
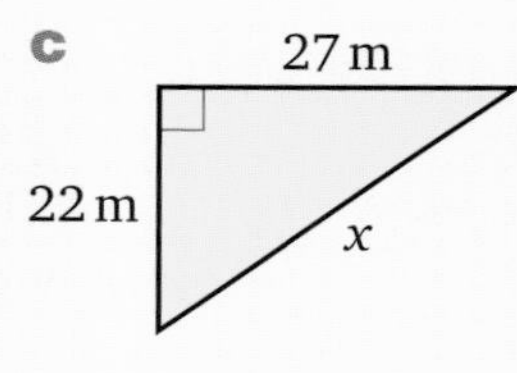

d
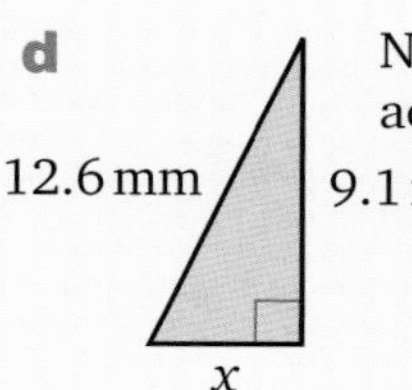

Not drawn accurately

C

2 An equilateral triangle has sides of length 15 cm.

Work out the perpendicular height of the triangle.

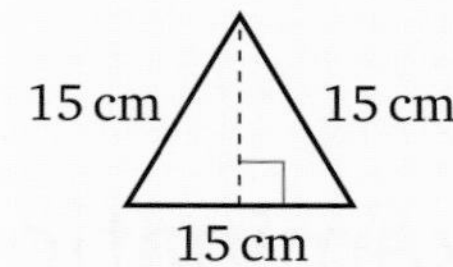

Not drawn accurately

3 A field is in the shape of a rectangle of length 37 m and width 29 m.
A path runs diagonally from one corner of the field to the opposite corner.

What is the length of the path?

4 A telegraph pole is held vertically by wires of length 10 metres and x metres fixed to the ground.

Calculate:

a the height, h, of the telegraph pole

b the length, x, of the second wire.

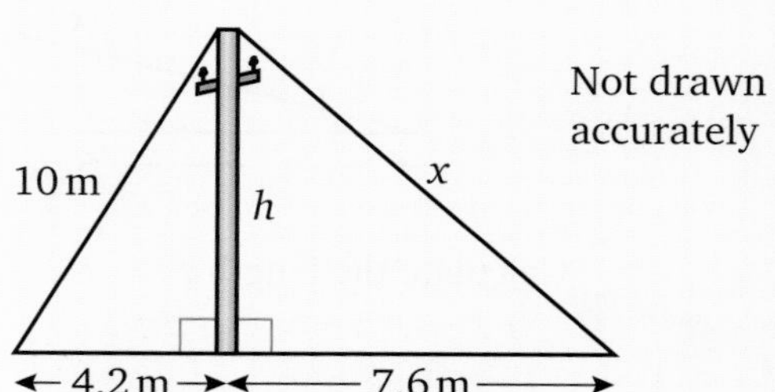

Not drawn accurately

5 Rob walks from his house 3 km due south then 2 km due west.

How far is Rob now from his house?

B

6 Find the distance between the following pairs of points.
Give each answer correct to one decimal place.

a $A(3, 6)$ and $B(5, 9)$

b $C(-2, 4)$ and $D(3, 1)$

c $E(2, 1)$ and $F(15, 8)$

7 Work out the perimeter of this quadrilateral.

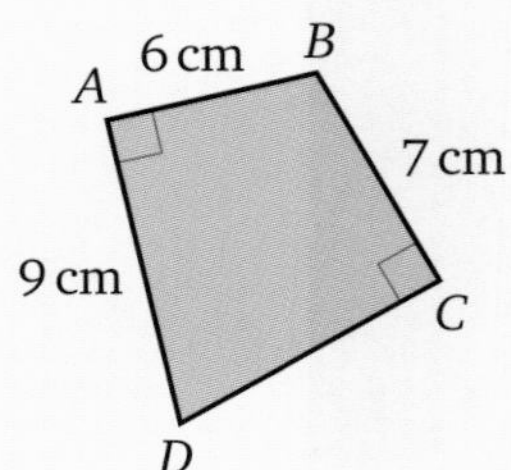

Not drawn accurately

A

8 Work out the length of the longest straight rod that could fit into:

a a cube of side 6 cm

b a cuboid of length 12 cm, width 10 cm and height 8 cm.

> **AQA Examiner's tip**
> Draw a sketch of each solid.

9 A roof is 14 m long and 3.5 m wide (i.e. $AX = 3.5$ m).
The top ridge, XY, is 2.5 m above the base.
Calculate the length of:

a AY **b** AC **c** BC

Give your answers to three significant figures.

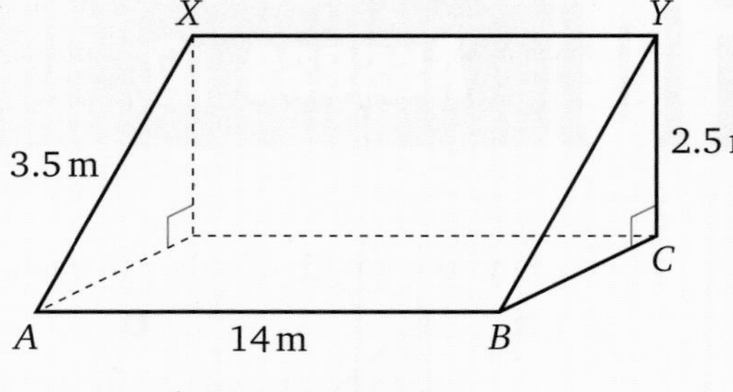

Not drawn accurately

AQA Examination-style questions

1 The diagram shows a right-angled triangle.

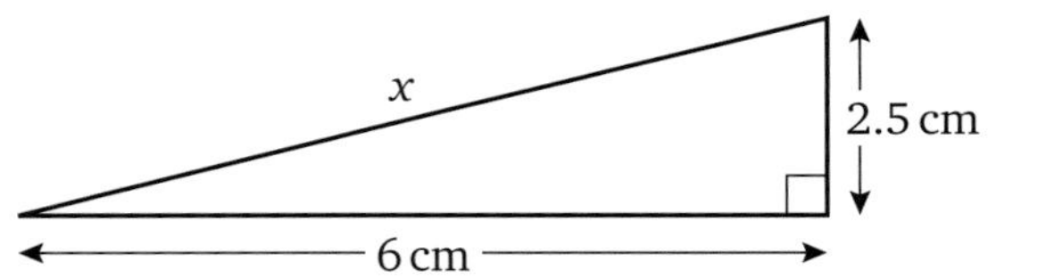

Not drawn accurately

Calculate the length x. *(3 marks)*

AQA 2009

2 A ladder of length 5 m rests against a wall.
The foot of the ladder is 1.7 m from the base of the wall.

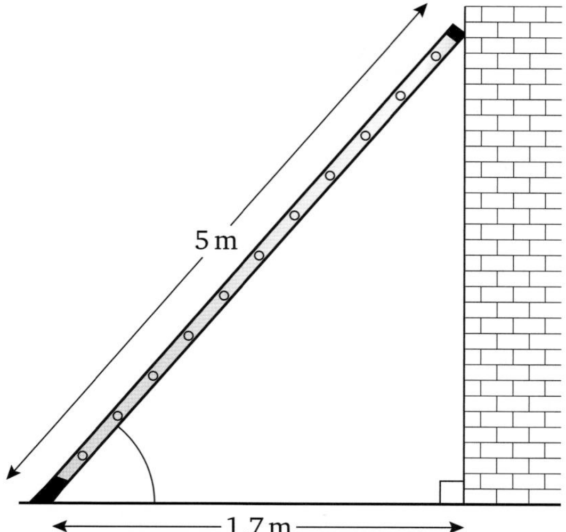

How far up the wall does the ladder reach? *(3 marks)*

AQA 2008

3 Three circles fit inside a rectangle as shown.
Two of the circles are identical and the third is larger.
The circles have radii 9 cm, 9 cm and 25 cm.

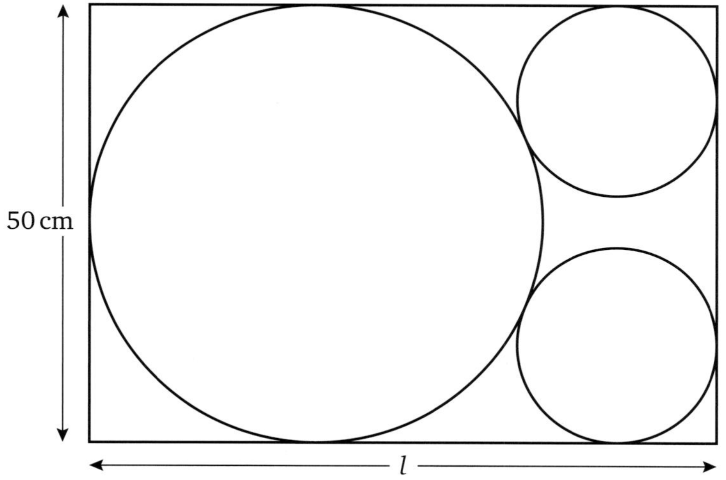

Calculate the length, l, of the rectangle. *(6 marks)*

AQA 2009

4 The diagram shows an isosceles triangle *ABC*.

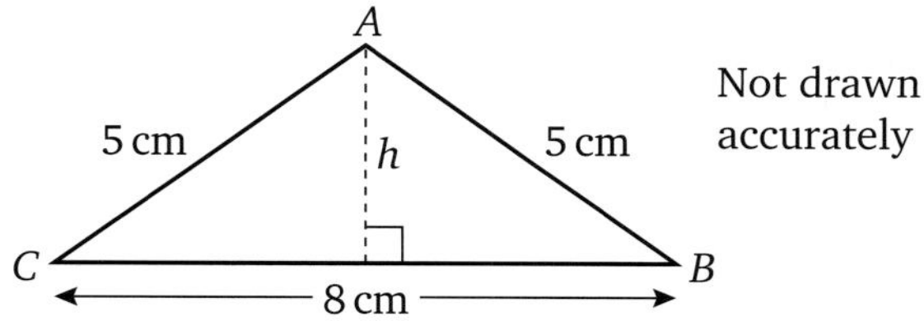

Calculate the area of the triangle *ABC*.
Show your working.
State the units of your answer. *(6 marks)*

AQA 2007

11 Properties of circles

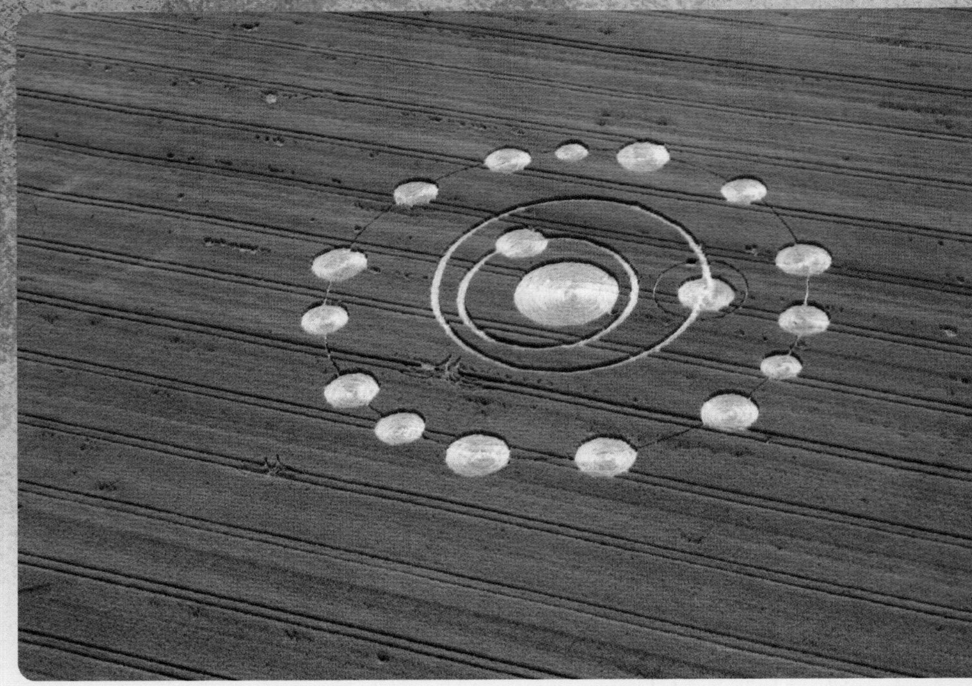

Objectives

Examiners would normally expect students who get these grades to be able to:

B

know the angle and tangent properties of a circle

A

understand the angle and tangent properties of a circle

understand the alternate segment theorem.

Did you know?

Crop circles

Crop circles are patterns made in fields of wheat, barley or corn. When crop circles were first spotted, they were normally quite simple patterns of different-sized circles. More recent crop circles are much more detailed and involve complex geometric patterns. There have been many theories about the origins of crop circles ranging from works of art to paranormal activity involving messages from extraterrestrial beings. Whatever the answer, there is no doubt that crop circles involve a lot of mathematics.

You should already know:

- ✔ the definition of a circle and the meaning of terms including centre, radius, chord, diameter, circumference, tangent, arc, sector and segment
- ✔ basic proofs such as the angle sum of a triangle is 180°, the exterior angle of a triangle is equal to the sum of the interior opposite angles
- ✔ and understand angle properties of parallel lines including corresponding angles and alternate angles on a transversal.

Key terms

angle subtended
arc
chord
cyclic quadrilateral
supplementary angles
tangent
alternate segment

Learn... 11.1 Angle properties of circles

There are four angle properties of circles you need to know.

- The **angle subtended** by an **arc** (or **chord**) at the centre of a circle is twice the angle subtended at any point on the circumference.
- The angle subtended at the circumference by a semicircle is a right angle.
- Angles subtended by the same arc (or chord) are equal.
- The opposite angles of a **cyclic quadrilateral** add up to 180°.

The angle subtended by an arc (or chord) at the centre of a circle is twice the angle subtended at any point on the circumference

The angle subtended by the arc PQ at O (the centre) is twice the angle at R (the circumference).

angle $POQ = 2 \times$ angle PRQ

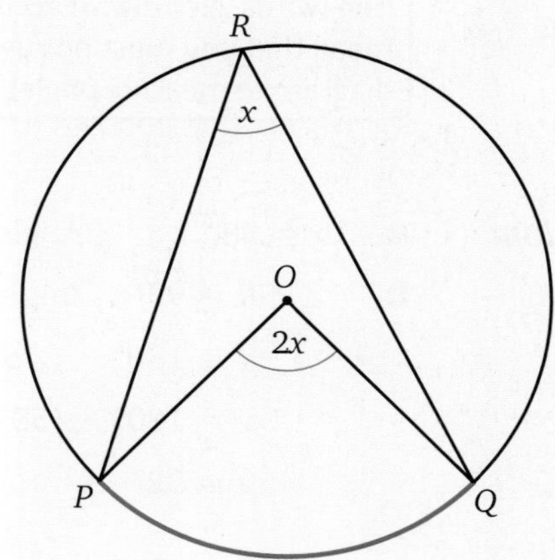

The angle subtended at the circumference by a semicircle is a right angle

The angle subtended by the arc PQ at O (the centre) is twice the angle at R (the circumference).

The angle at O is 180° so the angle at R (the circumference) is 90°.

Angle $PRQ = 90°$

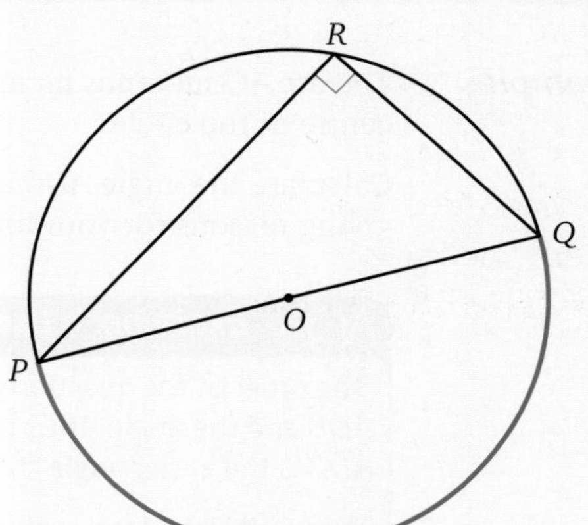

Angles subtended by the same arc (or chord) are equal

The angle at D and the angle at C and the angle at E are subtended by the same arc AB. Therefore, these angles are the same.

angle ADB = angle ACB = angle AEB

The angle at A and the angle at B are subtended by the same arc EC. Therefore these angles are the same.

angle EAC = angle EBC

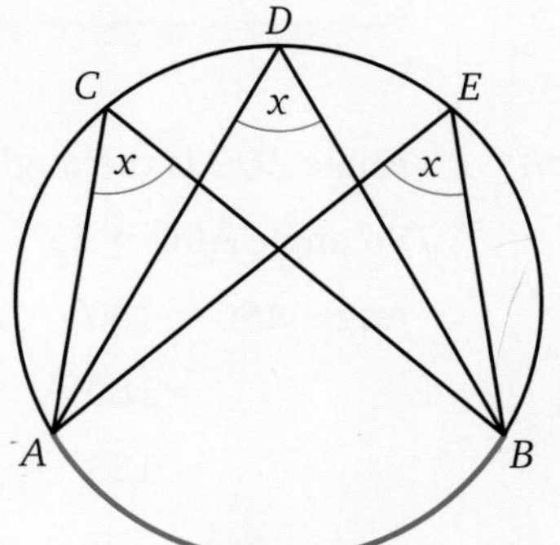

The opposite angles of a cyclic quadrilateral add up to 180°

A cyclic quadrilateral is any quadrilateral where the four vertices lie on the circumference of a circle.

The opposite angles of a cyclic quadrilateral add up to 180° (i.e. they are **supplementary angles**).

For the cyclic quadrilateral $EFGH$

$e + g = 180°$

$f + h = 180°$

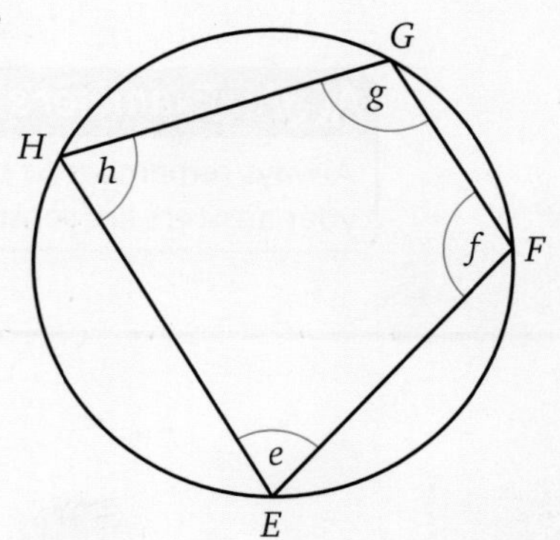

Example: P, Q, R and S are points on the circumference of a circle.

PR is a diameter and angle $SQR = 58°$

a Write down the value of x.

b Calculate the value of y.

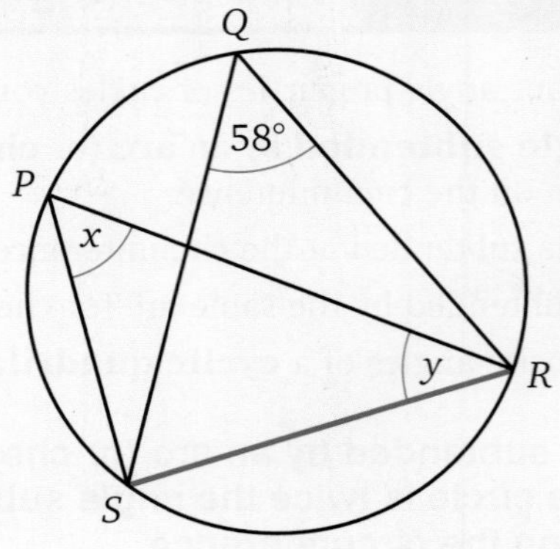

Not drawn accurately

AQA Examiner's tip

The words 'Not drawn accurately' mean that you must not use the diagram to measure angles.

Solution:

a $x = 58°$ Angles subtended by the same chord RS are equal.

b $\angle PSR = 90°$ Angle subtended at the circumference by a semicircle is a right angle.

$y = 180° - (\angle SPR + \angle PSR)$ Angles of a triangle add up to 180°.

$y = 180° - (58° + 90°)$

$y = 32°$

Example: The arc AC subtends an angle of 110° at the centre of the circle.

Calculate the angle ADC and the angle ABC giving reasons for your answers.

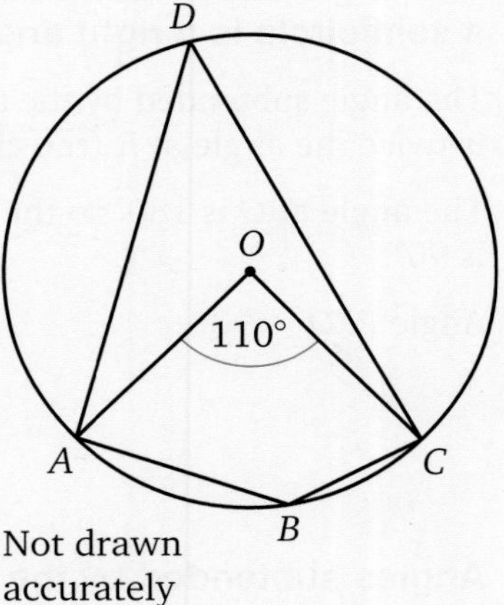

Not drawn accurately

AQA Examiner's tips

The order of the questions (i.e. calculate the angle ADB and the angle ABC) usually suggests that angle ADB is the easier angle to find first.

The word 'calculate' in this type of question means that you must work out the angles and **not** measure them.

Solution: Angle ADC is the angle at the circumference.

So angle $ADC = 55°$ angle at centre = 2 × angle at circumference

Angle $ABC = 180° - \angle ADC$ Opposite angles of a cyclic quadrilateral add up to 180°.

$= 180° - 55°$

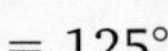

$= 125°$

Hint

Alternatively, you could use the fact that the angle at centre = 2 × angle at the circumference.

The reflex angle $AOC = 250°$ so angle $ABC = 125°$

AQA Examiner's tip

Always remember to check that your answers are sensible.

Practise... 11.1 Angle properties of circles

D C B A A*

B

The shapes in these exercises are not drawn accurately.

1 Calculate the marked angle on the following diagrams giving a reason for your answer.

The centre of the circle is marked O.

a

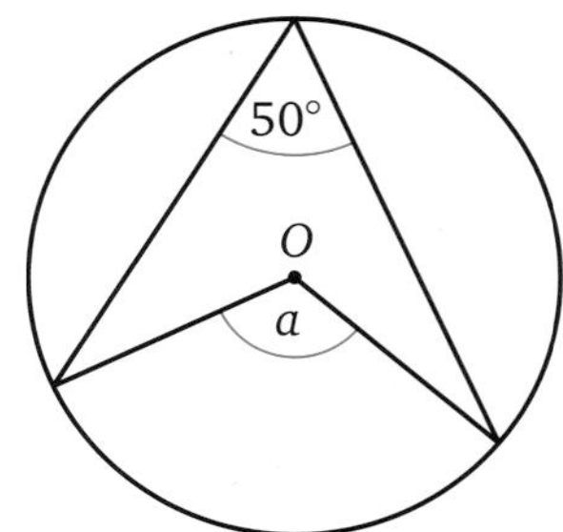

f

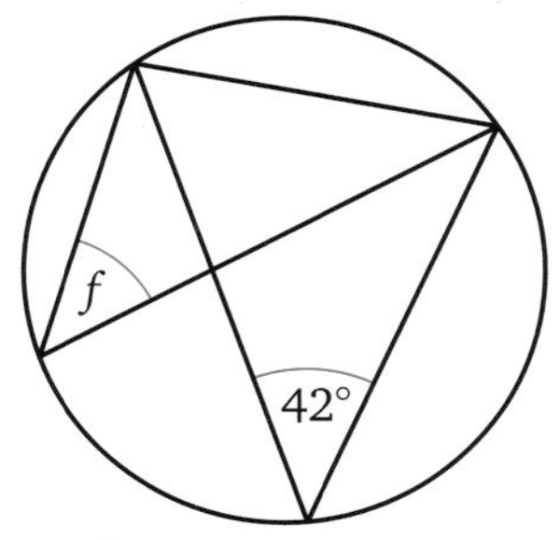

k

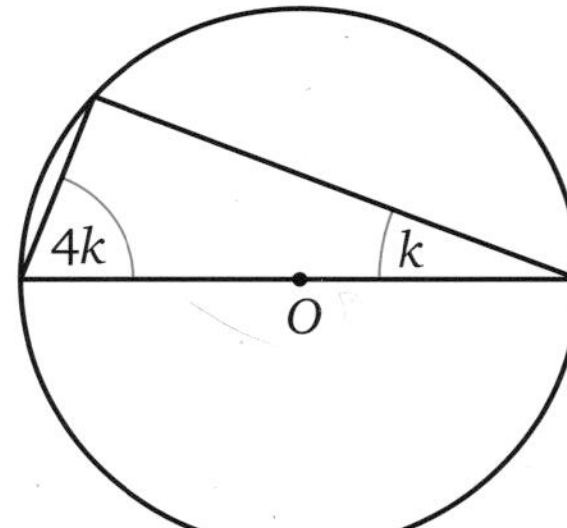

b

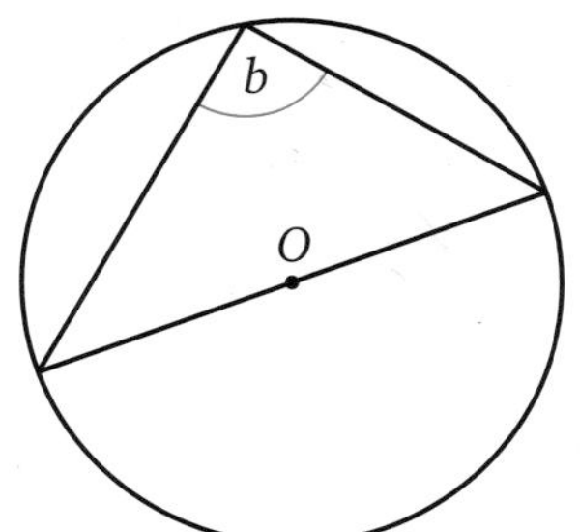

g

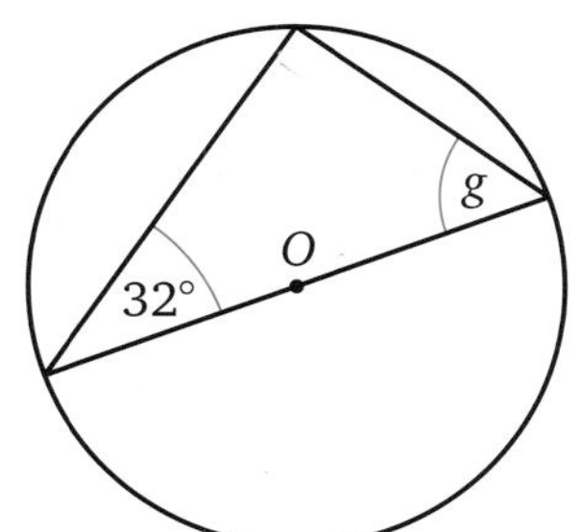

l

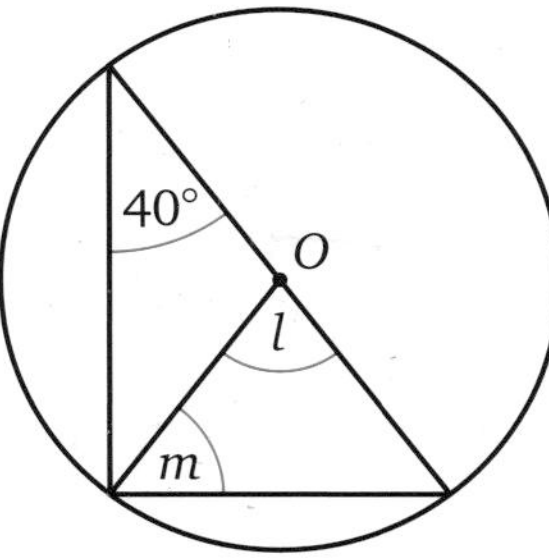

c

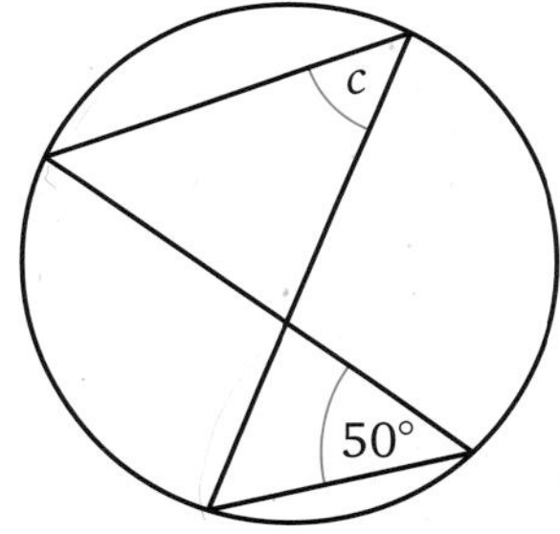

h

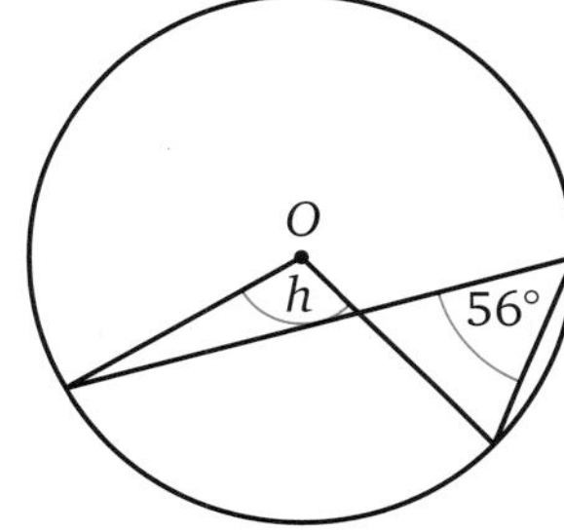

m

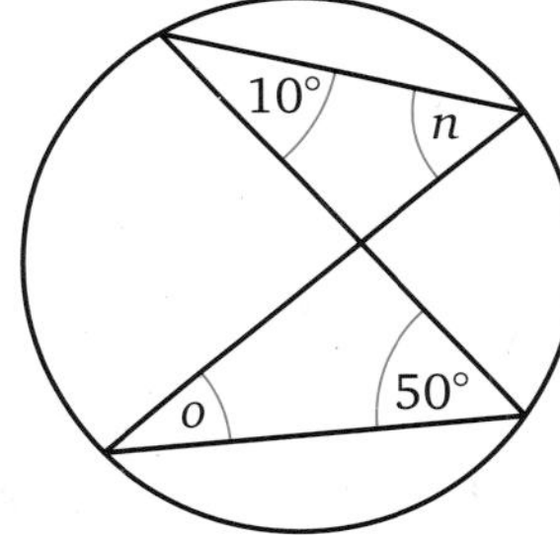

d

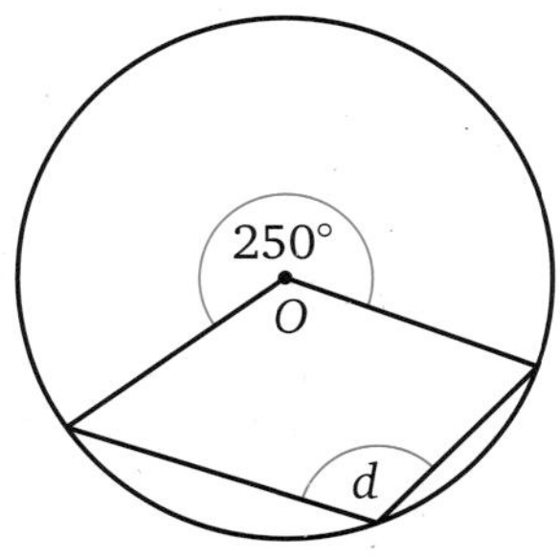

i

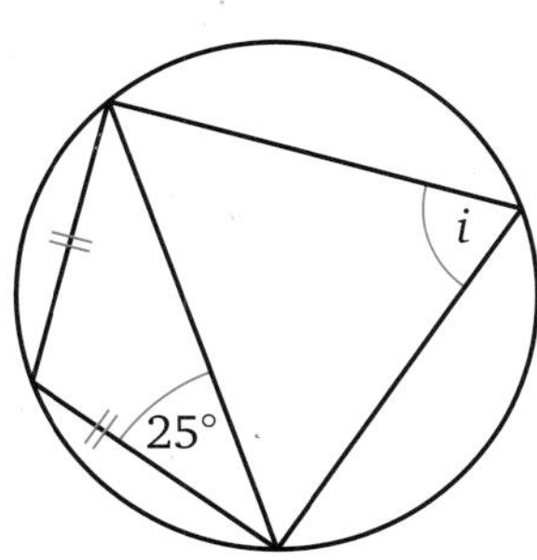

n

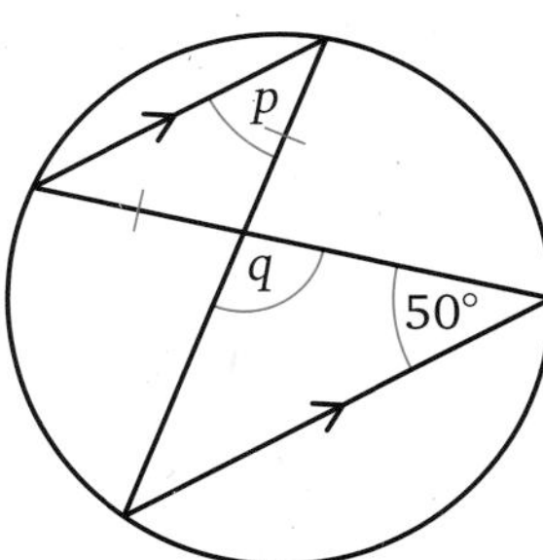

e

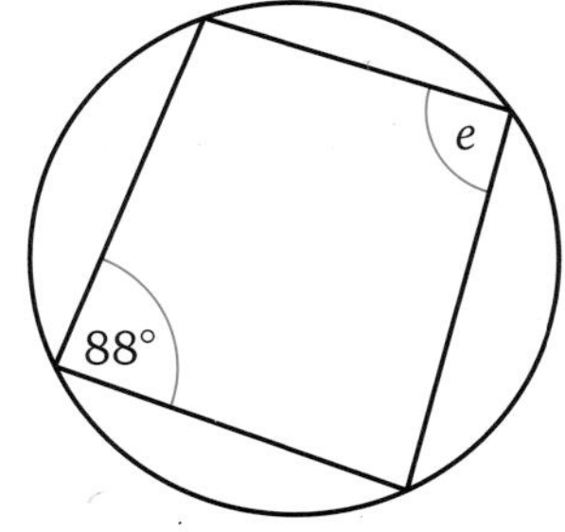

j

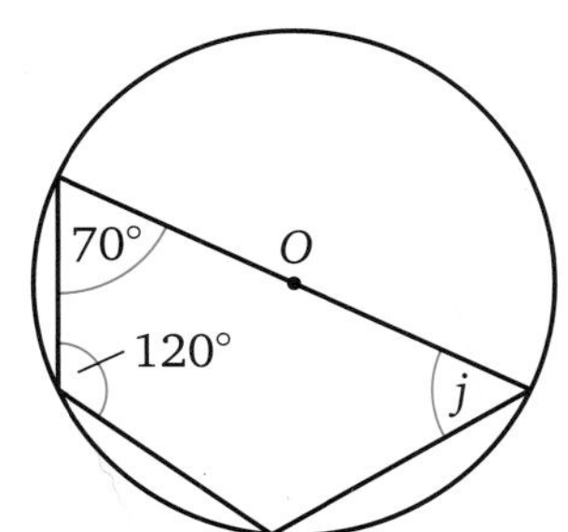

o

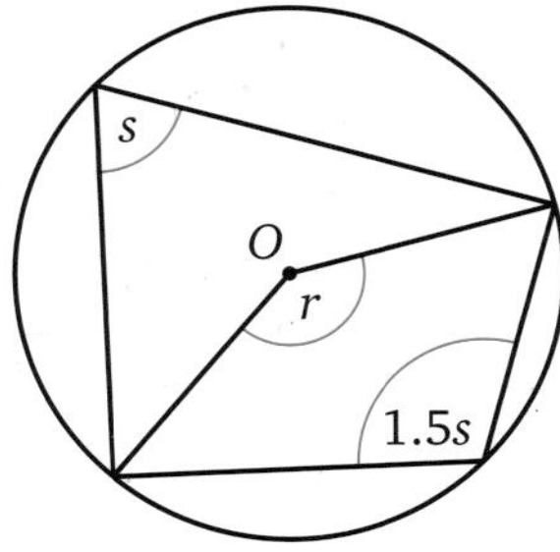

B

2 Write down whether the following are True or False.

a Angles subtended by the same arc of a circle at points on the circumference are equal.

b The angle subtended by a chord at the centre of a circle is twice the angle subtended at the circumference.

c The opposite pairs of angles of a quadrilateral both add up to 180° (they are supplementary).

d The angle subtended at the circumference by the diameter of a circle is a right angle.

e Adjacent angles in a cyclic quadrilateral are both complementary (i.e. they both add up to 90°).

f The angle subtended by a chord at the centre of a circle is half the angle subtended by the same chord at the circumference.

3 Sarah says that a trapezium cannot be a cyclic quadrilateral.

Is Sarah correct?
Give a reason for your answer.

4 *D* is the centre of the circle shown.

Assad thinks angle *ADC* is 65° because the angle at the centre is twice that subtended at the circumference.

Show Assad is incorrect by working out angle *ADC*.

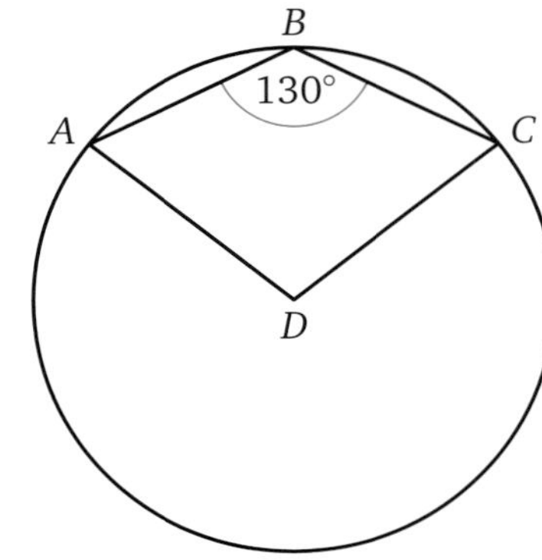

A

5 The lines *PR* and *QS* pass through the centre of the circle at *O*.

Angle *QRP* = 52°

Write down the angle *QSP* and angle *QOR*.
Give reasons for your answers.

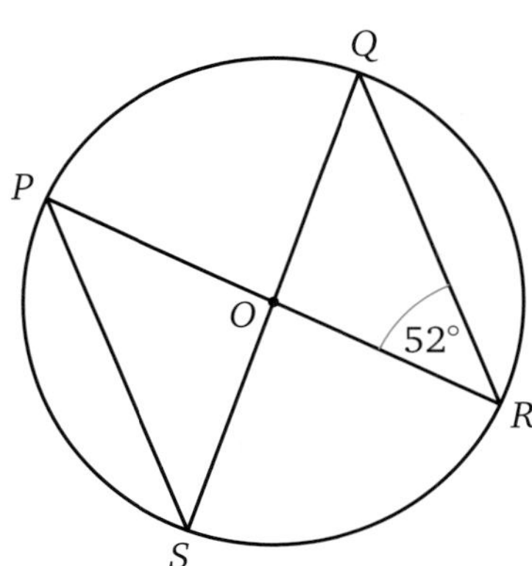

6 *LN* is the diameter of the circle.

Angle *NLM* = x° and angle MNL = $2x$°

Work out the value of x.

What is the size of angle *MNL*?

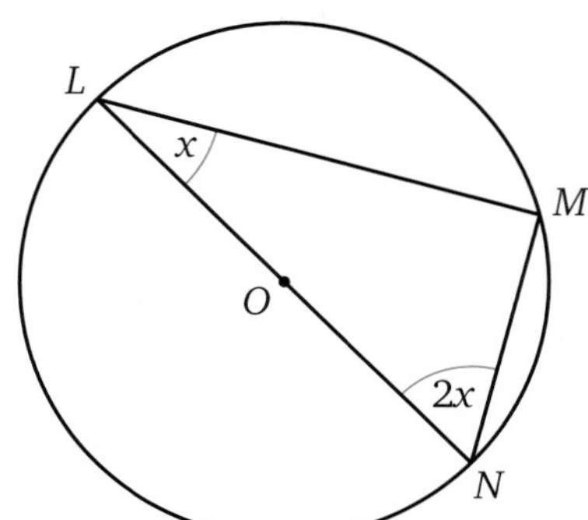

7 **a** Write down the value of x.

b Calculate the value of y.

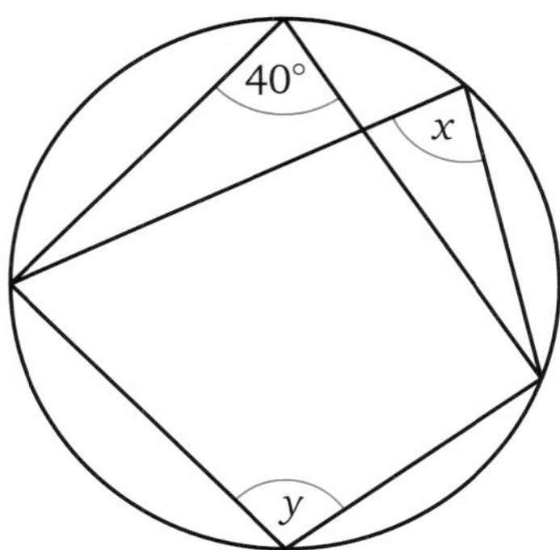

8 $PQ = PS$

Angle $QPS = 100°$

Show that angle $SRP = 40°$

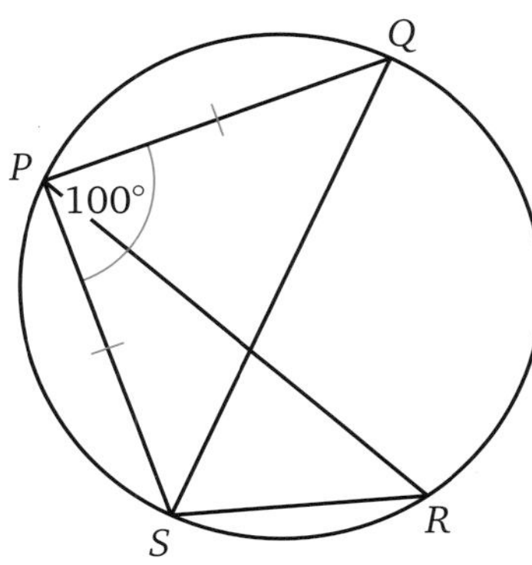

9 A quadrilateral $PQRO$ is drawn inside a circle centre O.

Angle $OPQ = 65°$ and angle $POR = 150°$

Calculate the other two angles of the quadrilateral. Show all your working.

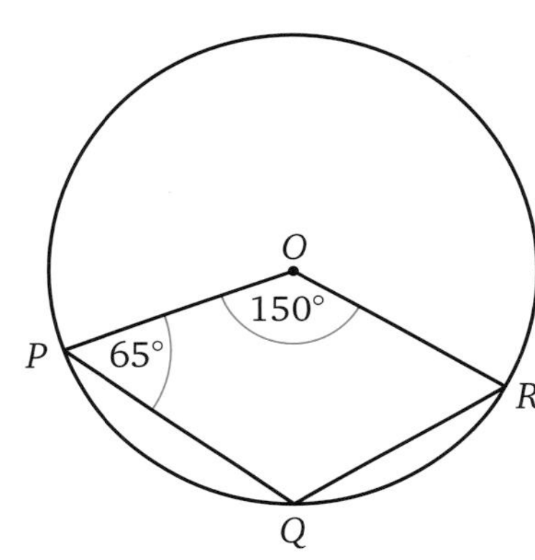

10 Prove the following circle properties.

a The angle subtended by an arc at the centre of a circle is twice the angle subtended at any point on the circumference.

b The angles subtended by the same arc (or chord) are equal.

c The angle subtended at the circumference by a semicircle is a right angle.

d Opposite angles of a cyclic quadrilateral add up to 180 degrees.

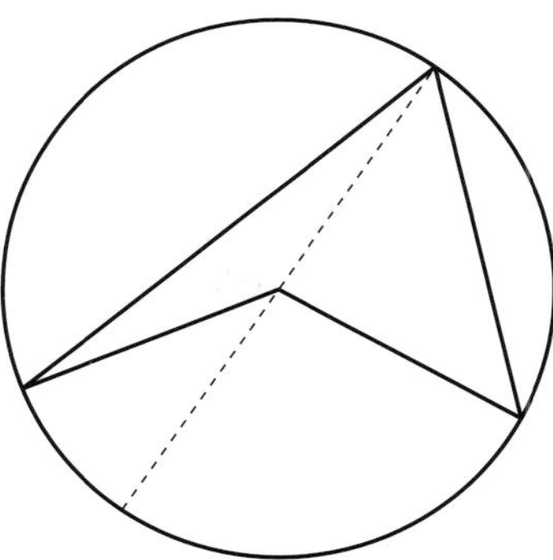

11 Work out angles a and b.

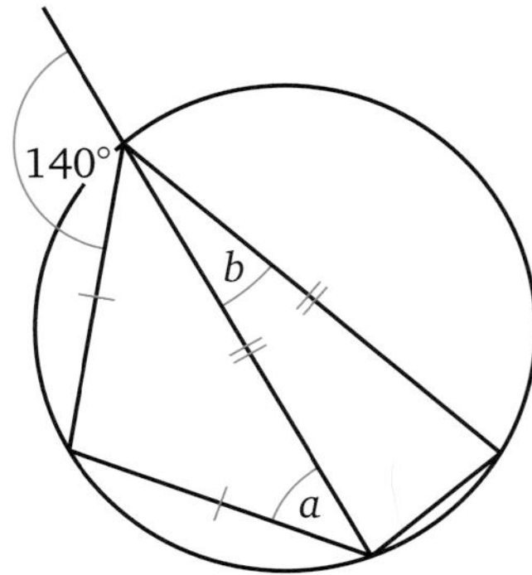

12 Work out angle x.

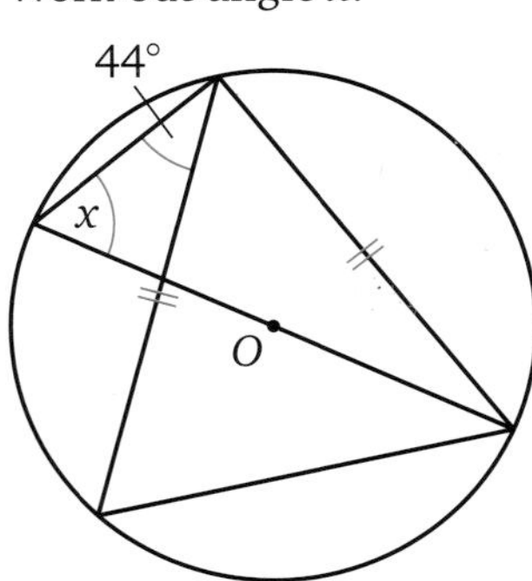

Learn... 11.2 Tangents and chords

Tangent

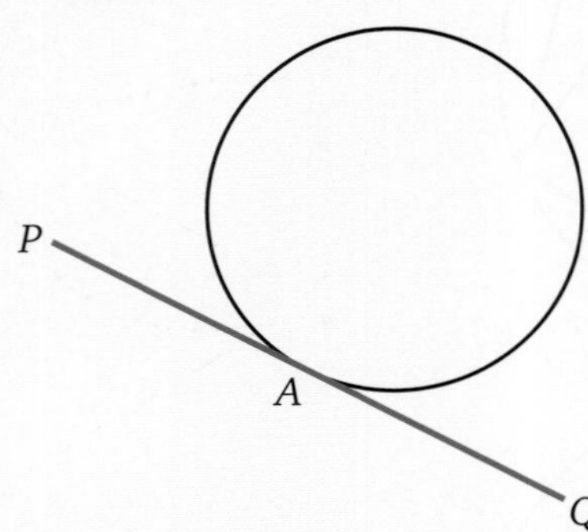

A **tangent** is a straight line that touches a circle at one point only.
PQ is a tangent to the circle.
It touches the circle at *A*.

Chord

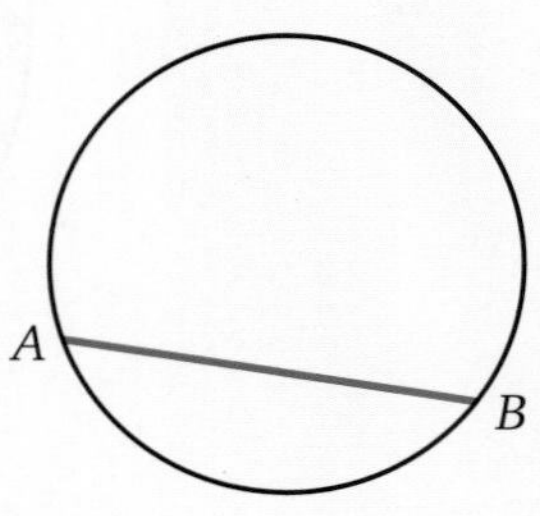

A chord is a straight line joining two points on the circumference of a circle. *AB* is a chord of the circle.
A diameter is a chord that passes through the centre of the circle.

Tangent properties

You need to know the following tangent properties.

- The tangent at any point on a circle is perpendicular to the radius at that point.
- Tangents from an external point are equal in length.

The tangent at any point on a circle is perpendicular to the radius at that point.

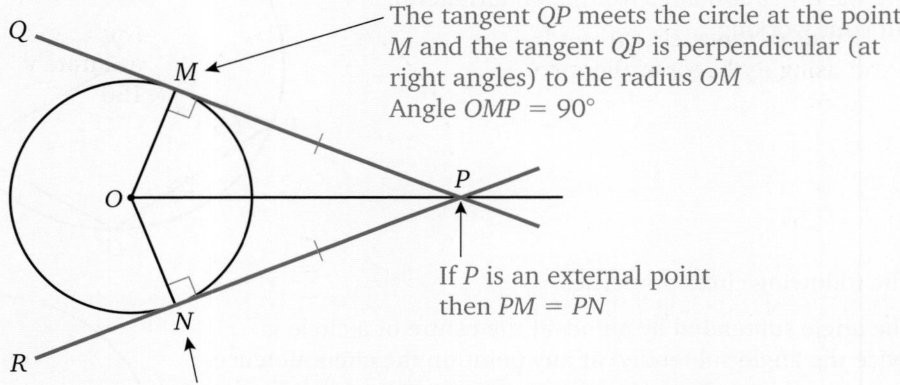

Q
M
O
P
N
R
The tangent QP meets the circle at the point M and the tangent QP is perpendicular (at right angles) to the radius OM
Angle OMP = 90°
If P is an external point then PM = PN
Also, the tangent RP meets the circle at the point N and the tangent RP is at right angles to the radius ON
Angle ONP = 90°

Tangents from an external point are equal in length.

So $PM = PN$ where *P* is the external point.

Chord properties

You need to know the following chord properties.

- The perpendicular line from the centre of the circle to a chord bisects the chord.
- The perpendicular bisector of any chord passes through the centre.

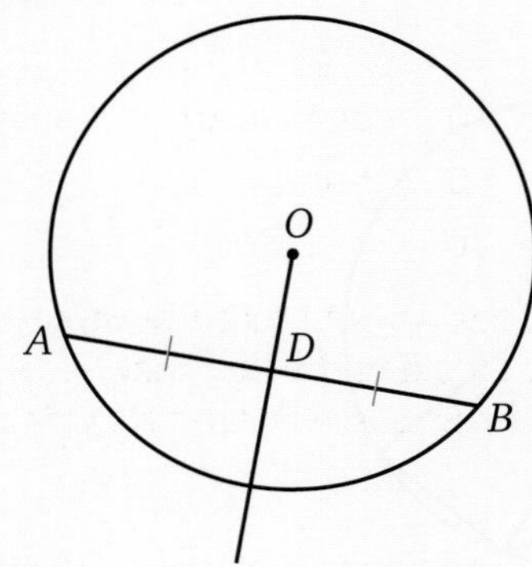

Alternate segment theorem

The **alternate segment** theorem says that the angle between the tangent and the chord is the same as the angle in the alternate segment.

The tangent PQ touches the circle at A. The angle between the tangent and the chord AC is the same as the angle subtended by the chord AC.

angle PAC = angle CBA

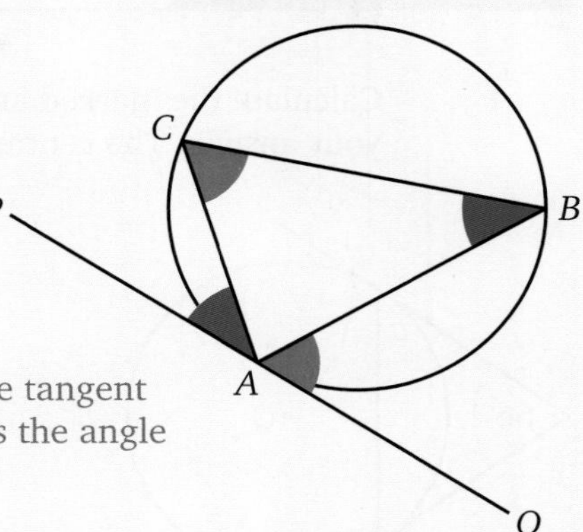

Similarly, the angle between the tangent and the chord AB is the same as the angle subtended by the chord AB.

angle QAB = angle ACB

Example: A chord of length 16 cm is 6 cm from the centre of a circle.

Calculate the radius of the circle.

Solution: Drawing a diagram will help to answer the question.

The chord AB is 6 cm from the centre of a circle.

The line from the centre to the chord OP is perpendicular to the chord.

The perpendicular from the centre to a chord bisects the chord.

So $AP = PB = 8$ cm

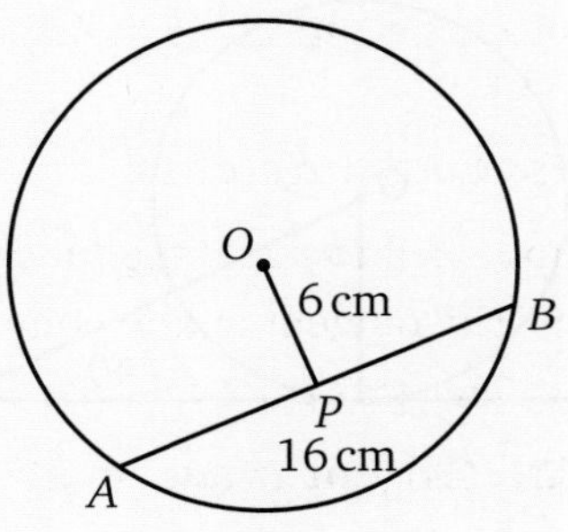

Not drawn accurately

The radius of the circle OA can be found using Pythagoras' theorem.

$OA^2 = OP^2 + AP^2$

$OA^2 = 6^2 + 8^2$

$OA^2 = 36 + 64$

$OA = \sqrt{100}$

$OA = 10$ cm

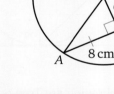

The radius of the circle is 10 cm.

Example: PQ is a tangent to the circle centre O.

$\angle CAP = 60°$

Calculate the angle ABC, angle AOC and the angle OAC.

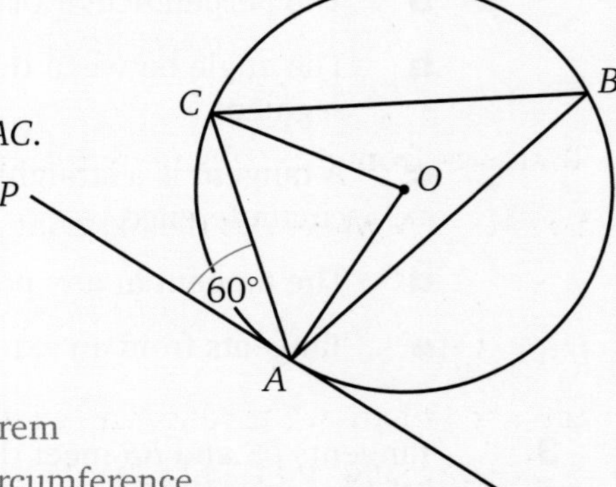

Solution:

angle $ABC = 60°$ using the alternate segment theorem

angle $AOC = 120°$ angle at centre $= 2 \times$ angle at circumference

angle $OAC = 90° -$ angle CAP Tangent is perpendicular to the radius.

angle $OAC = 90° - 60°$

angle $OAC = 30°$

Note that there are many other ways to find the value of angle OAC. For example, by using the fact that angle OAC is a base angle of the isosceles triangle OAC.

AQA Examiner's tip

Always remember to check that your answers are sensible.

Practise... 11.2 Tangents and chords 

B

1 Calculate the marked angle on the following diagrams, giving a reason for your answer. The centre of the circle is marked O. The diagrams are not drawn accurately.

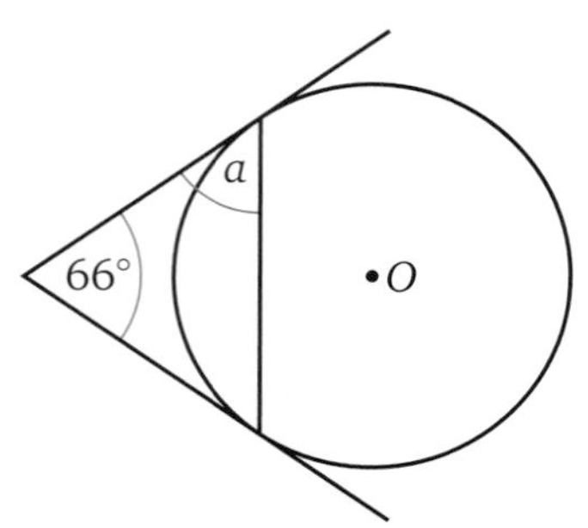

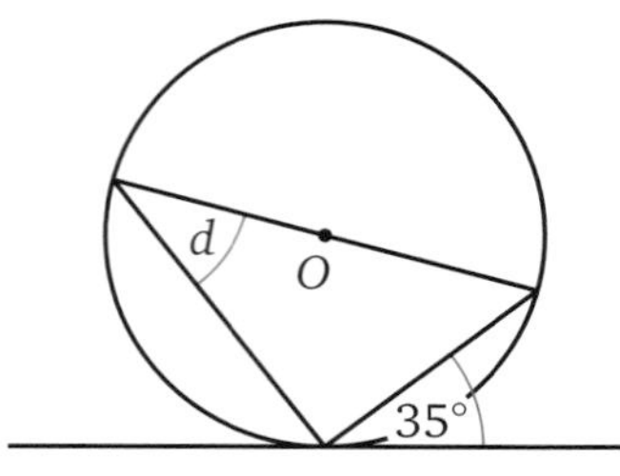

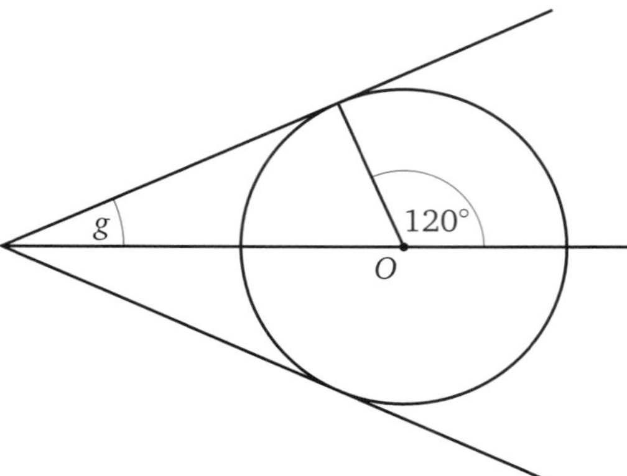

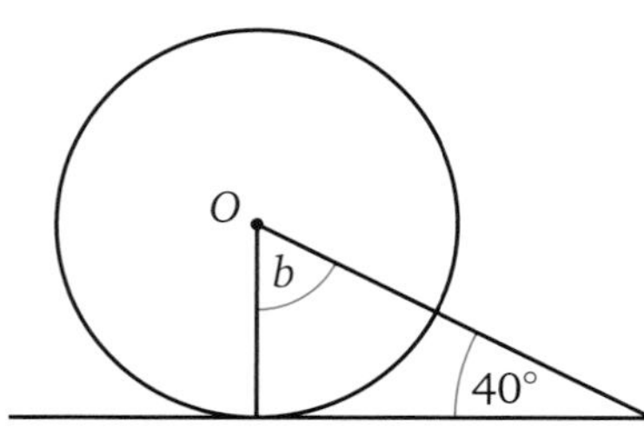

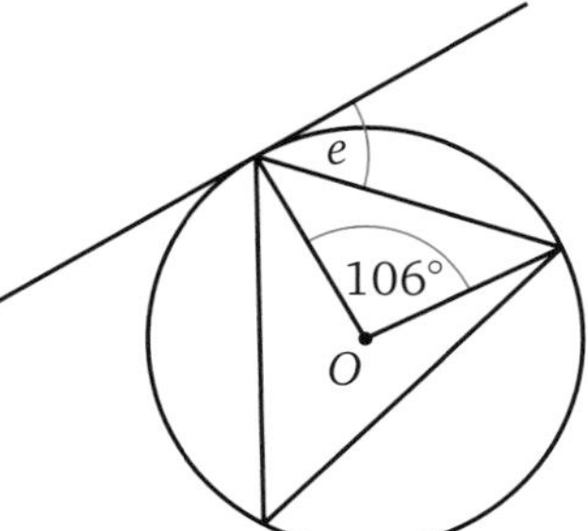

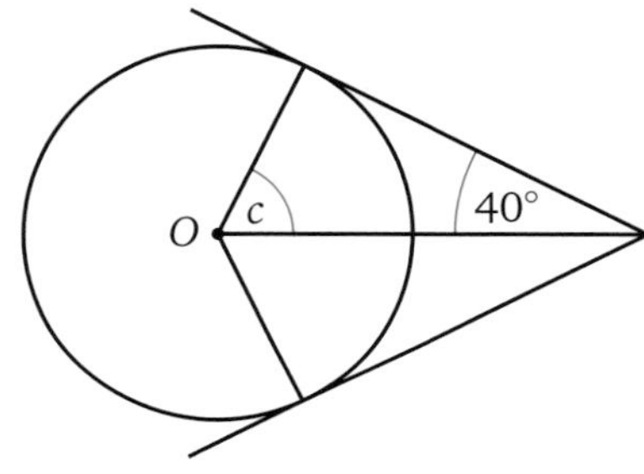

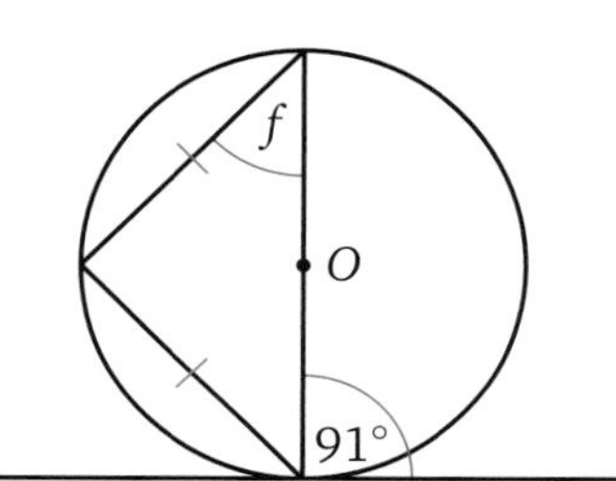

2 Write down whether the following are True or False.

- **a** The perpendicular bisector of a chord in a circle passes through the centre of the circle.
- **b** The angle between the tangent and the chord is the same as the angle in the alternate segment.
- **c** A tangent is a straight line that touches a circle at one or more points on the circumference.
- **d** The tangent at any point on a circle is perpendicular to the radius at that point.
- **e** Tangents from an external point to the circumference of a circle are equal in length.

3 Tangents PS and RS meet the circle at P and S respectively.
SOQ is a straight line passing through the centre of the circle O.
Angle $PSO = 22°$

Calculate:

- **a** angle RSO
- **b** angle POR
- **c** angle PQR.

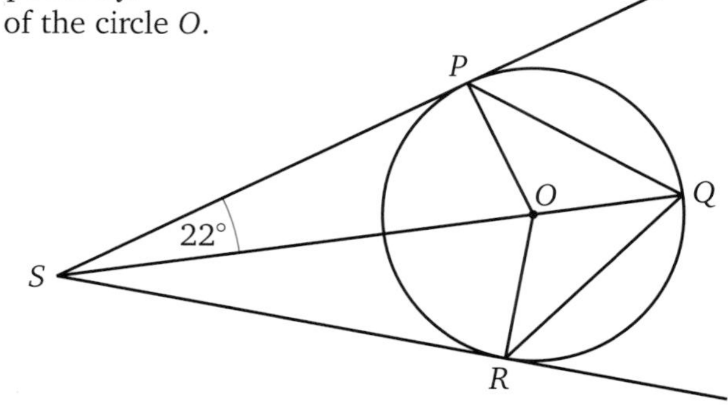

B

4 In the diagram below points Q and R lie on a circle centre O.
PQ is a tangent to the circle at Q. Angle $QPR = 40°$ and angle $QOR = 80°$

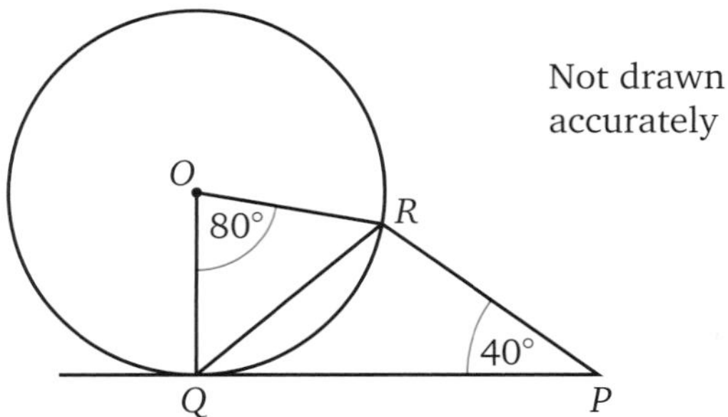

Prove that triangle QPR is isosceles.

5 A chord of length 24 cm is located 5 cm from the centre of a circle.

Calculate the radius of the circle.

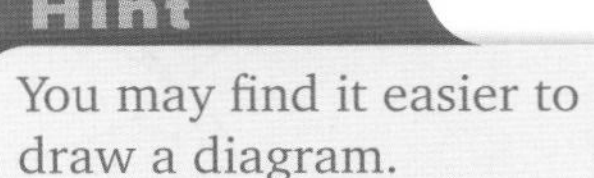

Hint
You may find it easier to draw a diagram.

6 A chord PQ is 12 cm from the centre of a circle of radius 15 cm.

What is the length of the chord?

A

7 XY is a tangent to the circle, touching at the point A.

$AB = BC$ and $CD = DA$

Angle $XAB = 48°$

Calculate:

a angle BCA

b angle ABC

c angle DAC.

Give reasons for your answers.

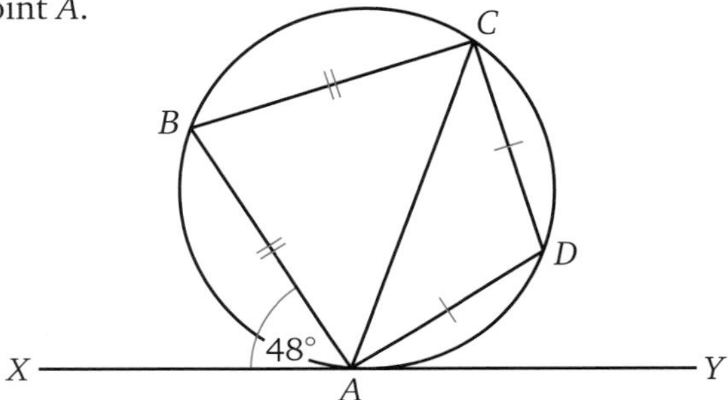

A*

8 PQ and QC are tangents.

The tangent PQ meets the circle at B.

The centre of the circle is O

Angle $PBA = 82°$ and angle $PQC = 40°$

Calculate the values of x and y.

Remember to show your working.

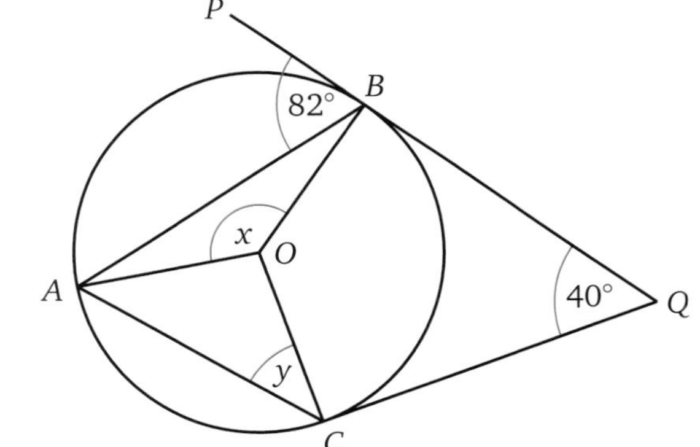

9 Tangents PR and PT meet the circle at Q and S respectively.

$WQ = WS$

Show that angle RQW = angle TSW

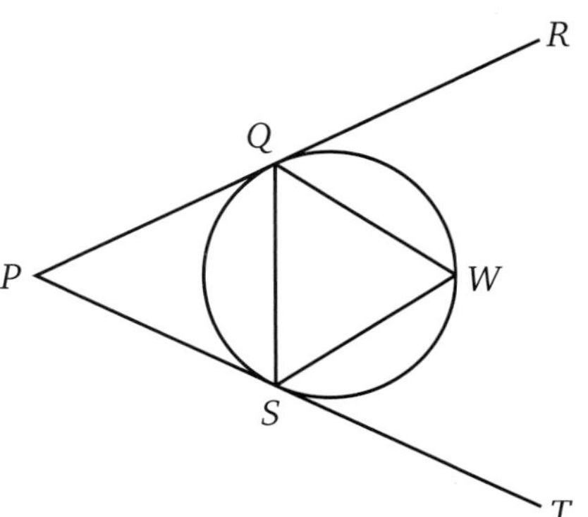

10 Prove that the angle between the tangent and the chord is the same as the angle in the alternate segment.

11 The perpendicular bisector of a chord passes through the centre of a circle.

Write down how you might use this fact to find the centre of any circle.

12 Work out the angle x, giving reasons for your answer.

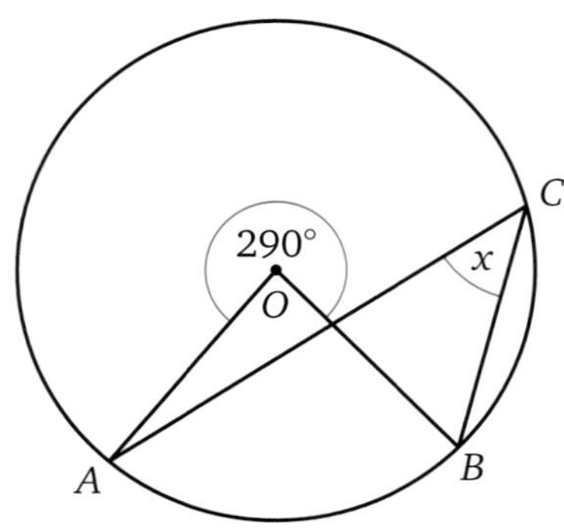

11 Assess

B

1 For each question part, calculate the marked angles.

The centre of the circle is labelled O.

a

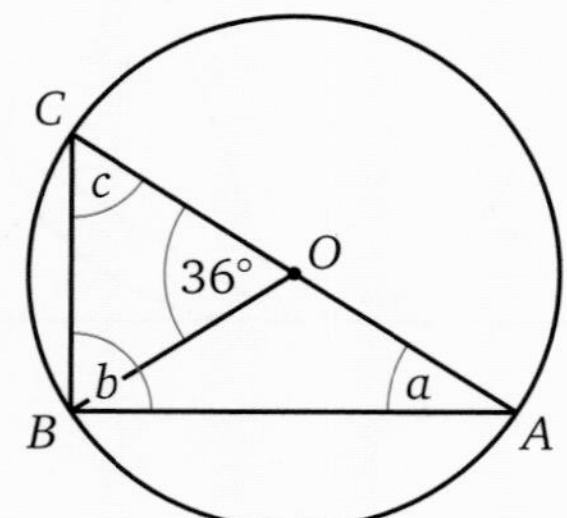

b

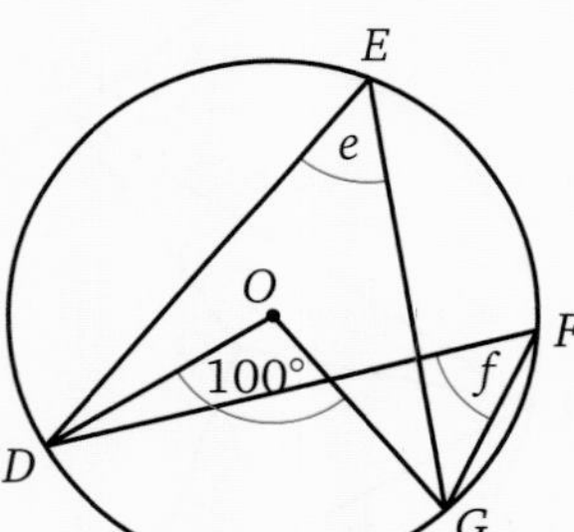

c

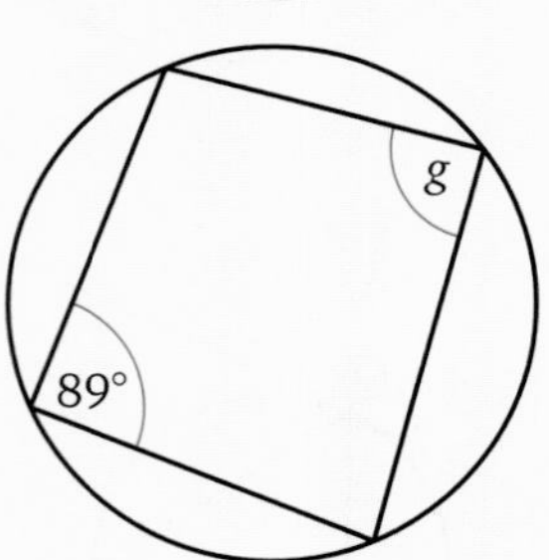

d

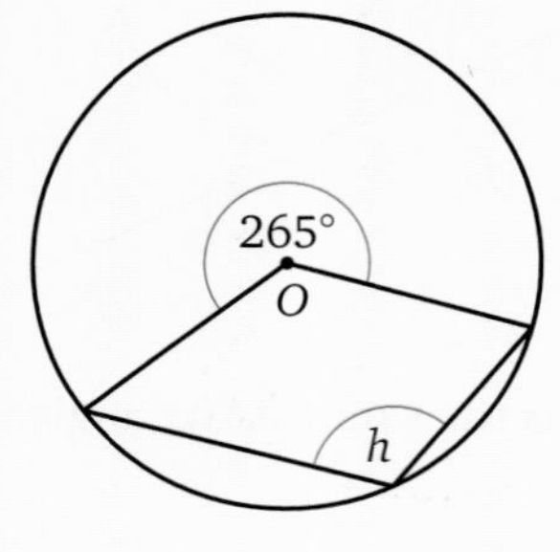

e

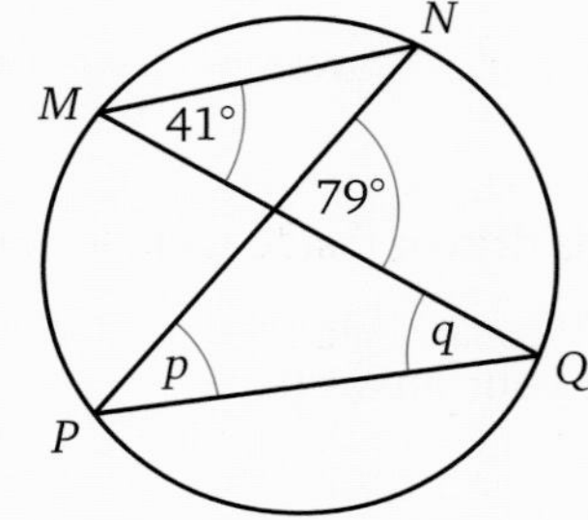

f

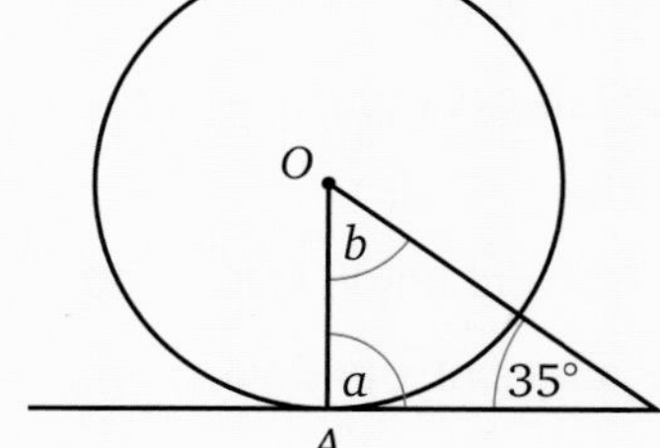

Not drawn accurately

g

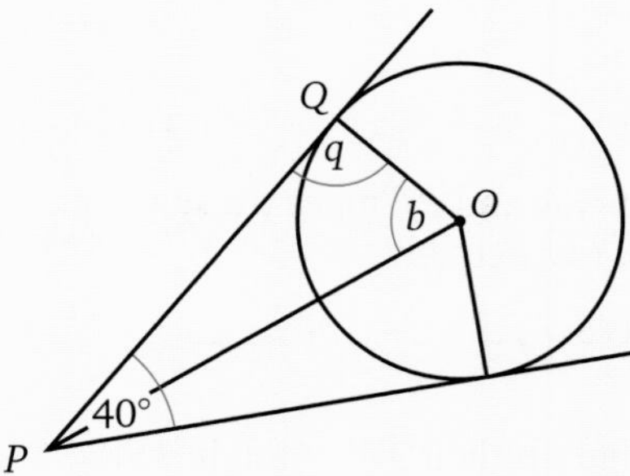

h

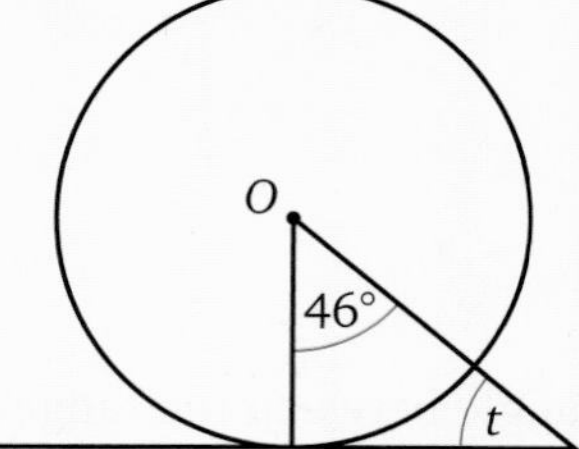

B

2 AD is a diameter of the circle centre O.
Angle $ADC = 60°$ and angle $ACB = 35°$

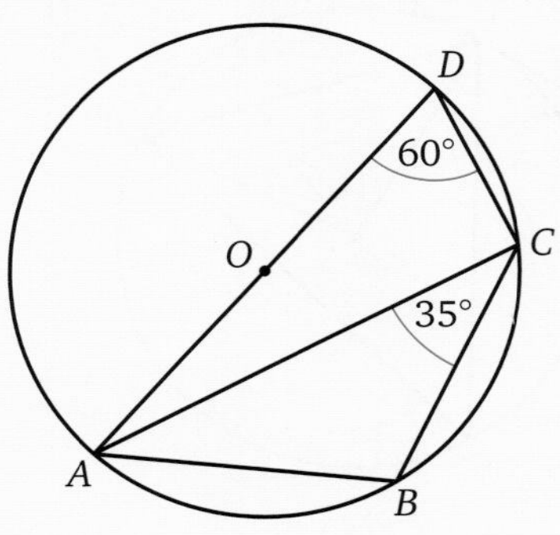

Not drawn accurately

Calculate:

a angle DCA

b angle DAC

c angle ABC

d angle DAB.

3 A chord AB is drawn on a circle centre O.
The length of the chord is 16 cm.
The radius of the circle is 10 cm.

Calculate the area of the triangle AOB.
State the units of your answer.

4 A quadrilateral is drawn inside a circle centre O.

Angle $OPQ = 55°$ and angle $POR = 150°$.

Calculate the other angles of the quadrilateral.
Remember to show all your working.

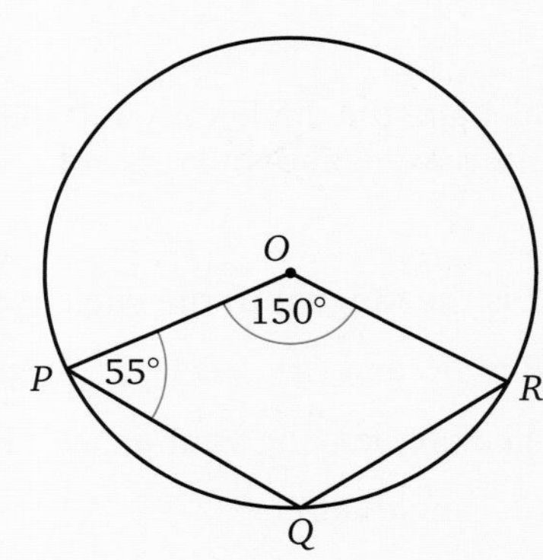

Not drawn accurately

A

5 For each question part, calculate the marked angles.

a

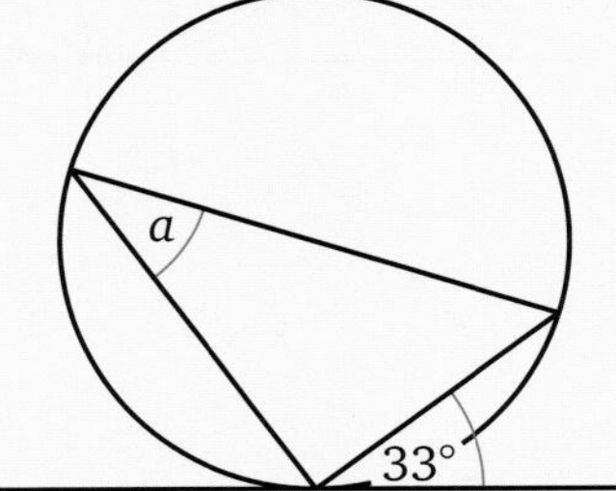

b

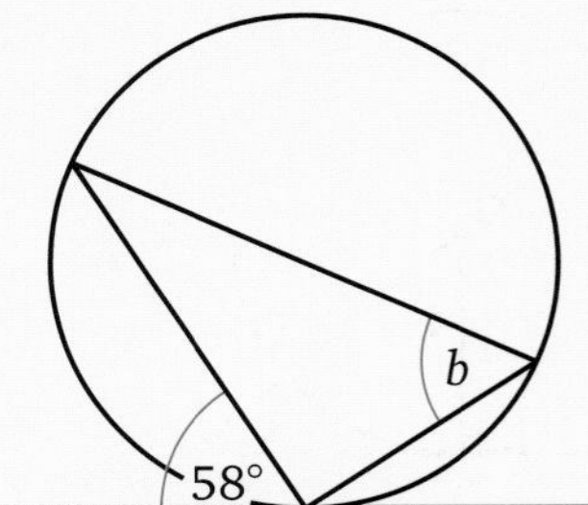

c

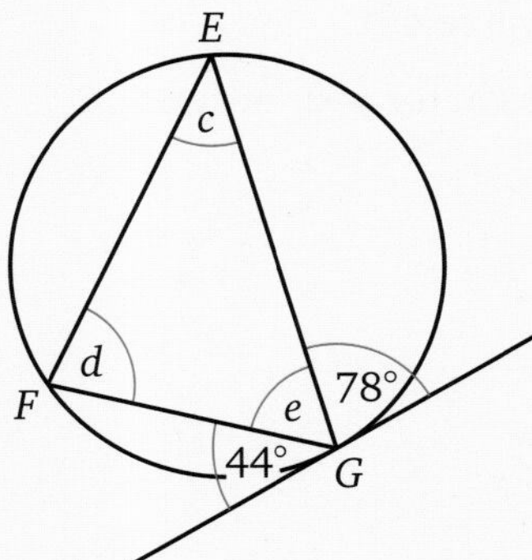

d

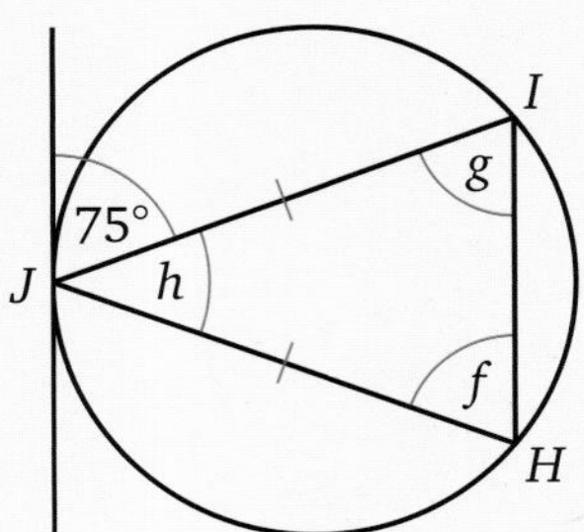

A

6 A, B and C are three points on a circle.

PQ is a tangent to the circle, which touches the circle at A.

The angle $AOB = 102°$

Calculate the angle QAB.

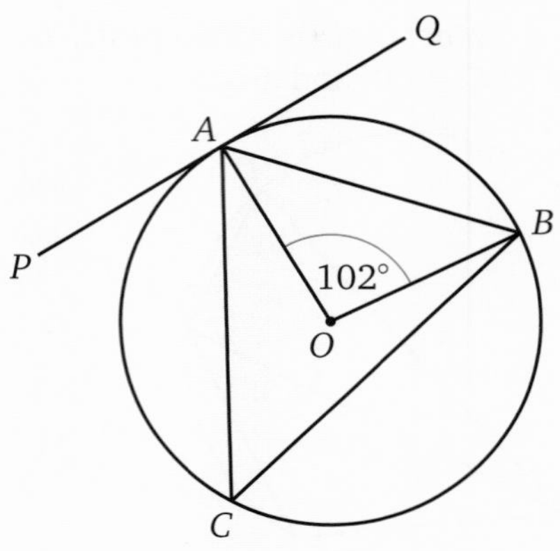

7 PA is a tangent to the circle centre O.
PB is a straight line passing through O.
Angle $PAC = 35°$

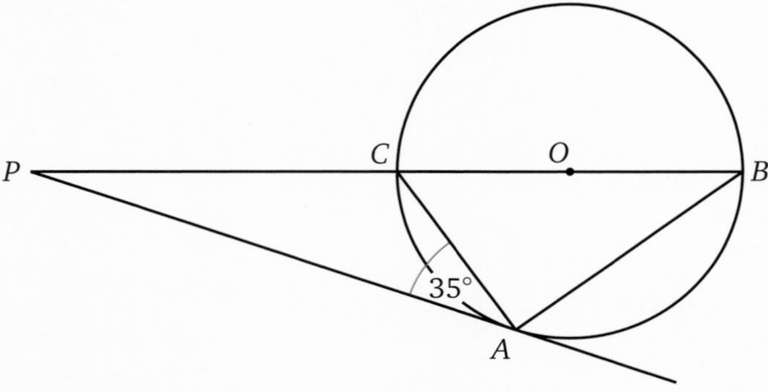

Not drawn accurately

Calculate the angle CBA and the angle CPA.
Give reasons for your answers.

A*

8 XY is the tangent to the circle touching at the point A.

$AB = BC$ and $CD = DA$ Angle $XAB = x$

Calculate, leaving your answer in terms of x:

a angle BCA

b angle ABC

c The angle $ADC = 102°$

Find the value of x.

Give a reason for your answer.

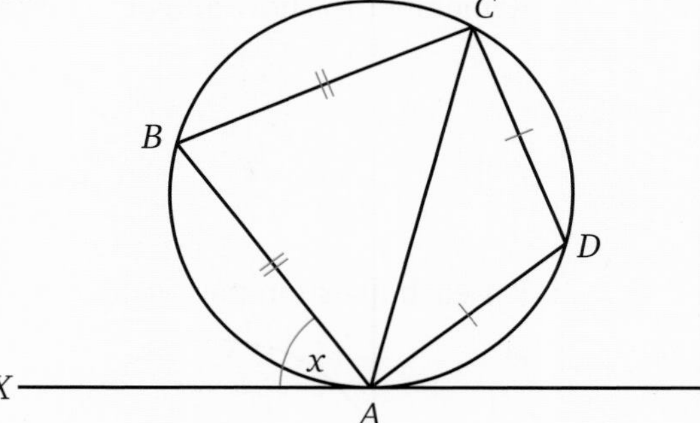

AQA Examination-style questions

1 **a** P, Q and R are points on the circumference of a circle, centre O.
PR is a diameter of the circle.

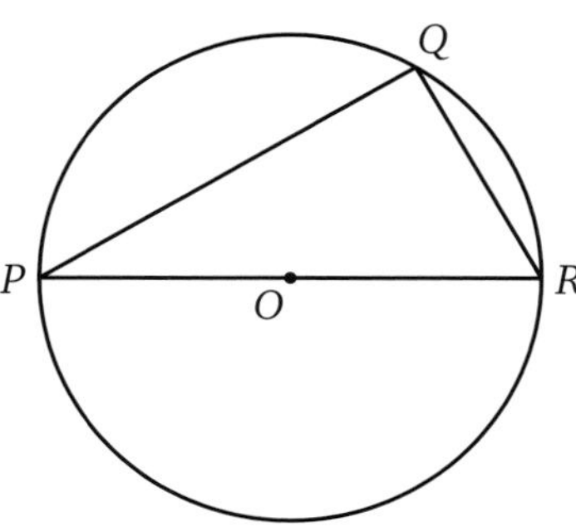

Not drawn accurately

Write down the size of angle PQR. *(1 mark)*

b T is also a point on the circumference of the circle in part **a**.
Angle $QTR = 27°$

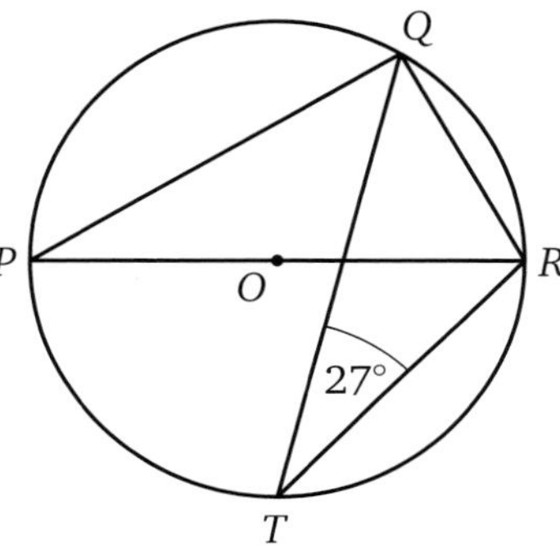

Not drawn accurately

i Write down the size of angle RPQ. *(1 mark)*

ii Work out the size of angle PRQ. *(1 mark)*

c S is another point on the circumference of the circle in part **a**.
QS is a diameter of the circle.
Angle $PRS = 38°$

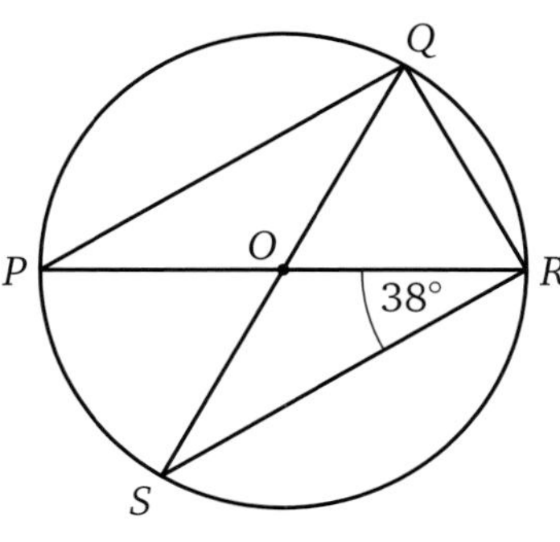

Not drawn accurately

Work out the size of angle SQR. *(1 mark)*

AQA 2008

12 Measures

Objectives

Examiners would normally expect students who get these grades to be able to:

D

calculate average speed

C

use compound measures such as speed

recognise that measurements to the nearest unit may be inaccurate by up to one half unit in either direction

B

use compound measures, such as density, for example find density given cross-sectional area, length and mass.

Did you know?

Archimedes

Part of Archimedes' fame comes from solving King Hiero's 'crown problem'. Legend has it that Hiero gave the goldsmith a quantity of gold to make a crown. The finished crown was the correct weight but Hiero suspected some silver had been substituted. The problem was brought to the attention of Archimedes.

Upon entering his bathtub, Archimedes noted that some water overflowed. This gave him an idea. A given weight of pure gold displaces a certain volume of water. The same weight of impure gold will displace a different volume of water. He could use this fact to tell whether the crown was made of impure gold. This proved bad news for the deceptive goldsmith.

Key terms

lower bound (or limit)
upper bound (or limit)
compound measures
density
mass

You should already know:

✔ how to convert between different units.

Learn... 12.1 Accuracy of measurement

When you measure to the nearest centimetre, any actual measurement, between 6.5 cm and up to but not including 7.5 cm, will round to 7 cm.

6 6.5 7 7.5 8

Any measurement in the shaded area of this scale rounds to 7.

The shaded area is from 6.5 up to 7.5

6.5 is the **lower bound**

7.5 is the **upper bound**

Notice: 'to the nearest centimetre' means that the actual distance could be any value from half a centimetre below to half a centimetre above.

Example: Zac measures the length of a shelf. He says it is 43 cm to the nearest cm.

What are the upper and lower bounds of Zac's measurement?

Solution: Zac's measurement rounds to 43 cm, so the actual length must be in the shaded area on the scale.

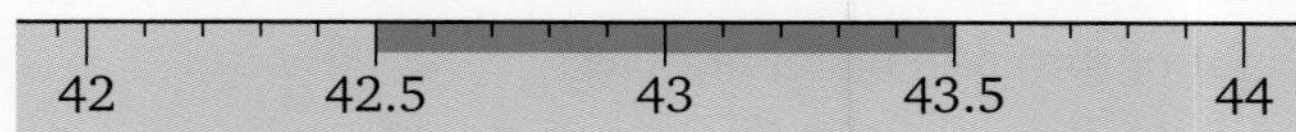

The shaded area is from 42.5 up to 43.5

The lower bound is 42.5 cm.

The upper bound is 43.5 cm.

Practise... 12.1 Accuracy of measurement 

D C B A A*

C

1 Round each of the following to the nearest unit.

- **a** 12.9 cm
- **b** 81.2 m
- **c** 49.9 miles
- **d** 6.5 g
- **e** 43.56 kg
- **f** 47.4999 litres

2 When rounded to the nearest centimetre each of these lengths rounds to 15 cm.

14.6 cm 15.4 cm 14.91 cm 15.49 cm

- **a** Write down some other lengths that also round to 15 cm to the nearest centimetre.
- **b** What is the minimum value that rounds up to 15 cm?
- **c** What is the maximum value that rounds down to 15 cm?

3 George measured the length of a post for a washing line in his garden.
He found it was 3 metres to the nearest metre.

What are the upper and lower bounds for his measurement?

4 Faye measured some of the crayons in her pencil case.
She found they were all 8 cm to the nearest centimetre.

Does this mean they were all the same length? Explain your answer.

C

5 Beshar measured the distance from his home in Windermere to his school in Kendal. He found it was 18 km to the nearest kilometre.

a What is the smallest distance that Beshar's measurement could be?

b What is the largest distance that Beshar's measurement could be?

6 The weight of a letter is 43 g to the nearest gram.

What are the maximum and minimum values for the weight of the letter?

7 A bridge has a sign stating the maximum weight allowed on it is 2 tonnes. A van driver knows his van weighs 2 tonnes to the nearest tonne.

Can the van driver be sure that it is safe for him to drive over the bridge? Give a reason for your answer.

8 The contents of a packet of crisps weigh 33 g to the nearest gram.

Which of the following could be the weight of the crisps?

a 33.2 g
c 33.49 g
e 32.5 g
b 33.6 g
d 33.94 g
f 32.29 g

B

9 Rachel weighs the books in her school bag.
All of her textbooks together weigh 3 kg to the nearest kg.
All of her exercise books together weigh 1 kg to the nearest kg.

What is the maximum weight of the contents of her bag?

10 George, Mildred, Henrietta and Dan all get into a lift.
To the nearest 1 kg, George weighs 80 kg, Mildred weighs 70 kg and Henrietta weighs 60 kg.
The lift states: Maximum weight 250 kg.

What is Dan's maximum possible weight if it is to be safe to use the lift?

Learn... 12.2 Compound measures

Compound measures combine two different units.

For example, density $= \frac{\text{mass}}{\text{volume}}$ and average speed $= \frac{\text{total distance}}{\text{total time}}$, population density $= \frac{\text{number of people}}{\text{area of land}}$,

fuel consumption $= \frac{\text{number of miles}}{\text{number of gallons used}}$

The units give you clues about what to divide by.

Density is **mass** per unit volume, kg/m^3 if the mass is in kg and the volume is in m^3.

The units tell you the formula, density = mass ÷ volume.

Speed is measured in km/hour. This tells you the formula, speed = distance ÷ time.

Population density is measured in people per square kilometre.

Fuel consumption is measured in miles per gallon or kilometres per litre.

AQA Examiner's tip

Make sure that you have the correct units for the problem. You may need to convert some units before you use them.

For km/h you need the distance to be in km and the time to be in hours.

For g/cm^3 you need the mass to be in grams and the volume to be in cm^3.

Example: Jamie is a runner in a club. He runs 200 metres in 32 seconds. Find his speed in:

i metres per second

ii kilometres per hour.

Solution: **i** Using the formula above:

average speed $= \frac{\text{distance}}{\text{time}} = \frac{200}{32}$ m/s = 6.25 m/s

ii 6.25 m/s means 6.25 metres every second.

In 60 seconds he runs 60 × 6.25 = 375 metres

In 60 minutes he runs 60 × 375 = 22 500 metres

To convert metres to kilometres, we need to divide by 1000 because there are 1000 metres in a kilometre, so:

22 500 metres = 22 500 ÷ 1000 = 22.5 km

Jamie runs at a speed of 22.5 km/h.

Example: **a** Work out the average speed in mph of a train that takes $1\frac{1}{2}$ hours to travel 102 miles.

b Nazneen cycled at a steady speed of 8 km/h.
How far did she travel in three and a half hours?

Solution: **a** average speed $= \frac{\text{distance}}{\text{time}} = \frac{102}{1.5}$ = 68 mph

b distance = average speed × time

distance = 8 × 3.5 = 28 km

AQA Examiner's tip

If time is in hours and minutes change it to hours and decimals of an hour before finding average speed. The unit for average speed is now mph or km/h.

Example: The density of lead is 11.4 g/cm^3.
A block of lead is 2 cm wide, 3 cm high and 6 cm long.

What is the mass of the lead in kilograms?

Solution: The volume of the block is 2 × 3 × 6 = 36 cm^3

Density is mass per unit volume.

density $= \frac{\text{mass}}{\text{volume}}$

$11.4 = \frac{\text{mass}}{36}$

11.4 × 36 = 410.4 = mass in grams

The block has a mass of 0.41 kg.

AQA Examiner's tip

You will get decimals and fractions as answers when you work with compound measures. When your answer is a recurring decimal, it is important to make sure that the decimal is rounded correctly.

Remember that $12\frac{2}{3} = 12.\dot{6} = 12.67$ (2 d.p.) and not 12.66

Practise... 12.2 Compound measures 

D

1 Work out the average speed for each of the following.
State the units of your answer.

- **a** A car that travels 200 metres in 8 seconds
- **b** A man that takes 28 seconds to run 200 metres
- **c** A train that takes 2 hours to travel 230 miles

2 Write each of these times using decimals of an hour.

a 30 minutes **b** 15 minutes **c** 4 hours 45 minutes

3 Write each of these times using hours and minutes.

a 2.5 hours **b** 3.25 hours **c** 1.75 hours

C

4 Find the speed in mph of:

- **a** a car that travels 85 miles in 1 hour and a quarter
- **b** a lorry that travels 75 miles in 1 hour 25 minutes.

5 A snail crawls at a speed of 5 cm every minute.

- **a** How far does it crawl in one hour?
- **b** How long does it take to crawl one metre?

6 Work out the time taken for each of these journeys. Give your answer in hours and minutes.

- **a** A car travels at 50 km per hour for 40 km.
- **b** A bus travels at 30 km per hour for 20 km.
- **c** A cyclist travels 45 km at 25 km per hour.

7 Work out the distance travelled for each of these journeys.

- **a** A person walks at 4 km per hour for 75 minutes.
- **b** A train travels at 110 km per hour for 90 minutes
- **c** A lorry travels for 45 minutes at 50 km per hour.

8 Jan drives her car 255 miles and uses 6 gallons of fuel.

- **a** What is her fuel consumption in miles per gallon?
- **b** How many gallons of fuel does Jan use for a journey of 400 miles?
- **c** The fuel tank in Jan's car contains 15 gallons of fuel when it is full.

Is it possible for Jan to travel 600 miles on one full tank of fuel?

9 An island has an area of 680 km^2 and a population of 3100 people.

What is the population density of the island?

B

10 A rock has a volume of 35 cm^3 and a mass of 266 g.

What is the density of the rock?

11 The density of gold is 19.3 g/cm^3. A gold bar has a volume of 5 cm^3.

What is the mass of the gold?

B

12 Four blocks of metal are shown below.
Each length has been recorded to the nearest mm.
Each weight has been recorded to 1 d.p.
Each density has been given to 1 d.p.

Use the information given in the table and on the diagrams to identify what each block is made from. Show working to justify your answers.

Metal	Density (g/cm³)
Copper	9.0
Iron	7.8
Lead	11.0
Gold	19.3

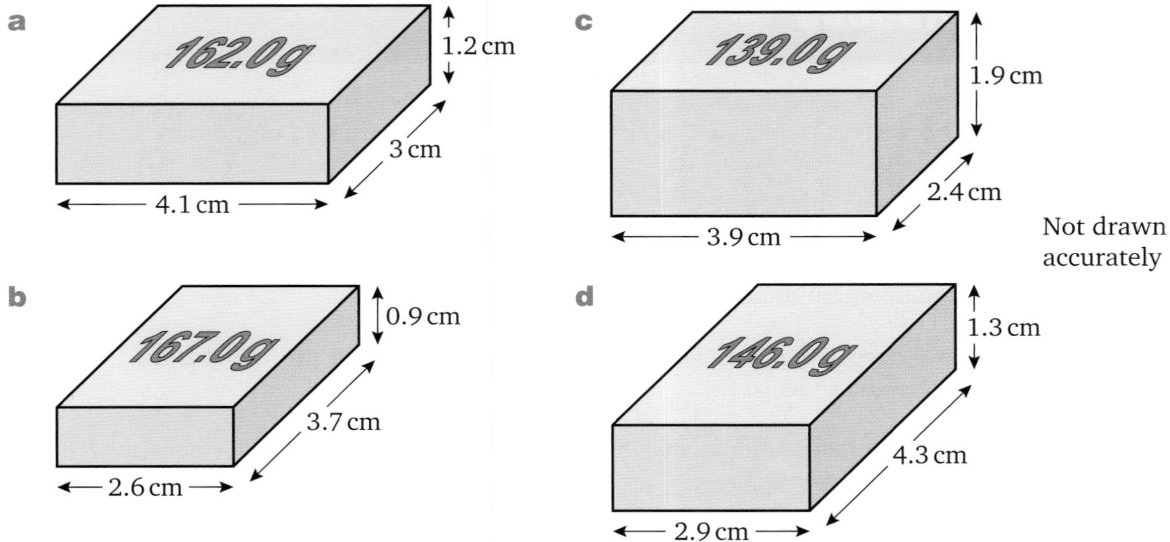

AQA Examiner's tip

When you have a lot of steps to do in a calculation, take them one at a time. When you think you have finished answering a question, remember to check that you have completed your solution. Many candidates lose marks for stopping part-way through and not checking their work.

A

13 The diagram shows a prism with a trapezium as its cross-section.

- **a** Find the volume of the prism.
- **b** The prism is made of gold. Gold has a density of 19.3 g/cm³. Find the mass of the gold.
- **c** Gold is worth £38.90 per gram. Find the value of the gold.

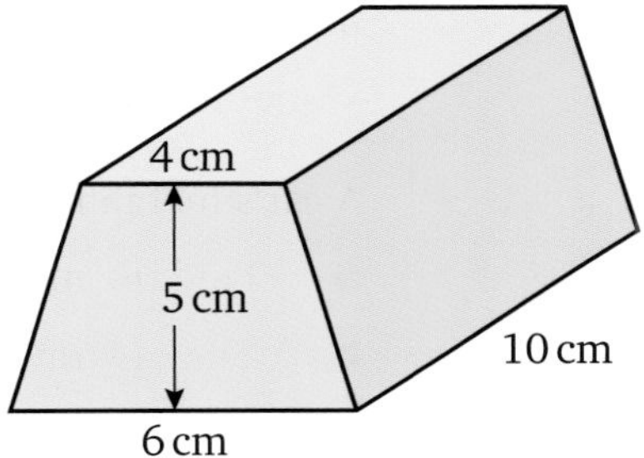

14 Mary has a solid ball which has a mass of 540 g.
The ball is made from material which has a density of 6 g/cm³.
Mary's little brother drops the ball in their fish tank.
The fish tank is in the shape of a cuboid 30 cm by 20 cm and 50 cm deep.

By how much does the depth of water in the fish tank rise when the ball is dropped in the tank?

15 John runs at 9 km per hour for 40 minutes. He then walks 2.5 km in 30 minutes.

What is John's average speed in km per hour?

16 Sam drives 265 miles from Kendal to Bristol at an average of 50 miles per gallon (mpg).
He then drives 221 miles from Bristol to Norwich at an average of 48 mpg.
He drives from Norwich back to Kendal, a distance of 278 miles, at an average of 47 mpg.
His fuel tank contains 15 gallons of fuel when full.

- **a** Is it possible for Sam to complete this journey on one tank of petrol?
- **b** What if all the quantities of petrol found in part **a** had been given to the nearest whole number?

? **17** Mary jogs 2 miles at 5 mph, then runs 2 miles at 7 mph.

- **a** Explain why her average speed is not 6 mph.
- **b** Mary wants do training runs of about 4 miles jogging at 5 mph and running at 7 mph. How far does Mary go at each speed in order to have an average speed of 6 mph?

12 Assess

D

1 A van travels 230 miles in 7 hours. What is the average speed of the van?

C

2 A car takes 45 minutes to travel 17 miles. What is the average speed of the car?

3 The temperature in a classroom is 18 °C to the nearest degree. What are the maximum and minimum temperatures?

B

4 A bar of chocolate has a mass of 250 g and a volume of 60 cm^3. What is the density of the chocolate?

5 A log has a cross-sectional area of 70 cm^2 and a length of 25 cm. The density of the log is 0.63 g/cm^3.

What is the mass of the log?

6 Harry is replacing the carpet in his rectangular front room. It is 4.5 m by 3.6 m. He pays £526.50 for the new carpet.

What is the cost per square metre of his new carpet?

7 Hamish drives 23 miles to the nearest mile in his car. The car's computer says it has used 0.4 gallons of fuel.

Work out the fuel consumption in mpg of Hamish's car.

B A

8 A car's fuel tank holds 60 litres of fuel when full. The density of the fuel is 925 kg/m^3.

- **a** Find the mass of the fuel in a full tank.
- **b** The car travels 375 miles at an average of 45 mph and 48 mpg.
  - **i** How long does the car take to complete its journey?
  - **ii** What is the mass of the fuel in the tank at the end of this journey? (1 gallon = 4.546 litres)

A

9 Darren has a spherical ball of modelling clay. It has a mass of 462 g. The density of the clay is 2.3 g/cm^3. He reshapes the clay into a cubical block with base 3 cm by 6 cm.

Find the height of the block.

AQA Examination-style questions 

1 Susan completes a journey in two stages.
In stage 1 of her journey, she drives at an average speed of 80 km/h and takes 1 hour 45 minutes.

- **a** How far does Susan travel in stage 1 of her journey? *(2 marks)*
- **b** Altogether, Susan drives 190 km and takes a total time of 2 hours 15 minutes. What is her average speed, in km/h, in **stage 2** of her journey? *(3 marks)*

AQA 2003

13 Trial and improvement

Objectives

Examiners would normally expect students who get these grades to be able to:

C

solve equations such as $x^3 + x = 12$ using systematic trial and improvement methods.

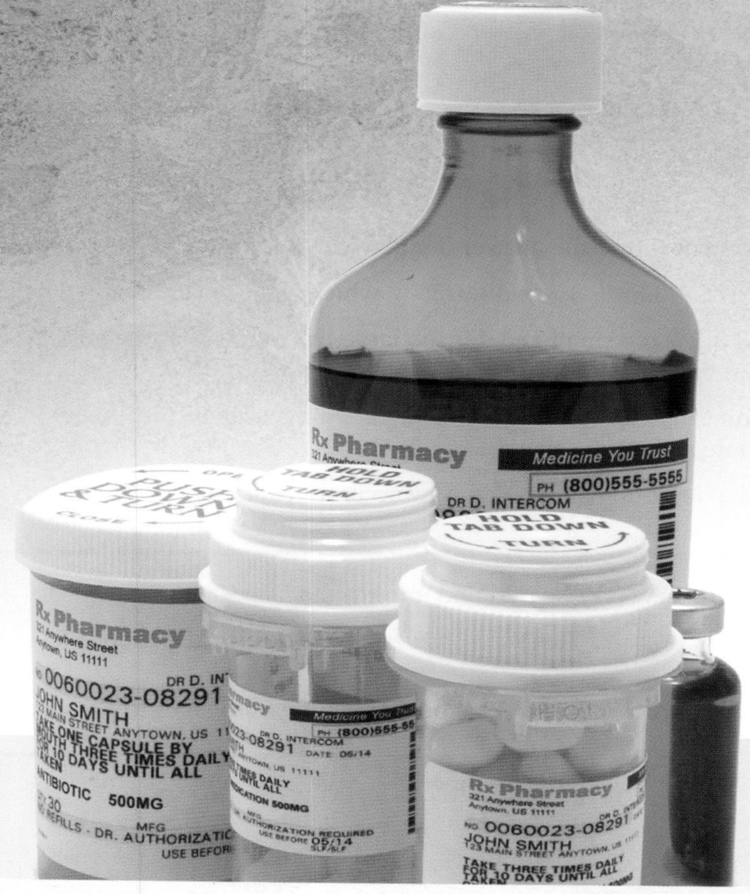

Did you know?

Trial and improvement

Trial and improvement (also known as trial and error) has many uses in the real world. It is often used by engineers when they develop complex equipment. For example, engineers will trial different fuel flow rates when determining the maximum thrust from a jet engine.

Also doctors may use trial and improvement to test different combinations of drugs for diabetes, epilepsy and high blood pressure. For the future, scientists are developing supercomputers which will act as 'virtual humans'. This will allow doctors to match different combinations of drugs to different patients.

Key terms

trial and improvement
decimal place

You should already know:

- ✔ how to substitute into algebraic expressions
- ✔ how to rearrange formulae
- ✔ how to use the bracket and power buttons on your calculator.

Learn... 13.1 Trial and improvement

Trial and improvement is a method for solving problems using estimations that get closer and closer to the actual answer. Trial and improvement is used where there is no exact answer, so you will be asked to give a rounded answer. On the examination paper, you will be told when to use a trial and improvement method.

If you are told that $x^3 + x = 50$ then you can work out that the answer lies between 3 and 4 because:

$3^3 + 3 = 30$ which is too small

and $4^3 + 4 = 68$ which is too large

You know that the answer lies between 3 and 4 so you might try 3.5:

$3.5^3 + 3.5 = 46.375$ which is too small

As 3.5 is too small and 4 is too large, you know that the answer lies between 3.5 and 4 so you might try 3.7 or 3.8...

You can keep going with this method to get an answer that is more and more accurate.

The question will tell you how accurate your answer should be.

> **AQA Examiner's tip**
> It is a good idea to lay your working out carefully. A table can be helpful.

Example: Use trial and improvement to solve $x^3 - x = 40$

Give your answer to one **decimal place**.

Solution: You can try out some different values to get you started.

Trial value of x	$x^3 - x$	Comment
1	0	too small
2	6	too small
3	24	too small
	← 40	
4	60	too large

The answer 40 lies between 24 and 60.

This tells you that x lies between 3 and 4.

You might try 3.5

3.5	39.375	too small
3.6	43.056	too large

Again, you can see that 40 lies between 39.375 and 43.056

This tells you that x lies between 3.5 and 3.6

You should try 3.55

The answer to 1 d.p. is either 3.5 or 3.6

Work out the value for 3.55 to see whether it is larger than 40.

If it is too large, then 3.55 is too large and the answer to 1 d.p. is 3.5

If it is too small, then 3.55 is too small and the answer to 1 d.p. is 3.6

3.55	41.188875	too large

You know that x lies between 3.5 and 3.55

But any answer between 3.5 and 3.55 is the same as 3.5 to 1 d.p.

The required answer is 3.5 to 1 d.p.

> **AQA Examiner's tip**
> On some examination questions you will be told where the answer lies. For example, you may be told that there is a solution between 2 and 3.

Practise... 13.1 Trial and improvement 

D C B A A*

C

1 Find, using trial and improvement, a solution to the following equations.
Give your answers correct to one decimal place.
You **must** show all your working.

a $x^3 + x = 10$

b $x^3 + x = 520$

c $x^3 - 5x = 400$

d $x - x^3 = -336$

2 Use trial and improvement to solve the equation $x^3 + x = 75$
Give your answer to one decimal place.
The table has been started for you

Trial value of x	$x^3 + x$	Comment
2	10	too low
3	30	too low
4	68	too low
5	130	too high

So now we know that the value lies between 4 and 5.

4.5	95.625	Too high
?		
?		

3 Use trial and improvement to find solutions to the following equations.
Give your answer to two decimal places.

a $a^3 - 10a = 50$ if the solution lies between 4 and 5

b $x^3 - x = 100$ if the solution lies between 4 and 5

⚠ **c** $5x - x^3 = 10$ if the solution lies between -2 and -3

4 Use trial and improvement to find solutions to the following.
Give your answer to two decimal places.

a $t^3 - 5t = 10$ if t lies between 2 and 3

b $x^3 - 5x = 60$ if x lies between 4 and 5

c $x(x^2 + 1) = 60$ if x lies between 3 and 4

⚠ **d** $p^3 + 6p = -50$ if p lies between -3 and -4

5 Use trial and improvement to find a negative solution of the equation $y^3 + 60 = 0$

B

6 The equation $x^3 - 4x^2 = -5$ has two solutions of x between 0 and 5.
Use trial and improvement to find these solutions.
Give your answer to three decimal places.

7 Use trial and improvement to find the value of $x^2 - \frac{1}{x} = 5$ where x lies between 2 and 3.
Give your answer to two decimal places.

⚠ 8 Solve the following equations using trial and improvement.
Give your answer to two decimal places.
You must show your working.

a $2^x = 20$

b $x^3 + \frac{1}{x^3} = 100$

c $x^3 - 2x^2 + x = 44$

d $x^4 - 3x = 99$

9 The difference between the square of a number and the cube of a number is 100.

Find the number to one decimal place.

10 The following solid consists of a central square and four equal arms.
The volume of the solid is 100 cm³.
Find the value of x correct to two decimal places.

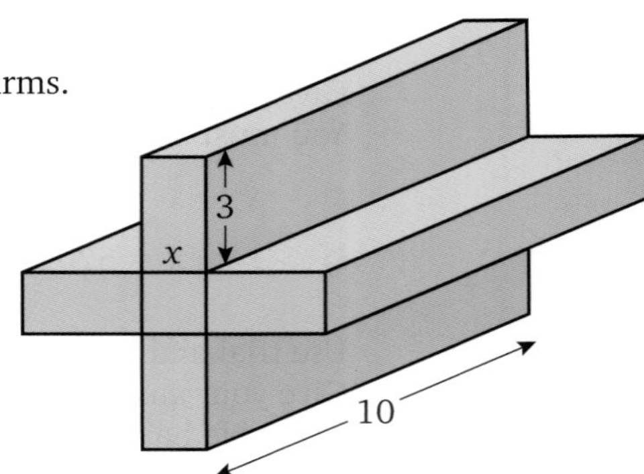

13 Assess

C

1 Use trial and improvement to solve the equation $x^3 - 4x = 100$
Give your answer to one decimal place.
The table has been started for you.

Trial value of x	$x^3 - 4x$	Comment
3	15	too small
4	48	too small
5	105	too large

2 The equation $x^3 + 8x^2 = 20$ has two negative solutions between 0 and −8.

Use trial and improvement to find these solutions.
Give your answer to two decimal places.

3 A cuboid measures $x \times x \times (x + 2)$.
The volume of the cuboid is 50 cm³.

Use trial and improvement to find x.
Give your answer to three decimal places.

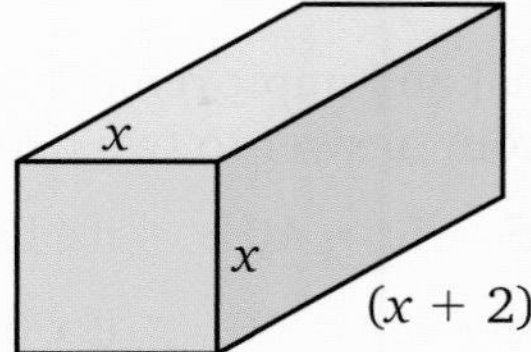

4 Use trial and improvement to find solutions to the following equations.
Give your answer to two decimal places.

a $x^3 + 5x = 50$ if the solution lies between 3 and 4

b $y^3 + 3y = 10$ if the solution lies between 1 and 2

c $x^3 - x = 100$ if the solution lies between 4 and 5

d $3x - x^3 = 25$ if the solution lies between −3 and −4

5 Use trial and improvement to find solutions to the following equations.
Give your answers to two decimal places.

a $x^3 + x = 60$ if x lies between 3 and 4

b $x^3 - 12x = 0$ if x lies between 2 and 6

6 Use trial and improvement to find the value of $8x - x^3 = 3$ where x is negative.
Give your answer to two decimal places.

7 Use trial and improvement to find the value of $t^3 + t^2 = 10$ where $1 \leqslant t \leqslant 2$
Give your answer to two decimal places.

B

8 Solve the following using trial and improvement methods.
Give your answer to two decimal places. You must show your working.

a $3^x = 15$ **b** $x^4 + x = 40$

9 $\frac{x^2 + 1}{x} = 10$ has two solutions greater than zero.

Use trial and improvement to find these solutions.
Give your answer to two decimal places.

10 $x^2 + \frac{2}{x} = 10$ has one negative solution.

Use trial and improvement to find this solution.
Work out x to one decimal place.

AQA Examination-style questions

1 Kerry is using trial and improvement to find a solution to the equation $8x - x^3 = 5$
Her first two trials are shown in the table.

x	$8x - x^3$	Comment
2	8	too high
3	−3	too low

Copy and continue the table to find a solution to the equation.
Give your answer to one decimal place. *(3 marks)*

AQA 2007

2 The sketch shows the graph of $y = x^3 - 3x - 8$
The graph passes through the points (2, −6) and (3, 10).

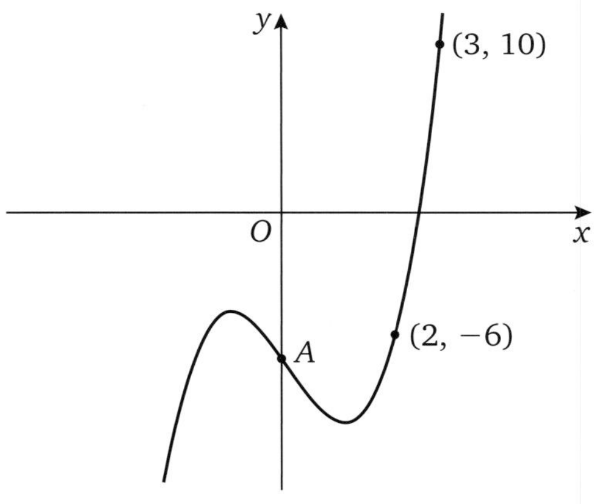

Not drawn accurately

a The graph crosses the y-axis at the point A.
Write down the coordinates of point A. *(1 mark)*

b Use trial and improvement to find the solution of:
$x^3 - 3x - 8 = 0$
Give your answer to one decimal place. *(4 marks)*

AQA 2008

3 Use trial and improvement to find the solution to the equation
$x^3 + 2x = 60$
Give your answer to one decimal place. You **must** show your working *(4 marks)*

AQA 2007

14 Enlargements

Objectives

Examiners would normally expect students who get these grades to be able to:

D

enlarge a shape by a positive scale factor from a given centre

compare the area of an enlarged shape with the original area

C

find the ratio of corresponding lengths in similar shapes and identify this as the scale factor of enlargement

use ratios in similar shapes to find missing lengths

B

enlarge a shape by a fractional scale factor

A

enlarge a shape by a negative scale factor

compare lengths, areas and volumes of enlarged shapes

use the effect of enlargement on perimeter, area and volume in calculations.

Did you know?

Ancient columns

This impressive building is the Colosseum in Rome built in ACE 70–82. It has always been recognisable by its shape. It is formed of layers that contain many striking columns.

Did you know that the three layers contain different types of columns? Did you know that for a column to be classed as Doric, Ionic or Corinthian requires a certain approximate height to width ratio?

Doric columns	8 : 1	strongest layer (bottom)
Ionic columns	9 : 1	
Corinthian columns	10 : 1	most slender layer (top)

Nelson's column in Trafalgar square is a Corinthian column. The height of the column itself is 46 m and the base diameter is 4.6 m.

More Corinthian columns can be found in the Pantheon in Rome. The columns are much smaller with heights of approximately 46 feet and base diameters of 4 feet 11 inches.

This means that all Corinthian columns are approximate enlargements of each other.

Key terms

enlargement	similar
transformation	ratio
image	scale factor
object	centre of enlargement
congruent	vertex, vertices

You should already know:

- ✔ how to plot coordinates in all four quadrants
- ✔ about units of length and how to use them
- ✔ about ratio and how to simplify a ratio
- ✔ how to use the vocabulary of transformations: object and image
- ✔ how to recognise and use corresponding angles
- ✔ how to find the area of simple shapes including a rectangle and a triangle.

14.1 Enlargement and scale factor

Enlargements are a type of **transformation**.

They are the only transformations learned at GCSE that change the size of a shape.

All the other transformations (reflections, rotations and translations) keep the **image** the same size as the **object** (the original shape). The shapes are **congruent**.

An enlargement changes the size of the object but not the shape.

All the lengths will be changed but all the angles will stay the same. The object and the image are **similar**.

The **ratios** of corresponding sides in similar shapes are all equal and are equivalent to the **scale factor** of enlargement.

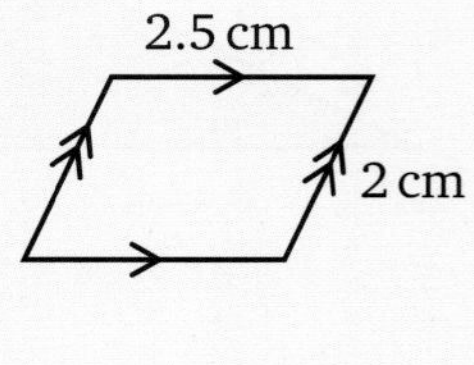

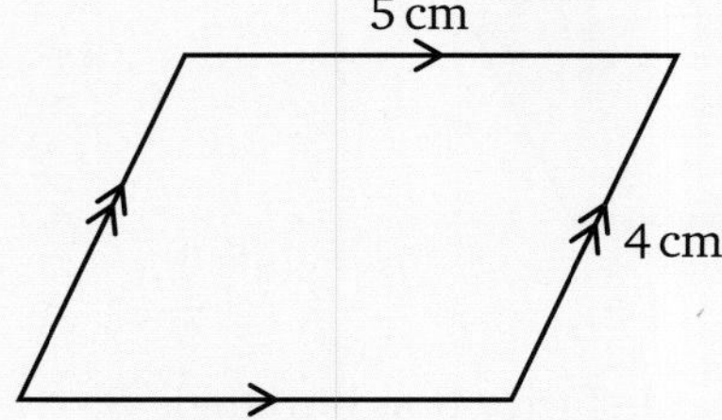

Not drawn accurately

$$\text{scale factor of enlargement} = \frac{\text{length on the enlarged shape (image)}}{\text{corresponding length on the original shape (object)}}$$

$\frac{5}{2.5} = 2$ and $\frac{4}{2} = 2$

The scale factor of enlargement is 2.

Example: The two doors are drawn on centimetre grids.
The doors are similar. One is an enlargement of the other.
All corresponding angles are equal.
All corresponding sides are in the same ratio.

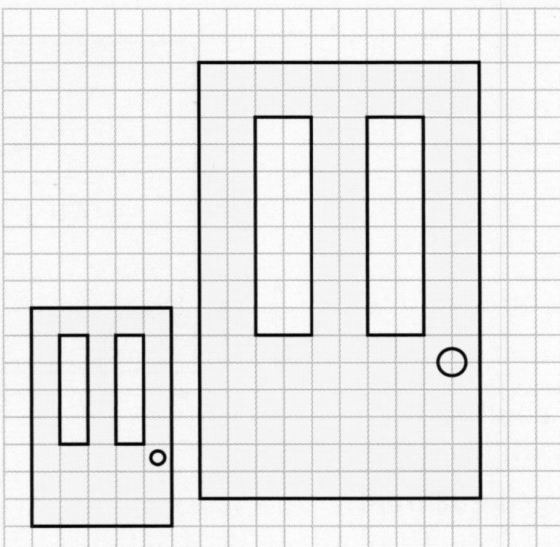

a What is the scale factor of enlargement?

b What is the area of one pane of glass in the small door (object)?

c What is the area of one pane of glass in the enlarged door (image)?

Solution:

a To find the scale factor, you need to take two corresponding sides: one on the original (object) and one on the new one (image).

$$\text{scale factor} = \frac{\text{length of door on enlargement}}{\text{length of door on the original}} = \frac{16}{8} = 2$$

b Area of pane of glass on the object $= 1 \times 4 = 4\,\text{cm}^2$

c Area of pane of glass on the image $= 2 \times 8 = 16\,\text{cm}^2$

Practise... 14.1 Enlargement and scale factor  

D

1 In these diagrams, A has been enlarged to give B.

a What is the scale factor of each enlargement?

b Work out the areas of the object (original) and the image (enlargement).

c What do you notice about the areas?

i

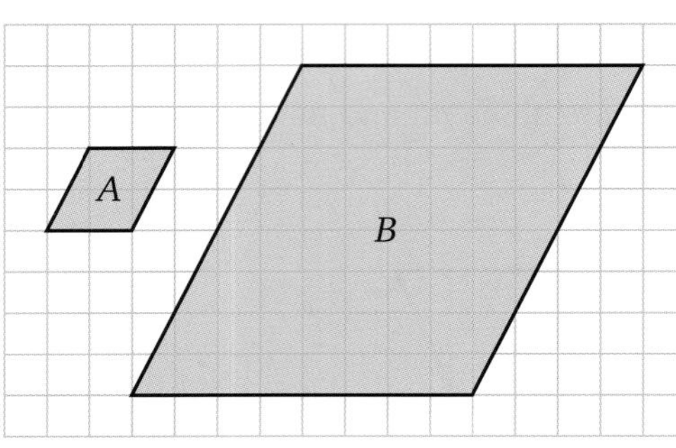

ii

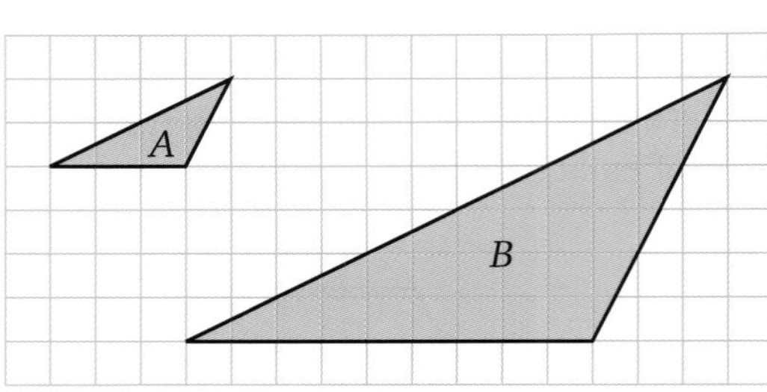

C

2 Triangle A is enlarged with scale factor 3 into triangle B.

a One side of the triangle B has length 7.5 cm.
What is the length of the corresponding side of triangle A?

b One angle of the triangle is 45°.
What is the size of the corresponding angle in triangle A?

3 Enlarge this shape by a scale factor of:

a 1.5

b 2.25

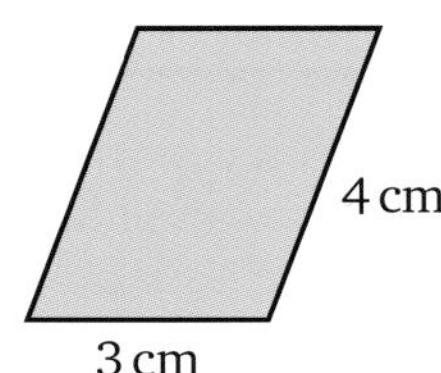

Not drawn accurately

4 The diagram shows a plan for a garden.
Quadrants of circles have been dug up to make a flower bed and a pond with a path beside it.
The radius of the pond is 2 metres.

3.6 m
lawn
10 m
?
2 m
5 m

Hint

A quadrant is a quarter of a circle.

a To form the enlarged quadrant (the pond and path), the pond has been enlarged with scale factor 1.5

i What is the radius of this enlarged quadrant?

ii Find the area of each of these two quadrants to two decimal places.

iii Using these two answers and the scale factor of enlargement, show that your answers are correct. Work to two decimal places.

b The flower bed is also an enlargement of the quadrant for the pond.

i What is the scale factor of enlargement?

ii Find the area of the flower bed.

c Find the area of the lawn.

d The gardener wants to copy this design for use in another garden. The new garden has dimensions that are exactly double this garden.
What will the area of the lawn be in the new garden?

AQA Examiner's tip

When you are working with areas or volumes, be careful that you are stating the units correctly, for example:
length cm   area cm^2 volume cm^3

5 **a** Julie has a photograph of her cat. She wants to have the photograph enlarged to put on the cover of her portfolio.

She wants the picture to fill the cover of the portfolio.

What is the scale factor of the enlargement she needs?

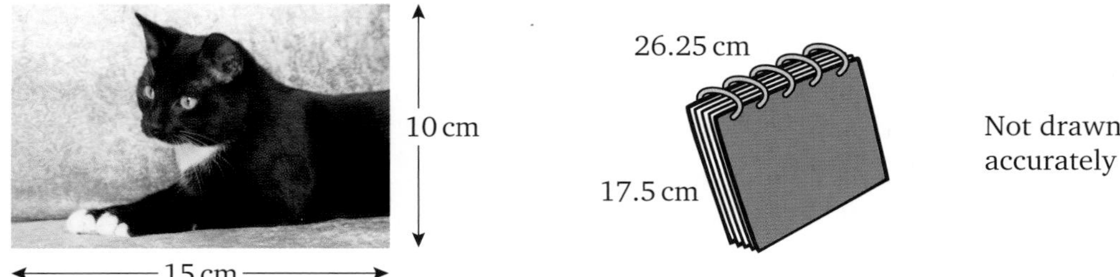

b Julie also wants to enlarge the photograph to fit a frame to go on her wall. The frame is 82.5 cm wide.

i Find the scale factor to be used to enlarge the original photograph to fit the frame.

ii Find the height of the frame.

iii If the photograph on the folder was enlarged to fit the frame, find the scale factor.

6 The area diagrams show a single square which has been enlarged by scale factor 2 and scale factor 3.

Areas

 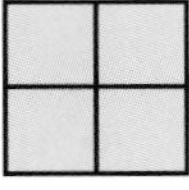 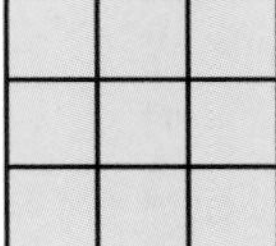

Volumes

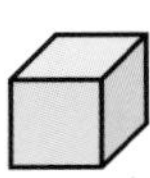  

a **i** Calculate the area of each of these shapes.

ii What do you notice about the areas compared with the original square?

b Now look at the diagrams for the volumes.

i Calculate the volume of each cube.

ii What do you notice about the volumes compared with the original cube?

Learn... 14.2 Centres of enlargement

When drawing enlargements you need some extra information.

An enlargement is defined by its **centre of enlargement** and **scale factor**.

The centre of enlargement can be outside, on the edge of, or even inside the object.

In each diagram the same scale factor has been used but the centre of enlargement, marked with a cross, is different.

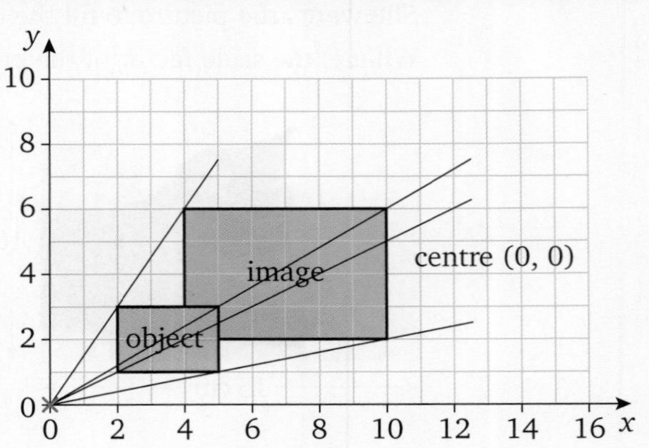

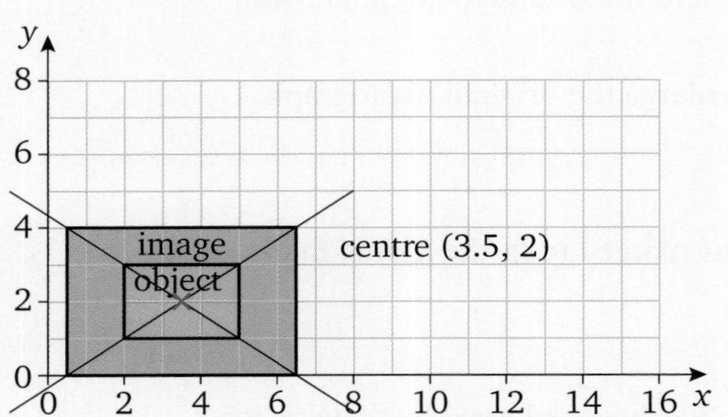

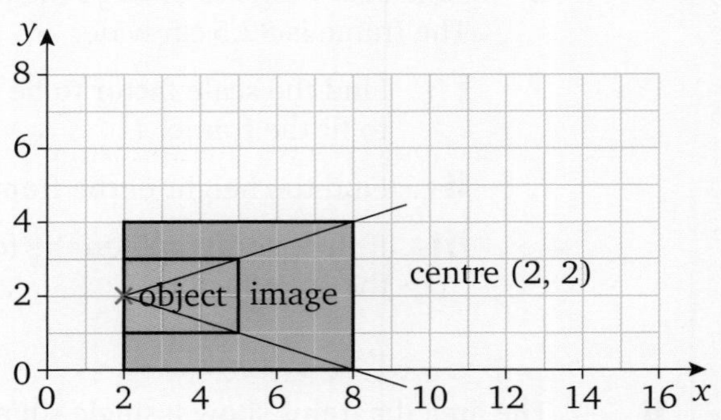

Example: Enlarge the rectangle A by scale factor 2, centre of enlargement (1, 1).

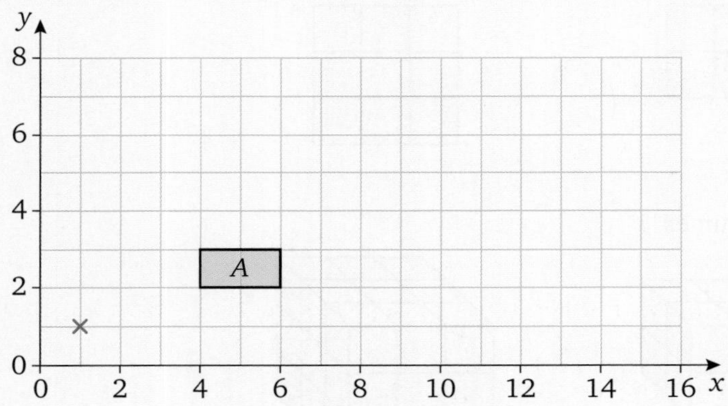

To enlarge a shape:

1 Plot the centre of enlargement on the grid with a cross.

2. Choose a **vertex** (corner) of the shape. Join the centre of enlargement to this vertex and extend the line past the vertex.

3. Measure the distance from the centre of enlargement to the vertex and multiply this by the scale factor. In this case, the scale factor is 2.

4. This is the new distance from the centre of enlargement to the corresponding vertex on the new shape. Measure this distance along the line you have drawn and mark the new point.

5. Repeat this for all the other **vertices**.

> **AQA Examiner's tip**
> Always use a sharp pencil and a ruler. Make sure that the lines are drawn exactly through the intersection of the lines of the grid.

> **AQA Examiner's tip**
> There are usually two vertices that are easier to draw because the construction lines do not cross over the shape itself.
> Do these first!

When the enlargement is finished, the distance from each vertex to the centre of enlargement will be twice as long as it was before.

Every length on the new rectangle will be twice as long as it was before.

The original rectangle was 2 by 1 units. The enlarged rectangle is now 4 by 2 units.

The rectangles are similar.

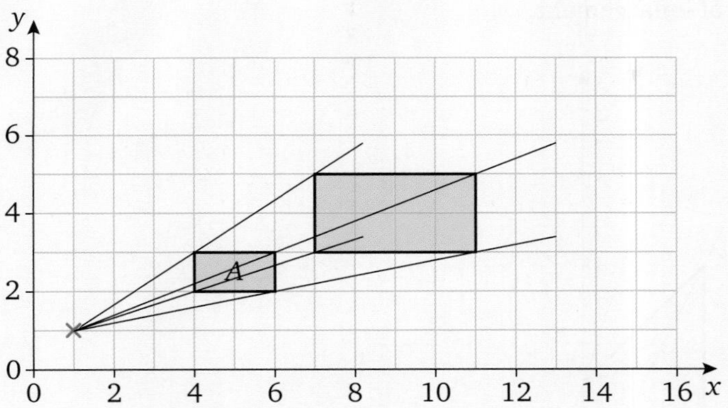

An alternative method would be to count the horizontal and vertical distance from the centre of enlargement to a vertex.

The new vertex would be at a point twice the horizontal distance and twice the vertical distance from the centre of enlargement.

Hint

You will know from the scale factor what the dimensions of the new rectangle should be. You can use this fact to help you draw an accurate diagram and to check your enlargement.

AQA Examiner's tip

Don't forget that you can use either of these methods to find the new position of a vertex. You can even use a combination of both!

Example: Enlarge triangle ABC with scale factor $\frac{1}{2}$ using O as the centre of enlargement.

Solution: A fractional scale factor, between 0 and 1, makes the object smaller rather than larger.

It reduces the size, but mathematically it is still called an enlargement.

Join each vertex in turn, from the object to the centre of enlargement.

OA' must be $\frac{1}{2}$ of the original distance OA, so measure OA and position A' accordingly.

Repeat for B and C.

If you have constructed this accurately, then the sides of $A'B'C'$ should be half the length of the sides of ABC.

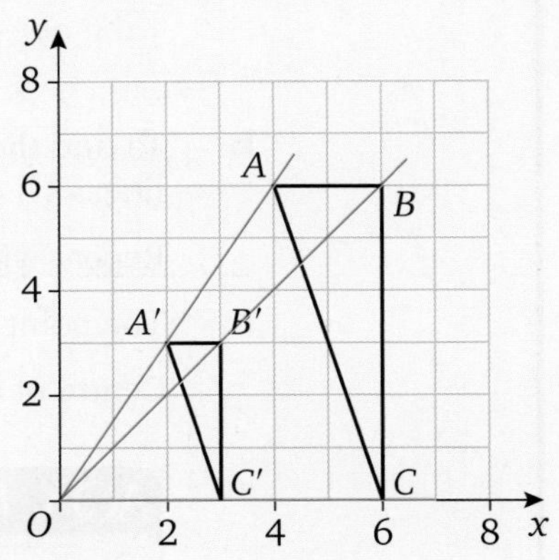

Example: The trapezium *RSTU* has been enlarged to give trapezium *VWYZ*.

Find:

a the scale factor of enlargement

b the centre of enlargement.

Solution:

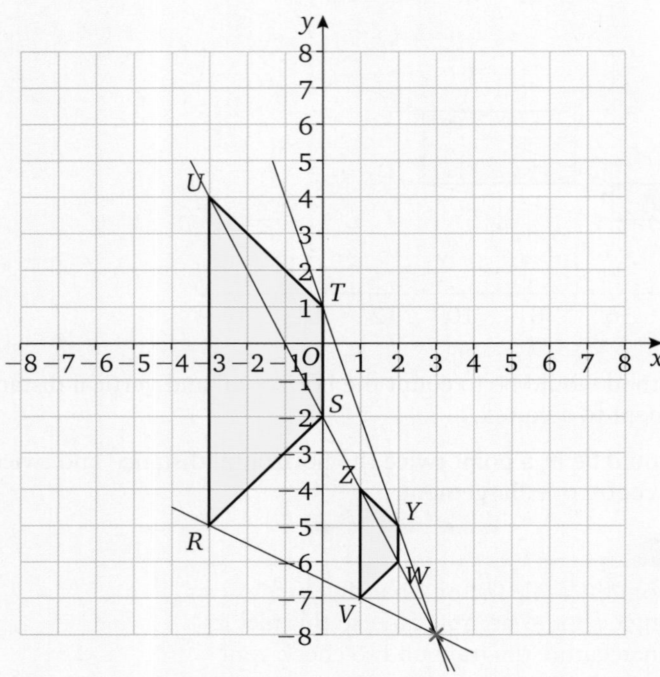

a Measure a side on the object *RSTU* (the original shape) and a corresponding side on the image *VWYZ* (the enlarged shape).

For example, $RU = 9$ units and $VZ = 3$ units

The scale factor $= \frac{VZ}{RU} = \frac{3}{9} = \frac{1}{3}$

Scale factor is $\frac{1}{3}$

The scale factor $= \frac{WY}{ST} = \frac{1}{3}$

Hint

It is a good idea to do this for two lengths on the object so that one can be used as a check.

b To find the centre of enlargement, join corresponding vertices *R* and *V*, and extend the line.

Repeat with the other three vertices.

The point where they meet is the centre of enlargement.

Centre of enlargement is $(3, -8)$.

AQA Examiner's tip

Once you have identified the centre of enlargement, choose a vertex.

Then check that:

the distance of the vertex on the image from the centre of enlargement = (scale factor) × distance of corresponding vertex on the object from the centre of enlargement.

Practise... 14.2 Centres of enlargement

D C B A A*

D

1 Copy the shape onto squared paper.

Enlarge the shape by a scale factor of 2.
Use the cross as the centre of enlargement.

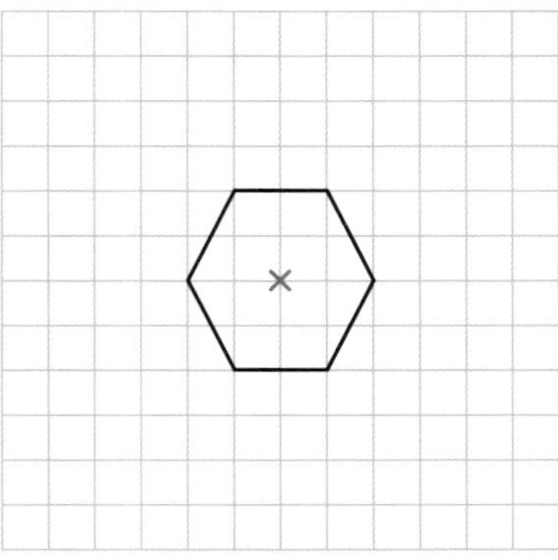

C

2 In this diagram, A has been enlarged to give B.

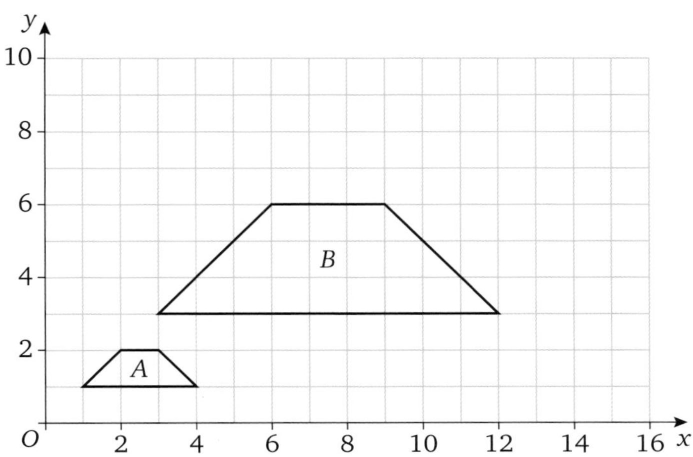

Find the scale factor and centre of enlargement.

3 **a** **i** Draw a pair of axes with x-values from 0 to 16 and y-values from 0 to 8.

ii Plot and label rectangle E with vertices at (6, 2), (6, 4), (10, 4) and (10, 2).

b Draw the image of rectangle E after an enlargement of scale factor 1.5 with centre (0, 0).
Label the image F.

c What are the coordinates of the vertices of F?

B

4 Copy each shape onto squared paper.

Enlarge **a** by a scale factor of $\frac{1}{2}$.

Enlarge **b** by a scale factor of $\frac{1}{4}$.

a

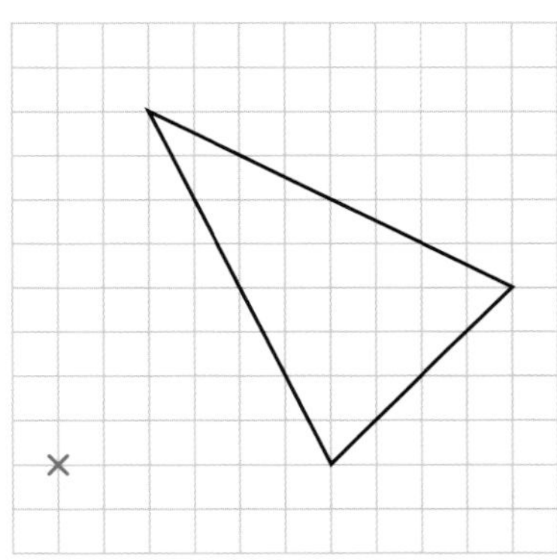

b

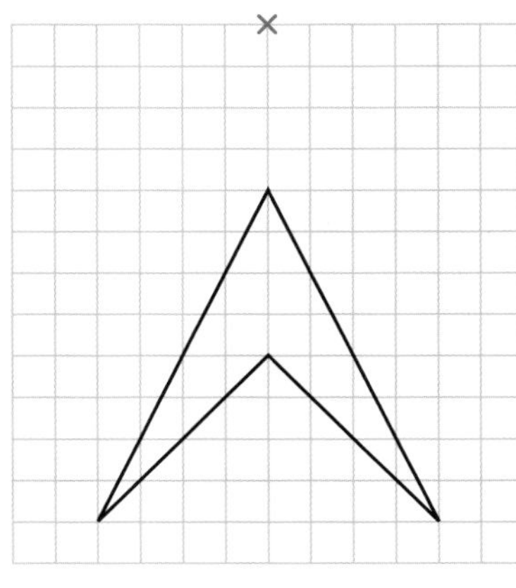

5 a i Draw a pair of axes with x-values from -2 to 12 and y-values from -4 to 6.

ii Plot and label parallelogram $ABCD$ with vertices at $(6, -1)$, $(10, 1)$ and $(10, 4)$ and $(6, 2)$.

b Draw the image of $ABCD$ after an enlargement of scale factor $\frac{1}{2}$ with centre $(-2, -4)$.
Label the image $A'B'C'D'$.

c What are the coordinates of the vertices of the image?

d i Measure the diagonals AC and $A'C'$.

ii What do you notice about these lengths?

e Repeat the process with the diagonals BD and $B'D'$.

f Relate your findings to the scale factor.

6 A jewellery designer was asked to make a pendant which perfectly matched a pair of earrings. In order for them to look similar, he decided that the pendant should be an enlargement of the earrings. He worked out that a scale factor of 3 would be most suitable.

Here is the plan of one of the earrings. Each square represents a square 0.5 cm by 0.5 cm.

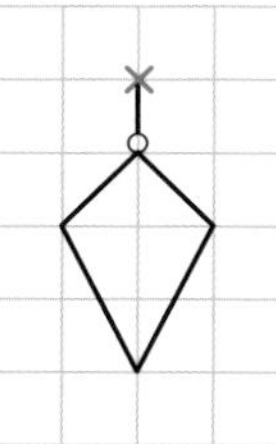

Copy this diagram onto squared paper and use this to draw the plan for the pendant.

The centre of enlargement is marked by a cross on the diagram.

7 a On an A4 piece of plain paper, construct an equilateral triangle of side 2 cm in the centre of the paper. Then construct another five similar triangles to form a regular hexagon. Label this $ABCDEF$.

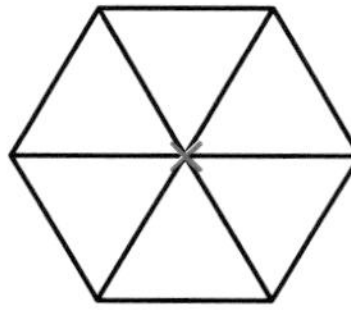

b Using the cross as the centre of enlargement and scale factor 1.5, draw the image of $ABCDEF$.
Label your image $A^1B^1C^1D^1E^1F^1$.

c Enlarge $A^1B^1C^1D^1E^1F^1$ with the same centre and scale factor to give $A^2B^2C^2D^2E^2F^2$.

d Repeat this process, always enlarging the latest shape you have drawn to get the next one.
Your diagram will then consist of several hexagons, each one outside the previous one.

e What do you notice about these hexagons?
Comment on their sizes. Give reasons for your answer.

Hint

When the second enlargement is formed, from $A^1B^1C^1D^1E^1F^1$ to $A^2B^2C^2D^2E^2F^2$, you were not enlarging the original. What would the scale factor of enlargement be for $ABCDEF$ to $A^2B^2C^2D^2E^2F^2$?

f Find the scale factors for each enlargement when $ABCDEF$ is always used as the object.

14.3 Negative scale factors

If the scale factor of an enlargement is negative, the object and the image are on opposite sides of the centre of enlargement.

Example: Copy the diagram onto squared paper. Enlarge the shape by a scale factor of −2, using the point marked with a cross as the centre of enlargement.

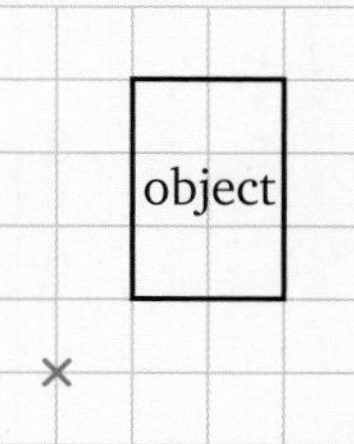

Solution: Label the object *ABCD*.

Join *A* to *X* and extend the line through *X* to the other side of the centre of enlargement.

Measure the distance *XA* and position *A*′ on this extended line such that $XA' = 2 \times XA$

Repeat this for all other vertices *B*, *C* and *D*.

Join *A*′, *B*′, *C*′ and *D*′ to form the image rectangle.

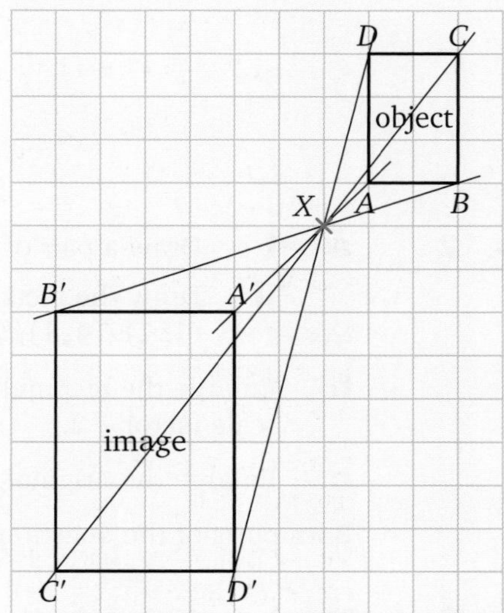

Hint

Do not forget that it can sometimes be helpful to use the patterns on the grid. To get from *X* to *A*, you move 1 to the right and 1 up. So to get from *X* to *A*′ you move 2 to the left and 2 down.

Example: Draw and label the *x*-axis from −8 to 12 and the *y*-axis from −6 to 10.

Draw the triangle *ABC* with coordinates (1, 2), (1, −4) and (10, −4).

Enlarge triangle *ABC* with centre (−2, 5) and scale factor $-\frac{1}{3}$

Solution: Draw the triangle and then join *A* to *X*, the centre of enlargement.

Extend this line through the point of enlargement, to form a line on the opposite side to *A*.

The scale factor is $-\frac{1}{3}$. The negative scale factor shows you to extend the line using the method shown in the previous example. The image point *A*′ will be a third of the distance from *X* as the object point *A* was from *X*.

You could also use the patterns on the grid. *X* to *A* involves moving 3 to the right and 3 down, so *X* to *A*′ will involve a journey of $\frac{1}{3}$ of this; 1 to the left and 1 up.

Repeat this process for *B* and *C*.

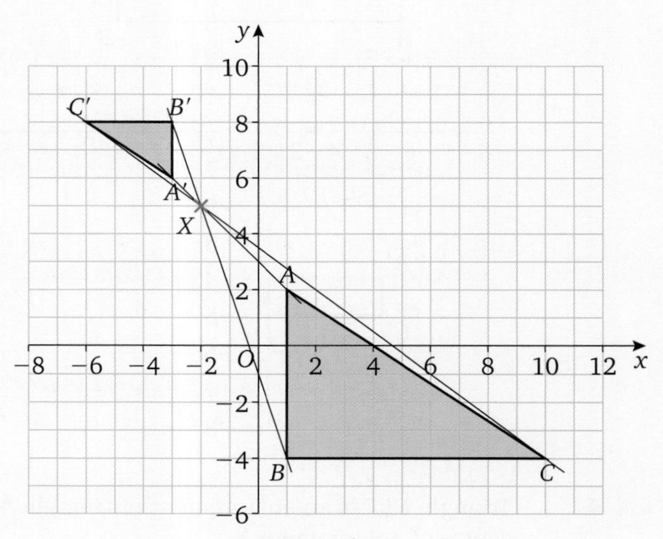

AQA Examiner's tip

An enlargement with scale factor $-\frac{1}{3}$ is equivalent to a rotation of 180° about the centre of enlargement, followed by an enlargement about the same centre, scale factor $+\frac{1}{3}$. Use this as a check.

Practise... 14.3 Negative scale factors 

A

1 Copy the axes and shape shown in the diagram.

Enlarge the shape from the centre (5, 5) with scale factor -3.

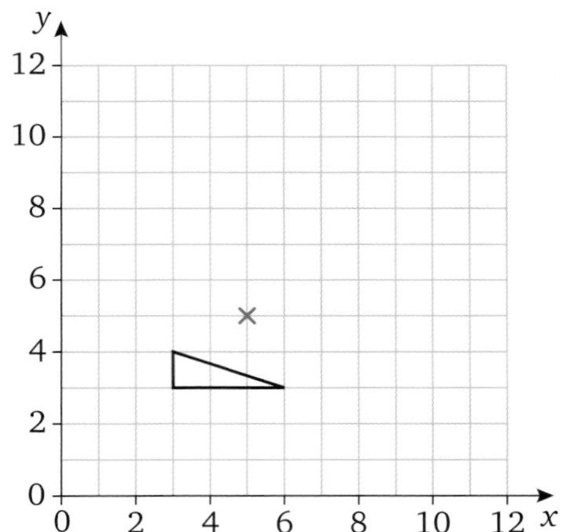

2 **a** **i** Draw a pair of x- and y-axes from -9 to 9.

ii Draw the rectangle with coordinates (1, 1), (4, 1), (4, 3) and (1, 3).

b Enlarge the rectangle with centre (0, 2) and scale factor -2.

c Enlarge the original rectangle with centre (1, -2) and scale factor $-1\frac{1}{3}$

Both parts of the question can be drawn on one set of axes.

Hint

For part **b**, use the pattern of squares as well as measuring the distance along the line.

3 The shape $ABCDEF$ has been enlarged to make the shape $A'B'C'D'E'F'$.

Find the scale factor and centre of enlargement.

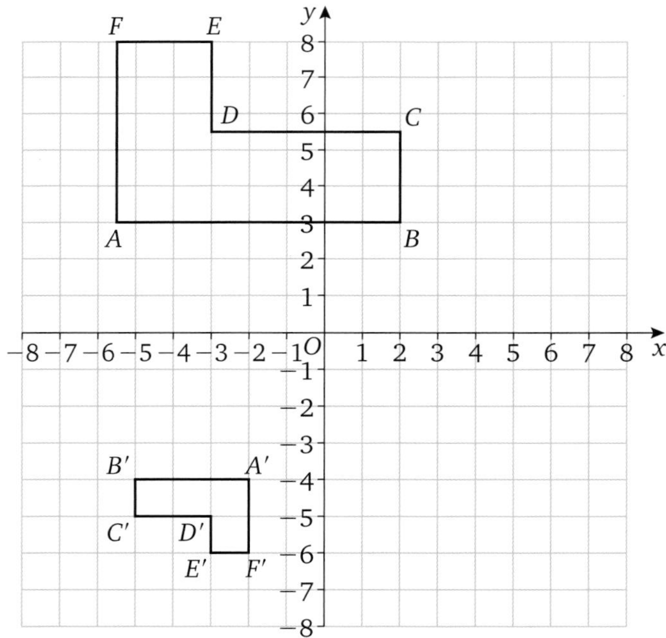

 **4** Triangle DEF is an enlargement of triangle ABC, centre of enlargement X.

a If $BX = 2EX$, what is the scale factor of enlargement?

b If $FE = 2.9$ cm and $AB = 8.2$ cm, find the lengths of DE and BC.

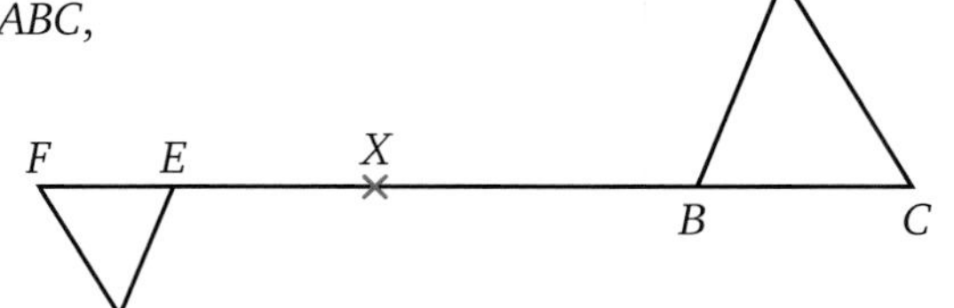

5 **a** **i** Draw a pair of x- and y-axes from -8 to 8.

ii Plot the trapezium $ABCD$ with these points: $A(2, 2)$, $B(5, 2)$, $C(5, 5)$, $D(2, 4)$.

b Enlarge $ABCD$ with scale factor -2 and centre $(1, 1)$.
Label the image $A^1B^1C^1D^1$.

c Enlarge $A^1B^1C^1D^1$ with scale factor -1 and centre $(-4, -2)$.
Label the image $A^2B^2C^2D^2$.

d Translate $A^2B^2C^2D^2$ with vector $\begin{pmatrix} 9 \\ 0 \end{pmatrix}$ to form $A^3B^3C^3D^3$.

e Find the transformation that takes $A^3B^3C^3D^3$ back to the original $ABCD$.

6 The following diagram shows a plan for making a two-ended measuring spoon.
The centre of the enlargement is marked with a cross.

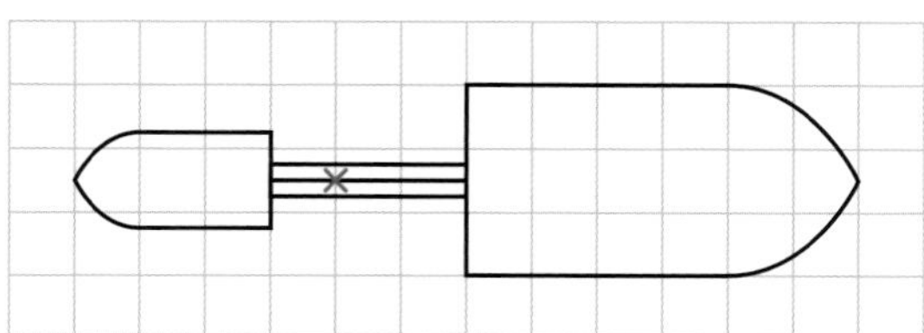

a Find the scale factor of the enlargement.

b Both spoons are the same depth but have the cross-sectional areas as shown.
The smaller spoon can contain 2.5 ml of liquid.

Explain why the larger spoon contains 10 ml.

c If the depth of the 2.5 ml spoon was also enlarged by the same scale factor to make the larger spoon, what would be the capacity of the larger spoon in ml?

Learn... 14.4 Similar shapes and scale factors

The following triangles are **similar**.

Triangle ABC is an enlargement of triangle DEF with a scale factor of 2.

This means that **every** side on triangle ABC is twice the length of the corresponding length on triangle DEF.

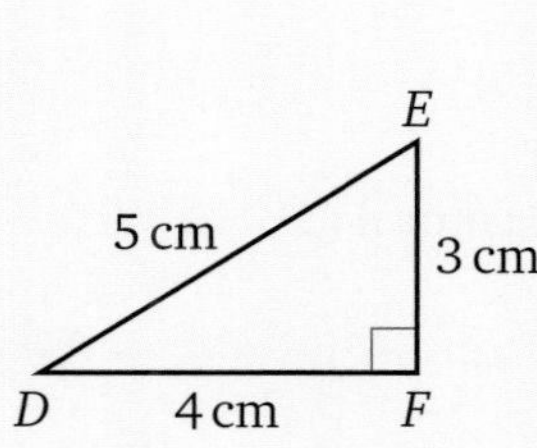

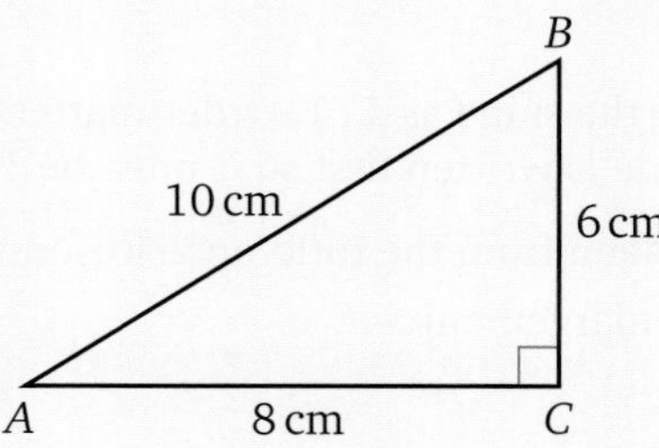

Not drawn accurately

The corresponding sides are all in the same ratio $2:1$

$AC:DF = 8:4 = 2:1$

$BC:EF = 6:3 = 2:1$

and $AB:DE = 10:5 = 2:1$

Note that the perimeters are also in the same ratio.

perimeter of ABC = 24 cm   perimeter of DEF = 12 cm

perimeter of ABC : perimeter of DEF

$= 24:12$

$= 2:1$

This is because perimeter is a 'length'. In similar shapes the perimeters will be in the same ratio as any other lengths.

Now look at the areas of the triangles:

area of triangle $ABC = \frac{1}{2} \times 8 \times 6 = 24\,\text{cm}^2$ area of triangle $DEF = \frac{1}{2} \times 4 \times 3 = 6\,\text{cm}^2$

area of triangle ABC : area of triangle DEF

$$= 24\,\text{cm}^2 : 6\,\text{cm}^2$$
$$= 24 : 6$$
$$= 4 : 1$$
$$= 2^2 : 1^2$$

In similar figures, the ratio of areas = (ratio of length)^2

or scale factor of enlargement for area = (scale factor of enlargement for length)^2

This can be extended into three dimensions.

In similar solids, the ratio of volumes = (ratio of length)^3

or scale factor of enlargement for volume = (scale factor of enlargement for length)^3

Example: Rectangles $ABCD$ and $EFGH$ are similar.

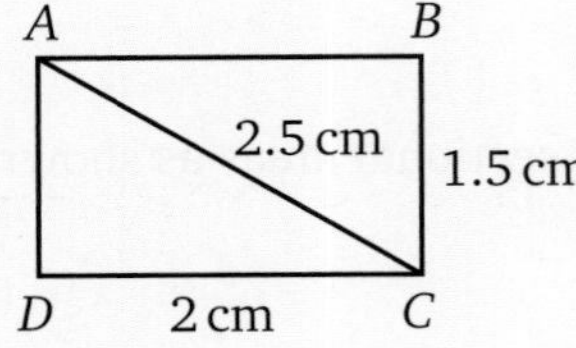

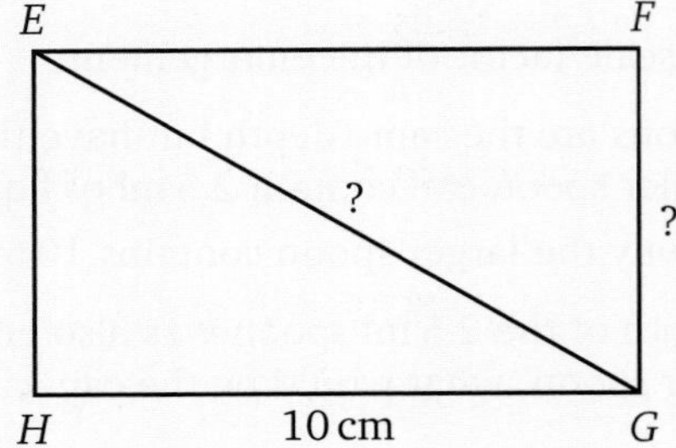

Not drawn accurately

a Write down the ratio of $DC : HG$ in its simplest form.

b What is the scale factor of the enlargement?

c Find the length of FG.

d Find the length of the diagonal EG.

Solution:

a $DC = 2\,\text{cm}$ $HG = 10\,\text{cm}$

$DC : HG = 2 : 10$

$= 1 : 5$

The ratio $1 : 5$ is not the same as $5 : 1$. Order matters.
Here the smallest side is written first so it must be $1 : 5$

b The scale factor is taken from the ratio once it is in the form $1 : n$ or $n : 1$

Scale factor of the enlargement = 5

c $BC = 1.5\,\text{cm}$

$FG = 1.5 \times 5 = 7.5\,\text{cm}$

d $AC = 2.5\,\text{cm}$

$EG = 2.5 \times 5 = 12.5\,\text{cm}$

Example: These cuboids are similar.

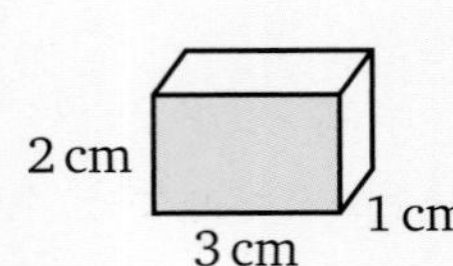

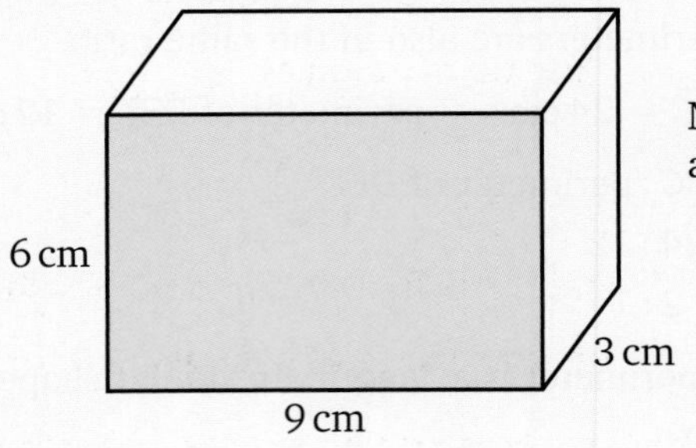

Not drawn accurately

a What is the ratio of the lengths and then the widths of the facing (shaded) rectangles?

b Show that the ratio of the perimeters of these rectangles is the same as the ratio of the lengths.

c Find the ratio of the depths of the cuboids.

d Find the ratio of areas of the two facing rectangles.

e Find the ratio of the volumes of the cuboids.

Solution:

a Length of small facing rectangle = 3 cm

Length of large facing rectangle = 9 cm

So ratio of lengths = 3 cm : 9 cm

= 1 : 3

As long as the units, for example, cm, are the same on both sides of a ratio, they can be left out of the following working.

Width of small facing rectangle = 2 cm

Width of large facing rectangle = 6 cm

So ratio of widths = 2 : 6

= 1 : 3

When a ratio is in the form $1:n$ or $n:1$, the scale factor of enlargement n can be read off easily.

Here the scale factor is 3.

b Perimeter of small rectangle = 2(3 + 2) cm = 10 cm

Perimeter of large rectangle = 2(9 + 6) cm = 30 cm

So ratio of perimeters = 10 : 30

= 1 : 3 as required

c Depth of small cuboid = 1 cm

Depth of large cuboid = 3 cm

So ratio of depths = 1 : 3

These depths are also in the same ratio, which confirms that the cuboids are similar.

d Area of small rectangle = $2 \times 3 = 6\text{ cm}^2$

Area of large rectangle = $9 \times 6 = 54\text{ cm}^2$

So ratio of areas = 6 : 54

= 1 : 9

Note that $1:9 = 1^2:3^2$

In similar figures, the ratio of areas = (ratio of lengths)^2

or scale factor of enlargement for area = (scale factor of enlargement for length)^2

e Volume of small cuboid = $3 \times 2 \times 1 = 6\text{ cm}^3$

Volume of large cuboid = $9 \times 6 \times 3 = 162\text{ cm}^3$

So ratio of volumes = 6 : 162

= 1 : 27

Note that $1:27 = 1^3:3^3$

In similar figures, the ratio of volumes = (ratio of lengths)^3

or scale factor of enlargement for volume = (scale factor of enlargement for length)^3

Example: Two similar glasses have heights of 15 cm and 5 cm. When full, the larger glass contains 540 ml of liquid.

How much liquid does the smaller glass contain when full?

Solution: Height of large glass = 15 cm

Height of small glass = 5 cm

The heights are both lengths.

Ratio of lengths = 15 : 5

= 3 : 1

The capacity of something is a measure of volume.

Because ratio of volumes = (ratio of lengths)^3

then ratio of capacities = (ratio of lengths)^3

$= 3^3 : 1^3$

$= 27 : 1$

This means that:

the capacity of the larger glass is 27 times the capacity of the smaller glass

or

the capacity of the smaller glass is $\frac{1}{27}$ of the capacity of the larger glass.

Capacity of large glass = 540 ml

Capacity of small glass = $\frac{1}{27} \times 540$ ml

= 20 ml

AQA Examiner's tip

Always re-read the question again at the end. Were you asked to find the smaller or larger value of something? Did your answer come out smaller or larger? Should you have multiplied or divided?

Practise... 14.4 Similar shapes and scale factors

1 Triangles *ABC* and *DEF* are similar.

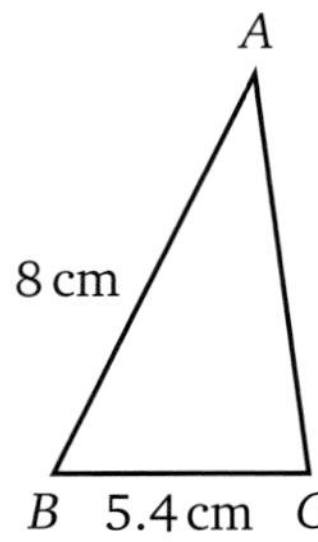

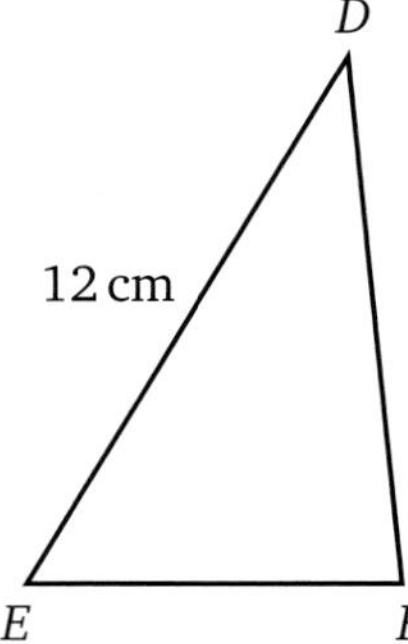

Not drawn accurately

a Work out the ratio of the corresponding lengths *AB* and *DE*.

b Simplify the ratio into the form $1 : n$

c Find the length of the side *EF*.

C

2 One triangle is an enlargement of the other.

Find the missing lengths.

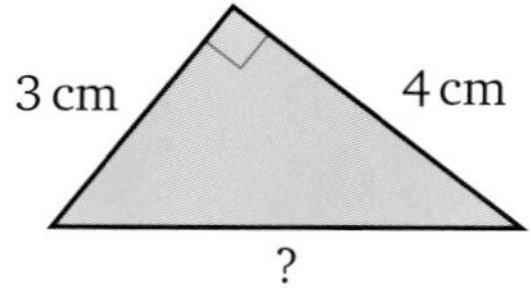

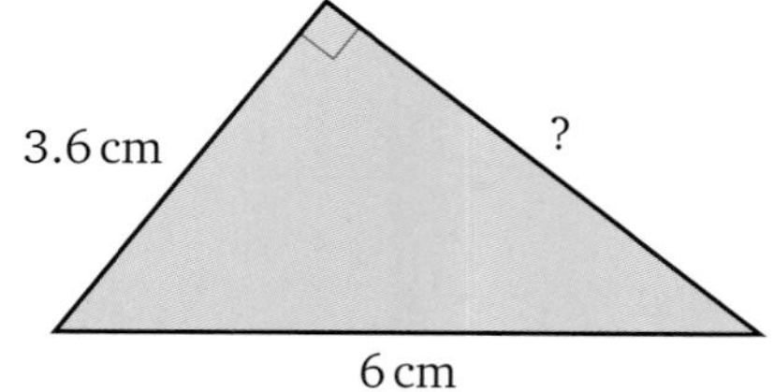

Not drawn accurately

B

3 The larger trapezium is an enlargement of the smaller one.

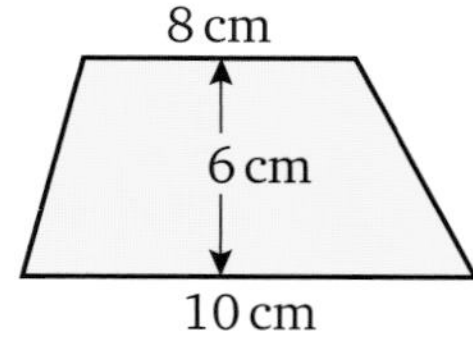

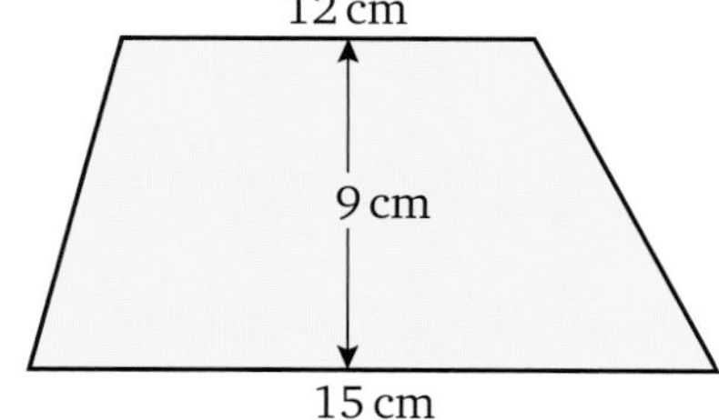

Not drawn accurately

a Find the ratio of lengths.

b **i** Find the area of each trapezium.

ii From these, find the ratio of areas.

iii Now rewrite this ratio in the form $1 : n$

iv What is the scale factor of enlargement for area?

4 These cylinders are similar.

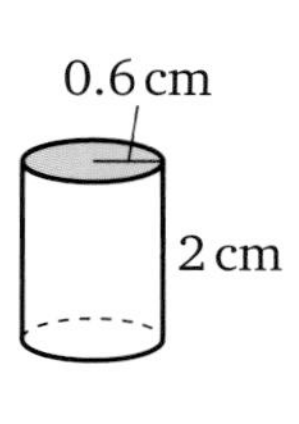

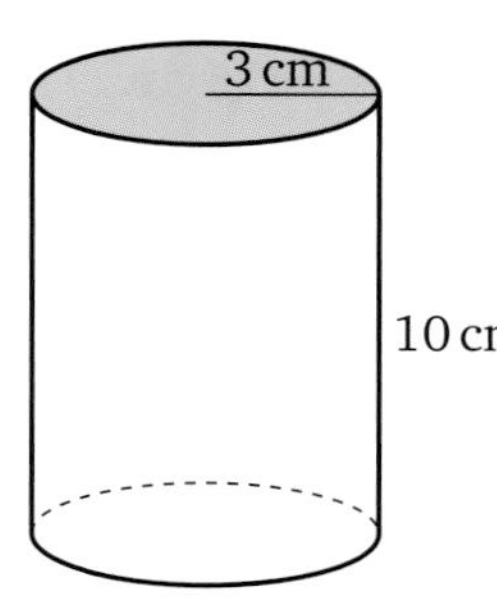

Not drawn accurately

a Find the ratio of heights.

b Find the ratio of radii.

c **i** Find the area of each cross-section shaded in the diagrams.

ii From these, find the ratio of the areas.

d Find the ratio of the volumes of the cylinders.

Hint

Area of circle $= \pi r^2$
Volume of cylinder $= \pi r^2 h$

B

5 In each of the following pairs of diagrams, decide whether the shapes shown are similar or not similar.

a 9 cm, 6 cm

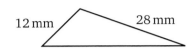

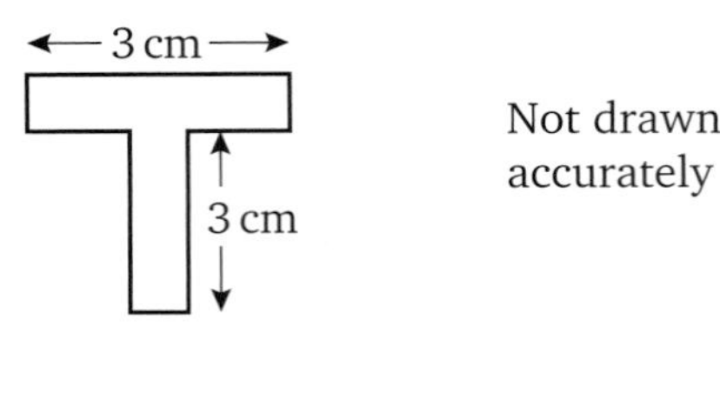

b 12 mm, 28 mm; 18 mm, 32 mm

Not drawn accurately

c

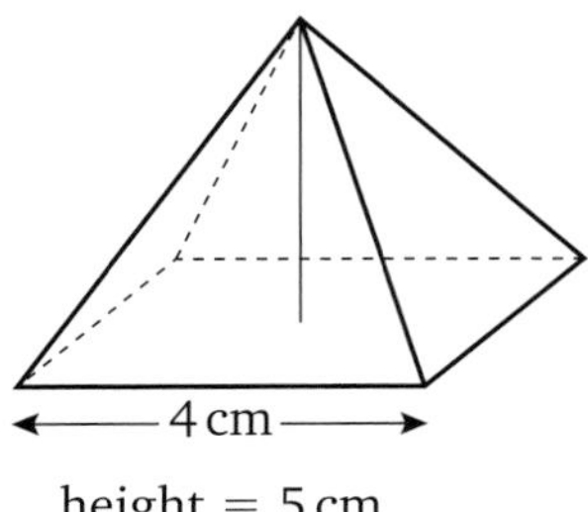

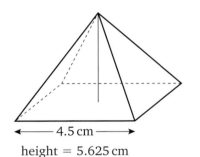

4.5 cm
height = 5.625 cm

Not drawn accurately

d

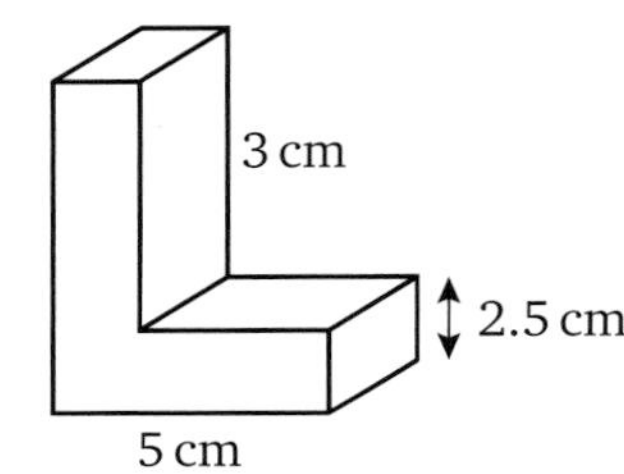

12 cm, 7.5 cm, 15 cm

Not drawn accurately

A

6 The area of a shape is 7 cm². Each side of the shape is enlarged by a scale factor of 5.
What is the area of the enlarged shape?

7 The volume of a shape is 4 cm³. Each side of the shape is enlarged by a scale factor of 5.
What is the volume of the enlarged shape?

8 A cuboid is enlarged. The original length was 3 cm and the enlarged length is 18 cm.
If the original volume was 24 cm³, what is the volume of the enlarged cuboid?

9 Two similar flower vases have heights of 18 cm and 28.8 cm.
It takes 0.5 litre of water to fill the smaller vase.
How much water does it take to fill the larger one?

10 A supermarket sells tinned fruit in two different sizes. The containers are similar.
The larger catering-size tin contains 4 kg of fruit and the smaller tin contains 500 g of fruit.
If the height of the smaller tin is 14 cm, what is the height of the larger tin?

⚠ 11 Two ice-cream cones are similar. The curved surface area of the large cone is 65π cm² and the curved surface area of the smaller cone is 16.25π cm².
If the volume of the small cone is 12.5 cm³, find the volume of the large cone in cm³.

12 **a** Twenty-seven small steel spheres are melted down and recast into one larger one.
The original radius of each sphere was 1 cm. What is the radius of the new steel sphere?

b This new steel sphere is now compared with an even larger one. These two spheres have curved surface areas of $36\pi\,\text{cm}^2$ and $144\pi\,\text{cm}^2$ respectively.
Write down in ascending order the ratio of the radii of all three different-sized spheres.

14 Assess

D

1 Copy this shape and enlarge it with scale factor 2, using the centre of enlargement shown.

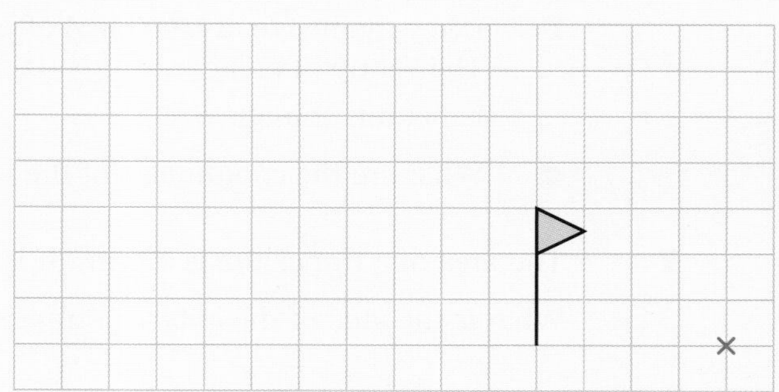

D C

2 **a** **i** Draw a pair of axes with x- and y-values from 0 to 16.
ii Plot and label triangle A with vertices (6, 7), (6, 11) and (10, 11).

b Draw the image of triangle A after an enlargement of scale factor 3.
The centre of enlargement is (7, 10).
Label this triangle B.

c What are the coordinates of the vertices of B?

C

3 The diagram below shows two different sizes of wheels. The dimensions labelled are the diameters of the alloy and of the wheel plus the alloy. These are still measured in inches. An alloy is sometimes referred to as the hub cap.

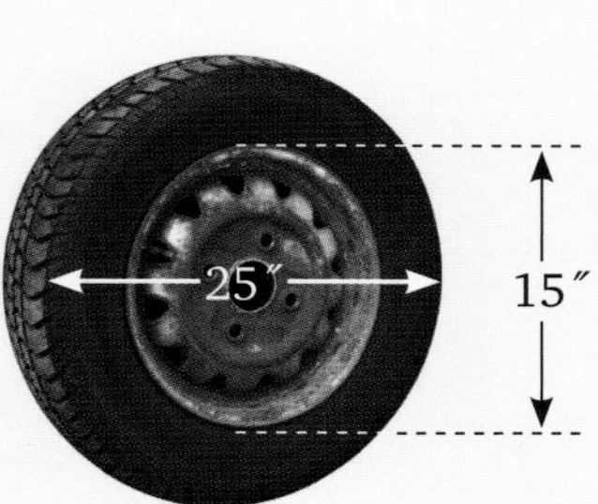

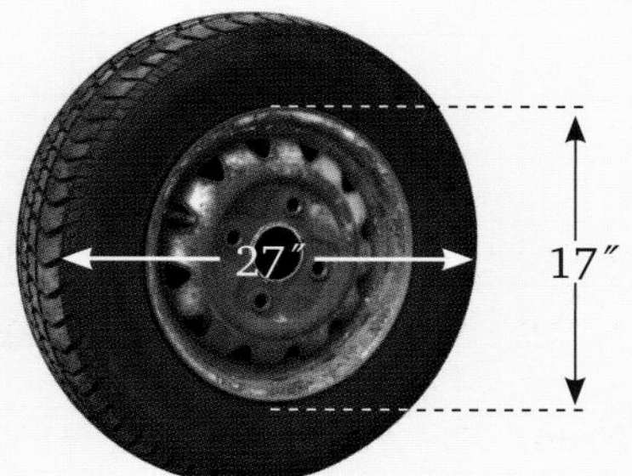

a What is the ratio of the corresponding diameters of the alloys?
Give the ratio in the form $1 : n$. Give your answer to two decimal places.

b What is the ratio of the corresponding external diameters? (alloy + tyre)

c Is the larger wheel an enlargement of the smaller one?

C B

4 The following are similar triangles.

Not drawn accurately

a Find the ratio of the corresponding sides in the form $1 : n$

b Find the missing lengths.

c Find the area of each of the triangles.

d What is the ratio of areas in the form $1 : n$
Give your answer to two decimal places.

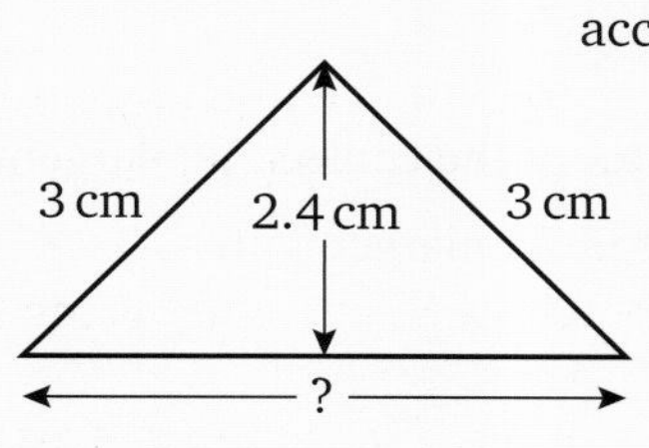

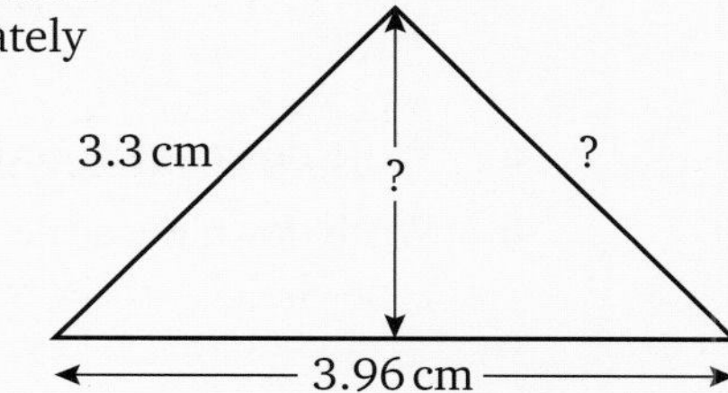

B

5 **a** **i** Draw a pair of axes with x- and y-values from -10 to 6.

ii Plot and label triangle A with vertices at $(-4, -2)$, $(-4, -10)$ and $(-8, -2)$.

b Draw the image of triangle A after an enlargement of scale factor $\frac{1}{4}$ with centre (4, 2).

Label the image B.

c What are the coordinates of the vertices of B?

A

6 **a** **i** Draw a pair of axes with x-values from -10 to 5 and y-values from -6 to 5.

ii Plot and label triangle A with vertices (1, 2), (3, 2) and (3, 3).

b Draw the image of triangle A after an enlargement of scale factor -3. The centre of enlargement is (0, 1).

Label this triangle B.

c What are the coordinates of the vertices of B?

7 The area of a trapezium is $12\,\text{cm}^2$. It is enlarged by a scale factor for length of 4.

What is the area of the enlarged trapezium?

AQA Examination-style questions

1 In the diagram, shape B is an enlargement of the shaded shape A.

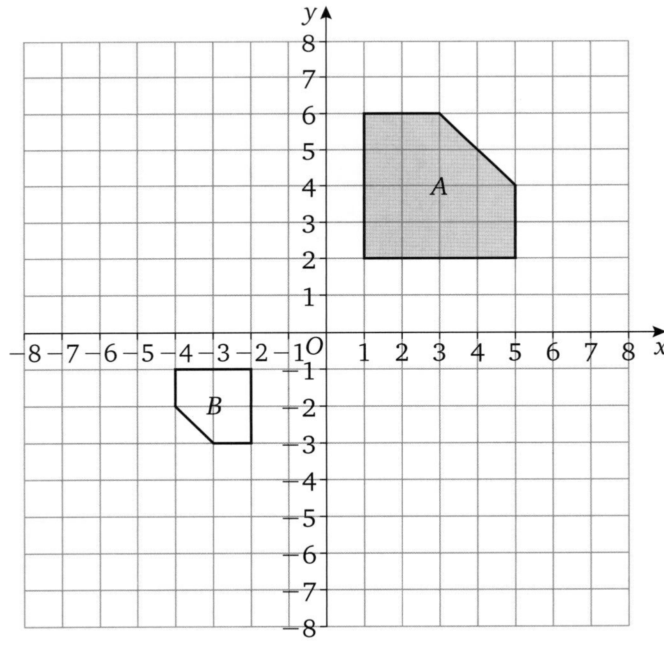

a Write down the coordinates of the centre of enlargement. *(1 mark)*

b Write down the scale factor of enlargement. *(1 mark)*

AQA 2005

2 Triangle PQR is an enlargement of triangle ABC with scale factor $\frac{5}{4}$

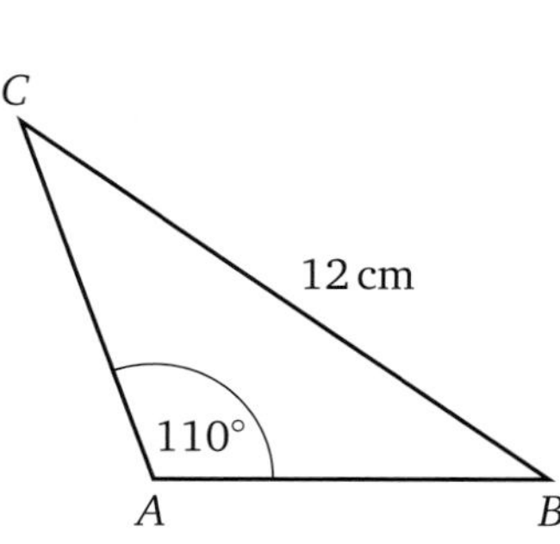

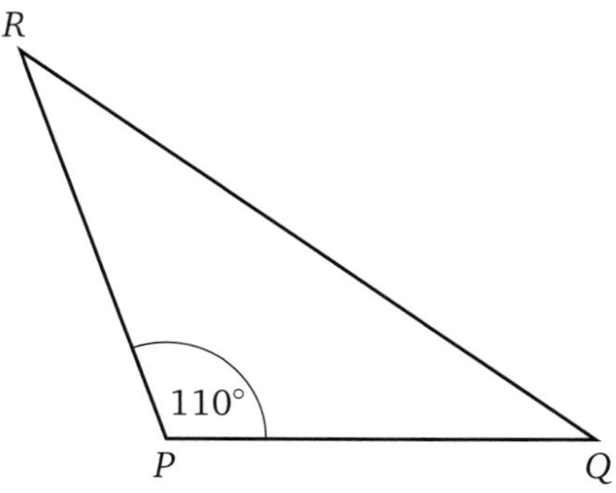

Not drawn accurately

Calculate the length of RQ.

(2 marks)

AQA 2007

3 The diagram shows three mathematically similar containers.

small

medium

large

Not drawn accurately

The table shows some information about the containers.

	Height (cm)	Area of top of container (cm^2)	Volume (cm^3)
small	12	X	400
medium	24	500	
large	36		Y

Calculate the missing entries X and Y.

(4 marks)

AQA Examiner's tip

Make sure that you show all the working on the answer page. Don't be tempted to write just the answers for X and Y in the table.

AQA 2007

15 Construction

Objectives

Examiners would normally expect students who get these grades to be able to:

D

draw a quadrilateral such as a kite, parallelogram or rhombus with given measurements

understand that giving the lengths of two sides and a non-included angle may not produce a unique triangle

C

construct perpendicular bisectors and angle bisectors

match one angle and one side of congruent triangles given some dimensions

B

construct perpendicular lines from a point to a line, perpendicular at a point on a line and an angle of 60°

match sides and angles of similar triangles

A

prove two triangles are congruent

prove construction theorems.

Key terms

construction	similar
equilateral	corresponding
perpendicular	scale factor
bisector	congruent

Did you know?

This type of mathematics is part of geometry and is very old

The ancient Greek mathematician Euclid is the inventor of geometry. He did this over 2000 years ago, and his book *Elements* is still the ultimate geometry reference. He used construction techniques extensively. They give us a method of drawing things when simple measurement is not appropriate.

You should already know:

- ✔ how to use scales and scale diagrams
- ✔ how to recognise enlargements
- ✔ how to find scale factors
- ✔ facts about angles in circles
- ✔ facts about angles in parallel lines
- ✔ angle properties of polygons
- ✔ how to use compasses
- ✔ how to measure an acute and an obtuse angle with a protractor.

15.1 Drawing triangles accurately 

There are different ways to draw a triangle accurately. The method you use depends on what you know about the triangle.

Drawing a triangle when all three sides are known

The following example shows how to draw a triangle when all three sides are known.

Example: Draw a triangle with sides 4.2 cm, 5.3 cm and 6 cm.

Solution: First draw a sketch to see what the triangle looks like.

Now, draw your **construction**. Draw and measure the longest side using a ruler and pencil. You should clearly mark end points on the line.

Open your compasses to the same length as one of your other sides. It does not matter which one you draw first. Put the point of the compasses on one end of the line and draw an arc.

Do the same from the other end of the line, with the compasses open to the length of the other side. Make sure that your arcs cross each other.

The arcs cross at the position of the third vertex. Join this point to the ends of the lines and label the diagram with the side lengths.

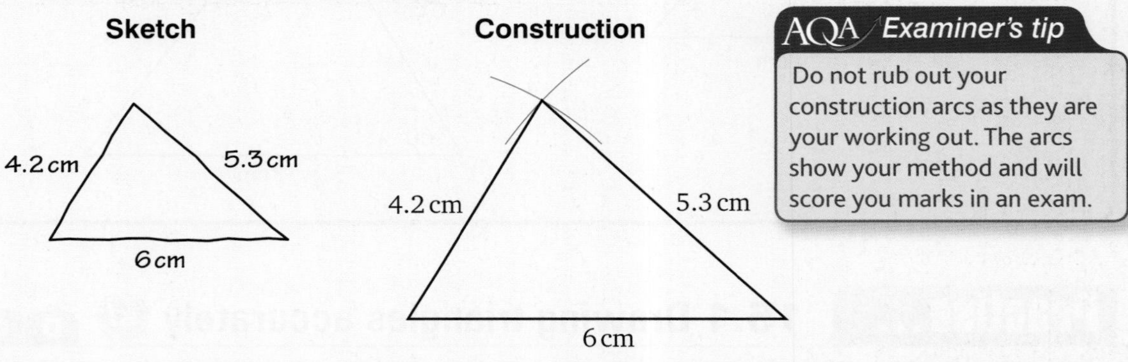

AQA Examiner's tip

Do not rub out your construction arcs as they are your working out. The arcs show your method and will score you marks in an exam.

Drawing a triangle when one side and two angles are known

The following example shows how to draw a triangle when one side and two angles are known.

Example: Construct a triangle with base 8 cm and two angles at the base of 50° and 40°.

Solution: Start by drawing the base, then measure one of the given angles carefully. It does not matter which one you measure first.

Mark a point on the protractor's scale at the value of the first angle and draw in the line. Extend it beyond the point you have marked. It is much better to have a line that is too long than too short.

Repeat for the other angle.

Extend this line to meet the line you have already drawn.
Label the triangle with two known angles and one known side.

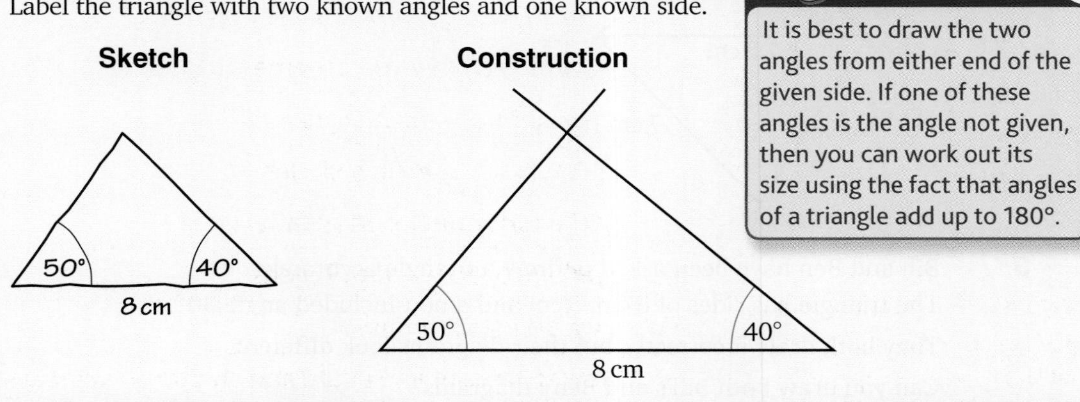

AQA Examiner's tip

It is best to draw the two angles from either end of the given side. If one of these angles is the angle not given, then you can work out its size using the fact that angles of a triangle add up to 180°.

Drawing a triangle when one angle and two sides are known

The following example shows how to draw a triangle when one included angle and two sides are known.

Example: Construct a triangle with two sides 6 cm and 7 cm, and an included angle of 60°.

Solution: Start with the longest side, then measure the angle given, using a protractor.
Draw in the side. Make it long enough as before; it is better too long than too short.
Now measure this side carefully. Mark on the line the point where it should end.
Join from this point to the end of the starting line.
Label the triangle.

Sketch

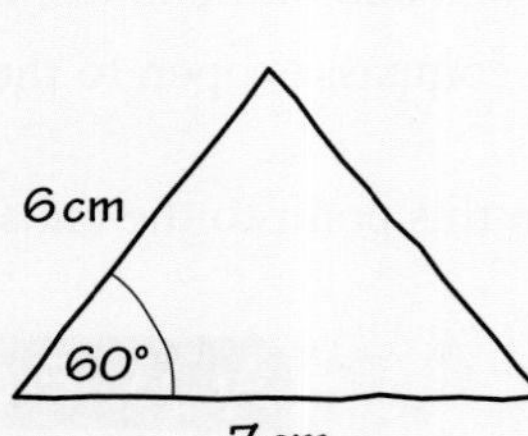

Construction

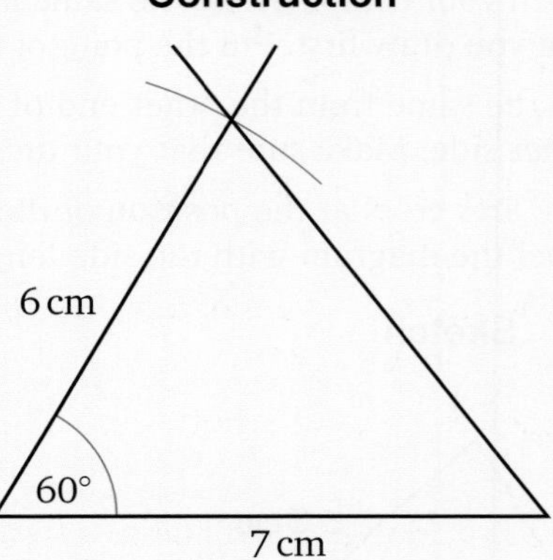

Practise... 15.1 Drawing triangles accurately

The shapes in these exercises have not been drawn accurately.

D

1 Draw these triangles accurately.

a

6 cm 8 cm 10 cm

b

50° 10 cm 40°

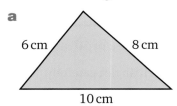

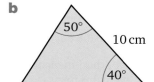

c

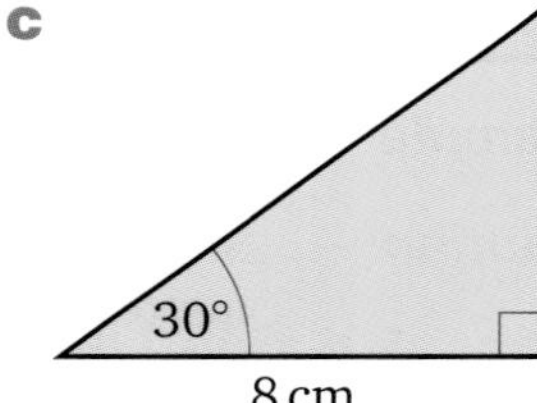

2 Draw this shape accurately.

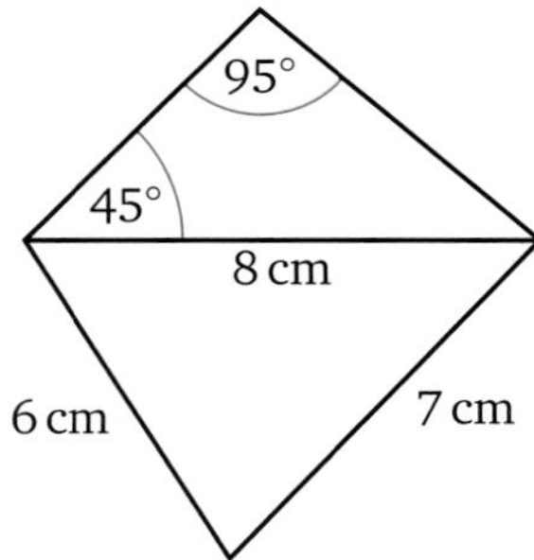

3 Bill and Ben have been asked to draw a triangle accurately.
The triangle has sides of 8 cm, 5 cm and a non-included angle 30°.
They both draw it correctly, but their diagrams look different.
Can you draw both Bill's and Ben's diagrams?

D

4 Draw a rhombus accurately, which has all sides equal to 6 cm and a shorter diagonal of 7 cm.

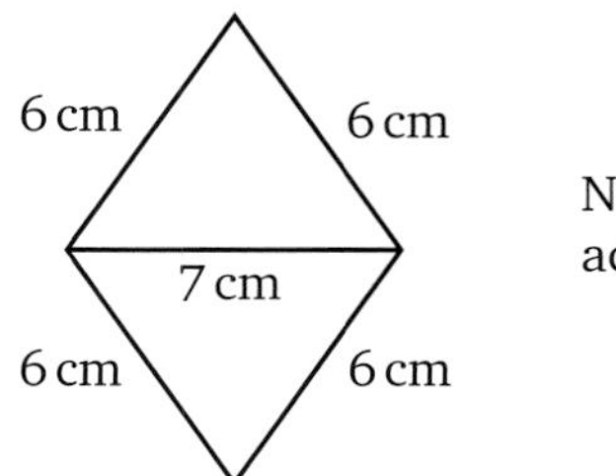

Not drawn accurately

5 John is making a triangular prism from card.

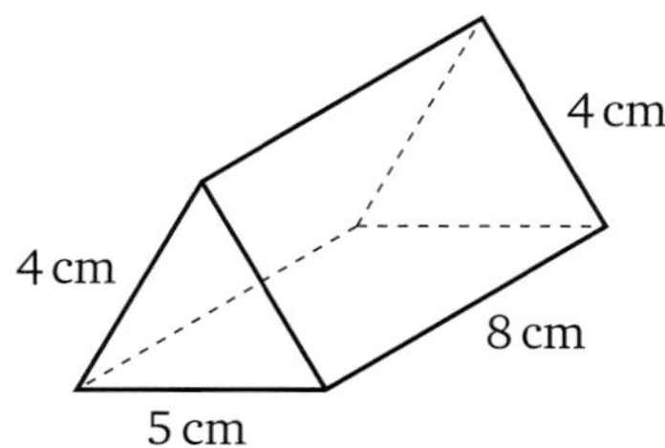

Not drawn accurately

Draw an accurate net for his prism.

6 Make an accurate drawing of a rectangle that has longer sides of 8 cm and diagonals of length 10 cm.

What is the length of the shorter sides?

7 A piece of material is cut to make a skirt.
The material is shown in this sketch and is in the shape of an isosceles trapezium.

Using a scale of 2 cm to 50 cm, make an accurate scale drawing of this piece of material.

Not drawn accurately

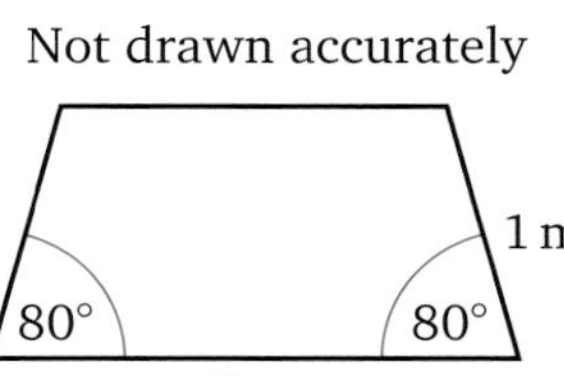

8 A field is in the shape of a pentagon as shown.

a Make a scale drawing of the field.

b Use your scale drawing to find the perimeter of the field.

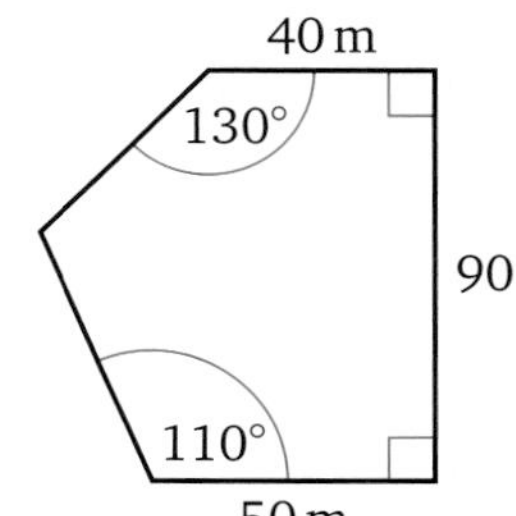

Not drawn accurately

9 On a map, two hotels (Sleep Inn and Stay Well Inn) are 8.8 cm apart. Sleep Inn is due north of Stay Well Inn. A third hotel (Lie Inn) is due east of Stay Well Inn and on a bearing of 143° from Sleep Inn.

a Draw the positions of the hotels accurately.

b If the map is drawn using a scale of 1 : 50 000, what are the actual distances between the hotels?

 10 This is a sketch of Hilary's bedroom.

She plans to reorganise her room, using a scale drawing to help decide where to put her furniture. She uses an A4 sheet of graph paper.

Draw an accurate plan of her bedroom using an appropriate scale.

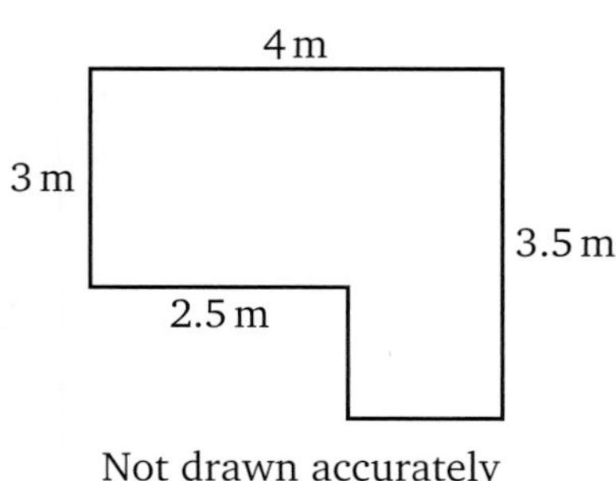

Not drawn accurately

11 The diagram shows the front elevation of Patrick's garage.

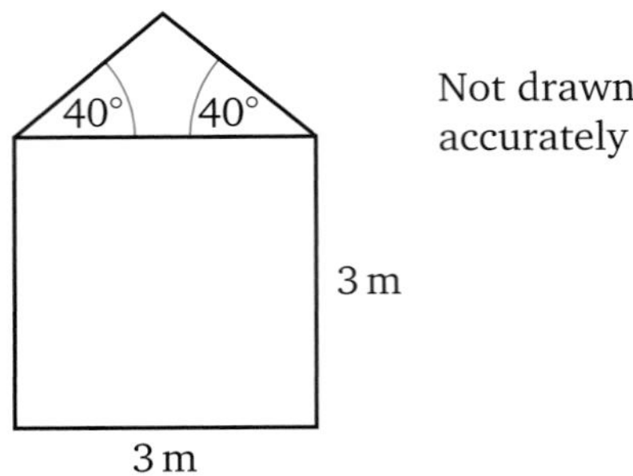

Patrick has a boat on a trailer, which he wants to keep in the garage.
The top of the mast is 4.4 metres above the ground.
Can Patrick keep the boat in the garage? Explain your answer.

Learn... 15.2 Constructions

Constructions are drawn using only a straight edge and a pair of compasses. You need to be able to construct **equilateral** triangles, the **perpendicular** bisectors of a line, and angle bisectors; perpendiculars from a point to a line, and perpendiculars at a point on a line.

A **bisector** is a line that cuts something into two equal parts. A line bisector cuts a line into two equal parts. An angle bisector cuts an angle into two equal parts.

Equilateral triangles

The following example shows how to construct an equilateral triangle.

Example: Construct an equilateral triangle.

Solution: Start with a line that will become one of the sides in your triangle, with a point P, where one vertex will be.

Open your compasses to the length of one side. With the point of your compasses on P draw a large arc that intersects the line at Q.

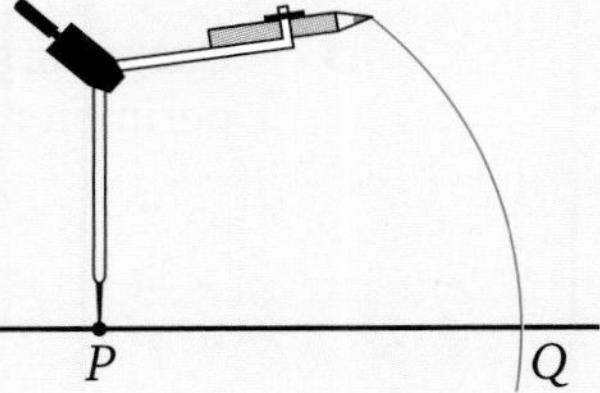

Keep the radius of your compasses the same. Put the point of the compasses on Q and draw an arc that passes through P and cuts the first arc at R.

Join P to R, and Q to R. You have now finished your construction.

You can use this technique to construct an angle of 60°. Follow the first two steps above, and then just join P to R (or R to Q).

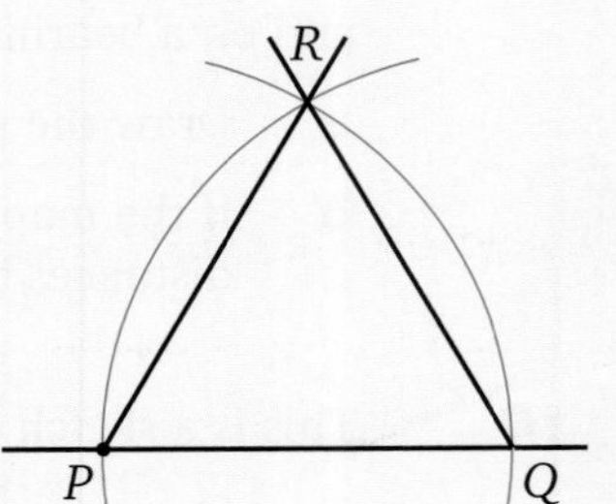

Remember to leave your arcs. They show your method and will score you marks.

Line bisectors

The following example shows how to construct the bisector of a line.

Example: Construct the bisector of line AB.

A ——————— B

Solution: Open your compasses to more than half of AB. Put the point on A and draw arcs above and below AB.

A ——————— B

Keep the radius of your compasses the same. Put the point of your compasses on B and draw two new arcs to cut the first two at C and D.

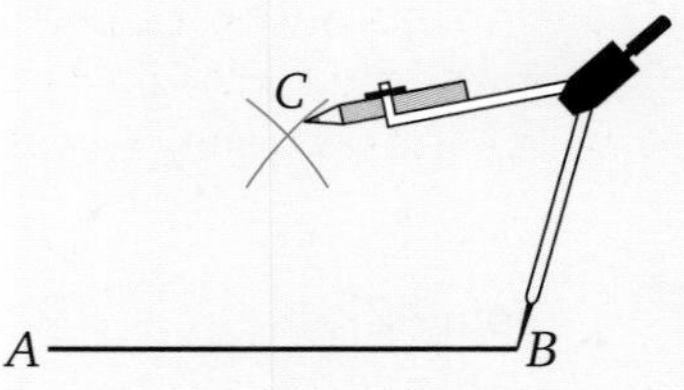

Join CD.

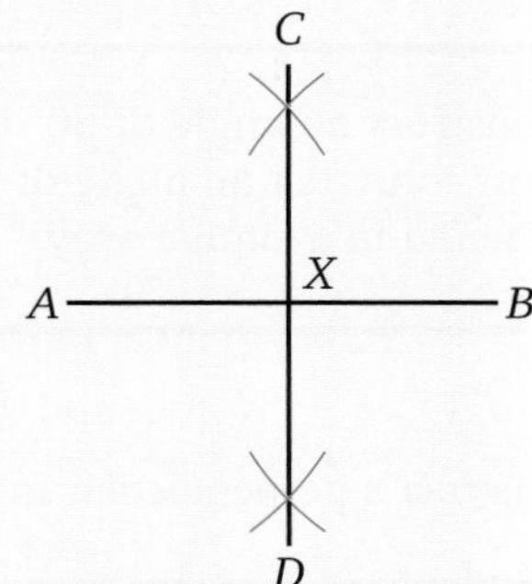

X is the midpoint of AB. CD not only bisects AB, it is called the perpendicular bisector of AB. This is because it meets AB at 90°.

AQA Examiner's tip

You have constructed a rhombus whose diagonals bisect each other at 90°.

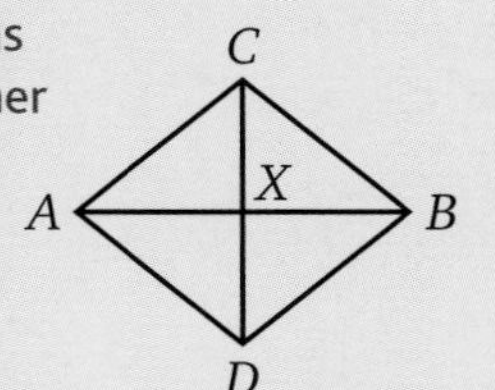

Bump up your grade

Learning to construct perpendiculars and bisectors is a Grade C skill. Learn to do these accurately to help you to get a Grade C in your exam.

Angle bisectors

The following example shows how to construct the bisector of an angle.

Example: Construct the bisector of angle BAC.

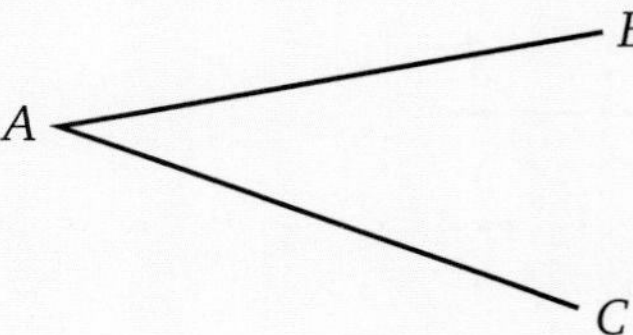

Solution: Open your compasses to less than the length of the shorter line. Put the point on A and draw arcs to cut AB at X and AC at Y.

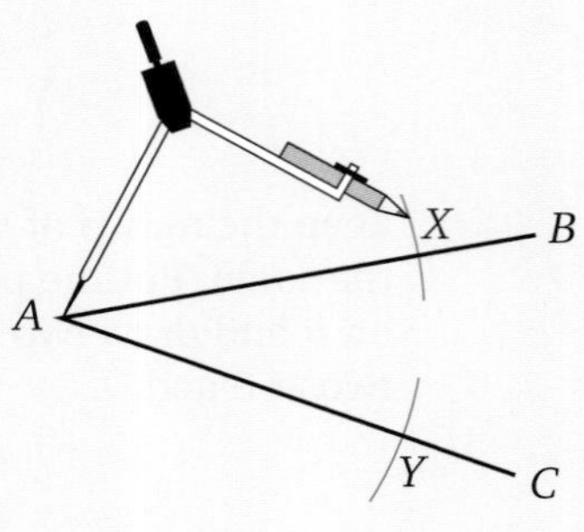

Keep the radius of your compasses the same. Put the point of your compasses on X and Y in turn and draw arcs that intersect at Z.

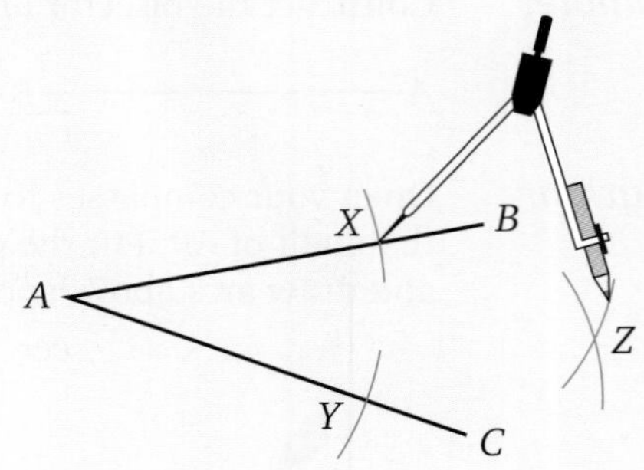

Join AZ.

AZ is the angle bisector of angle BAC.

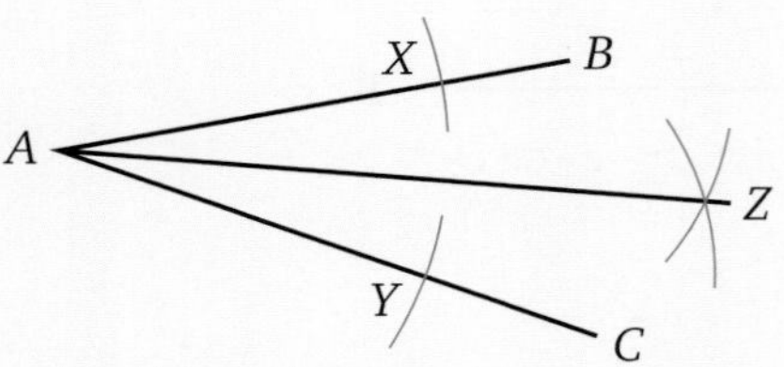

To construct an angle of 30° you first construct an angle of 60° (as part of an equilateral triangle), then bisect it. To construct an angle of 45° you construct an angle of 90°, and then bisect it. There are lots of other possible angles that can be constructed in a similar way.

Perpendiculars

The following examples show how to construct a perpendicular from a point to a line, and at a point on a line.

Example: Construct a perpendicular from a point to a line.

Solution: With the point of your compasses on P, draw two arcs that intersect the line at A and B.

P•

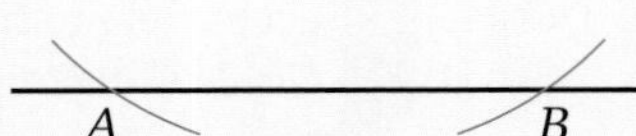

Put the point on A and B in turn and draw arcs that intersect at C.

P•

Join PC. This is the perpendicular to the original line from the point P.

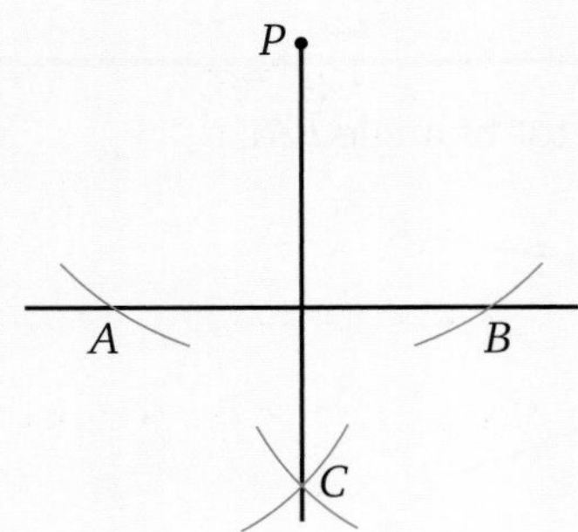

Example: Construct a perpendicular at a point P on a line l.

Solution:

P l

With the point of the compasses on P, draw two arcs to cut the line either side of P.

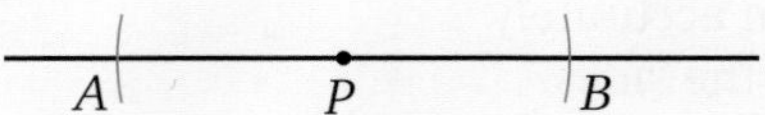

Make the radius of your compasses larger.
Put the point on A and B in turn drawing arcs that intersect.

Where the arcs intersect, label this new point C.

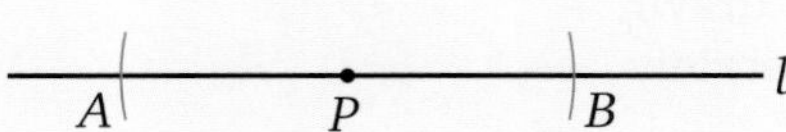

Join the point of intersection to P.

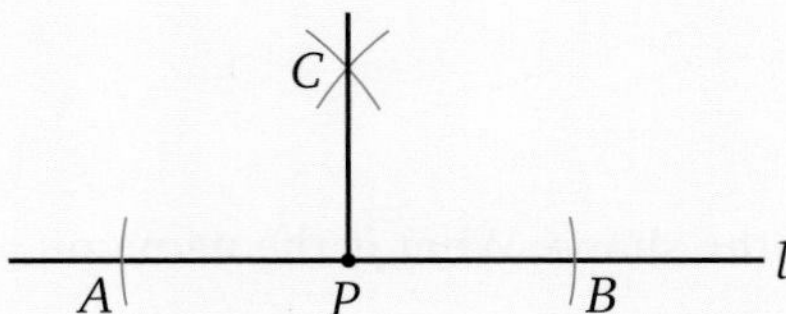

This method simply 'made' a line segment with P as the midpoint, and constructed the perpendicular bisector of it.

Practise... 15.2 Constructions

C

1 Construct the perpendicular bisector of a line 8 cm long.

2 Draw this rectangle accurately.

Using only a ruler and compasses, construct the perpendicular bisector of the diagonal BD.

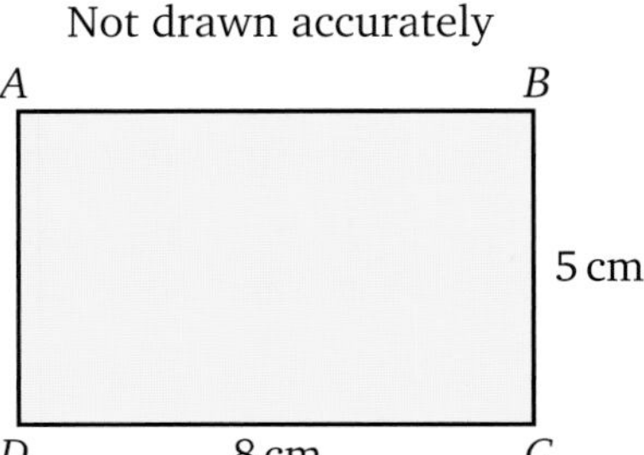

3 Construct an equilateral triangle of side 6 cm.

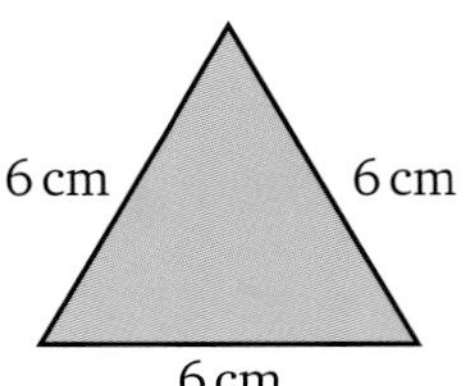

Not drawn accurately

4 Construct an angle of 60°. Bisect your angle to show an angle of 30°.

C

5 Draw a line segment 10 cm long.

a Construct the perpendicular bisector of your line segment.

b Your diagram shows four right angles. Bisect one of them to show an angle of 45°.

6 Draw a triangle with sides 8 cm, 9 cm and 10 cm accurately. Construct the perpendicular bisector of each of the sides.

What do you notice?

7 a Use only a ruler and compasses to construct this net accurately.

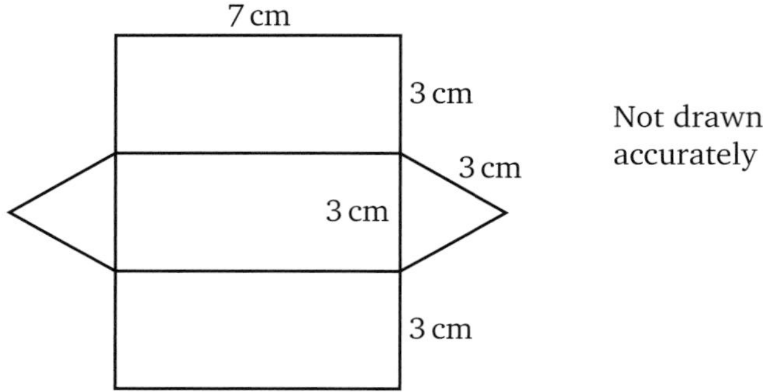

Not drawn accurately

b Put tabs on your net, cut it out and make the shape. What is the name of the shape you have made?

8 Accurately draw a triangle with sides 8 cm, 9 cm and 10 cm. Construct the angle bisector for each of the angles. What do you notice?

9 a Construct this rectangle.

b Label your diagram carefully and draw in the diagonal AC.

c Construct the perpendicular from vertex B to the diagonal AC.

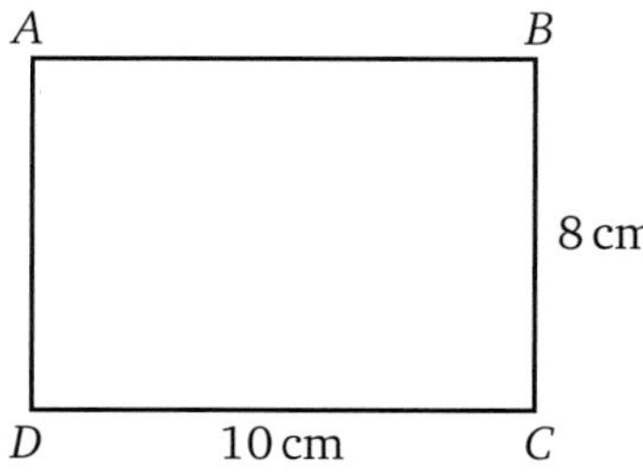

Not drawn accurately

10 Construct a right-angled triangle that has a hypotenuse 10 cm long and a shorter side 8 cm long.

Measure the third side. You may find it helpful to draw a sketch of the triangle first.

11 The diagram shows the roof truss design for a house.

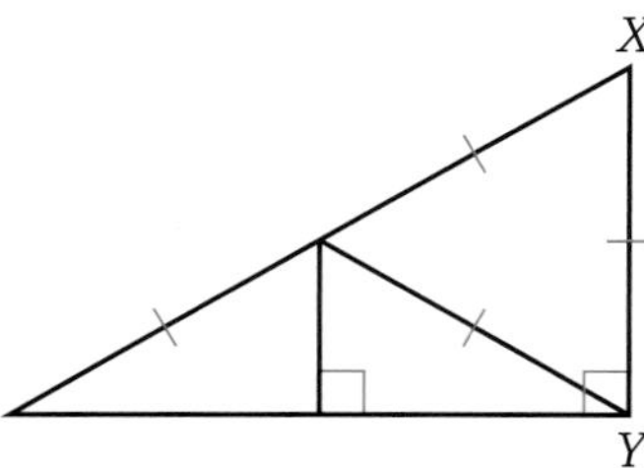

Construct the diagram starting with $XY = 5$ cm. Measure the width of the base of the roof truss to the nearest millimetre.

? **12** This diagram shows the space under the stairs in a house.

Jack wants to put a cupboard under the stairs.
The cupboard is 1.8 metres high, 40 centimetres deep and 1.2 metres wide. It has two doors on the front, each is 60 centimetres wide.

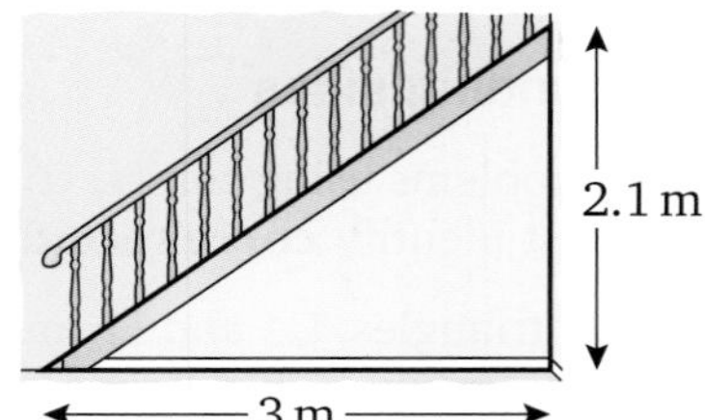

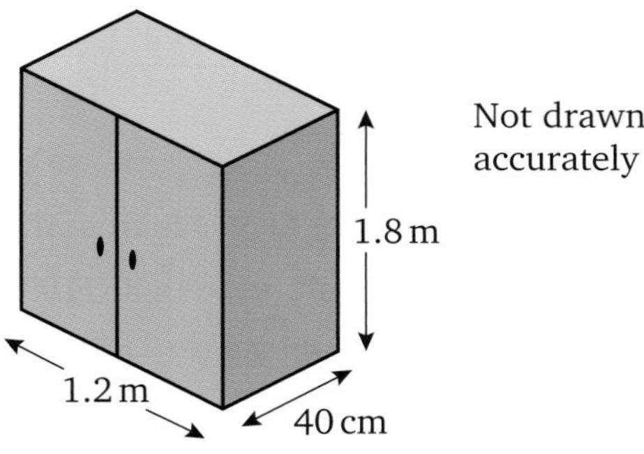

Not drawn accurately

Does this cupboard fit under the stairs? If so, can the doors open?
Show your working to justify your answers.

Learn... 15.3 Similar shapes

Two shapes are mathematically **similar** if they have the same shape but different sizes. That is, one shape is an enlargement of the other.

Shapes that are similar have the same angles

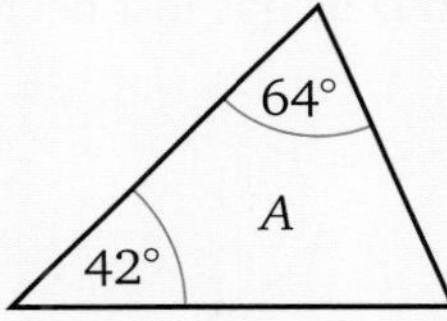

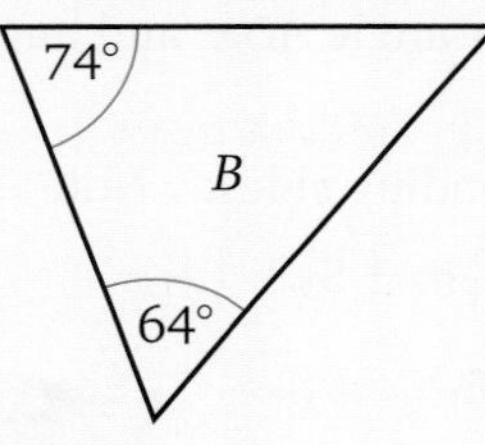

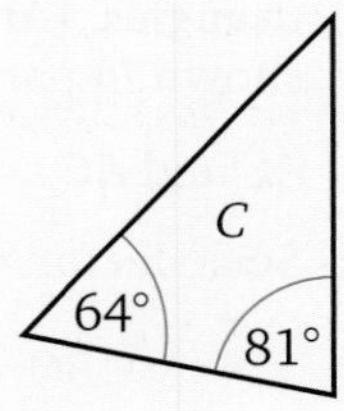

Not drawn accurately

These triangles are not all similar.

In triangle A the third angle $= 180° - (42° + 64°) = 74°$ So the angles are 42°, 64° and 74°.

In triangle B the third angle $= 180° - (74° + 64°) = 42°$ So the angles are 42°, 64° and 74°.

In triangle C the third angle $= 180° - (81° + 64°) = 35°$ So the angles are 35°, 64° and 81°.

So, triangles A and B are similar as they have the same angles.

Note: for shapes with more than three sides, the angles need to be the same and in the same order.

Consider the following two diagrams.

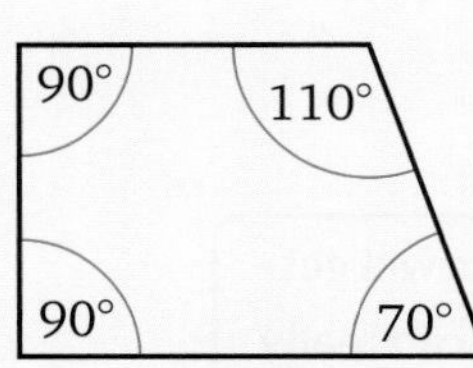

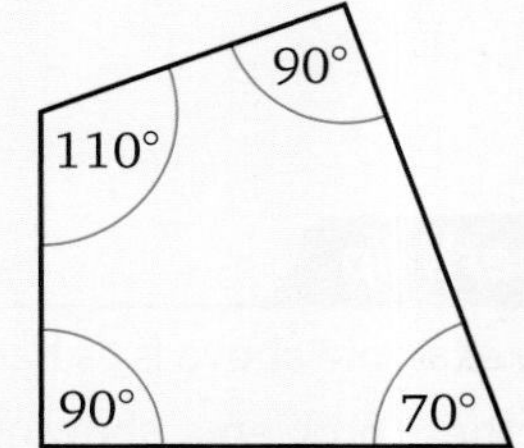

Not drawn accurately

These quadrilaterals have the same angles, but the angles are not in the same order.
They are therefore not similar.

Corresponding sides

To solve problems using similar triangles, you need to first identify **corresponding** sides.

In the two triangles, *CA* and *EF* are corresponding sides. This means they are in the same position in each triangle. They are both between the angles 42° and 64°.

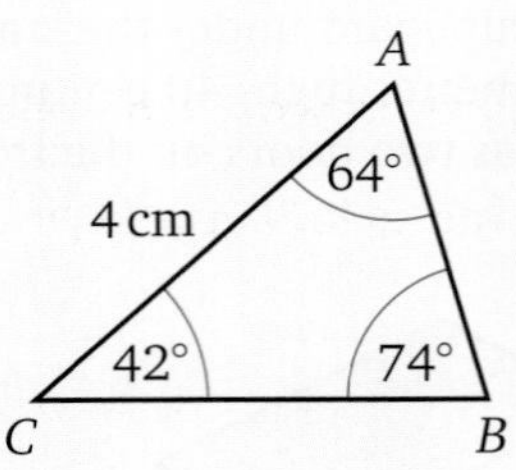

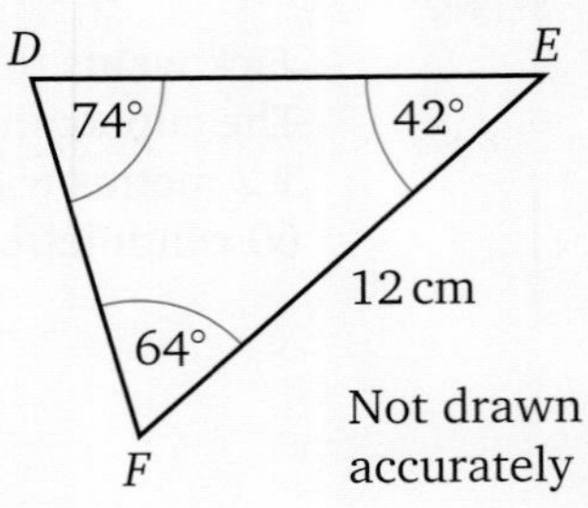

Scale factor

To find a **scale factor**, you divide one side by the corresponding side from the other triangle.

For the two triangles above, the scale factor from triangle *ABC* to triangle *FDE* is $\frac{12}{4} = 3$

You can use this to find the lengths of other sides, for example $ED = 3 \times CB$

Example: In the diagram *DE* and *BC* are parallel, *BC* is 12 cm, *AE* is 10 cm and *EC* is 5 cm.

a Explain why triangles *ADE* and *ABC* are similar.

b Find *DE*.

c If *DB* is 4 cm, find *AD*.

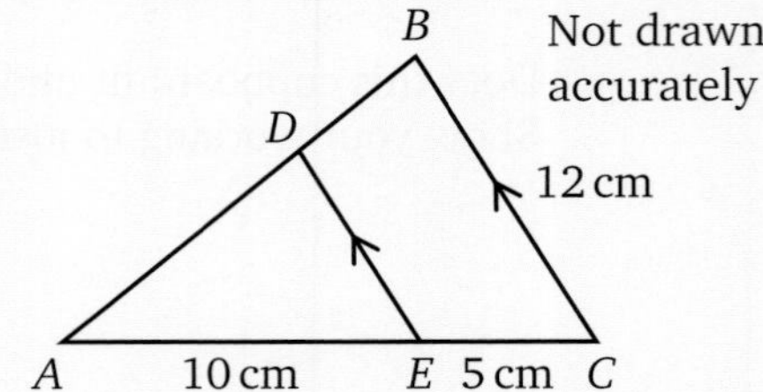

Solution:

a Angle *A* is common to both triangles.

Angle *ABC* = angle *ADE* as they are corresponding angles (*BC* and *DE* are parallel).

Angle *ACB* = angle *AED* as they are corresponding angles (*BC* and *DE* are parallel).

As both triangles have the same angles, they are similar.

b *DE* and *BC* are corresponding sides as they are in the same position in the similar triangles. (Angle *ABC* = angle *ADE*, and angle *ACB* = angle *AED* as has just been shown in part **a**.)

AE and *AC* are corresponding sides. Note $AC = 15$ cm $(10 + 5 = 15)$

Scale factor $= \frac{AC}{AE} = \frac{15}{10} = 1.5$

$BC = 1.5 \times DE$

$12 = 1.5 \times DE$

$DE = \frac{12}{1.5} = 8$ cm

See Chapter 6 for more information on how to solve equations.

c $AB = 1.5 \times AD$

We know $AB = AD + DB$, and $DB = 4$

So, $1.5 \times AD = AD + 4$

$0.5 \times AD = 4$ Subtract *AD* from each side, as you did with equations.

$AD = \frac{4}{0.5} = 8$

$AD = 8$ cm

AQA Examiner's tip

Finding *AD* in the example above is as hard as these questions will get. Learning to solve these problems, showing your working out, can really boost your marks.

Practise... 15.3 Similar shapes

D C B A A*

B

The shapes in these exercises are not drawn accurately.

1 Sort the following triangles into groups of similar triangles.

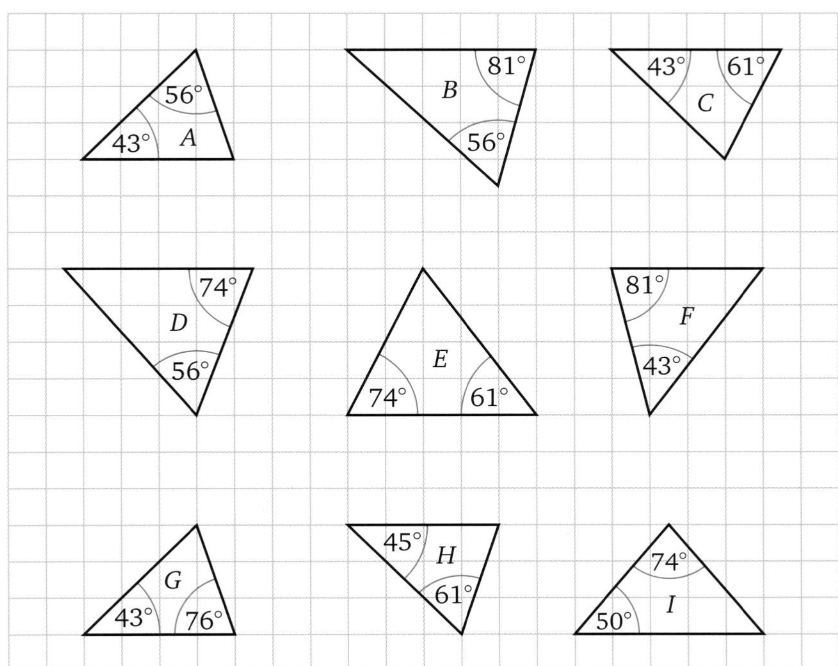

Not drawn accurately

2 Name three pairs of corresponding sides in these triangles.

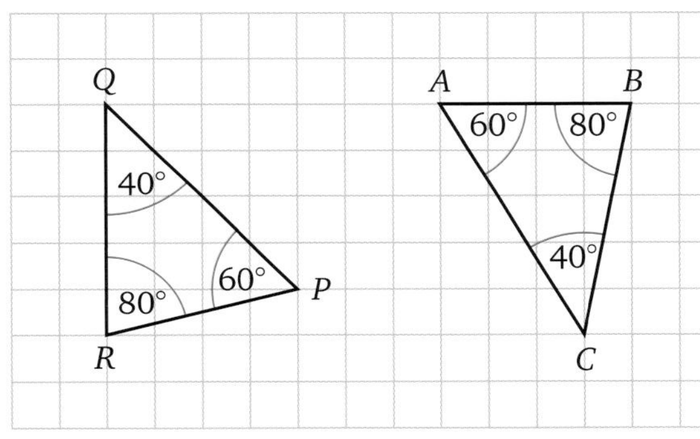

Not drawn accurately

3

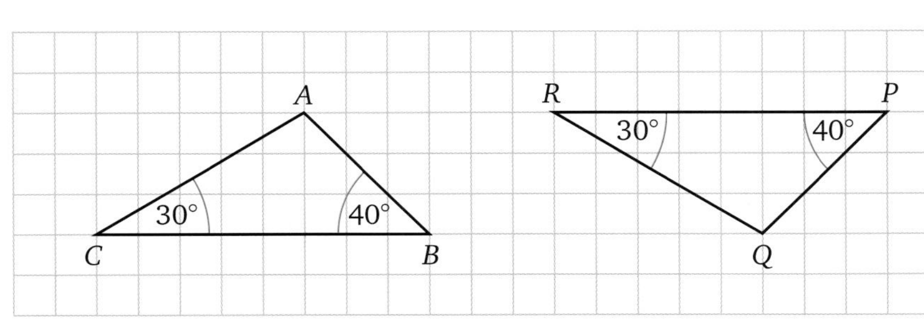

Not drawn accurately

a Show that these triangles are similar triangles.

b Name all the pairs of corresponding sides.

B

4 These pairs of shapes are all similar, but are not drawn accurately. Find the missing angles and sides.

a

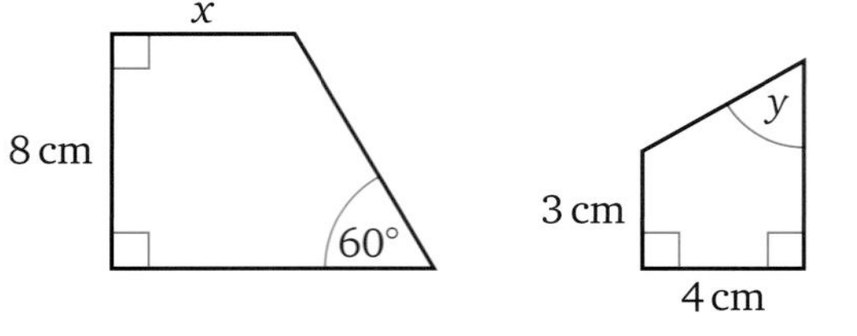

Not drawn accurately

b

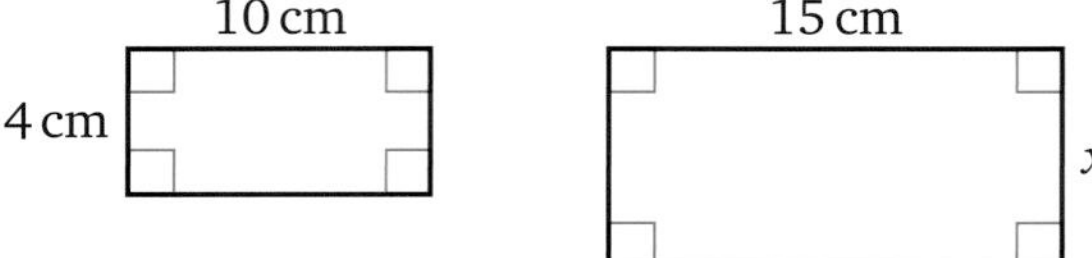

c

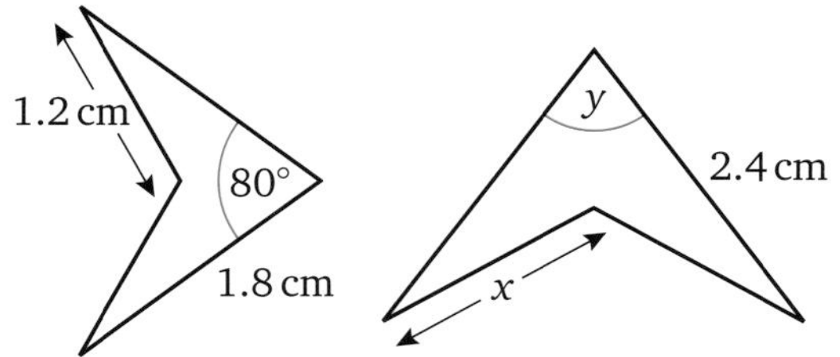

d

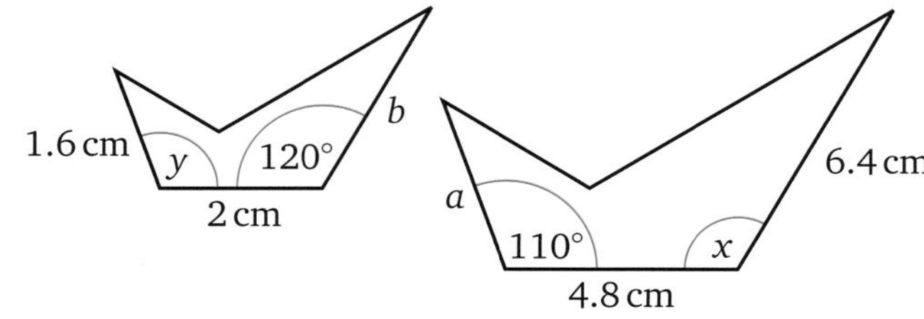

5 Each pair of triangles is similar.

Find the missing sides and angles.

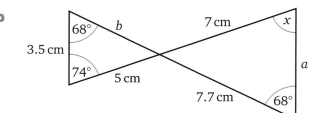

a

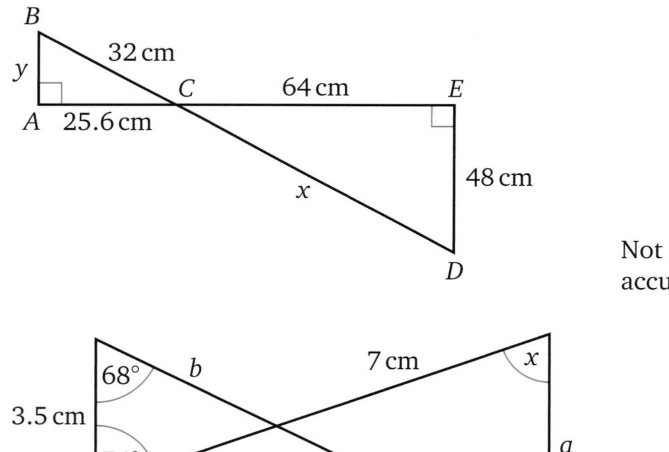

Not drawn accurately

b

B

6 Find the sides marked with the letters in these diagrams.

a

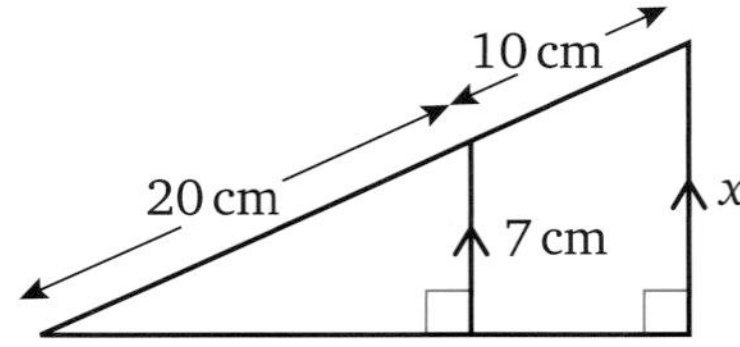

c

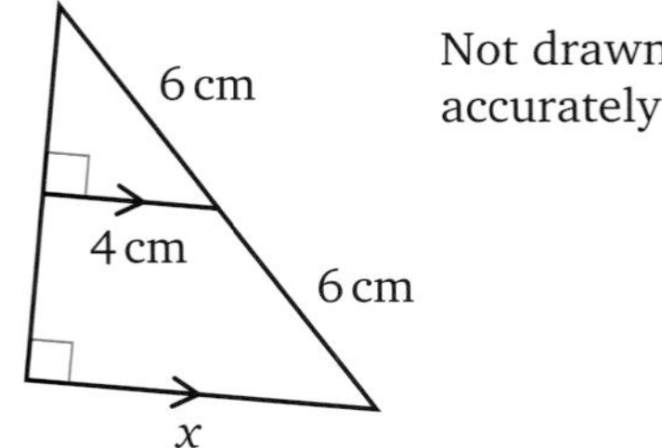

Not drawn accurately

b

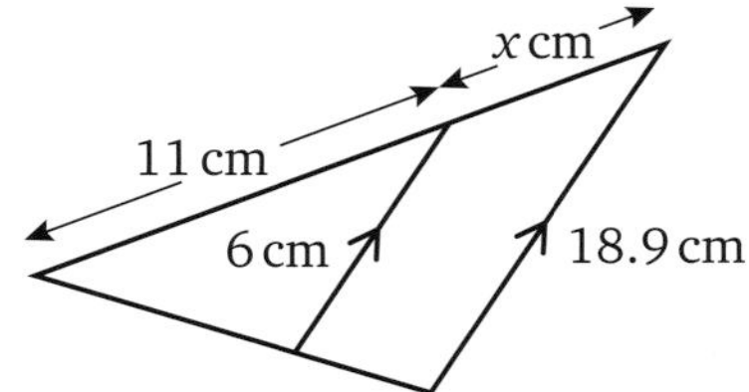

d

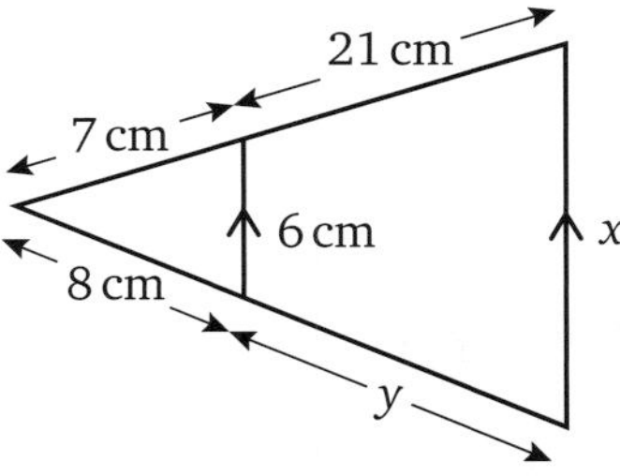

7 In this triangle *BC* and *DE* are parallel.

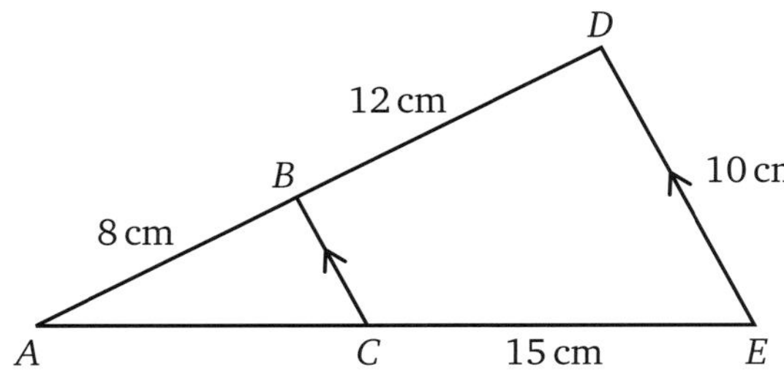

Not drawn accurately

a Explain how you can tell that triangle *ABC* and triangle *ADE* are similar.

b Work out: **i** *BC* **ii** *AC*

8 *ABCD* is a trapezium with *AD* parallel to *BC*.

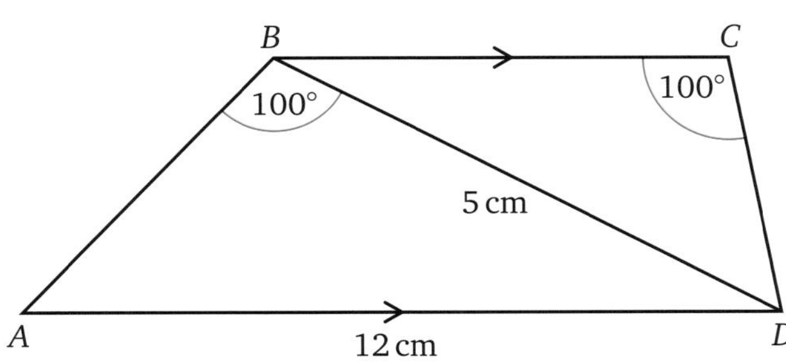

Not drawn accurately

a Explain why triangle *ABD* is similar to triangle *DCB*.

b Find the length of *BC*.

9 In the diagram *CB* is parallel to *ED*.

Find the lengths *AB* and *AC*.

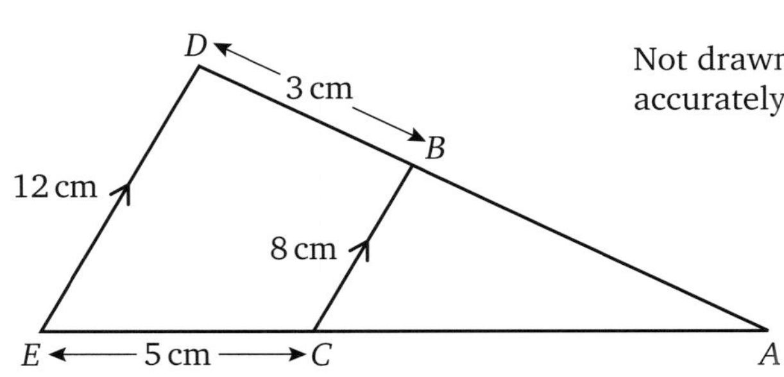

Not drawn accurately

10 **a** Explain a method that can be used to find an estimate of the height of the tree in the diagram.

Your method should include any measuring equipment you need.

b Using your method, find a tree, flagpole or tall building near you and calculate an estimate of its height.

Learn... 15.4 Congruent shapes

Two shapes are **congruent** if they have both the same shape and the same size. That is, the two shapes are identical.

When two triangles *ABC* and *DEF* are congruent you write $\triangle ABC \equiv \triangle DEF$. The '≡' sign means 'is congruent to'. The order of the letters is important, in this case it tells you that $\angle A = \angle D$, $\angle B = \angle E$, and $\angle C = \angle F$

You need to be able to prove that two triangles are congruent.

To do this, you need to show that one of these sets of conditions applies:

- Both triangles have three corresponding sides equal (SSS).
- Both triangles have two angles and one corresponding side equal (ASA).
- Both triangles have two corresponding sides and the angle between them equal (SAS).
- Both triangles have a right angle, hypotenuse and another side equal (RHS).

You need to give reasons for every fact you state.

Congruent triangles can be used to prove the constructions from Learn 15.2 in this chapter.

Example: **a** Prove that triangles *ABC* and *BCD* are congruent.

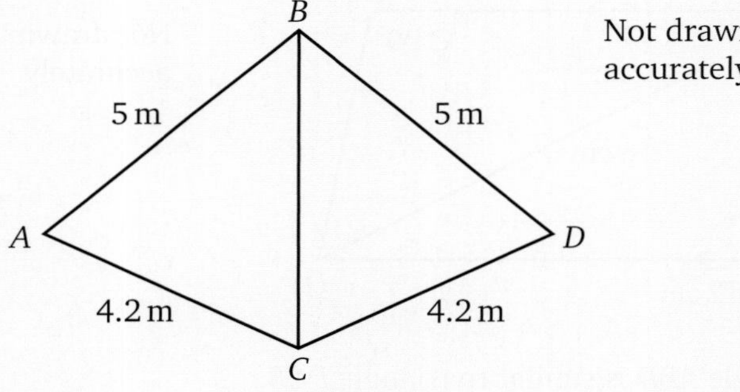

Not drawn accurately

b Prove that triangles *ABC* and *CDE* are congruent.

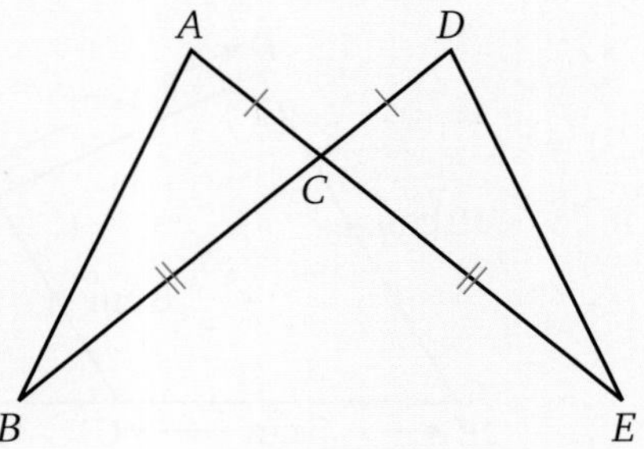

Solution: **a** In triangle ABC and triangle BCD:

$AB = DB$ (given)

$AC = DC$ (given)

$BC = BC$ (same line)

So triangle $ABC \equiv$ triangle DBC (SSS)

b In triangle ABC and triangle CDE:

$AC = DC$ (given)

$BC = EC$ (given)

angle ACB = angle DCE (opposite angles)

So triangle $ACB \equiv$ triangle DCE (SAS)

Example: Prove that the method for bisecting an angle does actually give the angle bisector.

Solution:

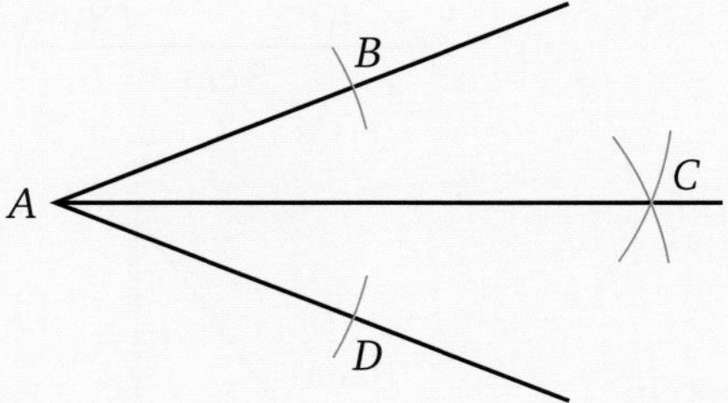

Think of this diagram as showing two triangles.

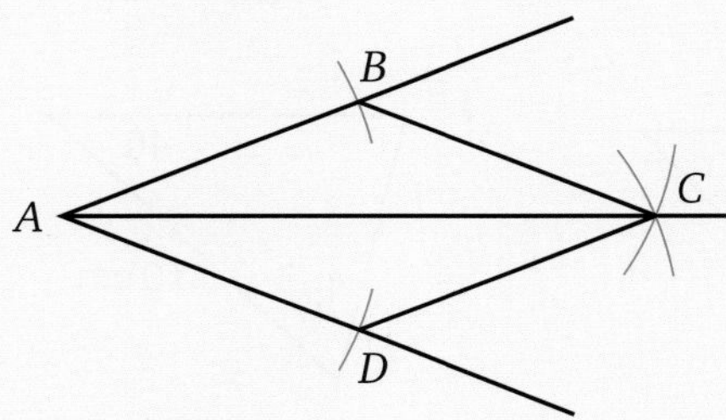

$AB = AD$ (given in method)

$BC = DC$ (given in method)

AC is common to both triangles.

So $\triangle ABC \equiv \triangle ADC$ (SSS)

and $\angle BAC = \angle DAC$

AQA Examiner's tip

When writing a proof, always give reasons clearly. Remember that the angles in a triangle **add up** to 180°. Do not say that they equal 180° because this implies that they are all 180°, which is incorrect.

Practise... 15.4 Congruent shapes 

B

1 State whether each of the following pairs of triangles is congruent.
For each pair that is congruent, give the condition that is being met.
The shapes are not drawn accurately.

a

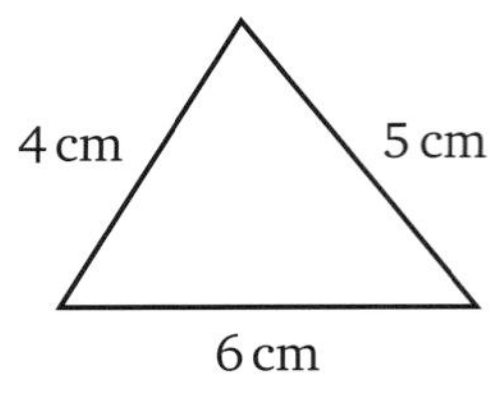

b

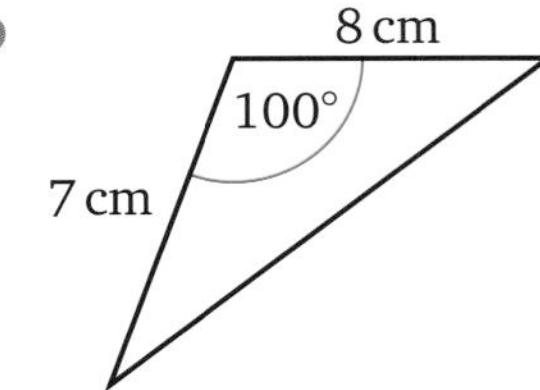

c

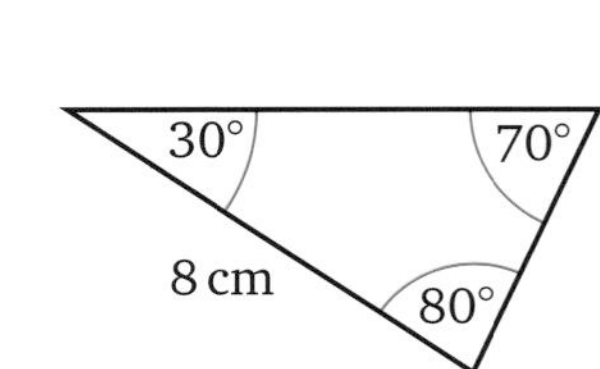

d

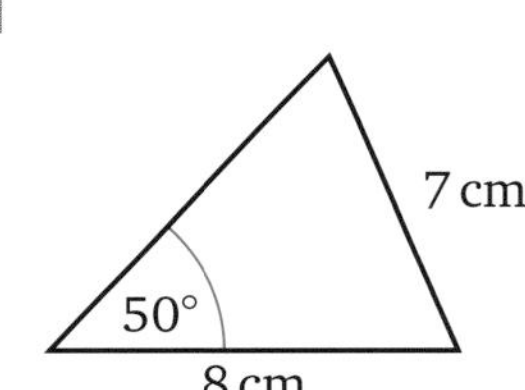

e

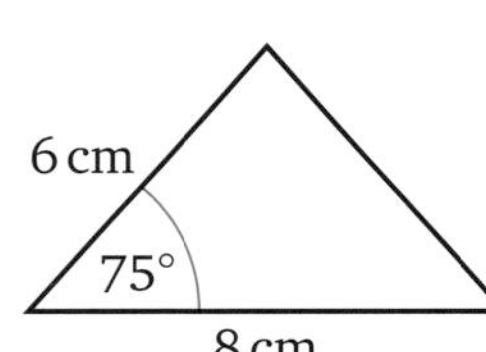

f

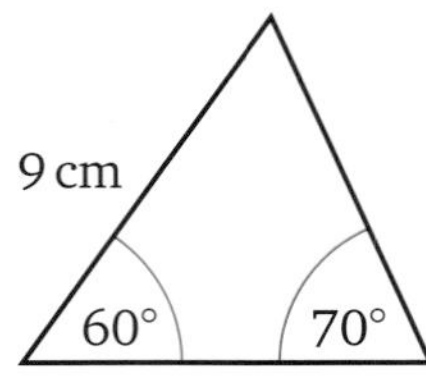

g

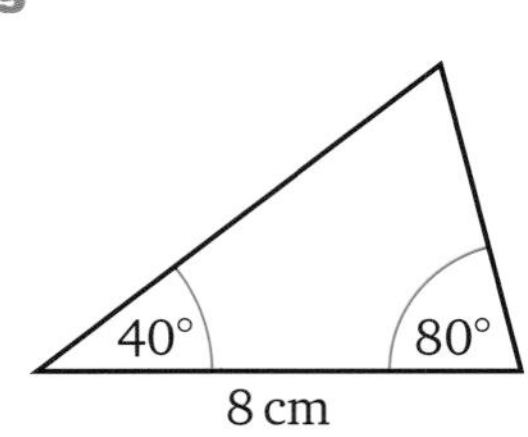

h

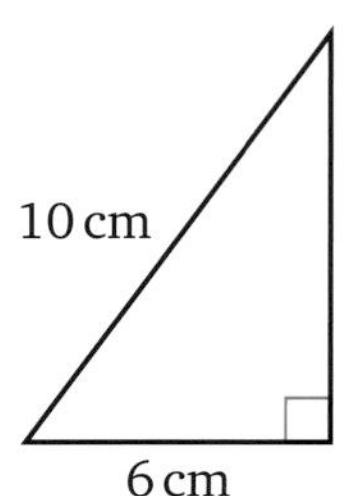

i

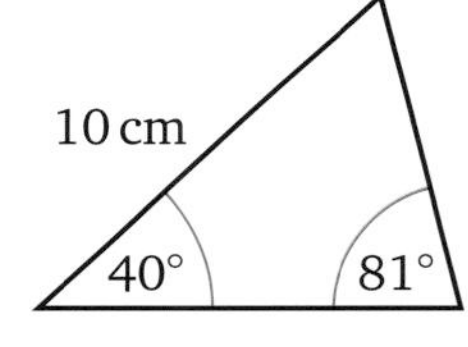

j

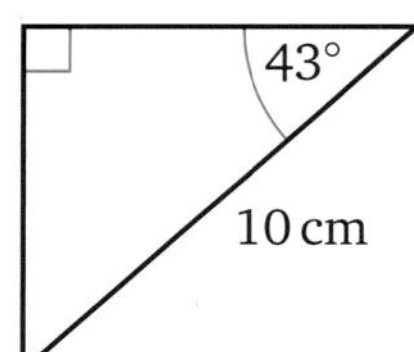

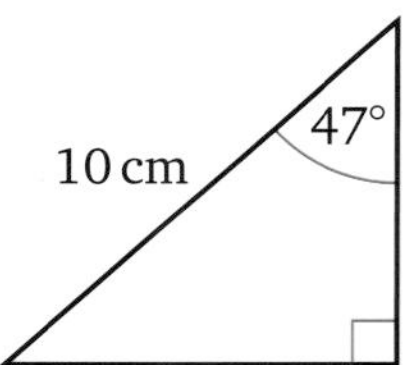

2 The triangles ABC and DEF are congruent.

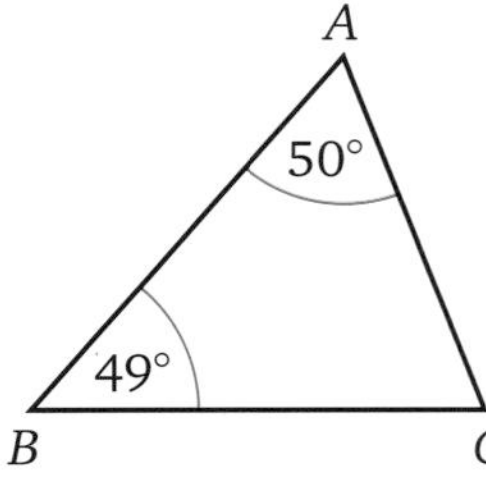

Not drawn accurately

a Find angles at **i** C **ii** D **iii** E **iv** F

b Which side in $\triangle ABC$ is equal to 8 cm?

3

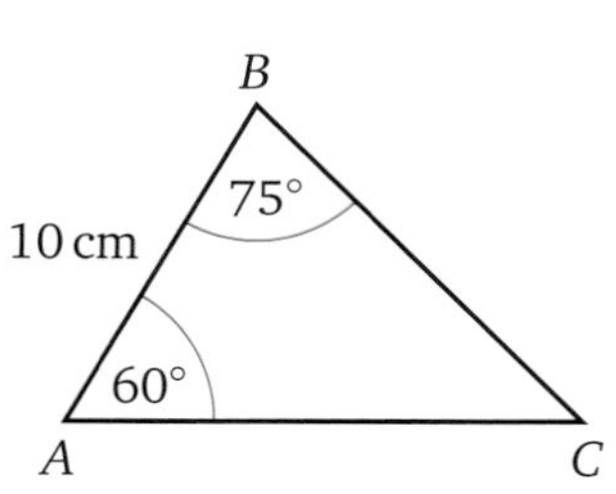

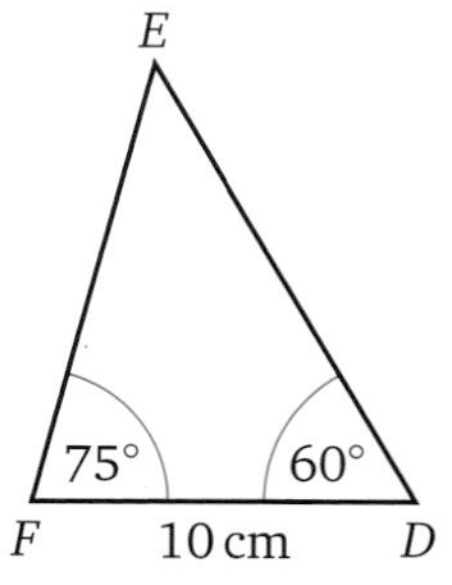

Not drawn accurately

Jen says that $AB = DF$, $\angle A = \angle D = 60°$, $\angle B = \angle F = 75°$, so $\triangle ABC$ is congruent to $\triangle DFE$ (ASA).

Malachie says Jen has made a mistake.

Who is correct? Explain your answer.

4 In this diagram $AB = DE$, $CD = 4$ cm and $CE = 9$ cm.

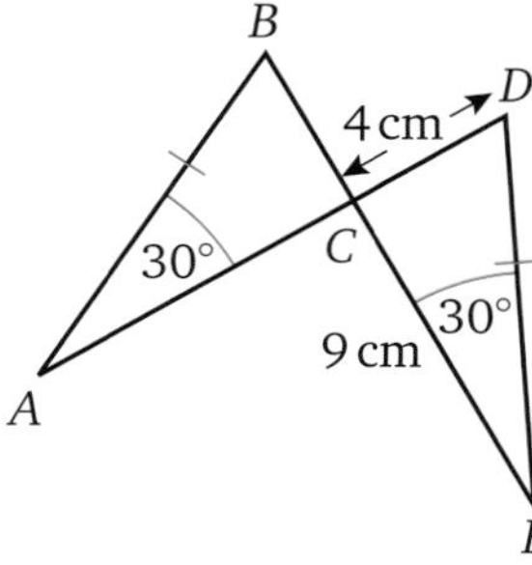

Not drawn accurately

Find the length of BC and AC.

5 Prove that each of the following diagrams contains a pair of congruent triangles.

a

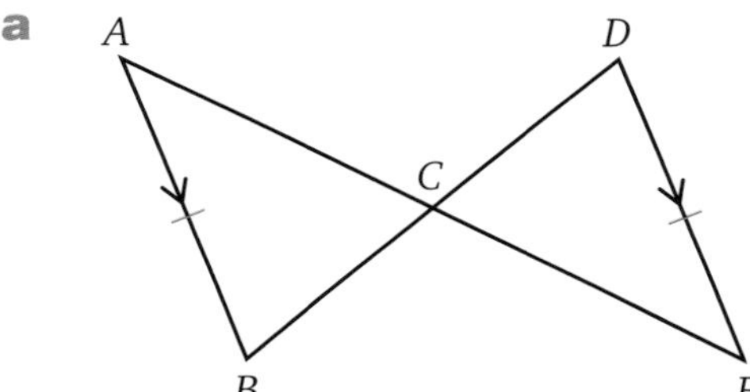

b

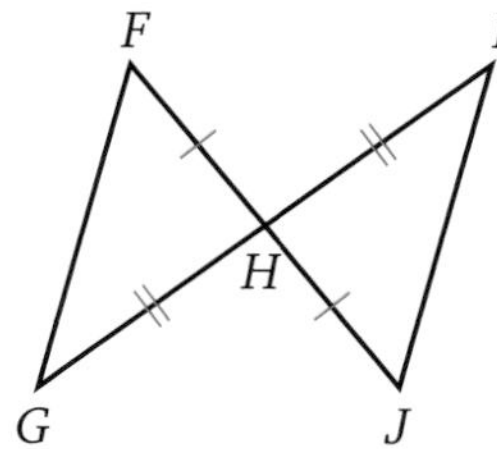

Not drawn accurately

c

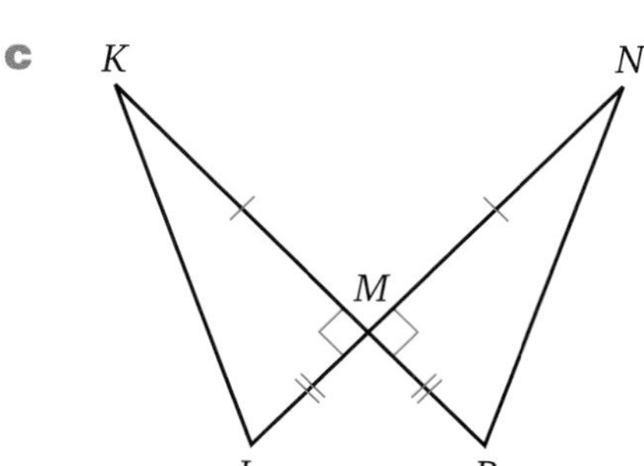

A

6 *ABCD* is a rectangle

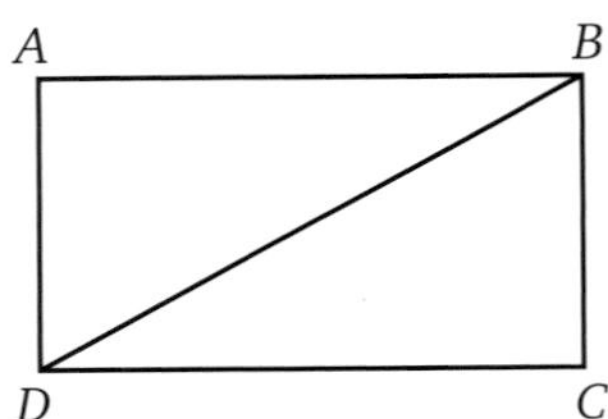

Prove triangles *ABD* and *CDB* are congruent.

A*

7 *ABCDEF* is a regular hexagon.

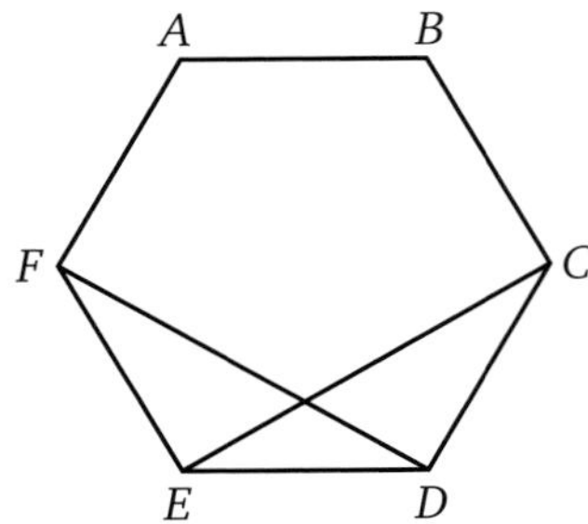

Prove triangles *DEF* and *CDE* are congruent.

8 Prove that the method of constructing a perpendicular at a point **on** a line does actually construct a perpendicular to the line.

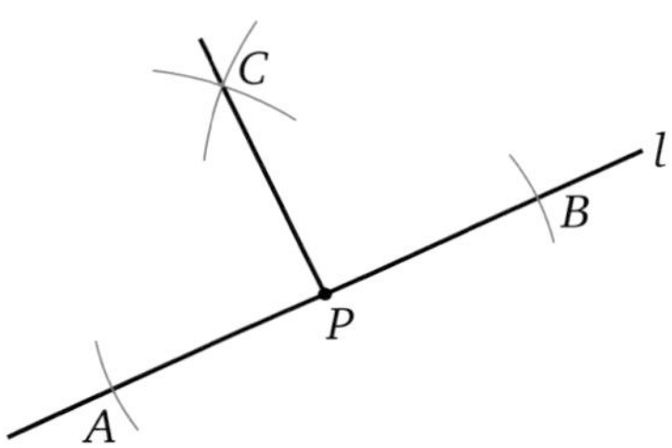

Hint

A completed diagram is shown to help you. Start by proving that triangle *APC* is congruent to triangle *BPC*, then use this to explain why *PC* is perpendicular to line *l*.

9 Prove that the method of constructing a perpendicular from a point **to** a line does actually construct a perpendicular to the line.

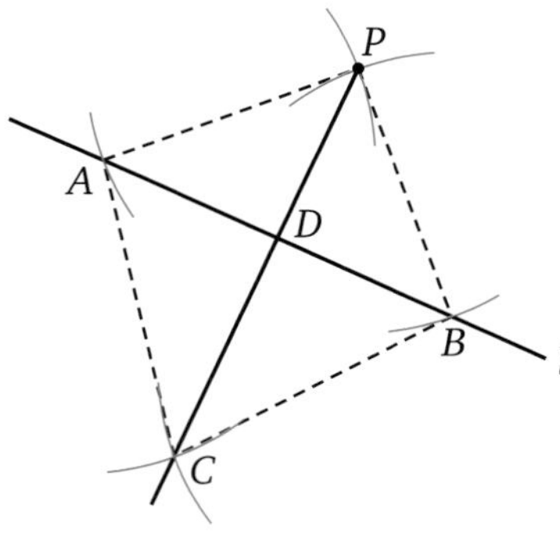

Hint

A completed diagram is shown to help you. Start by proving that triangle *APC* is congruent to triangle *BPC*. What does this tell you about the angles at *P*?

Then prove that triangle *APD* is congruent to triangle *BPD*. How does this tell you that *PD* is perpendicular to line *l*?

10 This is the standard construction of the perpendicular bisector of line *AB*. Draw and label the construction yourself.
Prove that the construction does in fact give the perpendicular bisector. You will need two steps, as in Question 9.

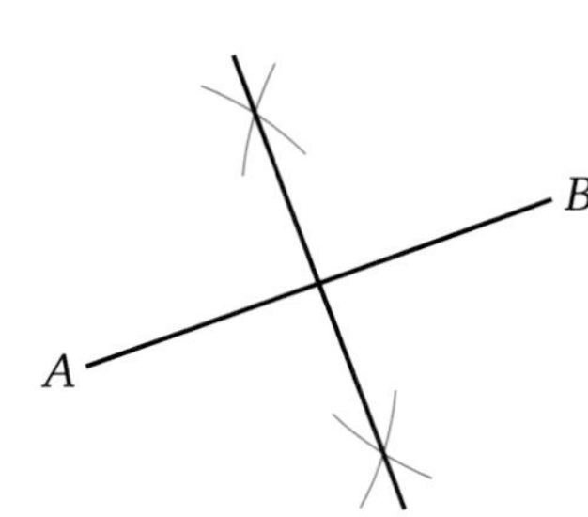

15 Assess

1 Draw this shape accurately.

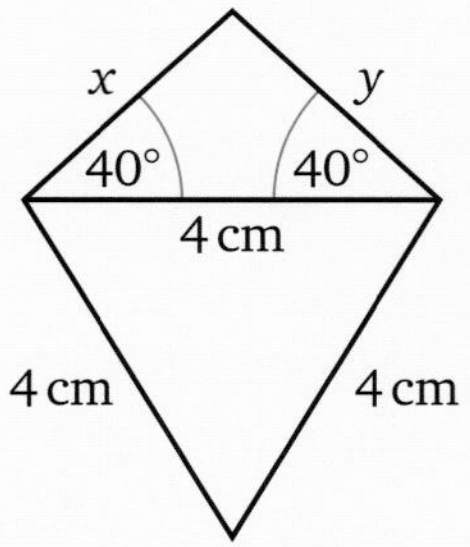

Not drawn accurately

2 Jamie was asked to draw this triangle accurately:

- a triangle of sides 6 cm and 8 cm and a non-included angle of 39°.

Nancy drew the same triangle but got a different diagram.
Both were correct.

a Explain how they can both have different but correct answers.

b Draw both their triangles.

3 Draw this parallelogram accurately.

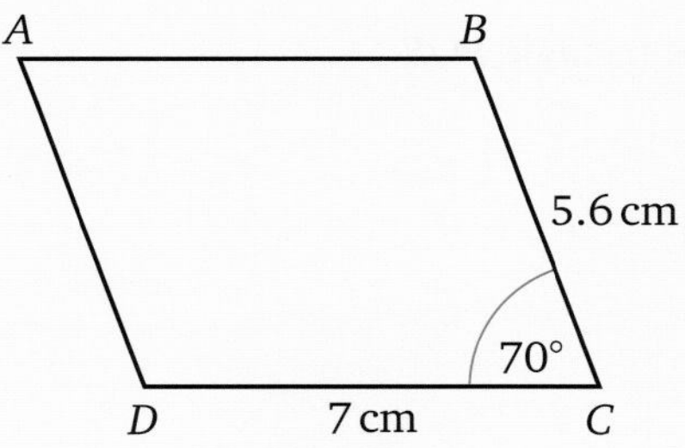

Not drawn accurately

Construct the angle bisector of angle D.

4 Find x in the following diagram.

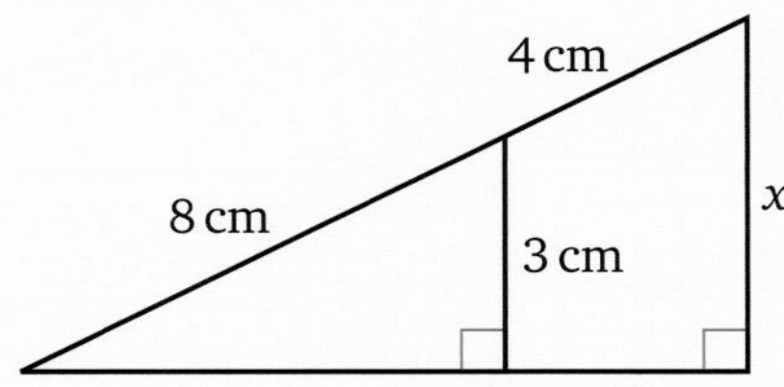

Not drawn accurately

5 In this diagram AB and CD are parallel.

a Prove that the two triangles are similar.

b Calculate the lengths x and y.

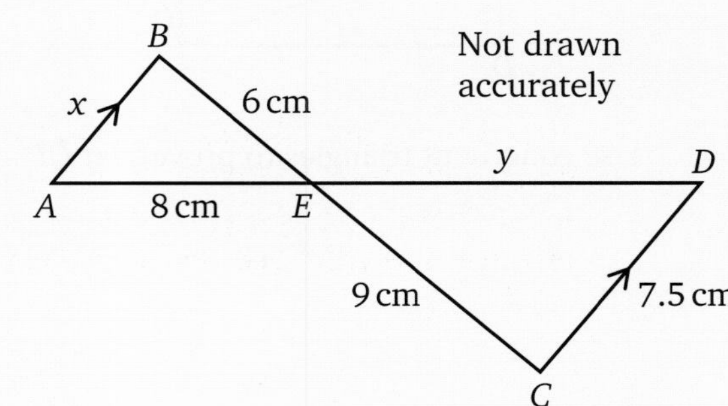

Not drawn accurately

B

6 In this diagram there are two parallel lines.

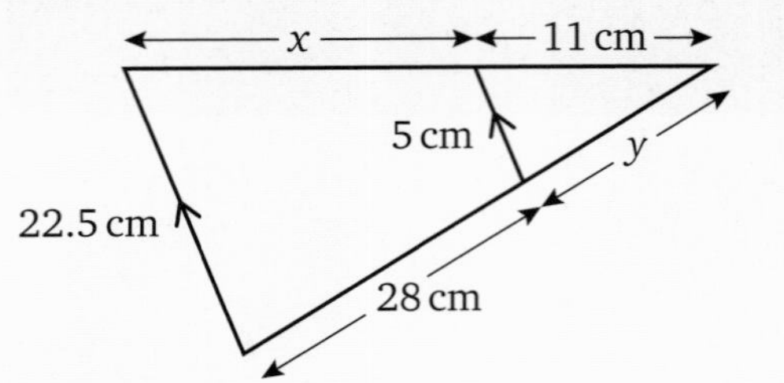

Not drawn accurately

Find the values of x and y.

A

7 In this diagram TA and TB are tangents to the circle with centre O.

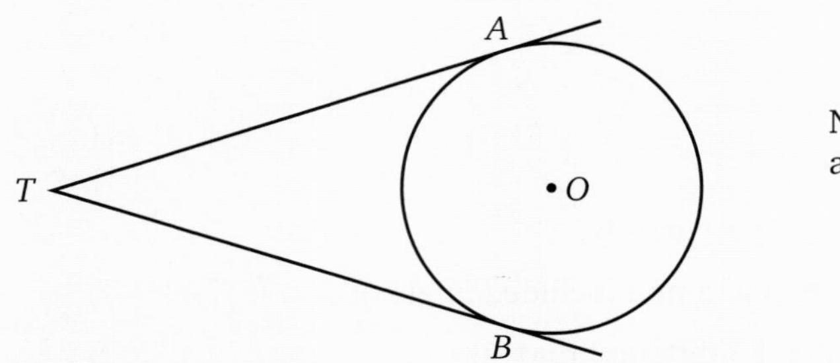

Not drawn accurately

Use your knowledge of angles to show that triangles TAO and TBO are congruent.

8 AXB and PXQ are two straight lines that bisect each other.

Prove that triangle $AXQ \equiv$ triangle BXP

A*

9 PQR is any triangle.
Equilateral triangles PAQ and PBR are drawn outside triangle PQR.

Prove that triangle $APR \equiv$ triangle QPB.

AQA Examination-style questions 

1 In the diagram $ABCD$ and $PQRC$ are squares.

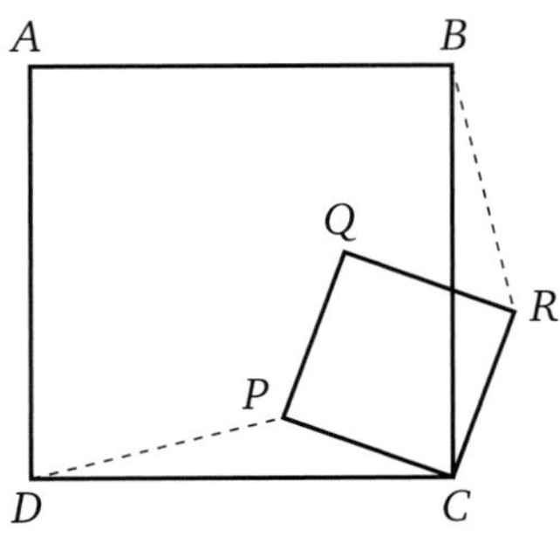

Use congruent triangles to prove that $DP = BR$ *(4 marks)*

AQA 2007

16 Loci

Objectives

Examiners would normally expect students who get these grades to be able to:

D

understand the idea of a locus

C

construct the locus of points equidistant from two fixed points

construct the locus of points equidistant from two fixed lines

Solve loci problems, for example find the points less than 3 cm from a point P.

Did you know?

Mobile phone masts

Some mobile phone masts have a range of 40 km.

Hills, trees and buildings all reduce this distance to as little as 5 km.

In some places, mobile phone masts are only 1 or 2 km apart. This is because they could not cope with the number of calls being made in the area on their own.

Key terms

locus, loci
perpendicular
bisect, bisector
equidistant

You should already know:

- ✔ how to construct the perpendicular bisector of a line
- ✔ how to bisect an angle
- ✔ how to construct and interpret a scale drawing.

Learn... 16.1 Constructing loci

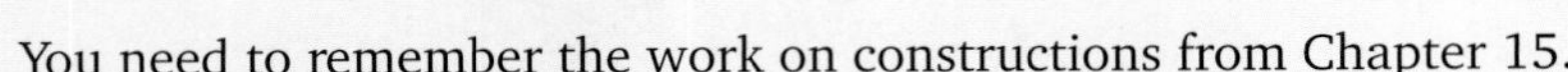

A **locus** can be thought of in two different ways.

It is the path that a moving point follows, or a set of points that follow a rule.

For example, a circle with a radius of 10 cm, centre C, can be thought of as all the points 10 cm from C, or as the path of a moving point which is always 10 cm from a fixed point, C.

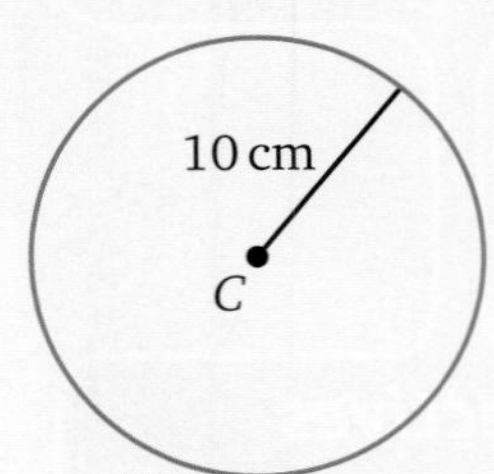

You need to remember the work on constructions from Chapter 15.

A **perpendicular bisector** of the line AB joins all the points **equidistant** (the same distance) from A and B.

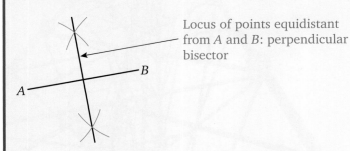

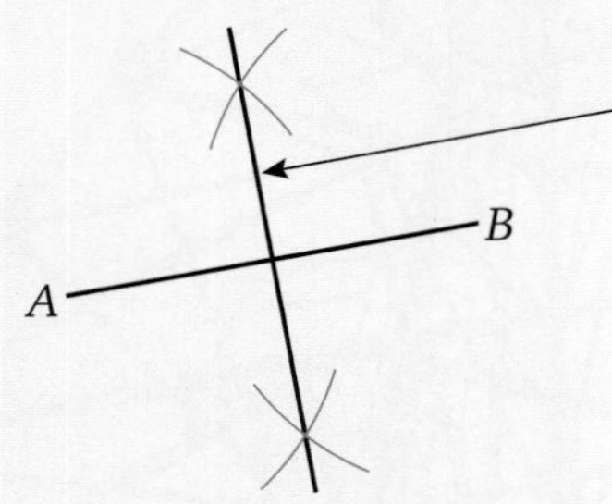

Draw four arcs of the same radius, two with centre at A, two with centre at B. Join the points of intersection.

An angle bisector joins all points equidistant from two lines.

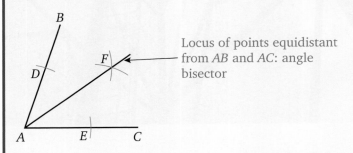

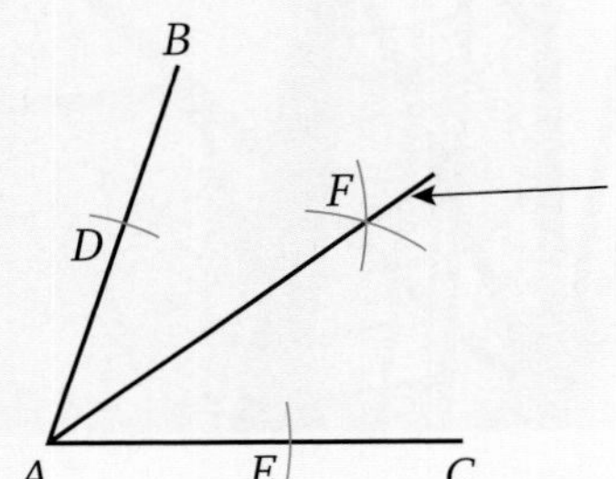

Draw two arcs of equal radius, centre A, to cut AB and AC at D and E.

Draw equal arcs from D and E to intersect at F.

Join AF and extend.

The perpendicular from a point to a line is the shortest distance from the point to the line.

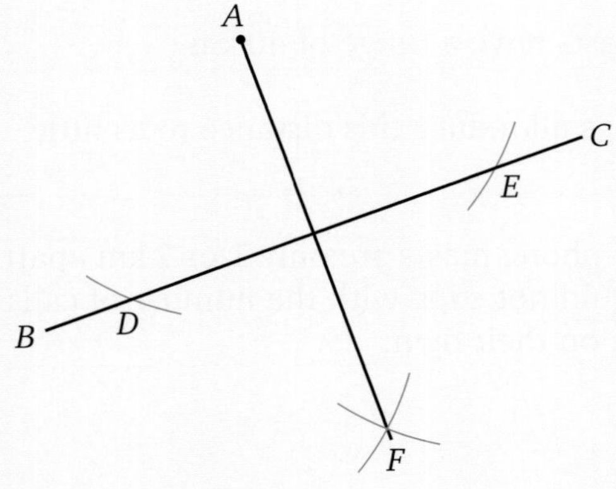

To draw the perpendicular from A to BC:

Draw two arcs of the same radius, centre A, to cross BC at D and E.

With D and E as centres, draw two more arcs of equal radius to meet at F.

AF is perpendicular to BC.

Example: An electricity pylon has to be placed so that it is equidistant from *AB* and *AC*, and no more than 200 m from *D*. It must be within the boundary.

Mark the points where the pylon could be placed.

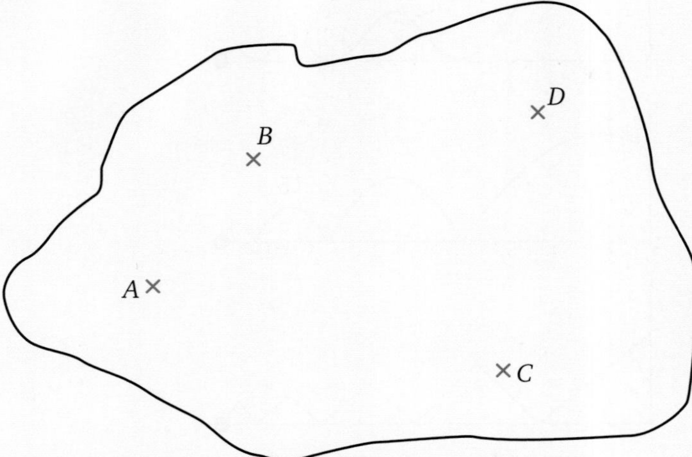

Scale: 1 cm = 100 m

Solution: Draw *AB* and *AC*.

Construct the angle bisector of *BAC*, as this marks the locus of points equidistant (the same distance) from *AB* and *AC*.

The points less than 200 m from *D* form a circle, radius 200 m.

With a scale of 1 cm to 100 m, this circle needs to have a radius of 2 cm.

The possible positions for the pylon are on the angle bisector and inside the circle.

This is shown in green.

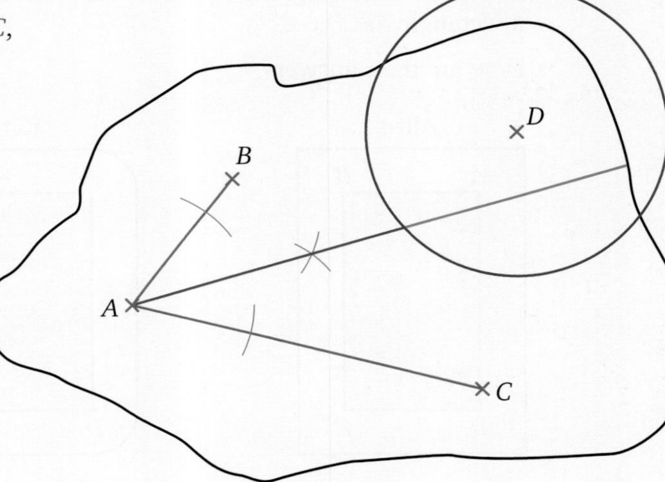

Scale: 1 cm = 100 m

AQA Examiner's tip

To get full marks, make sure that you leave your construction lines showing.

Practise... 16.1 Constructing loci

D

1 Most cars have one of the three arrangements of windscreen wipers shown below.

A

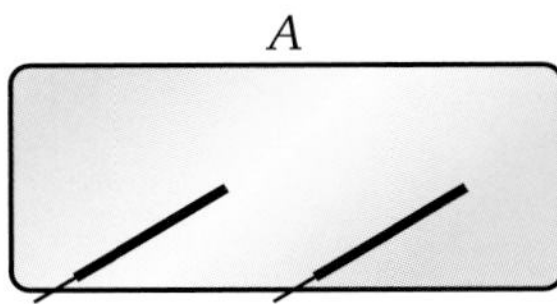

B

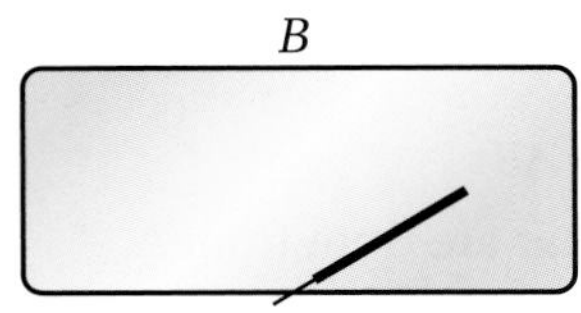

C

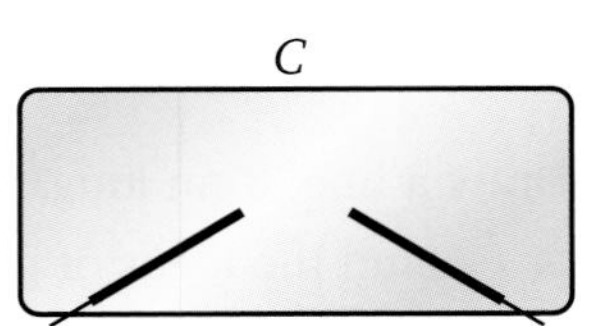

For each arrangement, sketch the area of windscreen that the wipers will clear.

Which is the best arrangement? Give a reason for your answer.

D

2 Bill throws a ball from his window.
Harry, Hope and Oli try to sketch the locus of the ball.

Harry draws this:

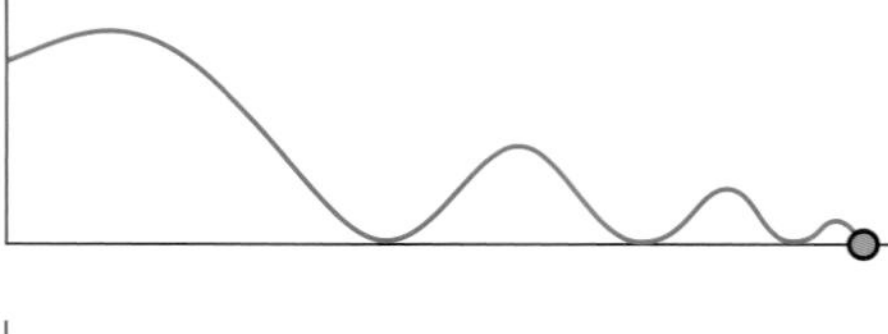

Hope draws this:

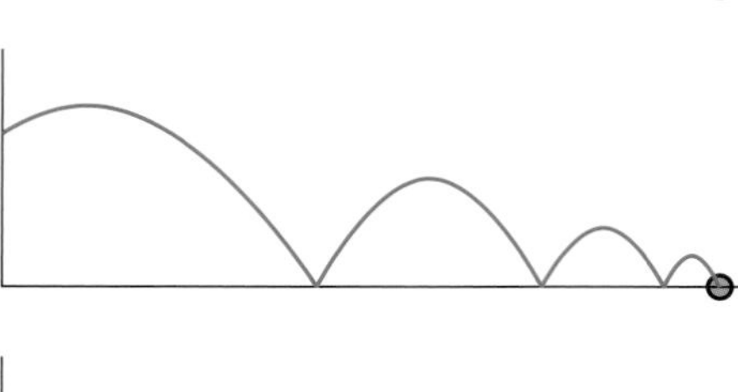

Oli draws this:

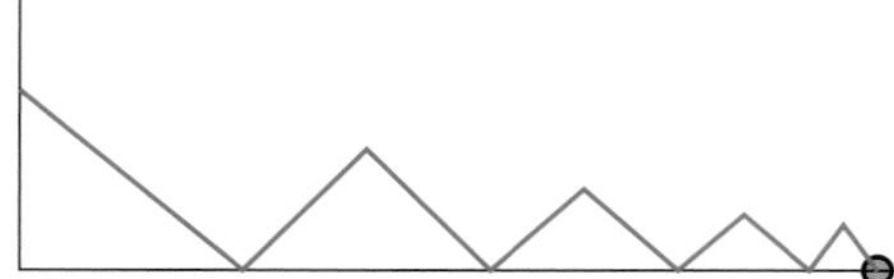

Who is correct?
Give a reason for your answer.

3 Alice, Kat and Becky tried to draw the locus of points a distance of 1 cm outside a rectangle *ABCD*.

Here are their answers.

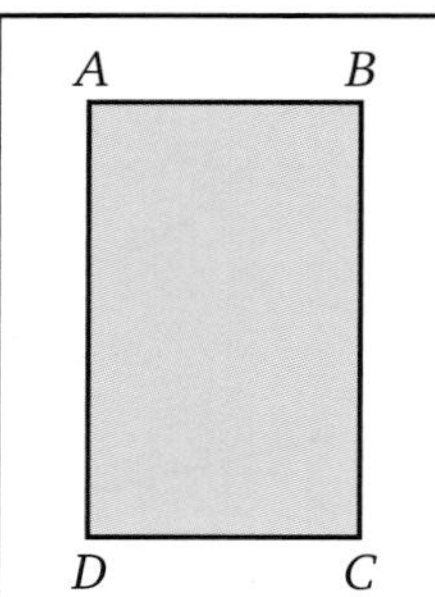

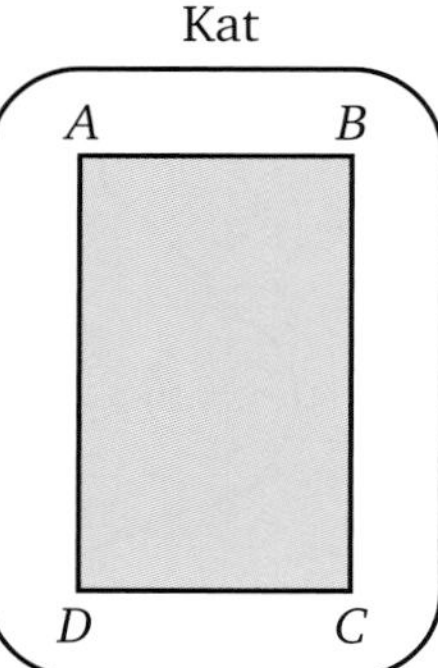

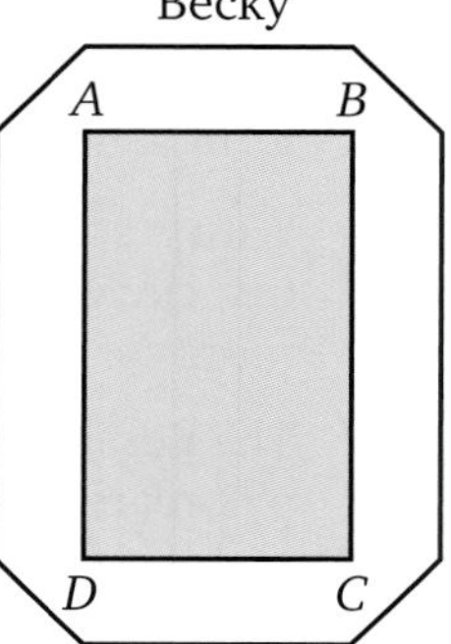

Who is correct: Alice, Kat or Becky?
Give a reason for your answer.

AQA Examiner's tip

You must explain answers when the question asks you to.
Just choosing the correct answer will not score full marks.

When explaining answers, use mathematical language. Answers do not have to be long, but must explain clearly why you are correct.

C

4 Draw a line, *AB*, 8 cm long.

Find the locus of points equidistant from *A* and *B*.

5 Draw an angle, *ABC*, of 70°.

Find the locus of points equidistant from *AB* and *BC*.

6 Draw a line, 8 cm long. Label it *AB*.

a Find the locus of points that are 6 cm from *A*.

b Find the locus of points that are 4 cm from *B*.

c Shade the area containing all the points that are less than 6 cm from *A* and less than 4 cm from *B*.

C

7 Draw a triangle ABC with sides at least 10 cm long.

a Draw the locus of points equidistant from A and B.

b Draw the locus of points equidistant from A and C.

c Mark the point that is equidistant from A, B and C. Label it X.

8 Three men discover an island.
Each wants to claim it as his own.
Each man plants a flag in part of the island.

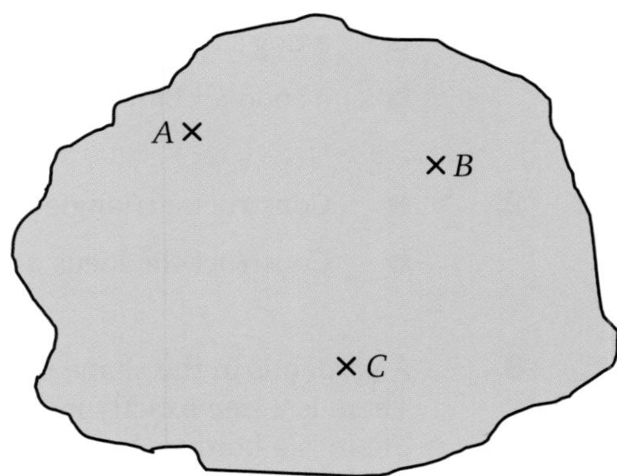

a Draw your own sketch of the island.

b Label three points, A, B and C.

c The men agree to divide the island between them, so that they keep the part of the island that is closer to them than to the others.

Show how the island can be divided up in this way.

9 Three telegraph poles, A, B and C, are on the corners of an isosceles triangle.
$AB = AC = 200$ m and $BC = 250$ m

a Using a scale of 1 cm to 20 m, make a scale drawing of where the poles are situated.

b A man wants to build a house within the triangle.
The house needs to be at least 90 m from each of the telegraph poles.

Shade the area on the plan where he can build the house.

c He wants the house to be rectangular, 50 m wide and 70 m long.

Can he fit a house this size in the shaded area?

10 Tommy has a rectangular garden, $ABCD$, 12 m by 8 m wide.
He wants to plant a tree in the garden.
He wants the tree to be at least 3 m from the edge CD of the garden.
It must be no more than 6 m from B.

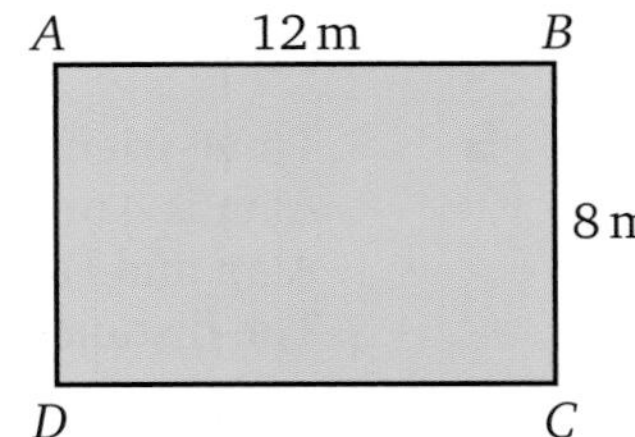

Using a scale of 1 cm to 1 m, make a drawing of the garden, and shade the region where Tommy can plant the tree.

11 In Suffolk, there are two mobile phone masts, 40 km apart.

a Make a map showing the two masts, using a scale of 1 cm to 4 km.
Label them A and B.

b Each mast has a range of 32 km if you are using a phone outside a building.
If you use a phone inside a building, the range drops to 24 km.

Find the locus of points where you can get a signal outside but not inside a house.

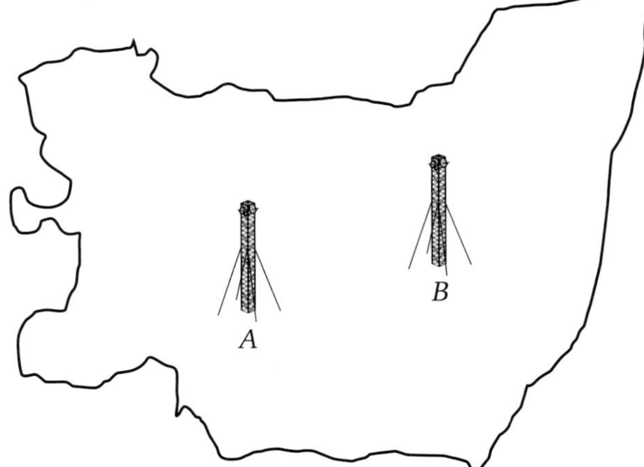

12 A rectangular lawn is 12 m long and 10 m wide.
A gardener waters the lawn with sprinklers, which spray water in a circle with a radius of 3 m.

Find the smallest number of sprinklers needed to water the entire lawn.

16 Assess

D

1 Sketch the locus of:

a a dog's nose as it chases its tail

b a snooker ball as it bounces off a cushion into a pocket.

C

2 a Construct a triangle ABC with $AC = 10$ cm, $AB = 8$ cm and $BC = 7$ cm

b Construct the locus of points equidistant from AB and AC.

3 A garden is in the shape of a rectangle, 16 m long and 10 m wide.
There is a tree exactly in the centre of the garden.
There is a border, 2 m wide, all around the garden, and a flower bed that includes all points up to 3 m from the tree.
The rest of the garden is lawn.

Use a scale of 1 cm to 1 m to make a scale drawing of the garden.

4 ABC is a triangle.
Angle $ABC = 90°$, $AB = 8$ cm and $BC = 6$ cm

a Make an accurate drawing of triangle ABC.

b Draw the locus of points equidistant from A and C.

c Draw the locus of points 6 cm from C.

d Mark the points that are equidistant from A and C and that are also less than 6 cm from C.

5 Three friends, Alan, Dave and Faye are sitting watching a DVD.
Alan is 4 m from the television, and Faye is 3.5 m from the television.
Alan and Faye are sitting 2.5 m apart.
The distance between Alan and Dave is the same as the distance between Alan and Faye.
Dave is exactly 3 m from the television.

Use a scale of 2 cm to 1 m to construct a plan of the room.

6 Draw a triangle ABC with sides at least 10 cm long.

a Draw the locus of points equidistant from AB and BC.

b Draw the locus of points equidistant from AB and AC.

c Mark the point that is equidistant from AB, BC and AC. Label it X.

AQA Examination-style questions

1 Two radio stations at A and B pick up a distress call from a boat at sea.
The station at A can tell that the boat is between 50 km and 70 km from A.
The station at B can tell that the boat is between a bearing of 060° and 070° from B.

Show clearly, using compasses and a protractor, the region where the boat will be found.

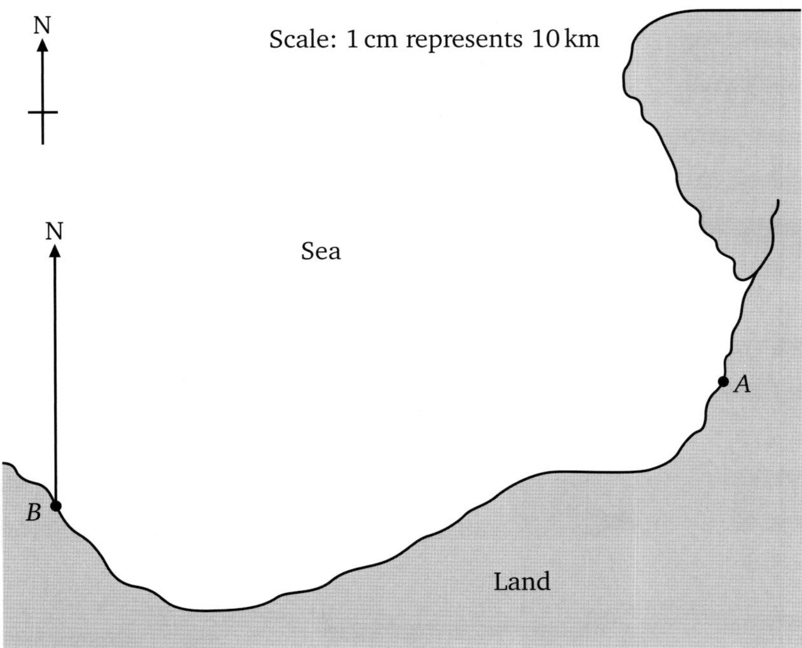

(3 marks)
AQA 2008

2 AB and AC represent two walls.
A pole must be erected that is

- equidistant from AB and AC
- between 40 m and 70 m from A.

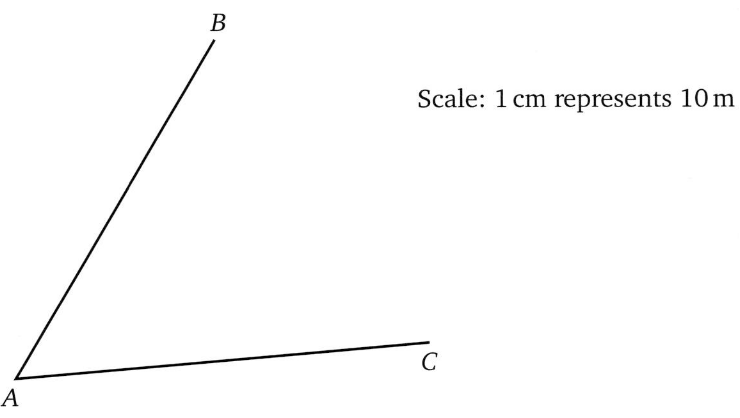

Show clearly all the possible positions of the pole. *(3 marks)*
AQA 2006

Consolidation 2

You have covered the following topics:

- Fractions and decimals
- Angles and areas
- Working with symbols
- Percentages and ratios
- Area and volume 1
- Equations and formulae
- Properties of polygons
- 3-D shapes, coordinates and graphs
- Reflections, rotations and translations
- Pythagoras' theorem
- Properties of circles
- Measures
- Trial and improvement
- Enlargements
- Construction
- Loci

All these topics will be tested in this chapter and you will find a mixture of problem solving and functional questions. You won't always be told which bit of maths to use or what type a question is, so you will have to decide on the best method, just like in your exam.

Example:

a Factorise $2r + \pi r$. *(1 mark)*

b Solve the equation $2r + \pi r = 21$
Give your answer to three decimal places. *(1 mark)*

c A prism with a semicircular cross-section is made by cutting out the shaded shape from an A4 piece of card. The dimensions of the card are 21 cm by 29.7 cm.

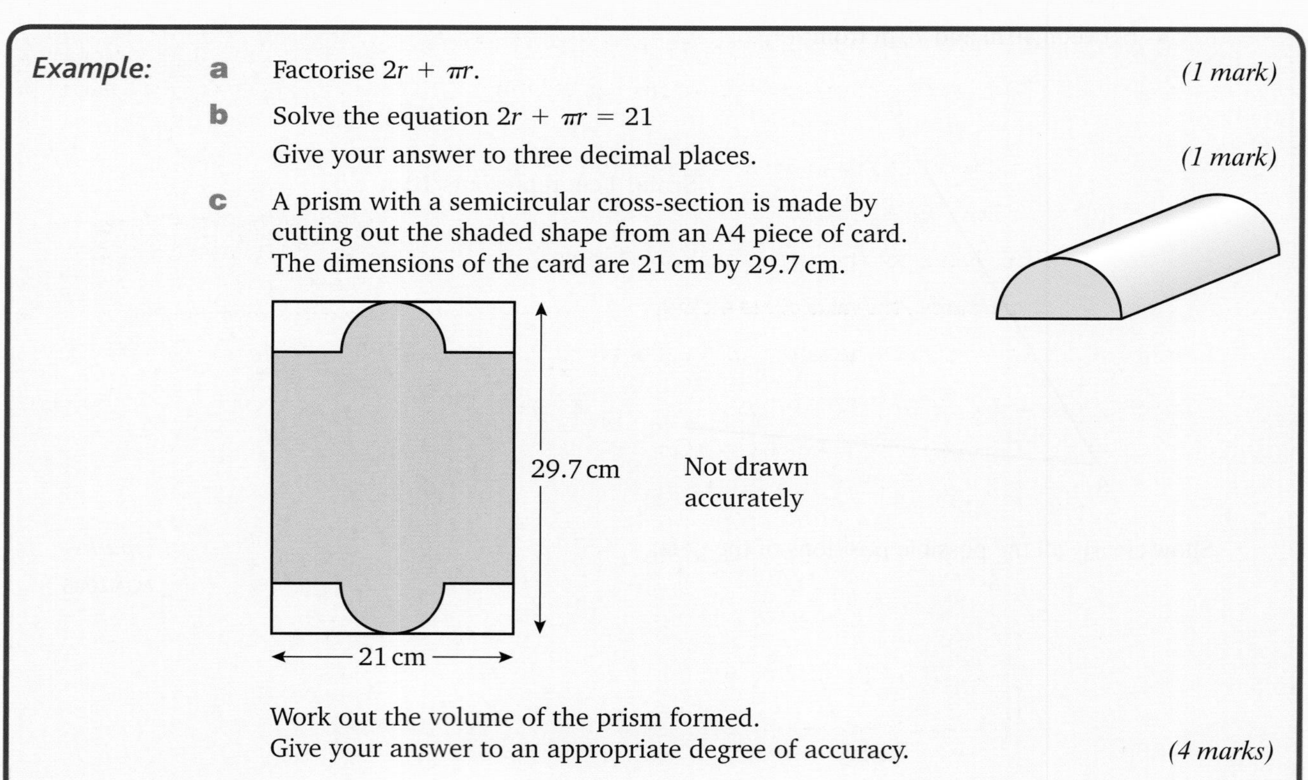

Work out the volume of the prism formed.
Give your answer to an appropriate degree of accuracy. *(4 marks)*

Solution:

a Always begin by looking for common factors.

Here, r is a common factor.

$2r + \pi r = r(2 + \pi)$

> **Mark scheme**
> - 1 mark for factorising $2r + \pi r$.

b Use your answer to part **a**.

$2r + \pi r = 21$

$r(2 + \pi) = 21$

$r = 21 \div (2 + \pi)$

$= 4.084337...$

$= 4.084$ (to 3 d.p.)

> **Mark scheme**
> - 1 mark for working out r.

c Here, think about how you work out the volume of a prism.

volume of a prism = area of cross-section × length

The cross-section is a semicircle.

volume of a semicircular prism $= \frac{1}{2}\pi r^2 \times l$

You need to be able to find r, the radius of the semicircle and l, the length of the prism to solve this problem.

Think about how the flat shape folds up to make the prism.

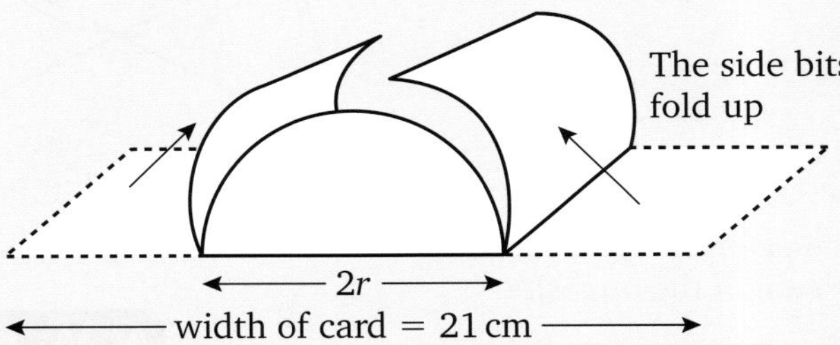

Together the side bits must be the same length as the curved part of the perimeter of the semicircle.

The curved part of the semicircle is worked out using πr.

Let the radius of the semicircle be r,

then $2r + \pi r = 21$

This is the equation you have solved in part **b**.

From part **b**, the value of r is 4.084

Now consider the length of the prism.

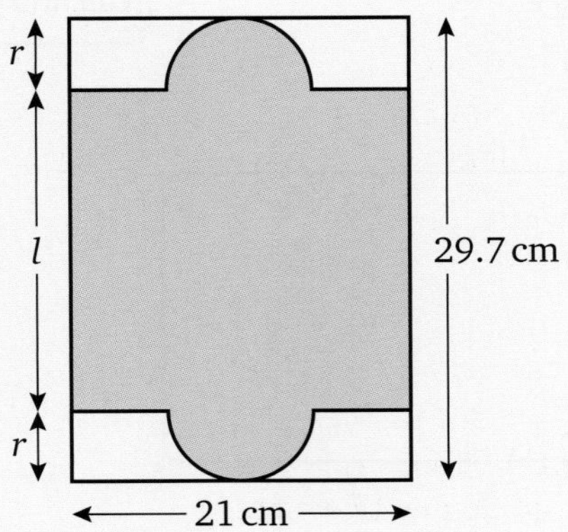

> **AQA Examiner's tip**
> One way of starting a problem is to try to work backwards.

> **AQA Examiner's tip**
> The formula of the volume of a prism would be given to you at the beginning of your exam paper.
> You need to learn the formulae to work out the area and perimeter of a circle and the area of a semicircle. From these, you can work out the area of a semicircle and the curved part of the perimeter of the prism in this question.

> **AQA Examiner's tip**
> When a long question is split into parts, keep looking back to the previous parts to see whether they can help.

Let ℓ be the length of the prism,

$\ell = 29.7 - 2r$

$= 29.7 - 2 \times 4.084$

$= 21.53$ cm

Volume $= \frac{1}{2}\pi r^2 \ell$

$= \frac{1}{2} \times \pi \times 4.084^2 \times 21.53$

$= 564.07$

$= 560$ cm^3 (to 2 s.f.)

AQA Examiner's tip

Remember that you will lose out on a mark if you do not round your answer to an appropriate degree of accuracy when the examiner has asked you to do so.

Mark scheme

- 1 mark for setting up the equation $2r + \pi r = 21$
- 1 mark for setting up the equation $l = 29.7 - 2r$
- 1 mark for using $\frac{1}{2}\pi r^2 \times l$.
- 1 mark for the correct answer to the appropriate accuracy.

Example: In the diagram *ABE*, *ADF*, *DCE* and *FCB* are straight lines.

A, *B*, *C* and *D* lie on the circle.

Angle *DCF* = 57°

Angle *CFD* is twice angle *BEC*.

Work out the value of *x*. *(4 marks)*

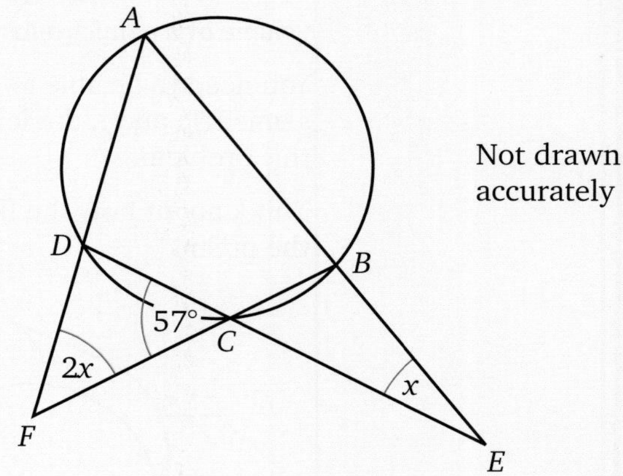

Solution: Use the fact that the exterior angle of a triangle equals the sum of the opposite two interior angles.

Angle *ADC* must be 57° + 2x

Angles on a straight line add up to 180°.

Angle *DCB* must be 180° − 57° = 123°

Opposite angles in a cyclic quadrilateral add up to 180°.

As *ABCD* is a cyclic quadrilateral, angles *DAB* and *DCB* must add up to 180° because they are opposite.

Angle *DAB* must be 180° − 123° = 57°

Now add up the angles in triangle *ADE*.

The sum of the angles must be 180°.

$57° + (57° + 2x) + x = 180°$

$114° + 3x = 180°$

$3x = 66°$

$x = 22°$

AQA Examiner's tip

Look at the diagram and shapes and think about which theorems are relevant.

Here we have a cyclic quadrilateral and triangles.

You know that:

- opposite angles of a cyclic quadrilateral add up to 180°
- angles of a triangle add up to 180°.

Try to see how many angles you can find in the diagram.

Mark scheme

- 1 mark for angle *ADC* (57 + 2*x*).
- 1 mark for angle *BAD* (57°).
- 1 mark for forming the equation.
- 1 mark for the correct final answer.

D

Consolidation

1 The diagram shows a centimetre grid with the points *A*, *B*, *C* and *D* marked.

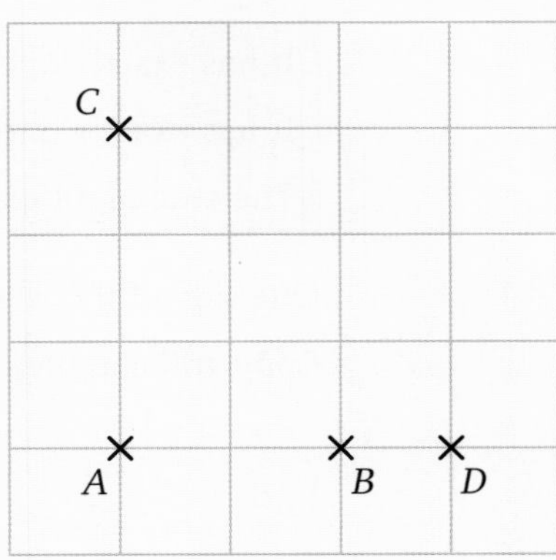

On a copy of the grid draw:

a the locus of the point that moves so that it is an equal distance from *A* and *B*

b the locus of the point that moves so that it is an equal distance from *A* and *C*

c the locus of the point that moves so that it is an equal distance from *C* and *D*.

2 Which of these two shapes has the bigger area? You must explain your answer.

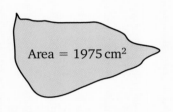

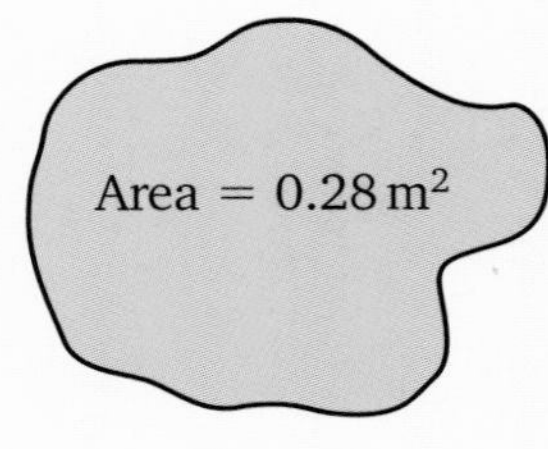

Shape A

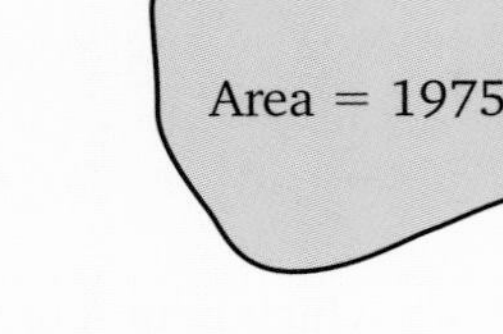

Shape B

Not drawn accurately

3 A quadrilateral has interior angles of x, $2x$, $(x + 35°)$ and $(x + 55°)$.

What is the value of x?

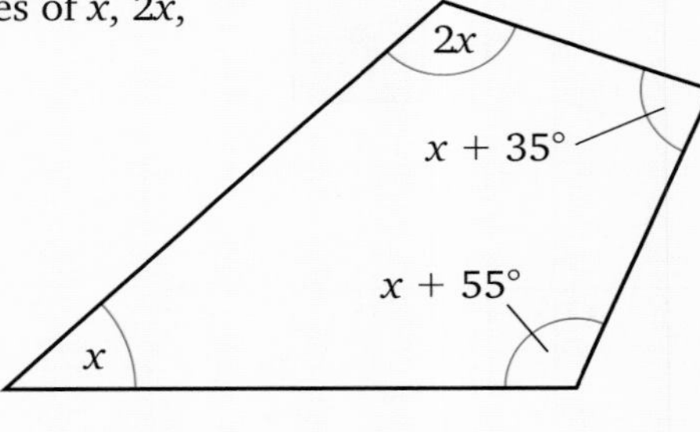

Not drawn accurately

4 The diagram shows a sketch of a triangular field, *ABC*.

a Using ruler and compasses only, construct an accurate scale drawing of the triangle.
Use a scale of 1 cm to 10 metres.

b In the field there is a telegraph pole. The telegraph pole is 44 metres from *A* and 63 metres from *C*.
How far is the telegraph pole from *B*?

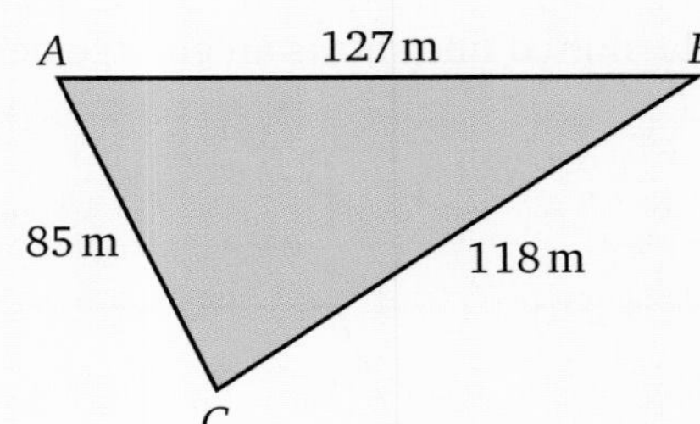

Not drawn accurately

5 These two right-angled triangles have the same area. What is the ratio $x : y$?

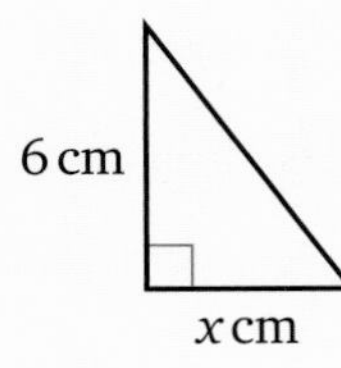

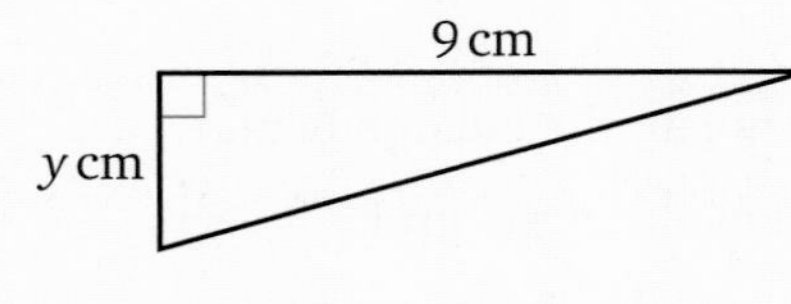

Not drawn accurately

D

6 A two-dimensional shape has the following properties.

It has exactly one line of symmetry.
It has exactly one reflex angle.
The sum of all its interior angles is 360°.

One side of the shape has been drawn on the grid below.

Copy and complete the shape.

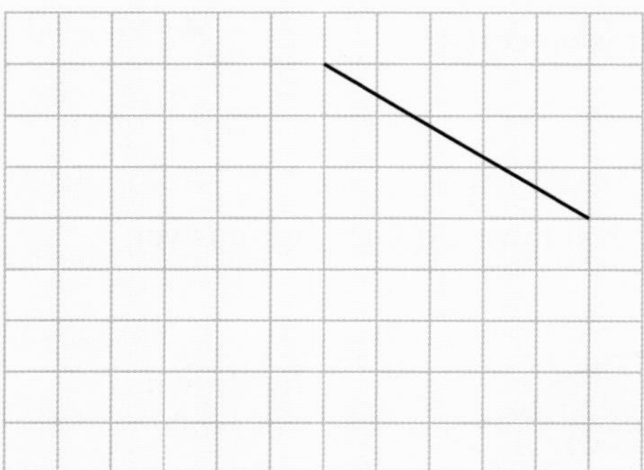

7 Write down two different transformations which map *A* onto *B*.

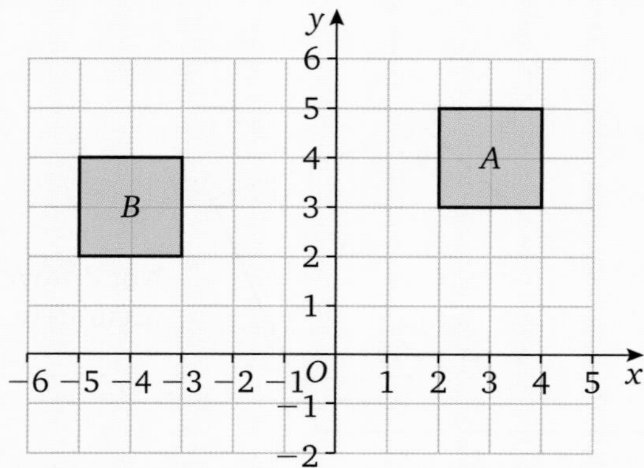

8 The diagram shows an irregular hexagon drawn on a centimetre square grid. The dotted line shows an enlargement of one of the sides of the hexagon.

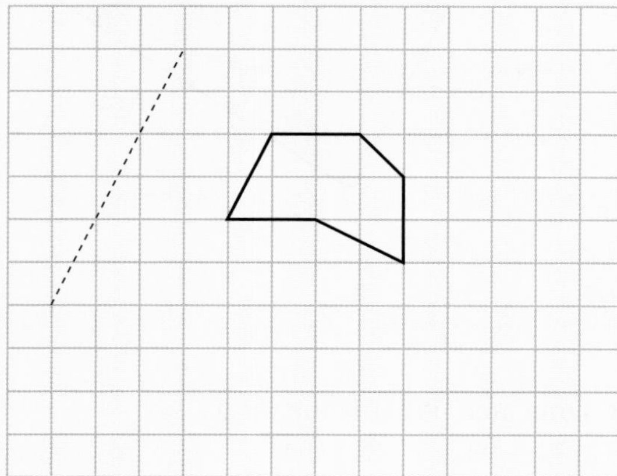

a Copy the diagram and complete the enlargement.

b Mark the centre of enlargement on the grid.

c Write down the scale factor of the enlargement.

9 **a** Show that the perimeter and the area of this triangle have the same numerical value. C

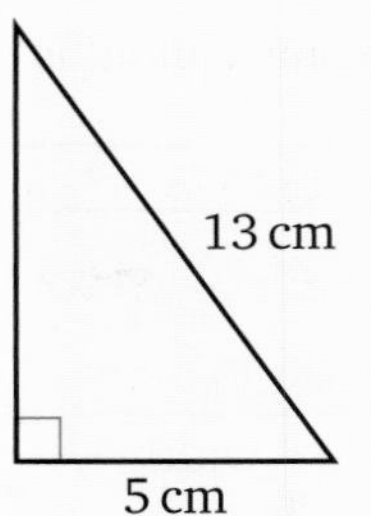

Not drawn accurately

b This triangle has integer lengths.
The numerical value of its perimeter and area are also the same.

Work out x and y. Show working to justify your answer.

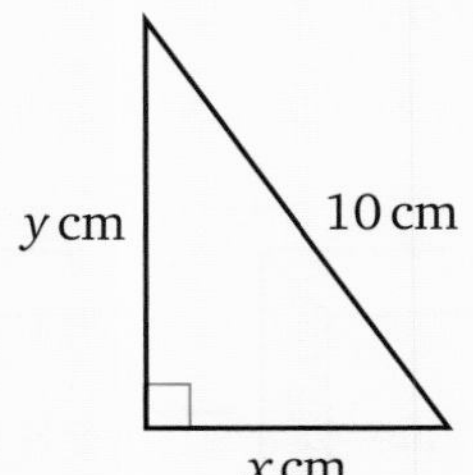

Not drawn accurately

10 Wayne is driving on a motorway. After $2\frac{1}{2}$ hours he has travelled 125 miles. D C

a What is his average speed?

b Wayne sees this sign.

SERVICES
20 miles

How long will it take Wayne to reach the services if he continues at his average speed?
Give your answer in minutes.

11 $AB = x\,\text{cm}$ C

BC is three times longer than AB.

CD is 2 cm longer than BC.

The perimeter of the shape is $2(4x + 3)$ cm.

Work out the length of AD in terms of x.

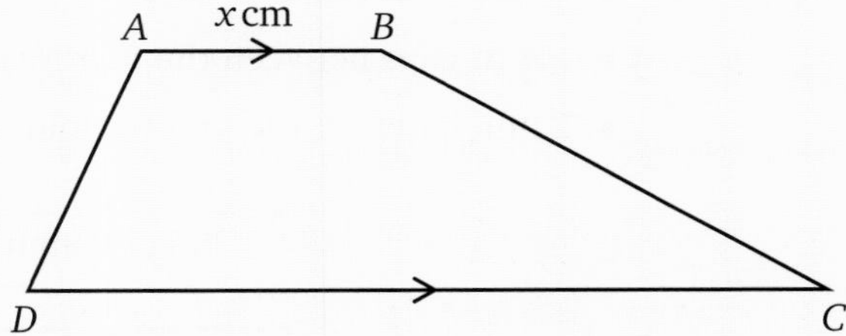

12 Oliver is downloading a file from the internet.
The file is 11.25 megabytes. The file downloads at a rate of 120 kilobytes per second.
There are 1024 kilobytes in a megabyte.

How long will it take to download the file?

13 **a** On the grid below, draw the line $y = x$

b Reflect the shape A in the line $y = x$

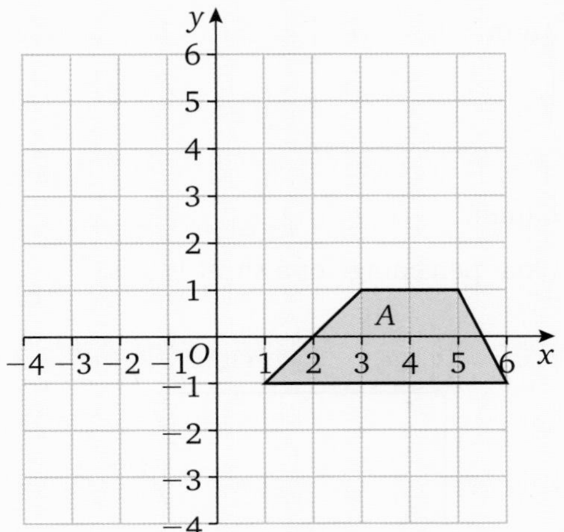

C

14 The volume of this cuboid is 480 cm³. Its height is 6 cm and its length is 5 cm.

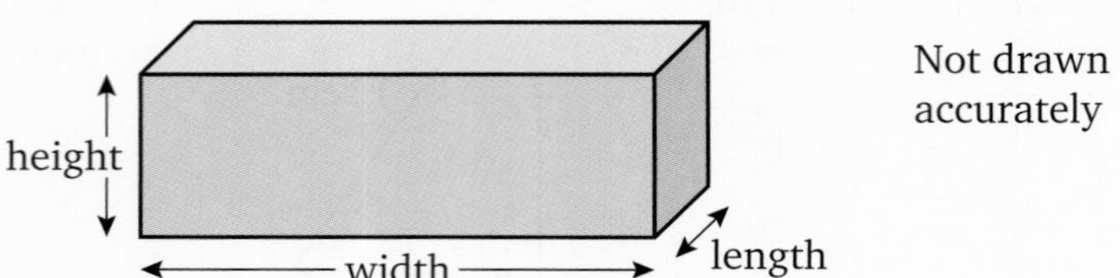

The shape is cut down the middle along the dotted lines and then glued together to make a new cuboid.

Not drawn accurately

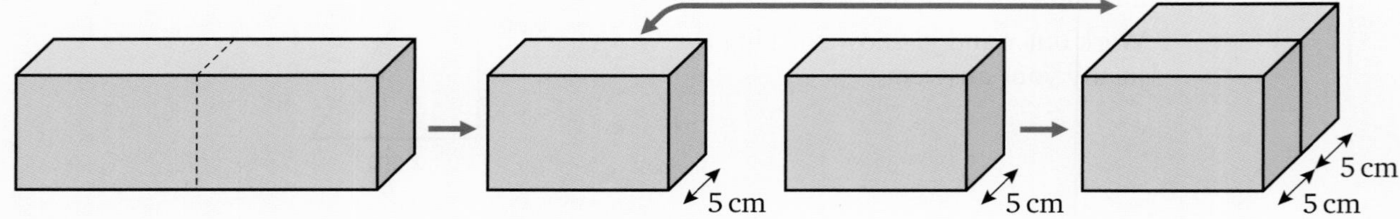

The surface area of the original cuboid is 412 cm².
Melissa says that the surface area of the new cuboid is the same as the original one.

Is Melissa correct? You **must** show all your working.

15 A machine cuts circular discs of diameter 5 cm from a sheet of rectangular plastic.
The dimensions of the sheet of plastic are 1.2 m by 0.8 m.
The machine leaves the following horizontal and vertical gaps:

- 2 mm between each disc
- 2 mm gaps between the first row of discs and the top of the sheet
- 2 mm between the first column of discs and the left side of the sheet.

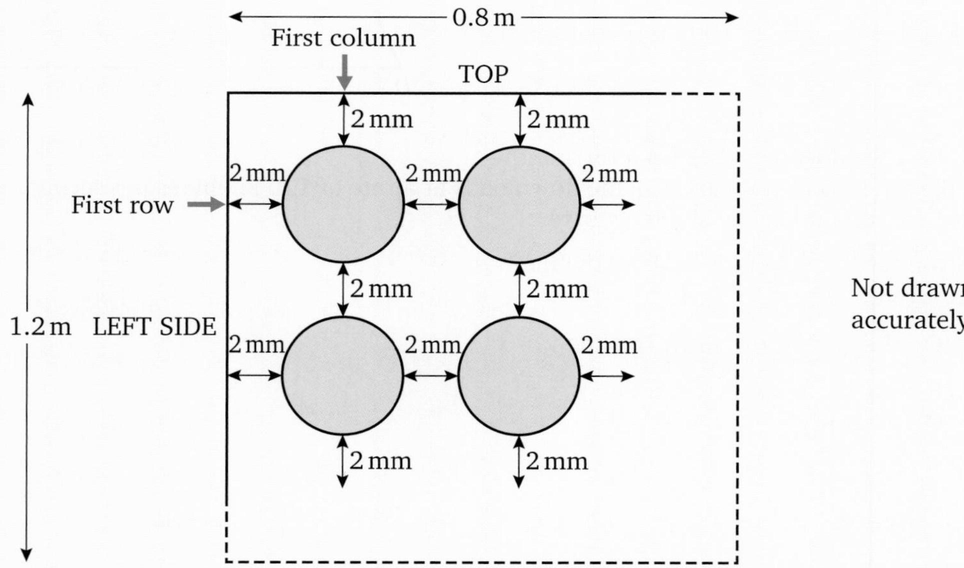

a Show that 23 discs can be cut in each column.

b The unused part of a sheet is recycled. What percentage of a sheet is this?

16 A rectangle has length x^2 and width $x + 3$.
The area of the rectangle is 40 cm².

Use trial and improvement to work out the value of x.

Give your answer to two decimal places.

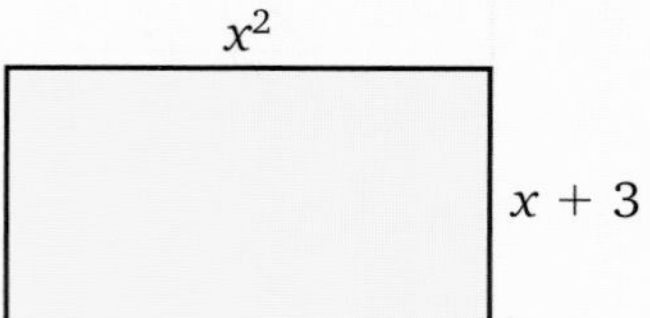

C

17 Tim and Sara are solving an equation using the method of trial and improvement. They find the solution lies between 2.641 and 2.648

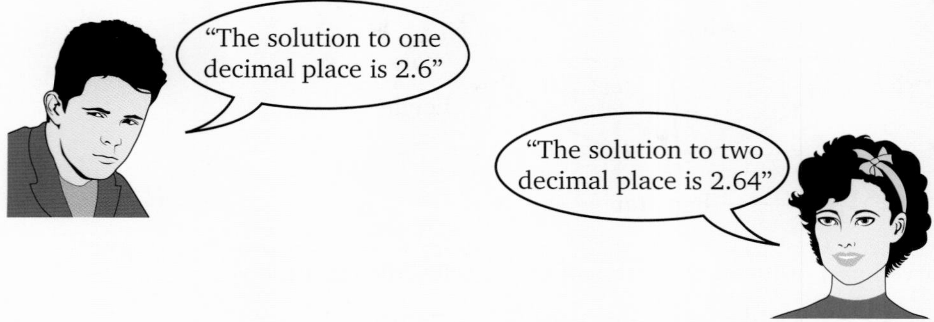

"The solution to one decimal place is 2.6"
"The solution to two decimal place is 2.64"

For each of Tim and Sara, which one of the following applies to their solutions?

- Definitely right
- Definitely wrong
- There is not enough information to tell

Give a reason for your answers.

B

18 Can either of these triangles be drawn inside a semicircle with the point X touching the semicircle as shown?

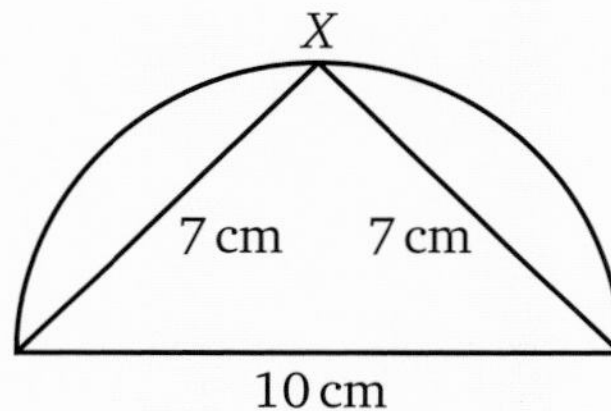

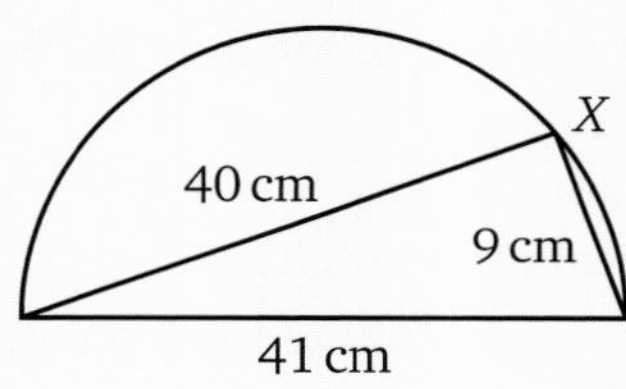

Not drawn accurately

You **must** show all your working.

19 In 2007 a bar of Yummichoc weighed 150 g and cost 85p.
In 2009 a bar of Yummichoc weighed 145 g and cost 95p.

a Work out the cost of 1 g of Yummichoc in 2007.

b What has been the percentage increase in the cost of a bar of Yummichoc, taking into account the increase in price and the reduction in weight?

20 A piece of wire 20 cm long is bent into a complete semicircle.

What is the area of the semicircle?

21 Each side of this hexagon is 6 cm long.
The angles BCD and EFA are right angles.

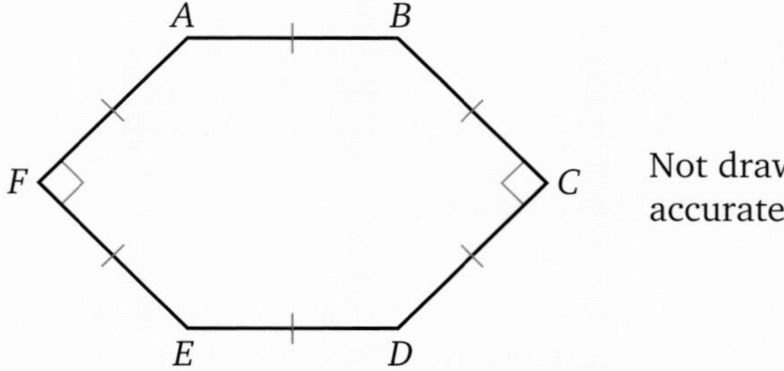

Not drawn accurately

Work out the area of the hexagon.

B A

22 The diagram shows the cross-section of a large vertical satellite dish.

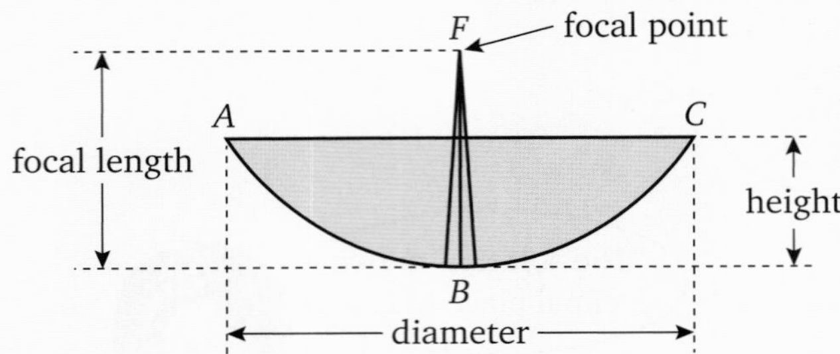

The curved part of the cross-section of the dish has the equation $y = \frac{(x - 10)^2}{25}$ where $0 \leqslant x \leqslant 20$ and x and y are measured in metres.

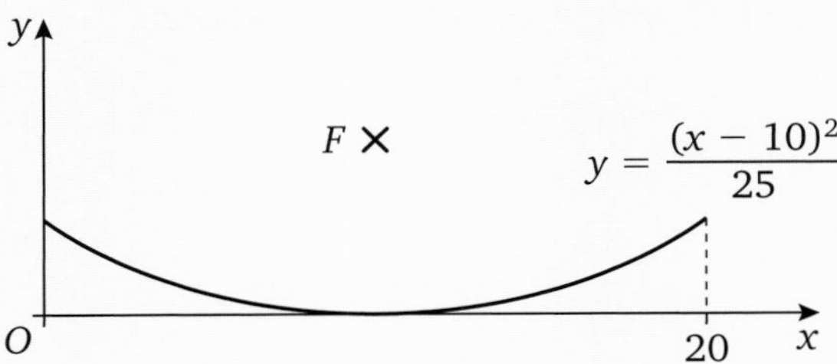

a Work out the height of the dish.

b Explain why the diameter of the dish is 20 m.

c The focal ratio of the dish is given by the formula:

$$\text{focal ratio} = \frac{\text{focal length}}{\text{diameter}}$$

The dish has a focal ratio of 0.3125

What are the coordinates of the focal point, F?

A

23 In the diagram ABP and DCP are straight lines.
$AC = AD$, angle $CAD = 28°$ and angle $BPC = 40°$

Work out the values of x and y.

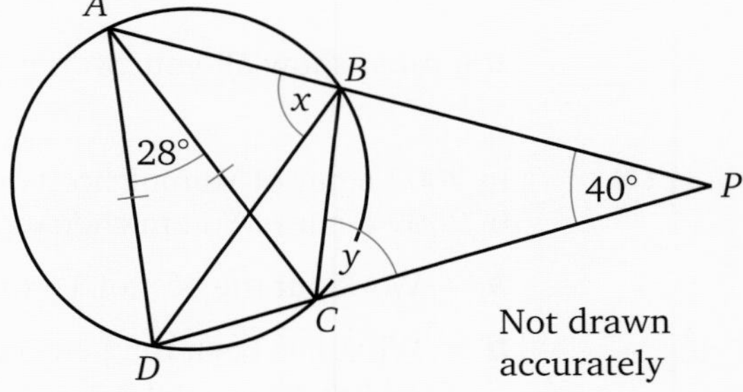

24 The formula $\frac{1}{u} + \frac{1}{v} = \frac{1}{f}$ is used with lenses.
u is the distance of an object from a lens.
v is the distance of the image from the lens.
f is the focal length of the lens.
All distances are in the same units.

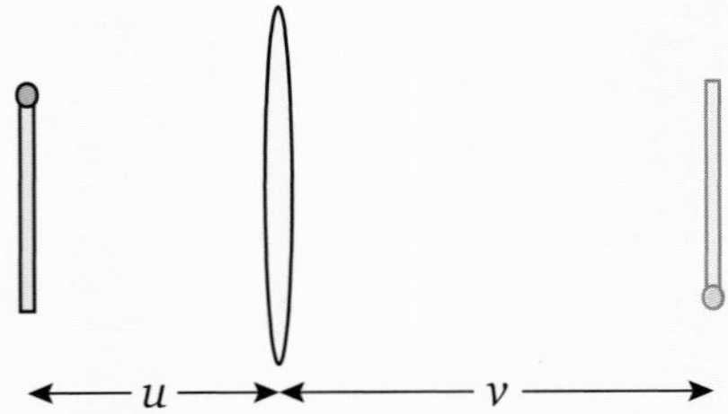

a Show that the formula can be rearranged into the form:

$$v = \frac{uf}{u - f}$$

b An object is placed in front of the lens at a distance that is twice the focal length. The distance of the object from the lens, u, is equal to $2f$.

Show that the distance of the image from the lens, v, is equal to u.

A

25 In 2010 the price of a ticket to a garden party was increased by 10% on its 2009 price.
The money raised from sales of tickets was the same as in 2009.

What was the percentage decrease in sales of tickets?

A A*

26 *ABCD* is a parallelogram.
P and *Q* are two points on *CD* so that $CP = CB$ and $DQ = DA$
Angle $DAQ = x$

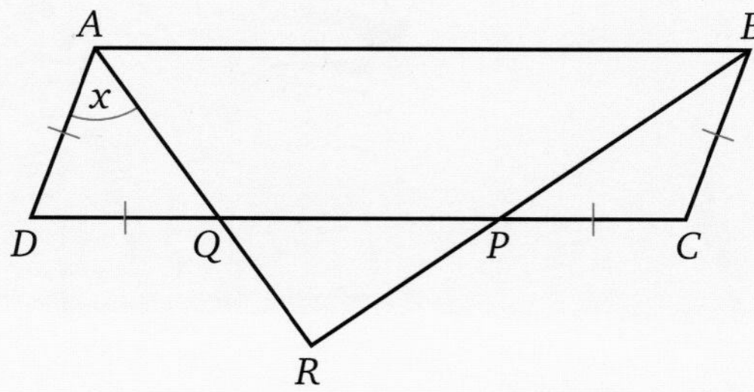

Not drawn accurately

a Express the following angles in terms of x.

i Angle *RAB*

ii Angle *BCP*

b Prove that angle *ARB* is a right-angled triangle.

AQA Examination-style questions

1 *ABCD* is a cyclic quadrilateral.
PCQ is a tangent at *C*.
O is the centre of the circle.
Triangle *ABC* is isosceles.

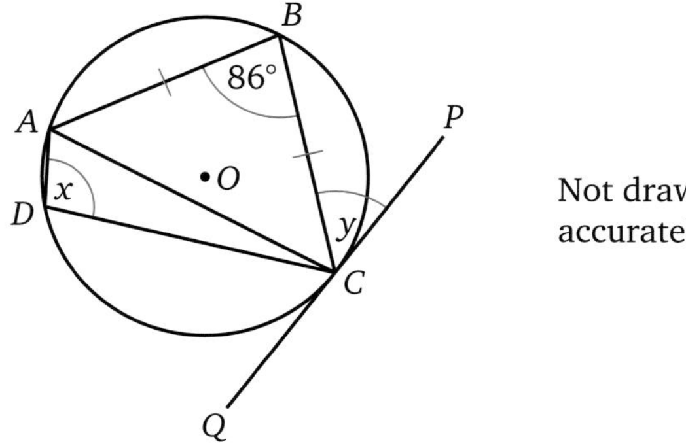

Not drawn accurately

a Work out the value of x. *(2 marks)*

b **i** Work out the value of y. *(3 marks)*

ii Write down the name of the circle theorem used in part **b i**. *(1 mark)*

AQA 2008

17 Quadratic functions

Objectives

Examiners would normally expect students who get these grades to be able to:

C

draw graphs of quadratics such as $y = x^2 + 2x + 1$

use a graph to estimate x- and y-values, giving answers to an appropriate degree of accuracy

draw graphs of harder quadratics such as $y = 2x^2 - 7x + 5$

B

factorise an expression such as $x^2 - 5x + 14$ or $x^2 - 9$

solve an equation such as $x^2 - 5x + 14 = 0$

A

factorise an expression such as $3x^2 + 7x + 2$ or $3x^2 - 27$

solve problems using equations that factorise such as $3x^2 + 7x + 2 = 0$

solve problems using equations such as $2x^2 - 6x + 1 = 0$ by using the quadratic formula

solve problems using equations such as $x^2 + 3x + 2 = 5$ by graphical methods

A*

solve problems using equations such as

$$\frac{3}{x-2} + \frac{4}{x-1} = 2$$

Did you know?

The Indian influence

Much of the mathematics learned today has taken many years to evolve. It can be traced back to several mathematicians in many different countries.

This is true of the development of the quadratic formula that you will meet later in the chapter.

The Egyptians, Babylonians, Greeks and Chinese all played their part in the development of the formula.

The most important of these mathematicians was an Indian astronomer by the name of Bhaskara, working in the 12th century ACE. This Hindu mathematician was the first to recognise that a positive number had two square roots: a positive one and a negative one. From this, other mathematicians were able to carry on with his good work and developed the general algebraic version, which we still use today.

Key terms

factorising	quadratic equation
quadratic expression	surd
expanding	quadrant
coefficient	parabola

You should already know:

- ✔ how to multiply two brackets together
- ✔ how to solve a linear equation
- ✔ how to simplify algebraic fractions
- ✔ how to take out a simple common factor
- ✔ the difference between a quadratic expression and a quadratic equation
- ✔ how to recognise a linear or a quadratic expression/equation.

Learn... 17.1 Factorising quadratic expressions

Factorising quadratic expressions is the inverse operation to **expanding**.

Make sure that the quadratic is arranged with descending powers of x. If not, rearrange it.

Begin by factorising the first term (x^2), then the last term (constant) and then look at all the pairs of possible brackets that could produce the middle term (x).

There may be many possible pairs of brackets. Find the correct one by multiplying out the brackets.

Example: Factorise $3x^2 + 8x + 5$.

Solution: In this example, the **coefficient** of x^2 does not equal 1, so the brackets will not both start with x.

The second sign is +, so both signs are the same. The first sign is a plus so they are both pluses.

$$(\ +\ )(\ +\ )$$

First term $3x^2$: to get $3x^2$, the brackets must start with $3x$ and x.

Last term $+ 5$: to get $+ 5$, the second terms must multiply to $+5$ so they must be $+1$ and $+5$.

Middle term $+ 8x$:

Now decide which of the two possible pairs of factors gives you the middle term $+ 8x$.

$(3x + 5)(x + 1)$

or $(3x + 1)(x + 5)$

Each pair of brackets multiplies to $3x^2 + \ldots + 5$

but only one multiplies to $3x^2 + 8x + 5$.

Check them by multiplying out.

The factorised expression is $(3x + 5)(x + 1)$.

Do not worry about not reaching an answer straight away.

Finding a selection of brackets that could give the answer is part of the method.

Sometimes you find the solution straight away and sometimes it happens to be the last pair of brackets that you write down.

As you practise factorisation, you will identify the correct answer more quickly.

> **AQA Examiner's tip**
> Always check your answer by multiplying it out to see if you arrive back at the quadratic.

> **AQA Examiner's tip**
> If the second sign in the unfactorised equation is a plus, both signs in the brackets will be the same. The first sign in the unfactorised equation tells you what these signs in the brackets will be.
> - If it is plus then the brackets are: (+) and (+)
> - If it is a minus then the brackets are: (−) and (−)
>
> If the second sign in the unfactorised equation is a minus, there will be one of each in the factorised brackets:
> **either** (+) and (−) **or** (−) and (+)

Example: The diagram shows a vegetable garden that consists of two identical plots for the vegetables completely surrounded by paths.
All dimensions shown are in metres.

a Write down an expression for the area of the whole garden.

b Write down an expression for the area of one of the vegetable plots.

c Find an expression for the area of the path, and simplify it.

d Factorise your expression for the area of the path.

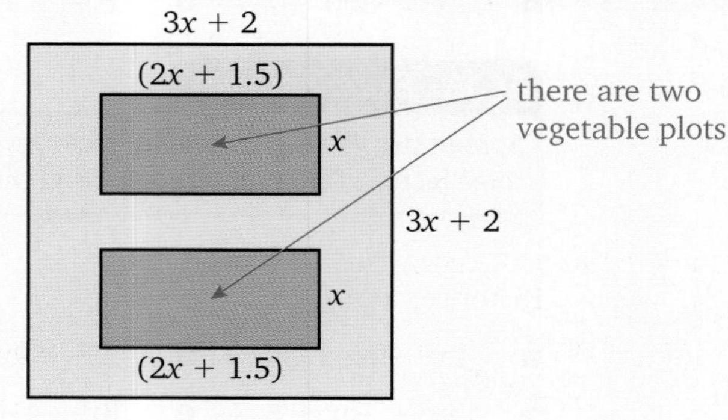

Solution:

a The area of the whole garden $= (3x + 2)^2$

$= (3x + 2)(3x + 2)$

$= 9x^2 + 6x + 6x + 4$

$= 9x^2 + 12x + 4 \quad \text{m}^2$

b The area of one vegetable plot $= x(2x + 1.5)$

$= 2x^2 + 1.5x \quad \text{m}^2$

c Area of path $= (9x^2 + 12x + 4) - (2x^2 + 1.5x) - (2x^2 + 1.5x)$

$= (9x^2 + 12x + 4) - 2(2x^2 + 1.5x)$

$= 9x^2 + 12x + 4 - 4x^2 - 3x$

$= 5x^2 + 9x + 4\,\text{m}^2$

The minus sign outside the bracket changes the signs inside a bracket.

d $5x^2 + 9x + 4$

The signs are both + so both brackets will have plus signs.

First term in expression $5x^2$: factors of $5x^2$ are $5x$ and x.

Last term in expression + 4: factors of 4 are 4 and 1 or 2 and 2.

Middle term in expression $9x$:

Possibilities are: $(5x + 4)(x + 1)$

$(x + 4)(5x + 1)$

$(5x + 2)(x + 2)$

The correct factors are $(5x + 4)(x + 1)$.

Practise... 17.1 Factorising quadratic expressions

D C B A A*

B

1 Factorise:

a $x^2 + 6x + 5$
b $x^2 - 10x + 21$
c $x^2 + 11x + 28$
d $x^2 - 13x + 40$
e $x^2 + 3x - 88$
f $x^2 - 10x + 24$
g $x^2 - 625$
h $4x^2 - 25$

2 Factorise:

a $x^2 - 1.44$
b $x^2 - 0.36$
c $x^2 - \frac{1}{25}$
d $x^2 - \frac{1}{4}$

B A

3 Factorise completely:

a $-2x + x^2$
b $2x^2 - 800$
c $-8x + 4x^2$
d $x^2 - 7 + 6x$
e $-12 - x + x^2$
f $-9x + x^2 + 14$
g $3x^2 - 6 + 3x$

AQA Examiner's tip

Sometimes you will be told to 'factorise completely' when there are three factors. Don't forget to write all three factors in your answer.

A

4 Factorise:

a $2x^2 + 5x + 3$
b $3x^2 + 16x + 5$
c $5x^2 - 7x + 2$
d $2x^2 - 9x - 5$
e $8x + 4 + 3x^2$
f $2x^2 + 13x + 6$
g $5x^2 - 27x + 10$
h $10x^2 - 51x + 5$
i $-7 + 6x^2 - x$
j $9 - 24x + 7x^2$
k $8x^2 - 10x + 3$
l $-51x - 13 + 4x^2$

5 Factorise completely:

a $2 + x - x^2$

b $-x^2 - 2x + 35$

c $28 + 10x - 2x^2$

d $25x^2 - 16$

e $1.44x^2 - 1.21$

f $18x^2 + 15x - 18$

g $30 + 4x - 2x^2$

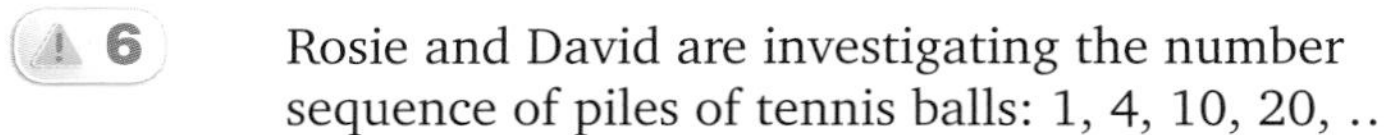

6 Rosie and David are investigating the number sequence of piles of tennis balls: 1, 4, 10, 20, …

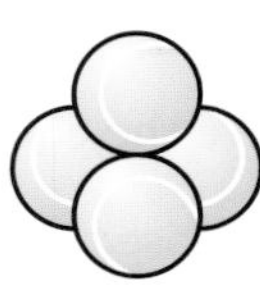
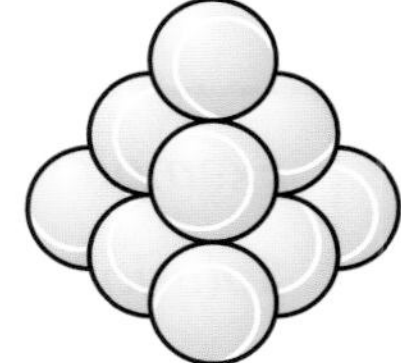

Rosie finds the expression for the *nth* term to be:

$$\tfrac{1}{6}n^3 + \tfrac{1}{2}n^2 + \tfrac{1}{3}n$$

David searches the internet and discovers that the numbers are called tetrahedral numbers and the *n*th term is:

$$\tfrac{1}{6}n(n + 1)(n + 2)$$

Rosie prefers David's expression.

Factorise Rosie's expression to give David's expression.

> **Hint**
> Write all the fractions with the same denominator.

7 A square is cut from another square to form the shaded region, as shown.

a Find a quadratic expression that represents the area of the shaded region.

b Factorise your quadratic expression.

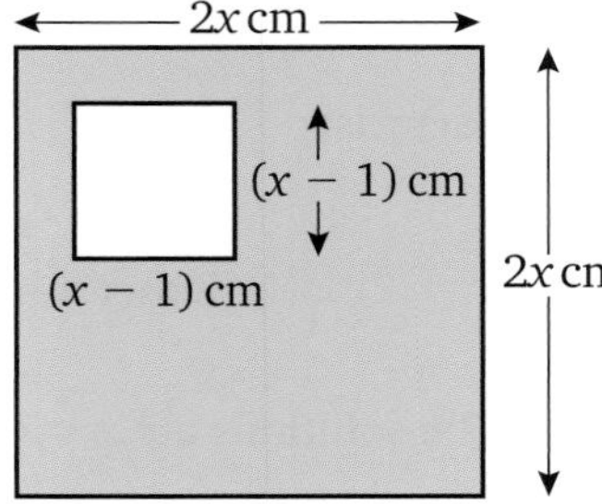

8 A hole is cut in a piece of metal that has dimensions $5x$ by $(3x + 1)$ cm.

The hole is in the shape of a parallelogram as shown.

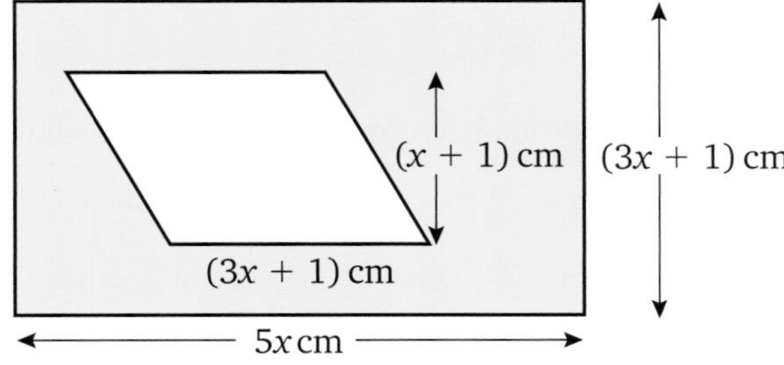

a Find an expression for the area of the rectangle.

b Find an expression for the area of the parallelogram.

c Find an expression for the shaded area.

d Factorise your answer to part **c**.

9 The factors of $x^2 + kx + 32$ are $(x + a)(x + b)$ where a and b are positive integers.

Show that there are three possible values of k.

10 One of the factors of $2y^2 - 7y + n$ is $(2y - 1)$.

Find the value of n.

Learn... 17.2 Solving quadratic equations by factorising

When solving a **quadratic equation**, it must first be rearranged into the form $ax^2 + bx + c = 0$

You then factorise the quadratic expression on the left-hand side of the equation.

The final steps relate to the two factorised expressions p and q.

If $p \times q = 0$ then either $p = 0$ or $q = 0$

For example, if $(x - 2) \times (x + 5) = 0$ then either $(x - 2) = 0$ or $(x + 5) = 0$ so x can be either 2 or -5.

Example: Solve the equation $2x^2 - 17x + 8 = 0$

Solution: Factorise: $(2x - 1)(x - 8) = 0$

Either $(2x - 1) = 0$ or $(x - 8) = 0$

$x = \frac{1}{2}$ $\quad$ $x = 8$

Hint

If $2x - 1 = 0$ then $2x = 1$ so $x = \frac{1}{2}$

The solutions are $x = \frac{1}{2}$ or $x = 8$.

Check: $2(\frac{1}{2})^2 - 17\times(\frac{1}{2}) + 8 = 2(\frac{1}{4}) - 8\frac{1}{2} + 8 = \frac{1}{2} - 8\frac{1}{2} + 8 = 0$ ✓

$2 \times 8^2 - 17 \times 8 + 8 = 128 - 136 + 8 = 0$ ✓

Example: Solve the equation $x^2 + 5x = 0$

Solution: Factorise: $x(x + 5) = 0$ Here the 'first bracket' is just x.

Either $x = 0$ or $(x + 5) = 0$

$x = -5$

The solutions are $x = 0$ or $x = -5$

Check: $0^2 + 5 \times 0 = 0$ ✓

$(-5)^2 + 5\times(-5) = 25 - 25 = 0$ ✓

Example: A rectangular garden is y metres wide.
The length of the garden is 7 metres more than the width.
The area of the garden is 120 m².

Form a quadratic equation in y and solve it to find the dimensions of the garden.

Solution: Width = y metres, so length = $y + 7$ metres

Area of garden = width × length

$= y(y + 7)$ Expand the brackets.

$= y^2 + 7y$

Area of the garden = 120 m²

So $y^2 + 7y = 120$

$y^2 + 7y - 120 = 0$ Take 120 from both sides.

$(y + 15)(y - 8) = 0$ Then factorise.

AQA Examiner's tip

Rearrange a quadratic equation in the form .... = 0 before you factorise.

Either $(y + 15) = 0$ or $(y - 8) = 0$

$y = -15$ or $y = 8$ Reject the negative value as y is a length.

So $y = 8$

Length = 8 + 7 = 15 m and width = 8 m

Check: area = 15 × 8 = 120 m² ✓

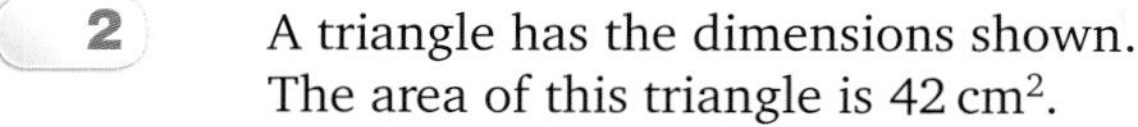

17.2 Solving quadratic equations by factorising

D C B A A*

B

1 Solve these equations.

a $x^2 + 17x + 72 = 0$

b $s^2 - 11s + 28 = 0$

c $z^2 + 10z - 24 = 0$

d $y^2 - 13y = 0$

e $b^2 - 2b = 15$

f $t^2 - 39 = 10t$

2 A triangle has the dimensions shown.
The area of this triangle is $42\,\text{cm}^2$.

a Write down a quadratic equation for the area of the triangle.

b Solve this equation to find the height and base length of the triangle.

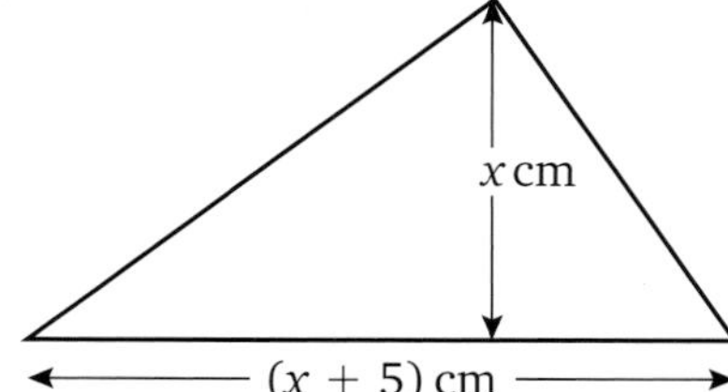

A A*

3 Solve these equations.

a $2x^2 + 9x - 18 = 0$

b $8y^2 - 18y + 9 = 0$

c $5p^2 - 23p + 12 = 0$

d $2x^2 - 0.5 = 0$

e $4y^2 - 8y - 8 = 13$

f $3p^2 + 16p + 7 = -14$

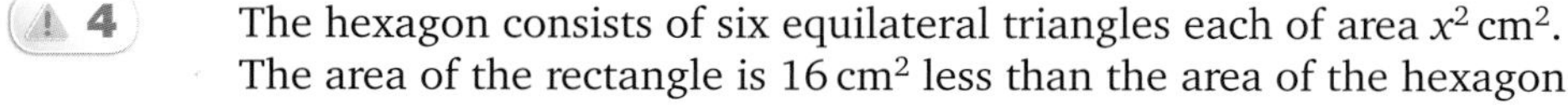

4 The hexagon consists of six equilateral triangles each of area $x^2\,\text{cm}^2$.
The area of the rectangle is $16\,\text{cm}^2$ less than the area of the hexagon.

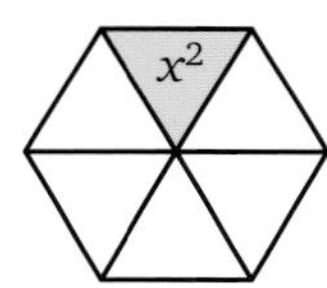

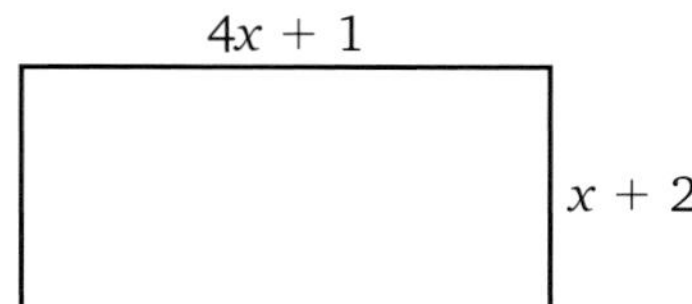

Form a quadratic in x and solve this to find the dimensions of the rectangle.

5 A box is made from a rectangular piece of card by cutting out a square from each corner.

a Write down expressions in terms of x for the length, width and height of the box formed from the sheet.

b The volume of the box is $144\,\text{cm}^3$.

Find the dimensions of the original piece of card.

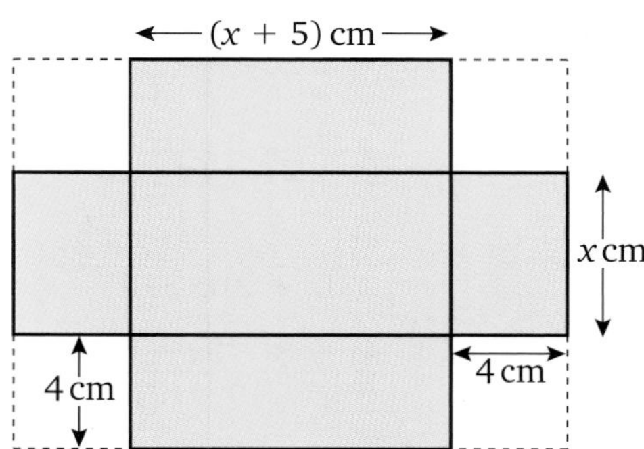

6 The length of a parallelogram is $(x - 3)$ cm and its perpendicular height is $(x + 3)$ cm.
The area is $16\,\text{cm}^2$.

Find the dimensions of the parallelogram.

7 A square of side $(x + 3)$ cm is cut from a square of side $(x + 4)$ cm.
The area remaining is $11\,\text{cm}^2$.

By forming an equation for the remaining area, find the side lengths of the two squares.

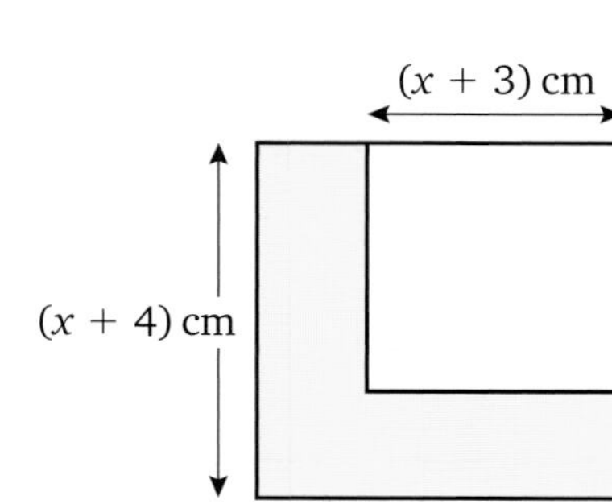

8 Pauline made a curtain for a concert hall. The curtain was a rectangle of length x m and width $(x - 7)$ m. The area of the material was $18\ \text{m}^2$.

a Find the dimensions of the curtain.

b The curtain was too short, so Pauline added a border on the top and the bottom of the curtain. Each border was 50 cm wide when finished.

What are the dimensions of the curtain now?

c Find the area of the new curtain.

9 **a** The quadratic equation $x^2 - 22x + 121 = 0$ has only one repeated solution. Explain why this is.

b Write down another quadratic equation that has only one solution.

10 Anna thinks of a number, squares it and then subtracts twice the original number. The result is 15.

Write down a quadratic equation and solve it to find two possible numbers that Anna could have started with.

Learn... 17.3 Solving equations with fractions

To solve an equation that has fractions:

- first put in any 'hidden brackets'
- give the fractions a common denominator then combine fractions and simplify
- then multiply both sides of the equation by the common denominator
- finally rearrange the terms and simplify/factorise.

Example: Solve the equation $\frac{4}{x+2} - \frac{3}{x-2} = 5$

Solution:

$\frac{4}{(x+2)} - \frac{3}{(x-2)} = 5$ First put in the 'invisible' brackets.

$\frac{4(x-2) - 3(x+2)}{(x+2)(x-2)} = 5$ Combine the fractions, giving them a common denominator: $(x + 2)(x - 2)$.

$\frac{4x - 8 - 3x - 6}{(x+2)(x-2)} = 5$ Multiply out the brackets in numerator.

$\frac{x - 14}{(x+2)(x-2)} = 5$ Collect like terms.

$x - 14 = 5(x + 2)(x - 2)$ Multiply both sides of the equation by the denominator.

$x - 14 = 5(x^2 - 4)$ Simplify.

$x - 14 = 5x^2 - 20$ Multiply out brackets.

$5x^2 - x - 6 = 0$ Rearrange the quadratic with 0 on right-hand side.

$(5x - 6)(x + 1) = 0$ Factorise.

Either $(5x - 6) = 0$ or $(x + 1) = 0$ Solve.

$x = \frac{6}{5} = 1.2$ or $x = -1$

Practise... 17.3 Solving equations with fractions

D C B A A*

A

1 Solve the equation $x = \frac{6}{x+1}$

2 Solve the equation $\frac{x}{x-2} - \frac{3}{x} = 1$

3 Solve the equation $\frac{3}{x} - \frac{2}{x+2} = 1$

A*

4 Solve the equation $\frac{6}{x-2} - \frac{4}{x+2} = 1$

5 Solve the equation $\frac{1}{x-3} - \frac{3}{x+2} = \frac{1}{2}$

6 A car is travelling at x km/h. It travels 60 km to its destination. On the way back, its speed is 20 km/h more, so the journey takes less time. If the times taken for each part of the journey are added together, the total journey time is $2\frac{1}{2}$ hours.

Hint

Remember that speed $= \frac{\text{distance}}{\text{time}}$

a By writing down expressions for the time taken for each part of the journey, form an equation for the total time taken.

b Solve this equation to find the speed for each part of the journey.

AQA Examiner's tip

Remember to check that your answer makes sense in the context of the question.

7 Shaun and Ben both play cricket. During the season they each score a total of 45 runs.

Shaun plays x games and Ben plays $x + 4$ games.

a Write down an expression for the batting average for each of them.

b There is a difference of 4 runs between their batting averages.
Using your answers to part **a**, write down an equation in x.

c Solve your equation to find out how many games Shaun played.

Learn... 17.4 Solving quadratic equations by using the quadratic formula

Quadratic equations may be solved by using the **quadratic formula**:

$$x = \frac{-b \pm \sqrt{b^2 - 4ac}}{2a}$$

where a, b and c are the **coefficients** of the terms in the quadratic equation $ax^2 + bx + c = 0$.

This formula is obtained from completing the square with the general quadratic $ax^2 + bx + c = 0$.

The proof is shown overleaf. However, you will only be expected to use the formula, not to prove it.

Proof of the quadratic formula

$ax^2 + bx + c = 0$

$x^2 + \frac{bx}{a} + \frac{c}{a} = 0$ Divide by a.

$\left(x^2 + \frac{b}{2a}\right)^2 = -\frac{c}{a} + \frac{b^2}{4a^2}$ Complete the square.

$\left(x + \frac{b}{2a}\right)^2 = \frac{-4ac + b^2}{4a^2}$ Combine the fractions on the right-hand side.

$x + \frac{b}{2a} = \pm\sqrt{\frac{-4ac + b^2}{4a^2}}$ Square root both sides.

$x + \frac{b}{2a} = \pm\frac{\sqrt{b^2 - 4ac}}{2a}$ Square root $4a^2$ and rearrange.

$x = -\frac{b}{2a} \pm \frac{\sqrt{b^2 - 4ac}}{2a}$ Subtract $\left(\frac{b}{2a}\right)$ from both sides.

$x = \frac{-b \pm \sqrt{b^2 - 4ac}}{2a}$ Combine the fractions on the right-hand side.

Example: Solve the quadratic equation $2x^2 - x - 9 = 0$.

Give your answers correct to two decimal places.

Solution: $2x^2 - x - 9 = 0$

First write down the values of a, b and c $a = 2$, $b = -1$ and $c = -9$

Using the formula $x = \frac{-b \pm \sqrt{b^2 - 4ac}}{2a}$

$x = \frac{-(-1) \pm \sqrt{(-1)^2 - 4(2)(-9)}}{2(2)}$ Substitute the values for a, b and c.

$x = \frac{+1 \pm \sqrt{1 + 72}}{4}$ Take care with the signs as you simplify this.

$x = \frac{+1 \pm \sqrt{73}}{4}$ Work out the values under the square root before splitting into two parts.

$x = \frac{+1 + \sqrt{73}}{4}$ or $= \frac{+1 - \sqrt{73}}{4}$

AQA Examiner's tip

If the number underneath the square root is negative, there are no solutions. If it is negative, check your working as it is likely you have made a mistake.

Store the value of the square root in the memory of your calculator so that you can use it in each part.

$x = \frac{1 + 8.5440...}{4}$ or $\frac{1 - 8.5440...}{4}$

$x = \frac{9.5440...}{4}$ or $\frac{-7.5440}{4}$

$x = 2.3860...$ or $-1.886...$

$x = 2.39$ or $x = -1.89$ (to 2 d.p.)

You will have to use the formula when you are asked to give an answer correct to a number of significant figures or decimal places. You would only be asked to leave your answer in **surd** form in Unit 2.

AQA Examiner's tip

The quadratic formula can be used to find the solution of a quadratic which has whole number answers but it is easier to factorise these quadratics without the formula.

So, before you use the quadratic formula, look at the quadratic you have been given to check whether it has whole number factors, e.g. $2x^2 + x - 3 = (2x + 3)(x - 1)$

Example: A window is 15 cm longer than it is wide. The area of the window is 225 cm^2.

a Form a quadratic equation in x for the area of the window.

b Solve the equation to find the value of x.

c What are the dimensions of the window? Give the answers correct to 2 d.p.

Solution:

a Let the width of the window be x cm so the length of the window is $(x + 15)$ cm.

x cm

$(x + 15)$ cm

Area $= x(x + 15)$

So $x(x + 15) = 225$

$x^2 + 15x - 225 = 0$

b In $x^2 + 15x - 225 = 0$, $a = 1$, $b = 15$ and $c = -225$

Using the formula $x = \dfrac{-b \pm \sqrt{b^2 - 4ac}}{2a}$

Substituting the values for a, b and c:

$$x = \frac{-15 \pm \sqrt{(15^2) - 4 \times 1 \times (-225)}}{2 \times 1}$$

$$= \frac{-15 \pm \sqrt{225 + 900}}{2}$$

$$= \frac{-15 \pm \sqrt{1125}}{2}$$

$$x = \frac{-15 + \sqrt{1125}}{2} \text{ or } \frac{-15 - \sqrt{1125}}{2}$$

Reject the negative value because x is a length.

$$x = \frac{-15 + 33.5410\ldots}{2}$$

$x = 9.27$ (to 2 d.p.)

c Width $= x = 9.27$ cm (to 2 d.p.)

Length $= x + 15 = 24.27$ cm (to 2 d.p.)

The dimensions of the window are 24.27 cm by 9.27 cm (to 2 d.p.).

> **AQA Examiner's tip**
>
> Watch out for solutions that do not fit the context of the question.
>
> They are usually negative solutions but not always.

Practise... 17.4 Solving quadratic equations by using the quadratic formula

A

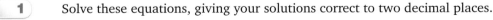

1 Solve these equations, giving your solutions correct to two decimal places.

a $x^2 + 8x + 9 = 0$

b $3z + z^2 - 5 = 0$

c $2x^2 - 7x + 4 = 0$

d $5y^2 - 15y + 2 = 10$

e $3x^2 + 6x - 2 = 0$

f $4p^2 - 6p = p - 2$

g $1.2x^2 + 0.8x - 14 = 0$

h $2 - 3x - x^2 = 0$

i $5x^2 = 4x + 2$

> **Hint**
>
> For part **g** either work with decimal values or multiply throughout by 10.

2 Solve the equation $\dfrac{2}{(x+1)} - \dfrac{3}{(2x+3)} = \dfrac{1}{2}$

3 Solve the equation $\dfrac{2}{(x+2)} = 4 - \dfrac{1}{(x+3)}$

> **Hint**
>
> Rearrange the equation so that both fractions are on the same side.

4 The surface area of an open cylinder is given by the formula $S = 2\pi rh$
The radius of the cylinder is x cm and its height is 9 cm more than the radius.

a If the surface area of the cylinder is $16\pi \text{ cm}^2$, find the value of x.

b What is the height of the cylinder?

5 A ball is thrown vertically upwards with a speed of 22 m/s.
After t seconds, its height above the ground, h metres, is given by the formula:

$$h = 22t - 4.9t^2$$

a Find the time taken for the ball to reach a height of 8 metres.
Give your answer to two decimal places.

b Explain why there are two valid answers to this problem.

6 A courtyard consists of a patio area with four **quadrant**-shaped flowerbeds at each corner.
They are all of radius x m. The patio has an area of 20 m^2.

a Form an equation in x, to find the radius of the flower beds.

b Calculate the dimensions of the courtyard.

c Use your answers to parts **a** and **b** to check your working.

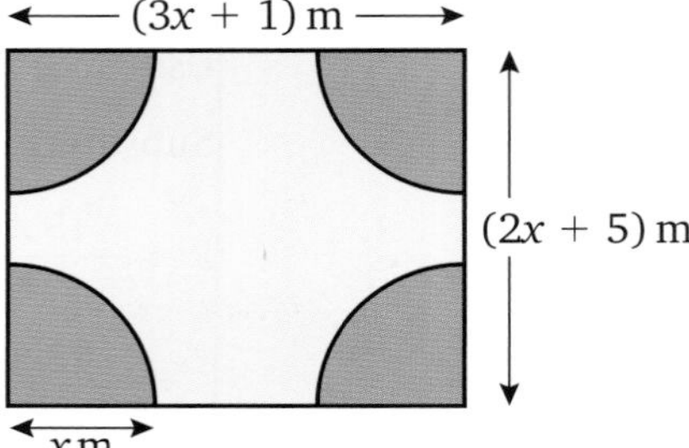

7 Explain why you cannot solve the equation $x^2 + 3x + 7 = 0$

8 Sam is using the quadratic formula to solve a quadratic equation. He writes his solution as:

$$\frac{(-3 + \sqrt{17})}{2}$$

Write down the quadratic equation Sam is trying to solve.

Learn... 17.5 Solving quadratic equations graphically

How to draw the graph of a quadratic

1. Complete or construct a table of values.

This is the table for the quadratic equation $y = x^2$

x	−3	−2	−1	0	1	2	3
y	9	4	1	0	1	4	9

When drawing the graphs of straight lines (linear graphs), you were taught to plot three points.

The graphs of quadratics are all curves, so you need more than three values in the table.

The y-values are found by substituting the x-values into the equation of the quadratic, e.g. for $y = x^2$

when $x = 2$ $\quad y = 2^2 = 4$

when $x = -3$ $\quad y = (-3)^2 = 9$

Sometimes you are asked to construct the table for yourself.

Make sure that you construct this table for the range of values given in the question.

Bump up your grade

To get a Grade C you need to be able to draw accurate graphs of quadratics. Ensure that you get plenty of practice drawing smooth, symmetrical, continuous curves.

2. **By looking at your table, find the smallest and largest y-values that you will need on the y-axis.**
3. **Draw a pair of axes for your graph and label them x and y.**
4. **From your table of values, plot the points on the graph as small crosses.**

   The points from this table would be $(-3, 9)$, $(-2, 4)$, $(-1, 1)$, $(0, 0)$, $(1, 1)$, $(2, 4)$, $(3, 9)$.
5. **Join the points with a smooth curve.**

   It should be smooth at the bottom, not 'pointy'.

   Always draw your line with a sharp pencil, making sure that it passes clearly through the centre of the points plotted.

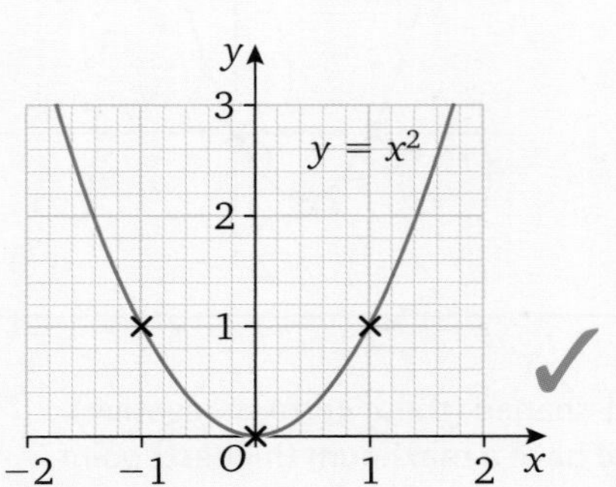

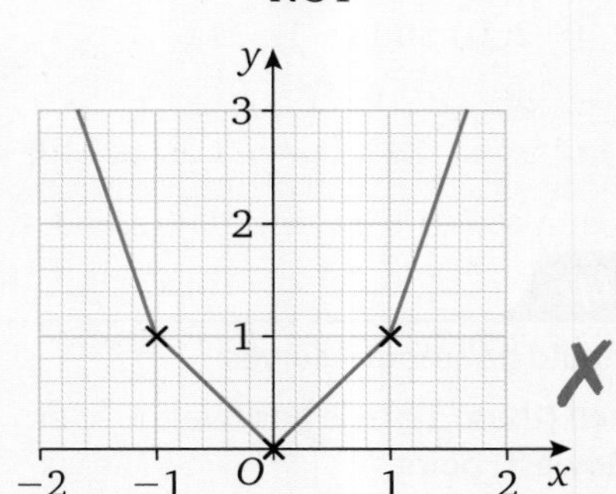

AQA Examiner's tip

Make sure that the curve passes through the middle of all the plotted points.

- Join the points with one continuous curved line.
- If you join the points with straight lines, you will lose marks.
- If you are right handed, you will find it easier to join the points by turning your page upside down.

The graph obtained is a curve. A curve of this shape is called a **parabola**.

$y = x^2$ is a U-shaped graph and has a minimum (lowest) point at $(0, 0)$.

All quadratics are parabolas. They are also symmetrical about a line.
Notice that this graph is symmetrical about the line $x = 0$, the y-axis.

You can sometimes see the pattern in the table. If you look back at the table for this graph, you will see that the values for $x = -1$ and 1 give the same y-values. (The points are equally spaced about $x = 0$.)

Graph of $y = x^2$

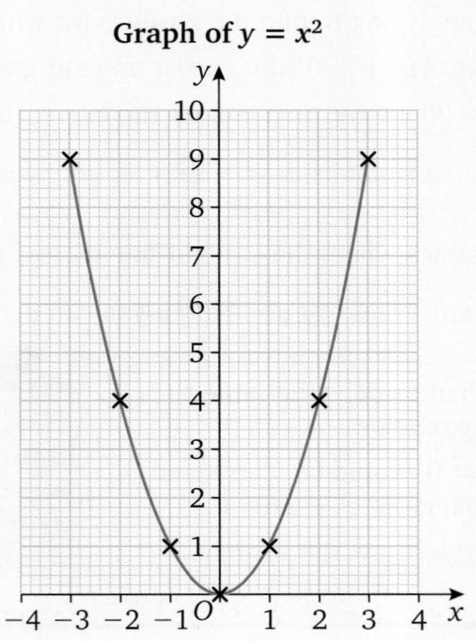

With more complicated quadratics, it is useful to construct a more detailed table.

When drawing the graph of $y = x^2 + 2x$, you would construct a table that shows a row for each term in the expression.

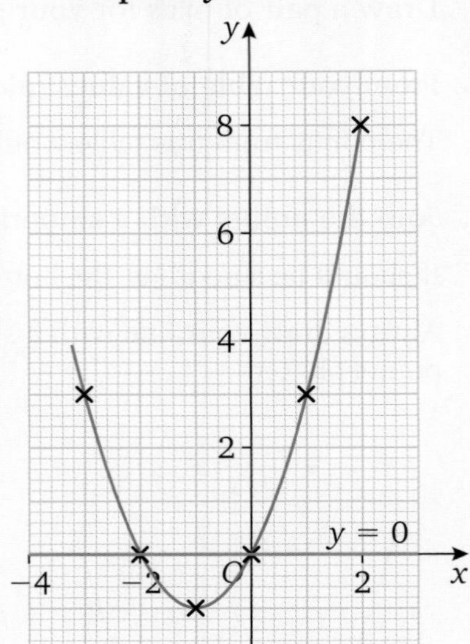

Add these two rows to get the y-values (rows x^2 and $+2x$)

x	−3	−2	−1	0	1	2
x^2	9	4	1	0	1	4
$+2x$	−6	−4	−2	0	2	4
y	3	0	−1	0	3	8

The points to be plotted are the x and y rows.

For example, $(-3, 3)$, $(-2, 0)$ and $(-1, -1)$.

AQA Examiner's tip

All quadratic graphs should be smooth curves.

They are either U-shaped (the x^2 term is positive) and have a minimum (lowest) point or hill-shaped (the x^2 term is negative) and have a maximum (highest) point.

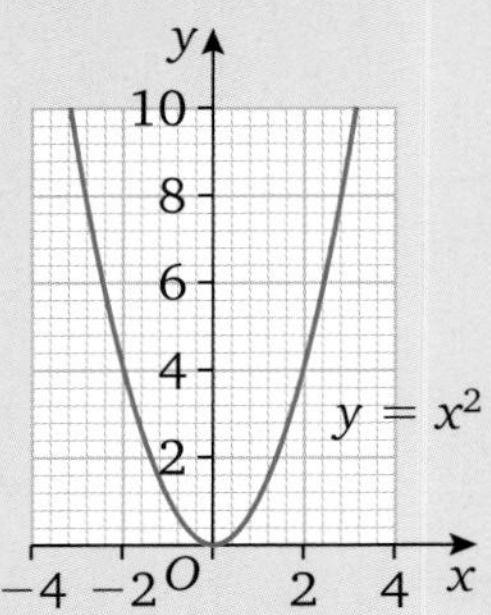

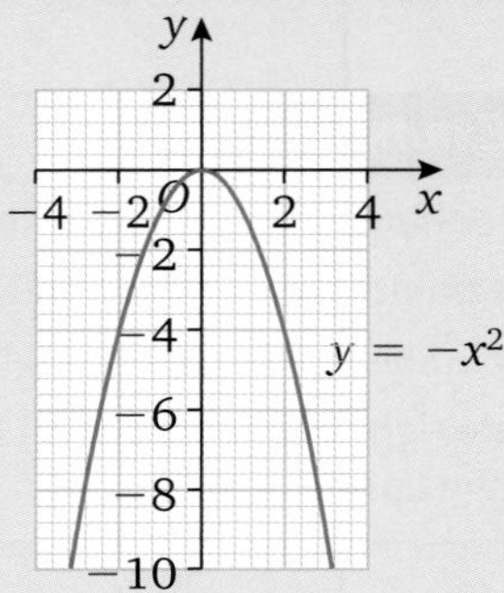

If your graph does not look like one of these, check your working in the table.

You can use these graphs to find the solution of quadratic equations.

If you were asked to find the solution to the equation $x^2 + 2x = 0$, you would do this by adding a straight line to your graph.

To get from $y = x^2 + 2x$ to $x^2 + 2x = 0$ $(0 = x^2 + 2x)$, the y has been replaced by 0.

The y-coordinate equals zero.

In solving the equation, you are trying to find the x-values for which $y = 0$.

To find these, draw the straight line $y = 0$ (the x-axis) on your graph.

The solutions are the x-values where this line crosses the curve, that is, $x = 0$ and $x = -2$

If the equation given equals a constant as in $x^2 + 2x = 4$, you would solve this by drawing the straight line $y = 4$ on your graph.

The solutions are still the x-values where the straight line crosses the curve.

Sometimes you have to rearrange the equation before you can decide which straight line to draw.

Given the equation $x^2 + 2x - 6 = 0$, you must rearrange the equation so that the left-hand side is the same as for the quadratic that you have drawn.

This means that $x^2 + 2x - 6 = 0$, would be rearranged by adding 6 to both sides which gives $x^2 + 2x = 6$.

To find the solutions you need to draw the straight line $y = 6$.

Link

Sometimes you have to rearrange equations further to give expressions on the right-hand side. This will be covered in Chapter 20, Simultaneous equations.

Example:

a Draw the graph of $y = 3 + 2x - x^2$ for values of x from $-2 \leqslant x \leqslant 4$

b Write down the coordinates of its maximum point.

c Use your graph to find the solutions of the equation $3 + 2x - x^2 = 0$

d Use the graph of $y = 3 + 2x - x^2$ to solve the following equations.

i $4 + 2x - x^2 = 0$ **ii** $2x - x^2 = -1$

Solution:

a

x	−2	−1	0	1	2	3	4
3	3	3	3	3	3	3	3
$+2x$	−4	−2	0	2	4	6	8
$-x^2$	−4	−1	0	−1	−4	−9	−16
y	−5	0	3	4	3	0	−5

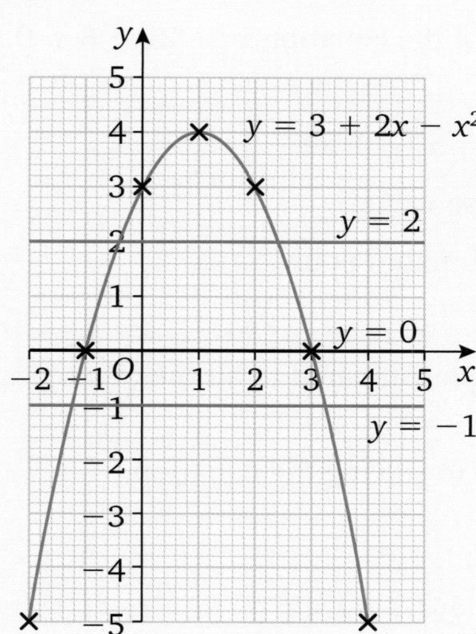

AQA Examiner's tip

In the exam you may be asked to work with the shortened table. Sometimes you may be able to choose which you use.

Whenever you use the shortened form, make sure that you show how you obtained each y-value.

b Maximum point = (1, 4)

c The solutions of the equation $3 + 2x - x^2 = 0$ are found where the line $y = 0$ (x-axis) crosses the curve.

Reading from the graph, these are $x = -1$ and 3.

d **i** To solve: $4 + 2x - x^2 = 0$

$4 - 1 + 2x - x^2 = -1$ Subtract 1 from both sides.

$3 + 2x - x^2 = -1$ The left-hand side of the equation now matches the equation of the line drawn.

Solutions are found where the line $y = -1$ crosses the curve.

Reading from the graph, these are $x = -1.2$ and 3.2 (to 1 d.p.).

ii To solve: $2x - x^2 = -1$

$1 + 2x - x^2 = 0$ Rearrange the equation into the same format as the original.

$1 + 2 + 2x - x^2 = 0 + 2$ Add 2 to both sides.

$3 + 2x - x^2 = 2$ The left-hand side of the equation now matches the equation of the line drawn.

Solutions are found where the line $y = 2$ crosses the curve.

Reading from the graph, these are $x = -0.4$ and 2.4 (to 1 d.p.).

Practise... 17.5 Solving quadratic equations graphically

C

1 **a** Copy and complete this table for $y = x^2 - 2x - 8$.

x	−2	−1	0	1	2	3	4	5
x^2	4	1	0		4		16	25
$-2x$	4		0	−2	−4	−6	−8	−10
-8	−8		−8		−8			−8
y	0		−8		−8			7

b Draw the graph of $y = x^2 - 2x - 8$ for values of x from −2 to 5.

c When $x = 3$, what is the value of y?

d Use your graph to find the solutions of the equation $x^2 - 2x - 8 = 0$.

2 **a** Construct a table for the quadratic $y = x^2 + 1$ for $-3 \leqslant x \leqslant 3$.

b Draw the graph of $y = x^2 + 1$ for these values.

c Write down the coordinates of the minimum point.

d Look at the graph.

i What is different about this graph compared to all the other graphs you have drawn so far?

ii Explain what relevance this has to solving the equation $x^2 + 1 = 0$.

B

3 **a** Copy and complete this table for $y = 2x^2 - 7x + 5$

x	0	0.5	1	1.5	2	2.5	3
$2x^2$	0	0.5		4.5			18
$-7x$		−3.5	−7		−14		−21
$+5$	+5	+5	+5	+5	+5	+5	+5
y		2					2

b Draw the graph of $y = 2x^2 - 7x + 5$ for values of x between 0 and 3.

c Find the coordinates of the minimum point.

d What are the values of x when $y = 4$?

A

4 **a** Draw the graph of $y = 2x^2 - 7x + 3$ for values of x from $0 \leqslant x \leqslant 3$ at intervals of 0.5

b It has a minimum when $x = 1.75$

Find the y-coordinate of this point.

c Use your graph to find the solutions of these equations.

i $2x^2 - 7x + 5 = 0$

ii $2x^2 - 7x + 3 = 0$

d Use your graph to explain why you cannot solve the equation $2x^2 - 7x + 7 = 0$.

5
a Draw the graph of $y = 1 + 4x - \frac{1}{2}x^2$ for values of x from $-1 \leqslant x \leqslant 9$

b Write down the coordinates of its maximum point.

c Use your graph to find the solutions of the equation $1 + 4x - \frac{1}{2}x^2 = 9$

d Find solutions to these equations.

i $1 + 4x - \frac{1}{2}x^2 = 7$ **ii** $1 + 4x - \frac{1}{2}x^2 = 0$

e What is the significance of the answers in part **d**?

6
A square of side x cm is removed from a triangle of card as shown.
The remaining area of card (shaded in the diagram) is 5 cm^2.

x, x, $x + 1$, $3x$

a Form an equation for the area of the shaded part.

b Simplify this equation into the form $ax^2 + bx + c = 0$.

c Construct a table for the quadratic, taking values of x from 0 to 3, at intervals of 0.5

d Draw the graph of the quadratic.

e Use your graph to find the x-value that satisfies the equation in part **b**.

f What are the dimensions of the triangle and the square?

7
A flower bed is in the shape of a semicircle as shown.
The quadratic $A = \frac{1}{2}\pi r^2$ gives the area for the flower bed in square metres.

r

a Copy and complete the following table, giving values of A to one decimal place.

r	1	2	3	4	5
$A = \frac{1}{2}\pi r^2$		6.3		25.1	39.3

b Draw the graph of $A = \frac{1}{2}\pi r^2$ using 2 cm to represent 1 m on the r-axis and 2 cm to represent 10 m^2 on the A-axis.

c Use your graph to estimate the area of the flower bed when the radius is:

i 1.5 m **ii** 3.8 m

d Use your graph to estimate the radius r, of the flower bed when the area is:

i 30 m^2 **ii** 16 m^2

17 Assess

1
Solve these equations.

a $x^2 + 3x + 2 = 0$

b $x^2 - 7x + 10 = 0$

c $x^2 - 3x - 28 = 0$

d $2x^2 - 5x - 18 = 0$

A*

2
Solve these equations by using the quadratic formula.

Give your answers as decimals to two decimal places.

a $x^2 + 8x + 6 = 0$

b $4x^2 - 3x - 2 = 0$

c $x^2 - 5x - 3 = 0$

d $5x^2 + 10x + 4 = 0$

e $3x^2 - 4x - 7 = 0$

A

3 Draw the graph of $y = x^2 - 2x - 3$ for x-values between -2 and 4.

Use your graph to solve the following equations.

Give your answers correct to two decimal places

a $x^2 - 2x - 3 = 0$ **b** $x^2 - 2x - 1 = 0$

4 Draw the graph of $y = x^2 - 5x + 6$ for x-values between -1 and 4.

Use your graph to solve the following equations.

Give your answers correct to two decimal places

a $x^2 - 5x + 6 = 0$ **b** $x^2 - 5x + 5 = 0$ **c** $x^2 - 5x + 2 = 0$

5 Solve the following equations.

a $\frac{x}{(x + 1)} - \frac{4}{(x - 1)} = 1$ **b** $\frac{12}{(x + 3)} + \frac{x}{(x + 5)} = 2$

6 The length of a garden is 5 metres longer than its width. The area is $84\,\text{m}^2$.

a Write down expressions for the length and width of the garden.

b Use these to form a quadratic equation for the area.

c Find the length and width of the garden.

7 A box is made from taking a rectangular piece of card and cutting out a square from each corner, as shown in the diagram.

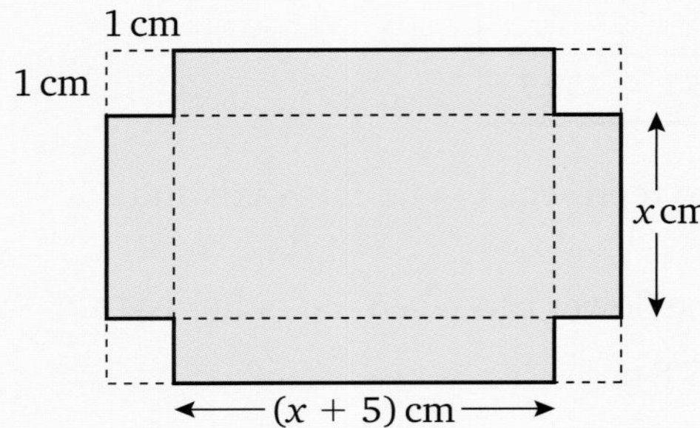

Not drawn accurately

a Write down expressions for the length, width and height of the box formed from the sheet.

b The volume of the box is $14\,\text{cm}^3$.
Find the dimensions of the original piece of card.

A*

8 A cyclist travels 100 km from her home to a seaside town. The next day she then cycles back home.

On the way there, her speed is 5 km/h quicker than on the way back.

There is a difference of 1 hour between the time to travel there and the time to travel back.

a Write down expressions for the time taken for each journey.

b Use these to form an equation for the time taken.

c Solve this equation to find the cyclist's speed both on the way there and on the way back.

AQA Examination-style questions

1 **a** Copy and complete the table of values for $y = x^2 - x - 5$.

x	-2	-1	0	1	2	3	4
y	1		-5	-5	-3	1	

(2 marks)

b On graph paper draw the graph of $y = x^2 - x - 5$ for values of x from -2 to 4. *(2 marks)*

c An approximate solution of the equation $x^2 - x - 5 = 0$ is $x = 2.8$
Explain how you can find this from the graph. *(1 mark)*

AQA 2008

2 The diagram shows a garden in the shape of a rectangle measuring 10 m by 8 m.
On two sides of the garden there is a path x metres wide.
The remaining area is covered by grass.

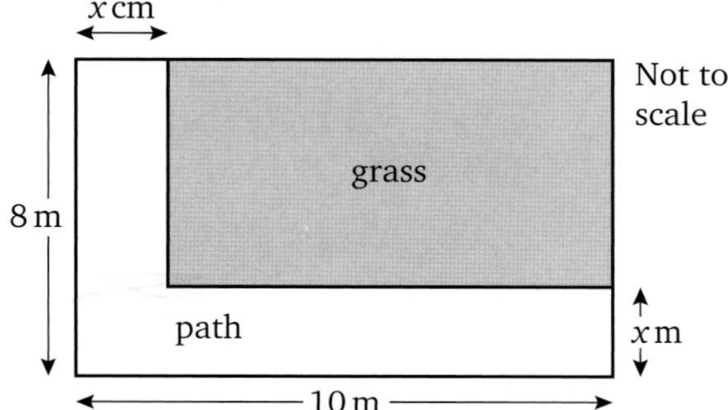

a The area covered by grass is $\frac{3}{5}$ of the area of the garden.
Show that x satisfies the equation $x^2 - 18x + 32 = 0$ *(3 marks)*

b Hence, or otherwise, find the width of the path. *(2 marks)*

AQA 2008

3 Solve the equation $2x^2 + 3x - 7 = 0$.
Give your answers correct to two decimal places.
You **must** show your working. *(3 marks)*

AQA 2008

18 Trigonometry 1

Objectives

Examiners would normally expect students who get these grades to be able to:

B

use sine, cosine and tangent to calculate a side in a right-angled triangle

use sine, cosine and tangent to calculate an angle in a right-angled triangle

A/A*

use trigonometry to solve problems, including those involving bearings

use trigonometry to find sides and angles in three dimensions.

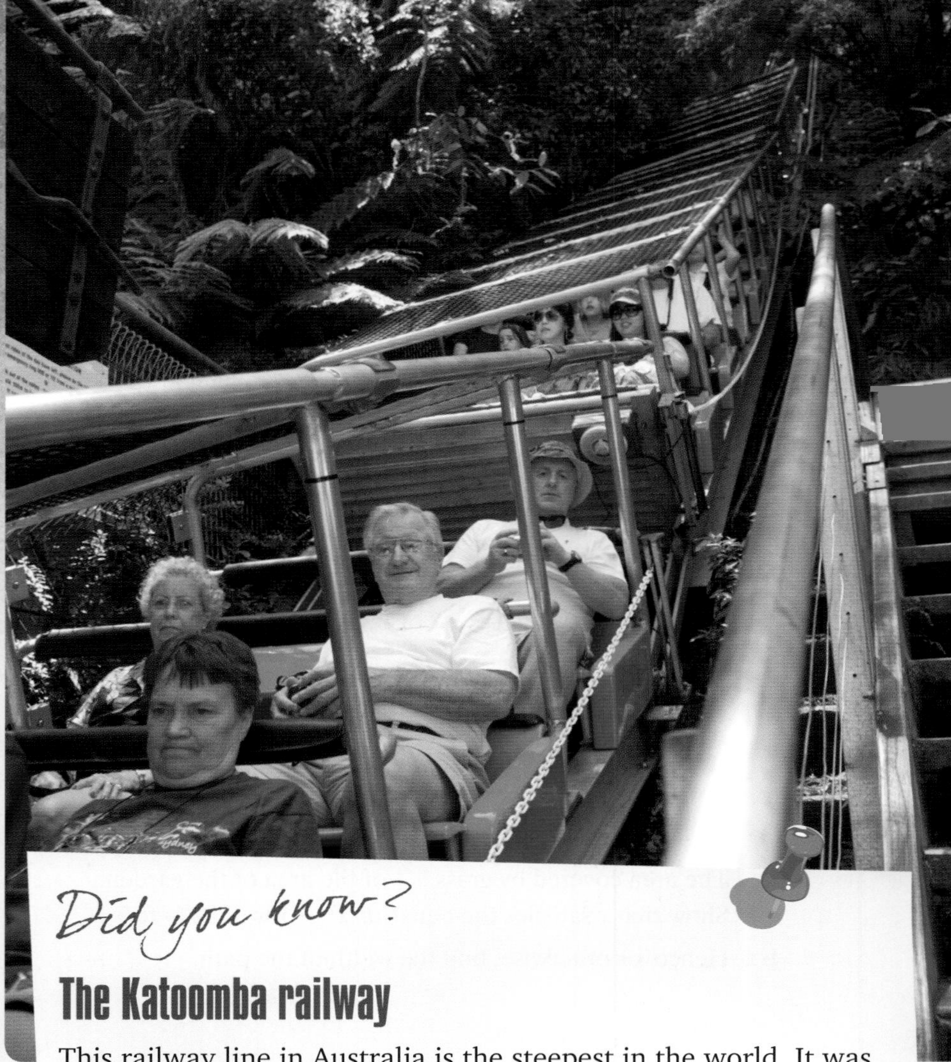

Did you know?

The Katoomba railway

This railway line in Australia is the steepest in the world. It was originally used for mining, but still runs as a tourist attraction.

The line is 415 m long, and the end is 178 m lower than the start.

At its steepest, it makes an angle of 52° with the horizontal.

You should already know:

- ✔ the sum of the angles in a triangle
- ✔ the sum of angles round a point
- ✔ symmetry and angle properties of isosceles triangles
- ✔ how to measure angle bearings
- ✔ ratio properties of similar triangles
- ✔ Pythagoras' theorem
- ✔ how to solve equations when the unknown is the denominator of a fraction.

Key terms

trigonometry
hypotenuse
opposite side
adjacent side
tangent
cosine
sine
angle of elevation

18.1 Calculating the side of a right-angled triangle using trigonometry

Trigonometry is the study of the relationship between sides and angles in triangles.

All **right-angled triangles** that contain the same angle (in addition to the right angle) are similar. For example, all right-angled triangles with an angle of 30° are similar.

This means their sides are in the same ratio (or proportion) regardless of the actual size of the triangle.

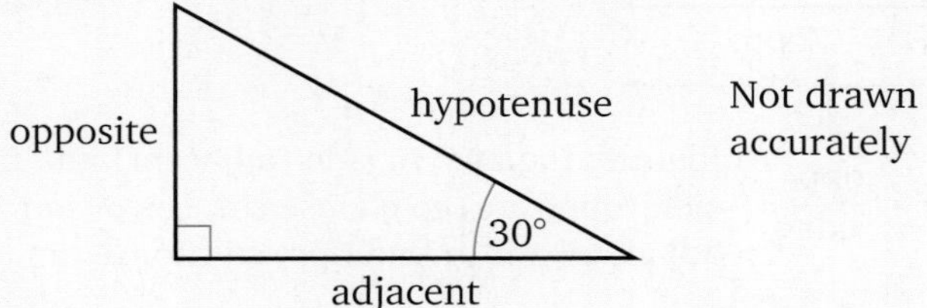

The longest side, opposite the right angle, is called the **hypotenuse**.

The side opposite the known angle is called the **opposite side**.

The side next to the known angle is called the **adjacent side**.

The length of the opposite side divided by the length of the adjacent side is called the **tangent** of the angle.

This is usually abbreviated to $\tan x$, where x is the size of the angle.

$$\tan x = \frac{\text{opposite}}{\text{adjacent}}$$

The length of the adjacent side divided by the length of the hypotenuse is called the **cosine** of x, or $\cos x$.

$$\cos x = \frac{\text{adjacent}}{\text{hypotenuse}}$$

The opposite side divided by the hypotenuse is called the **sine** of x, or $\sin x$.

$$\sin x = \frac{\text{opposite}}{\text{hypotenuse}}$$

For an angle x:

$$\sin x = \frac{\text{opposite}}{\text{hypotenuse}}, \quad \cos x = \frac{\text{adjacent}}{\text{hypotenuse}}, \quad \tan x = \frac{\text{opposite}}{\text{adjacent}}$$

Trigonometry questions will always involve two sides and an angle.

Identify the two sides, and then choose the ratio (sin, cos or tan) that includes those two sides.

Example: Calculate the length of BC.

Give your answer to an appropriate degree of accuracy.

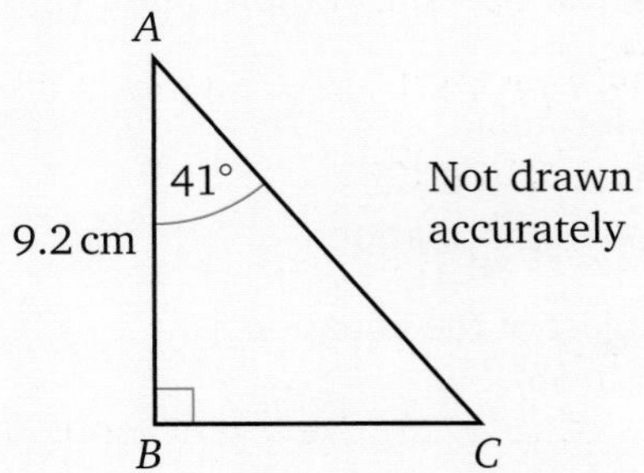

Solution: Copy and label the diagram with hyp, opp and adj.

Put $BC = x$

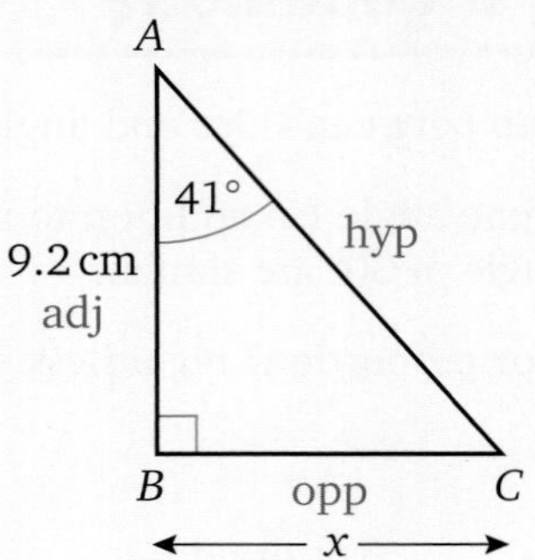

AQA Examiner's tip

Once you know which acute angle you are using (41°), mark the opposite side to that angle 'opp'. The side opposite the right angle is 'hyp', which leaves the remaining side 'adj'.

$\sin = \dfrac{\text{opp}}{\text{hyp}}$, $\cos = \dfrac{\text{adj}}{\text{hyp}}$, $\tan = \dfrac{\text{opp}}{\text{adj}}$ — Identify the known side (adjacent) and the required side (opposite) to choose sin, cos or tan. You know adj and want to find opp, so choose tan.

$\tan = \dfrac{\text{opp}}{\text{adj}}$ — Start with the formula.

$\tan 41° = \dfrac{x}{9.2}$ — Substitute known information.

$9.2 \times \tan 41° = x$ — Multiply both sides by 9.2

$x = 7.997437988\ldots$

$BC = 8.0\text{ cm}$ (to 2 s.f.) — Choose an appropriate degree of accuracy; in the question, AB was given to 2 s.f. (9.2 cm), so give the answer to the same degree of accuracy.

Example: Calculate the length of AB, giving your answer to an appropriate degree of accuracy.

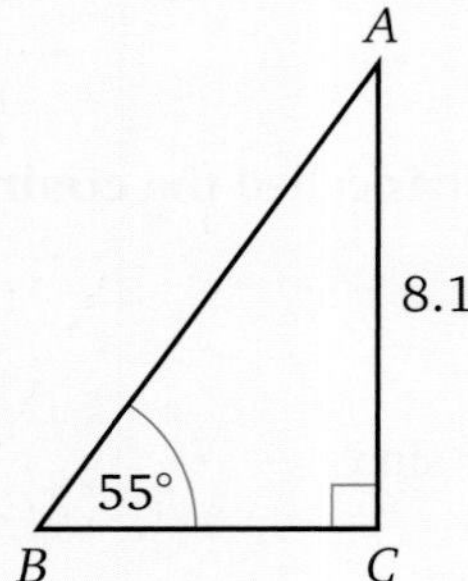

Not drawn accurately

Solution: Copy and label the diagram.

Put $AB = x$

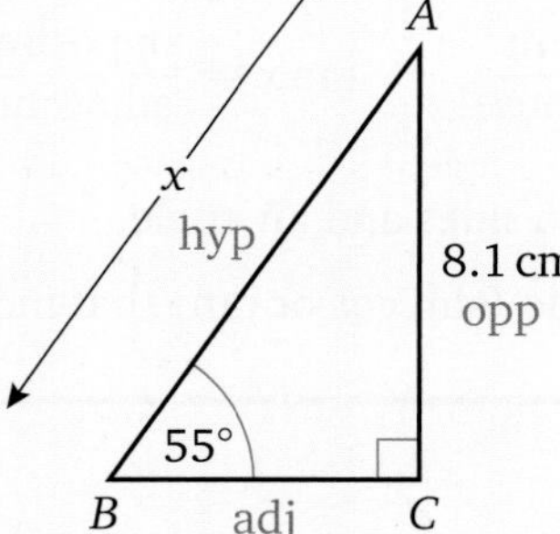

AQA Examiner's tip

Always complete the algebraic manipulation before using your calculator.

$\sin = \dfrac{\text{opp}}{\text{hyp}}$, $\cos = \dfrac{\text{adj}}{\text{hyp}}$, $\tan = \dfrac{\text{opp}}{\text{adj}}$ — You know opp and want to find hyp, so choose sin.

$\sin = \dfrac{\text{opp}}{\text{hyp}}$ — Start with the formula.

$\sin 55° = \dfrac{8.1}{x}$ — Substitute known information.

$x = \dfrac{8.1}{\sin 55°}$ — Make x the subject of the equation. Divide 8.1 by sin 55°.

$= 9.888274169\ldots$

$AB = 9.9\text{ cm}$ (to 2 s.f.)

Practise... 18.1 Calculating the side of a right-angled triangle using trigonometry

D C B A A*

The shapes in these exercises are not drawn accurately.

C

1 In the diagrams below, identify the opposite side, the adjacent side and the hypotenuse.

a

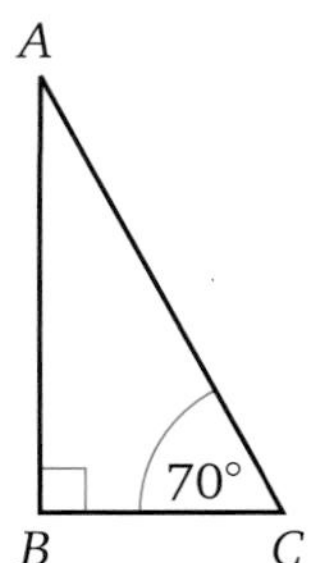

b

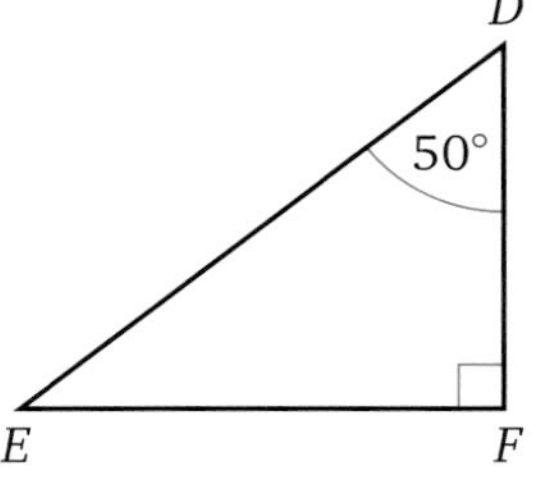

c

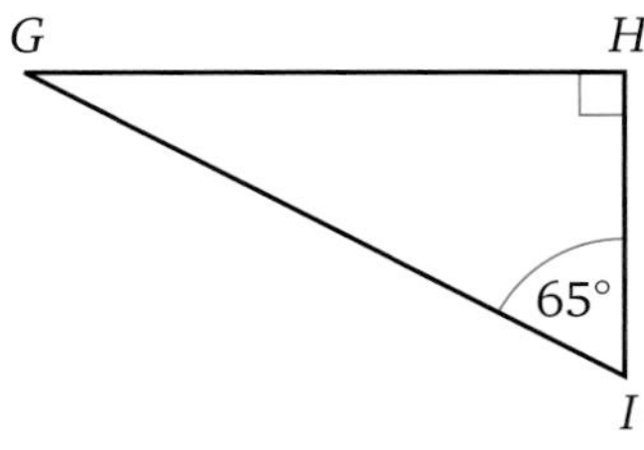

B

2 Copy these diagrams and label them.

Calculate the length marked x, giving your answers correct to one decimal place.

a

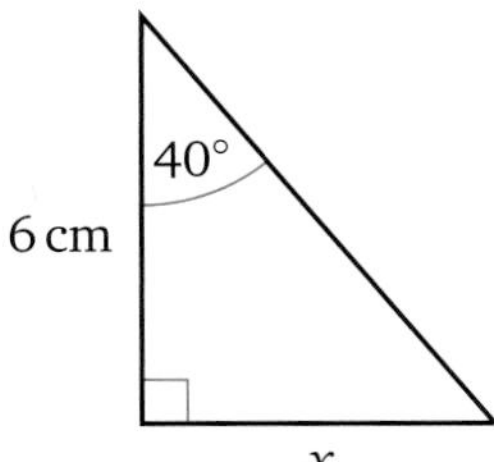

c

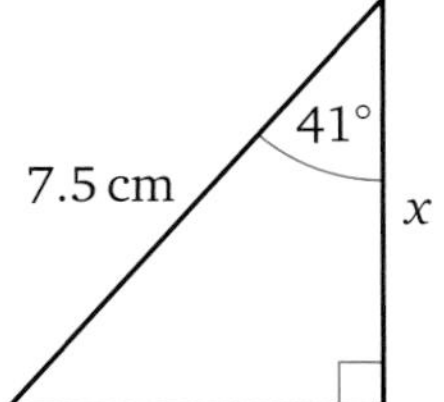

e

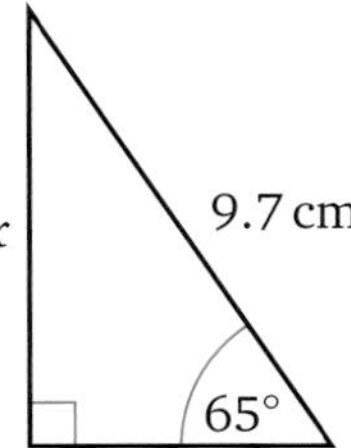

b

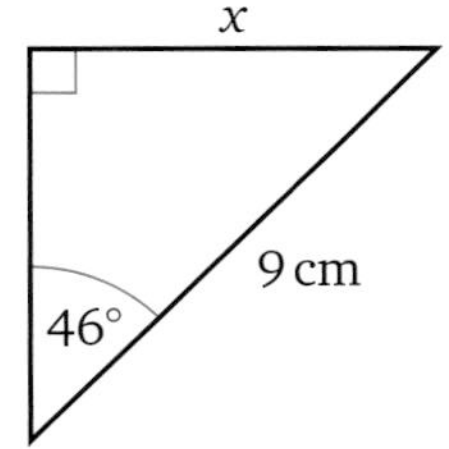

d

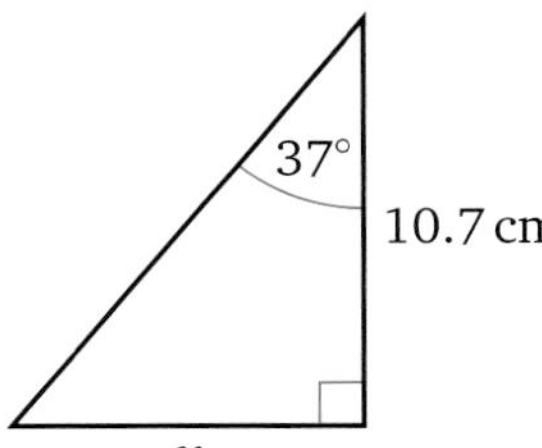

f

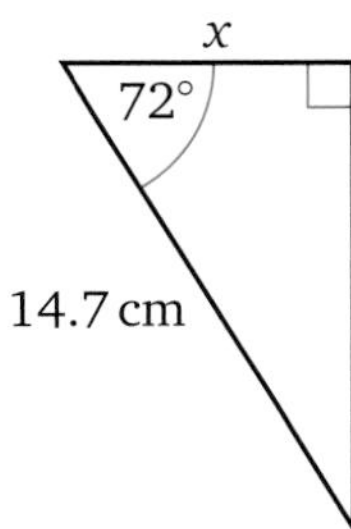

3 Copy these diagrams and label them.

Calculate the length marked x, giving your answers correct to one decimal place.

a

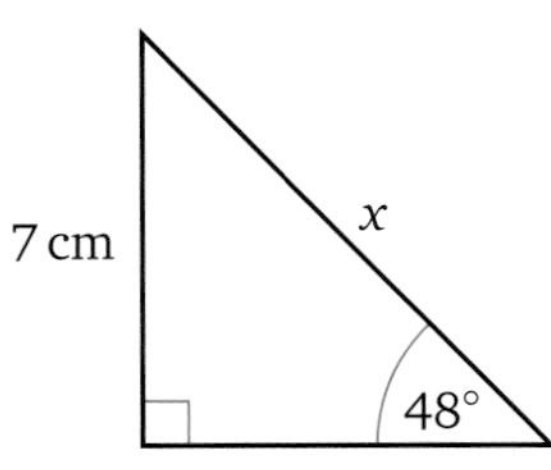

c

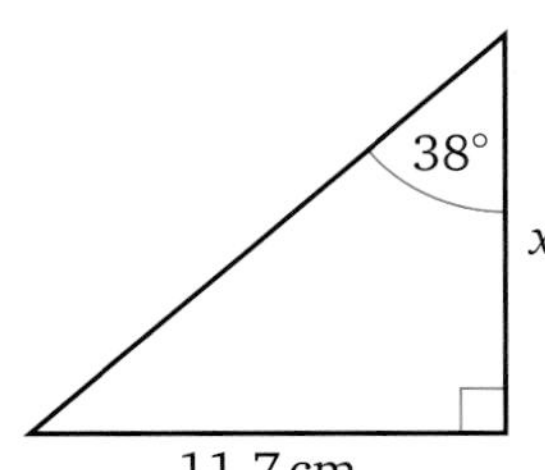

e

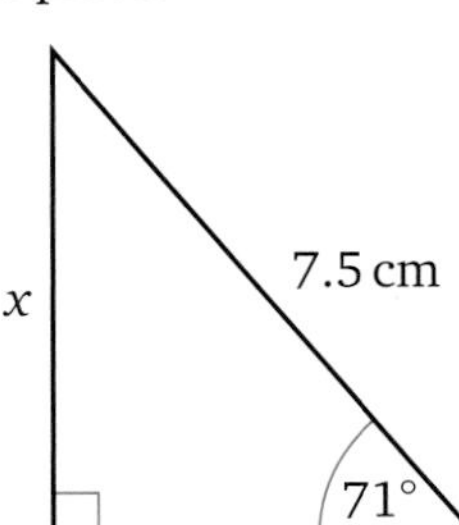

b

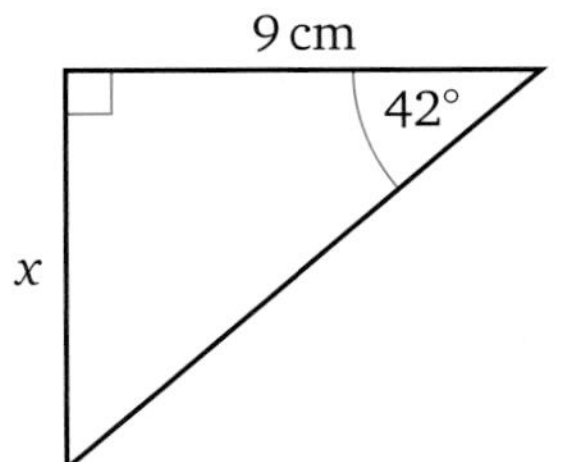

d

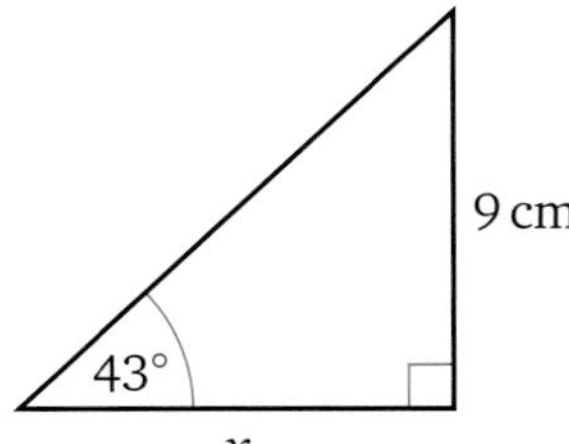

f

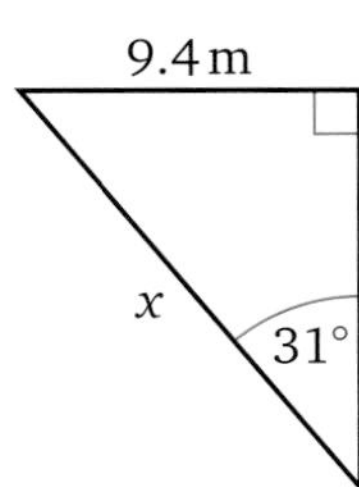

4 *ABCDE* is a pentagon made up of three similar right-angled triangles.

AB = 10 cm

Calculate *AE*.

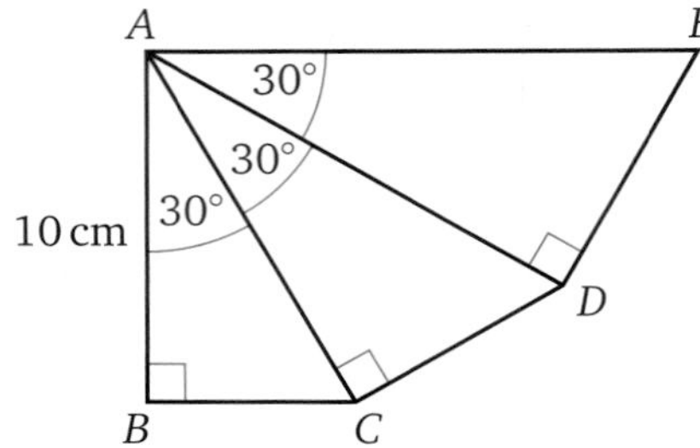

5 A 5 m ladder leans against a vertical wall, making an angle of 73° with the horizontal ground.

How far up the wall does it reach?

6 A shed roof has sloping sides at an angle of 30°. Each edge is 3 m long.

Calculate, x, the width of the shed.

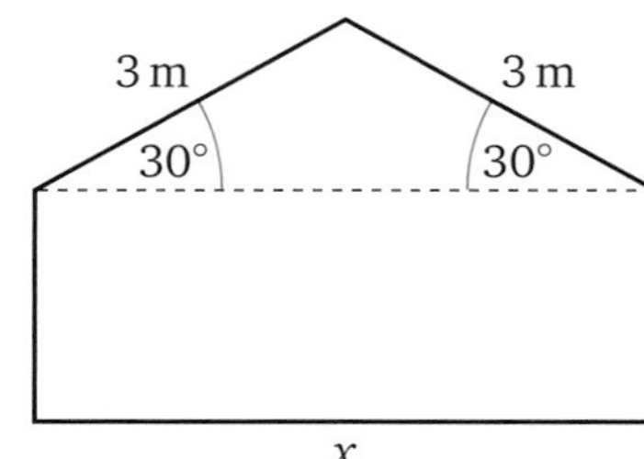

7 An aerial runway is to be built in an adventure playground.
It needs to have a slope of 5° to the horizontal.
The lower end must be 2 m high.
The runway is 30 m long.

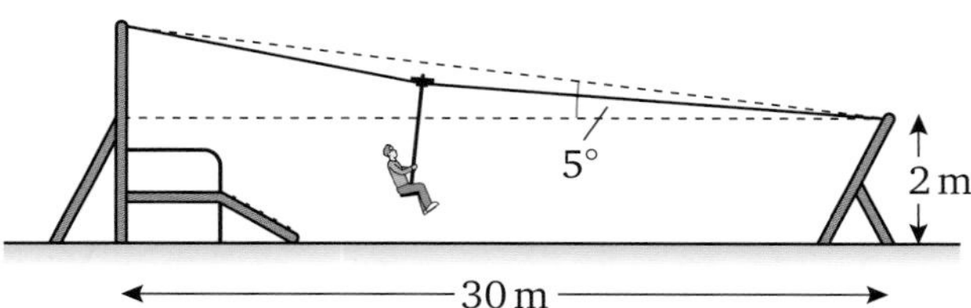

How high should the starting post be?

8 Dave is making some trestles.
Here is his design.

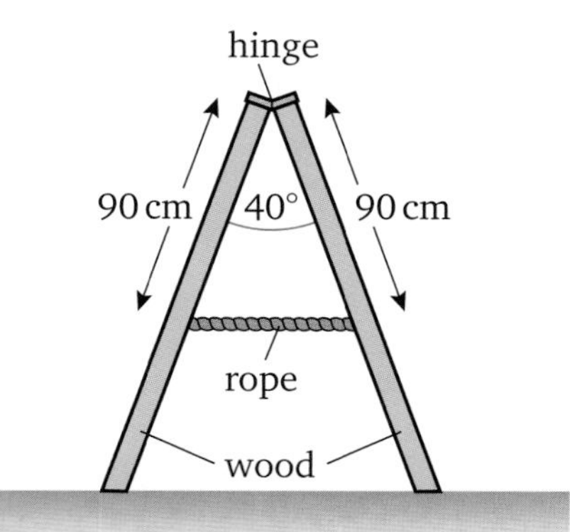

He wants the trestles to open to an angle of 40°.
The horizontal rope support needs to be fixed 90 cm away from the hinge.

How long should the rope be?

9 Holly is working out with a punchbag.
The distance from the end of the punchbag to its support is 130 cm.
It swings so it is 20° from the vertical.

a How far has the bottom moved in a horizontal direction (shown in red)?

b How far would it move horizontally if it swings at an angle of 40°?

c How far would it move horizontally if it swings at an angle of 100°?

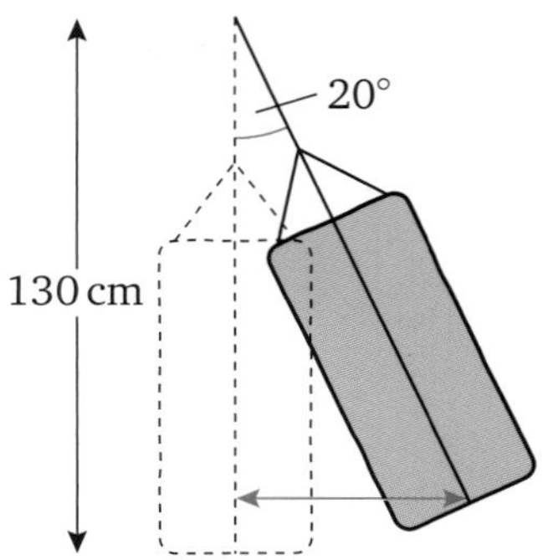

Learn... 18.2 Calculating angles using trigonometry

Your calculator can change the sine, cosine or tangent back to an angle.
The reverse of sin is written as $\sin^{-1}$, or inverse sin or arc sin.

On many calculators, you access it by pressing SHIFT, INV or 2ND FUNCTION before the sin key.

Using your calculator, $\sin 60° = 0.8660254038...$, and $\sin^{-1} 0.8660254038... = 60°$

Remember you can type in sin 60° to get the answer of 0.8660254038… and then SHIFT sin ANS to get back to 60°.

In the same way, SHIFT cos and SHIFT tan convert cosines and tangents into angles.

In the diagram, $\sin = \frac{\text{opp}}{\text{hyp}}$

$\sin x = \frac{6}{8} = 0.75$

So $x = \sin^{-1} 0.75$

SHIFT sin 0.75 = 48.59037789…°

Or $x = 48.6°$ (1 d.p.)

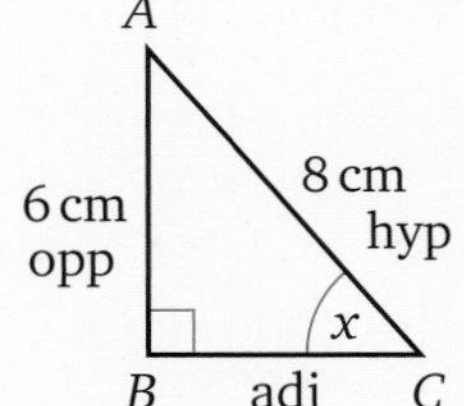

Not drawn accurately

AQA Examiner's tip

Make sure your calculator is in degree mode.

Example: *ABCD* is a rectangle.

Calculate the size of angle *BAC* in this diagram.

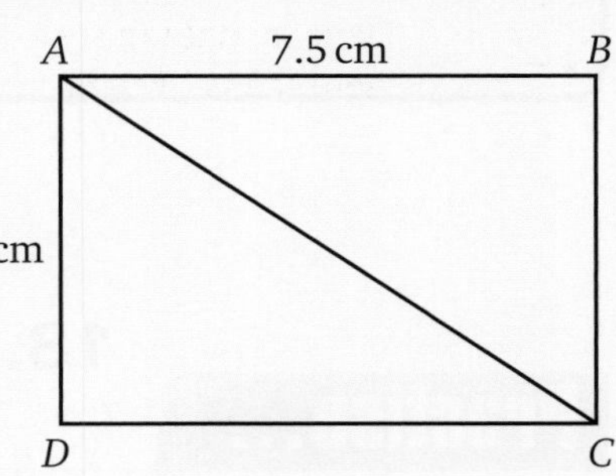

Not drawn accurately

Solution: First, copy and label the diagram.

Use x for angle *BAC*, and mark on 'opp', 'hyp' and 'adj'.

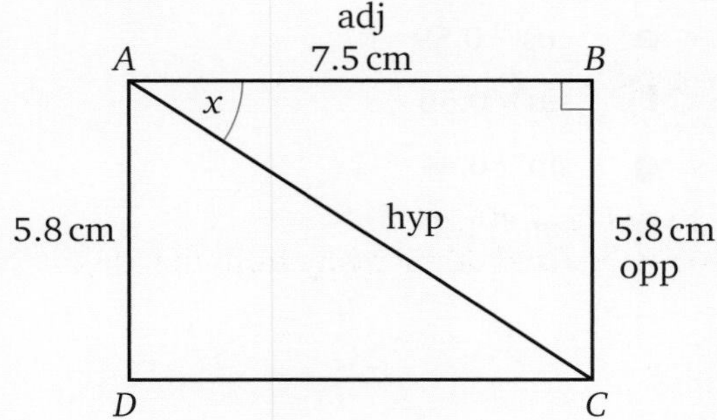

$\sin = \frac{\text{opp}}{\text{hyp}}$, $\cos = \frac{\text{adj}}{\text{hyp}}$, $\tan = \frac{\text{opp}}{\text{adj}}$

$\tan x = \frac{\text{opp}}{\text{adj}}$ — The two known sides are opposite and adjacent to angle x, so choose tan.

$\tan x = \frac{5.8}{7.5}$ — Substitute.

$x = \tan^{-1}\left(\frac{5.8}{7.5}\right)$

$= 37.71597635...°$ — Calculate $\tan^{-1}$.

Angle BAC = 37.7° (1 d.p.)

Example: Bayston is 8 km due west of Greenmore. Greenmore is 7 km due north of Sliddow.

Calculate the bearing of Bayston from Sliddow.

Solution: First, draw a sketch. Bearings are measured clockwise from north.

Identify the required bearing (marked in red).

Identify angle x.

Label the sides as hypotenuse, opposite and adjacent.

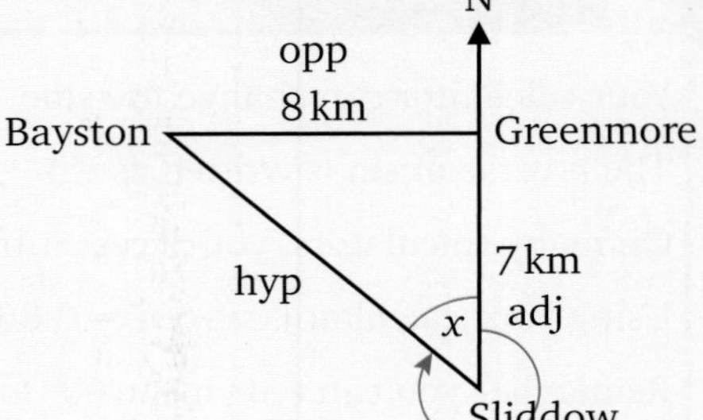

$\sin = \frac{\text{opp}}{\text{hyp}}$, $\cos = \frac{\text{adj}}{\text{hyp}}$, $\tan = \frac{\text{opp}}{\text{adj}}$ — You know the opposite and adjacent, so choose tangent.

To calculate angle x,

$\tan x = \frac{\text{opp}}{\text{adj}}$

$\tan x = \frac{8}{7}$

$x = \tan^{-1}\left(\frac{8}{7}\right)$

$x = 48.81407483...°$

$x = 48.8°$ (1 d.p.)

So the bearing is 360° − 48.8° = 311.2°

Practise... 18.2 Calculating angles using trigonometry

B

1 Write the angles given below to one decimal place.

a $\sin^{-1} 0.65$

b $\cos^{-1} 0.7$

c $\tan^{-1} 0.86$

d $\cos^{-1} 0.234$

e $\cos^{-1} 0.59$

f $\sin^{-1} 0.56$

g $\sin^{-1} 0.44$

h $\tan^{-1} 1.2$

B

2 Calculate the angles marked with letters in the diagrams below.

a

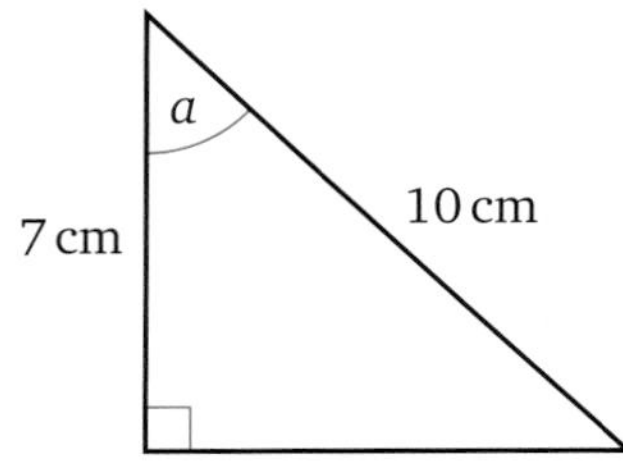

c

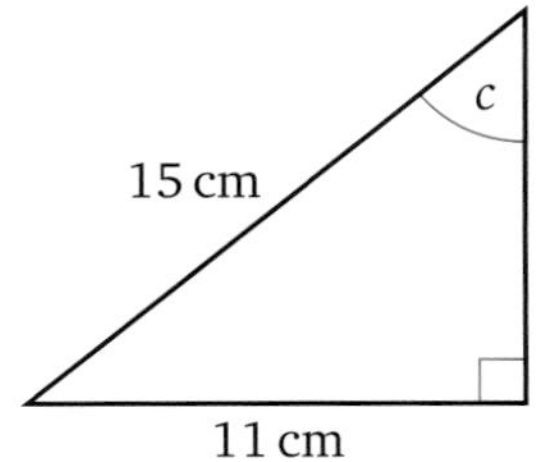

e

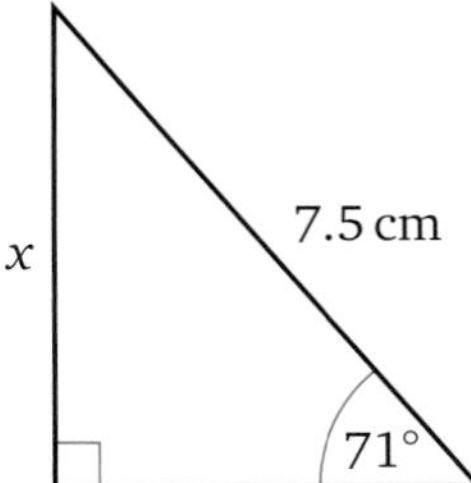

b

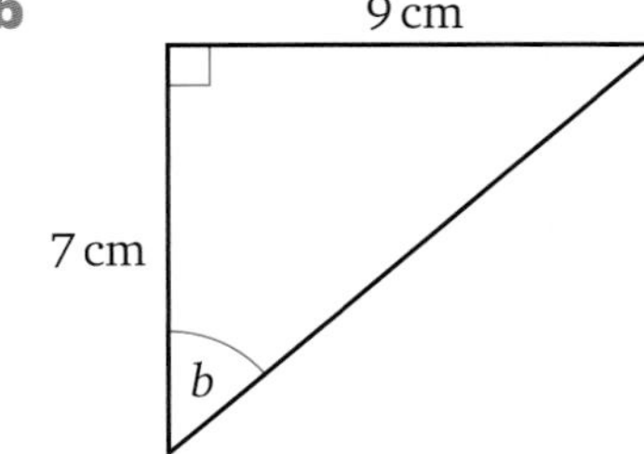

d

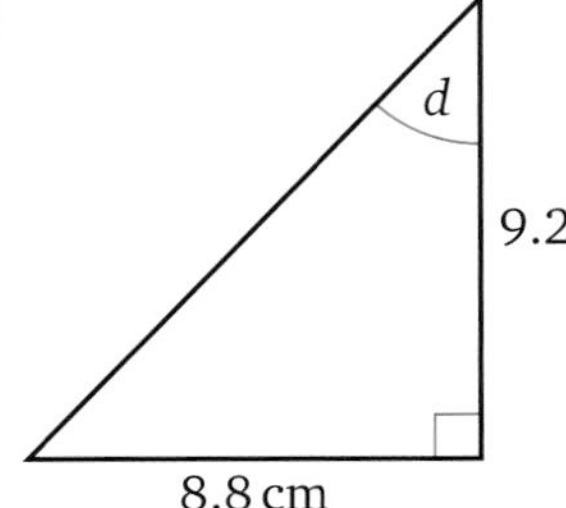

f

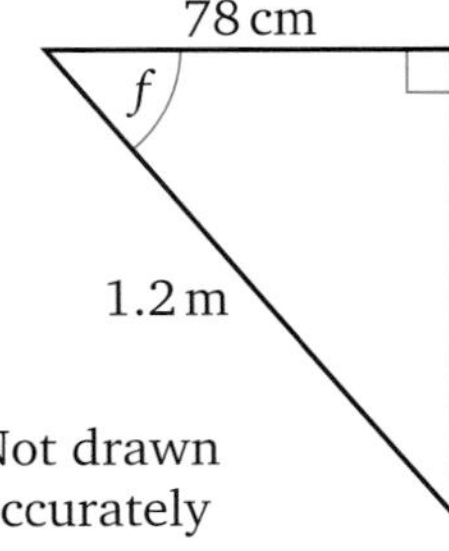

Not drawn accurately

AQA Examiner's tip

When you calculate the size of a missing angle, always check that your answer makes sense. The smallest angle is always opposite the smallest side; checking for this can indicate you have made an error.

A

3 In the triangles below, calculate all the missing sides and angles.

a

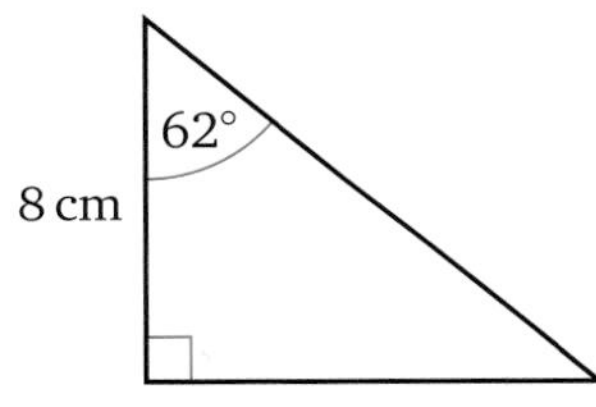

c

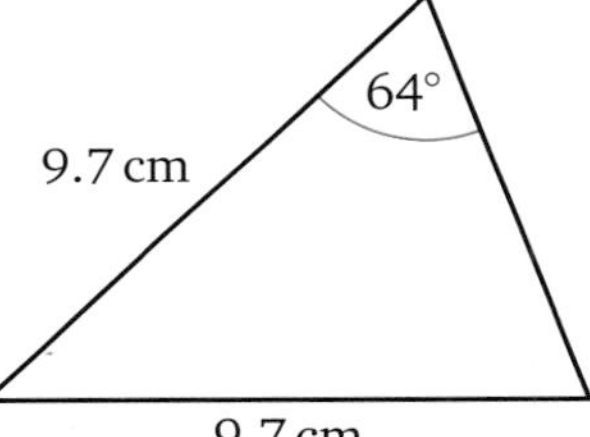

b

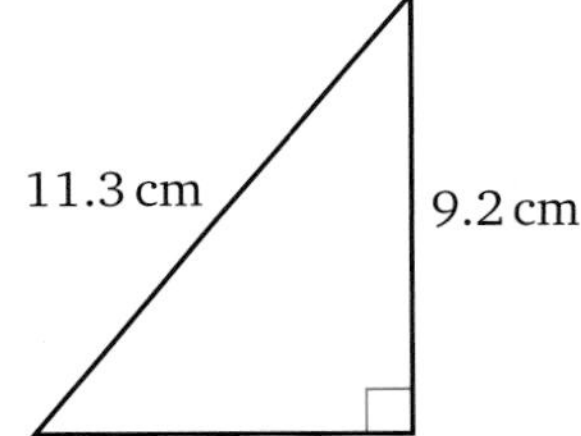

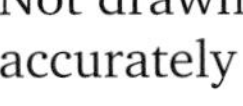

Not drawn accurately

4 *ABCD* is a rectangle inscribed in a circle with centre *O* and radius 8 cm.

AD = 11 cm

Calculate angle *ABD*.

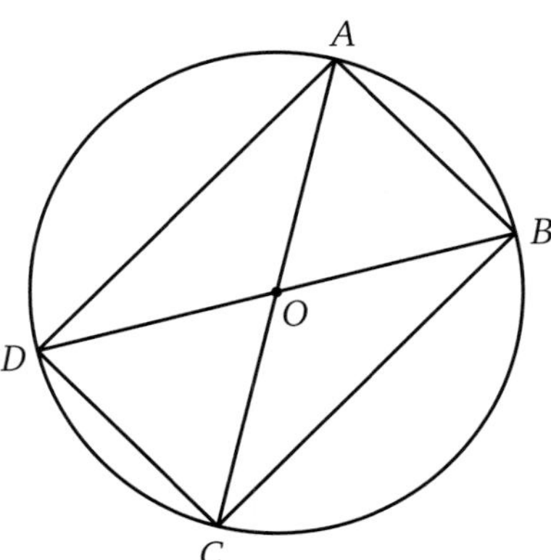

5 A ship sails 12 km due east, then 8 km due south. It then sails straight back to the starting point.

a On what bearing does it sail back?

b How far is the total journey?

6 a A train is climbing a hill with a gradient of 1 in 14.

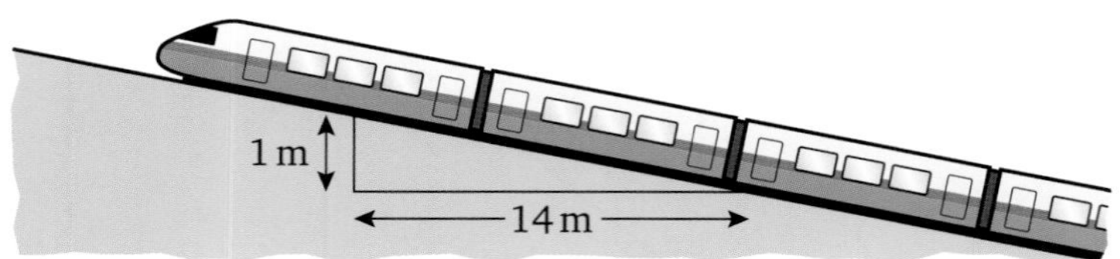

> **Hint**
> The **angle of elevation** is the angle with the horizontal.

Calculate the angle of elevation.

b When the gradient is measured, it is easier to measure the distance along the surface rather than the horizontal distance.
The real measurements taken are as shown.

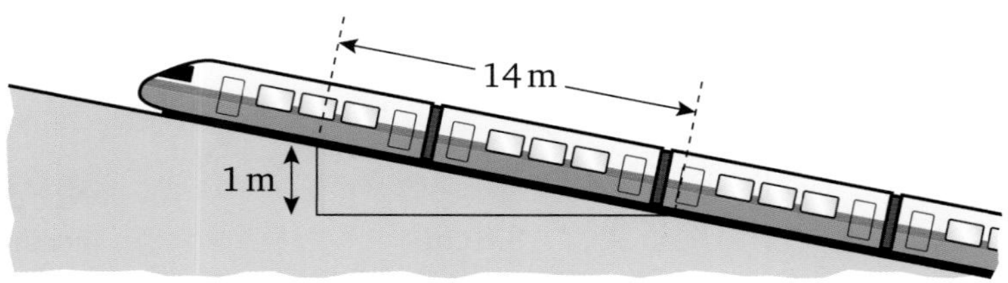

Calculate the angle of elevation now.

7 What angle has a cosine equal to sin 30°?

What angle has a cosine equal to sin 40°?

Write a general rule for this relationship.

Learn... 18.3 Trigonometry in three dimensions

Look at the cuboid *ABCDEFGH*.

It contains many right angles, for example angles *ADE*, *ADC* and *CDE*.

It is not always easy to tell whether angles are right angles.

For example, look at angles *ADH* and *FDH*.

Imagine turning the cuboid so that the face *DCHE* is at the bottom.

DH is then horizontal. *AD* is vertical. So *ADH* must be a right angle.

It is not possible to arrange the cuboid so that *DH* is horizontal and *DF* is vertical, so *FDH* cannot be a right angle.

Identifying right-angled triangles enables you to apply trigonometry and Pythagoras' theorem to three-dimensional problems.

The angle between a line and a plane

Look at the diagonal *AH*.

When questions ask about the angle between *AH* and the base *EFGH* it means the angle *AHF*.

Then the shortest distance from *A* to the base is *AF*.

AHF is a right-angled triangle, so trigonometry can be used.

Sometimes, it is necessary to use Pythagoras' theorem first.
For example, if you knew the lengths of *EH* and *EF*, you could use Pythagoras' theorem to calculate *FH*.

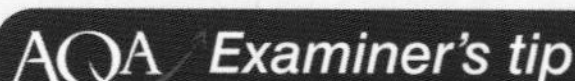

> **AQA Examiner's tip**
> Always draw out the right-angled triangle you are going to use, in this case triangle *AFH*.

Example: A pyramid has a square base with sides of 8 cm and a perpendicular height of 6 cm.

Calculate the angle between DE and the base.

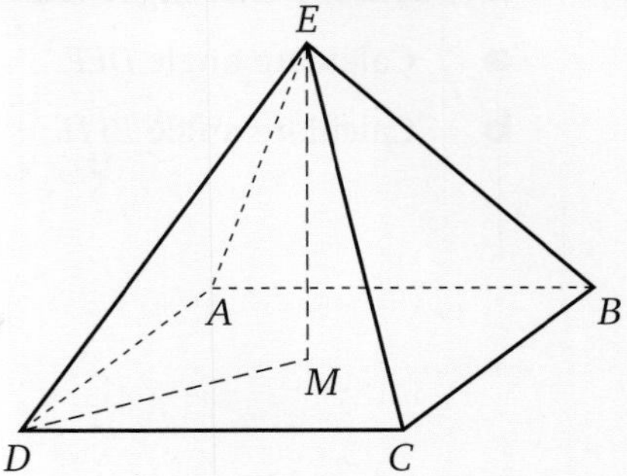

Solution: The view of the base from above looks like this.

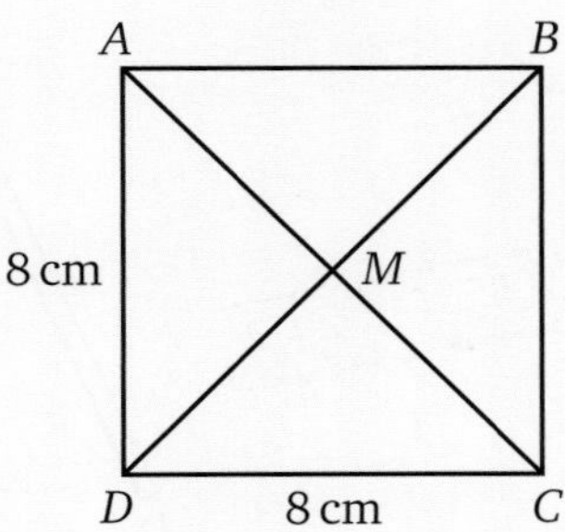

It is very important to draw two-dimensional diagrams to show the right-angled triangles correctly.

Using Pythagoras' theorem,

$$DC^2 + BC^2 = DB^2$$
$$8^2 + 8^2 = DB^2$$
$$64 + 64 = DB^2$$
$$DB^2 = 128$$
$$DB = \sqrt{128} = 8\sqrt{2}\text{ cm } (= 11.3137\ldots)$$ Using surds avoids any errors due to rounding.
$$DM = \tfrac{1}{2}DB = 4\sqrt{2}\text{ cm } (= 5.6568\ldots)$$

Now consider triangle EDM.

$$\tan x = \frac{\text{opp}}{\text{adj}}$$
$$\tan x = \frac{6}{4\sqrt{2}}$$
$$x = \tan^{-1}\left(\frac{6}{4\sqrt{2}}\right)$$
$$= 46.6861\ldots^\circ$$

Angle $EDM = 46.7°$ (1 d.p.)

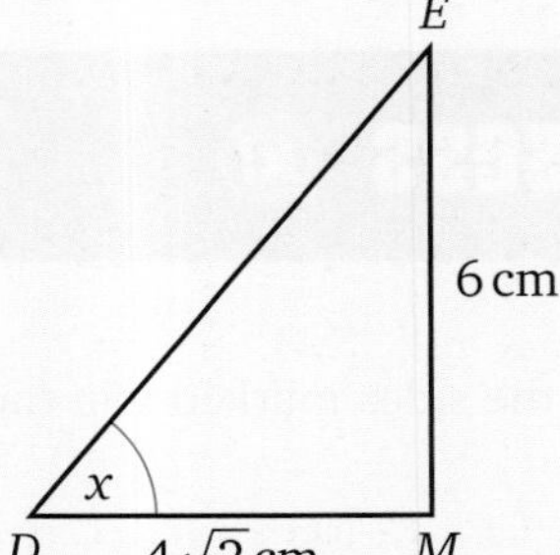

Practise... 18.3 Trigonometry in three dimensions

B

1 A 4 m flagpole leans at an angle of 84° with the ground.

Calculate the distance between the top of the flagpole and the ground.

2 A cone has a radius of 7 cm and a perpendicular height of 12 cm.

Calculate the angle ABC.

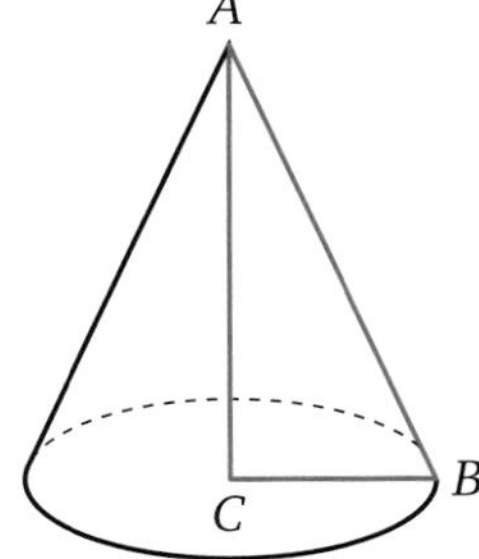

A*

3 A cuboid is 8 cm long, 6 cm wide and 5 cm high.

a Calculate angle DEF.

b Calculate angle BFH.

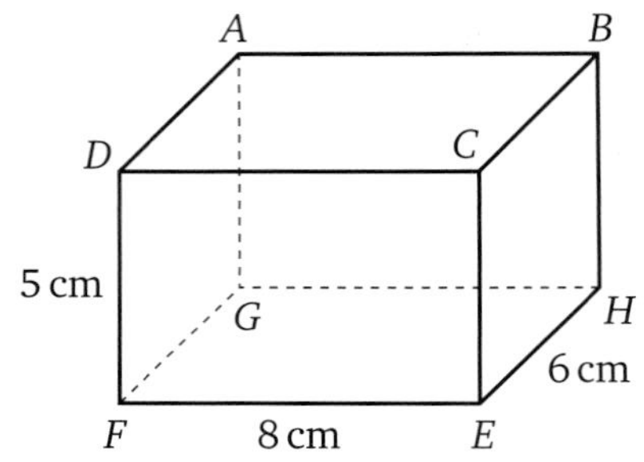

Not drawn accurately

4 Four 3-metre bamboo canes are used to make a support for growing some runner beans.
One end of each cane is stuck in the ground at the corners of a square of side 1 m. The top ends are tied together.

a Calculate the angle BAC.

b Calculate the angle BAD.

Hint

Drawing in the line of symmetry in an isosceles triangle makes two congruent right-angled triangles.

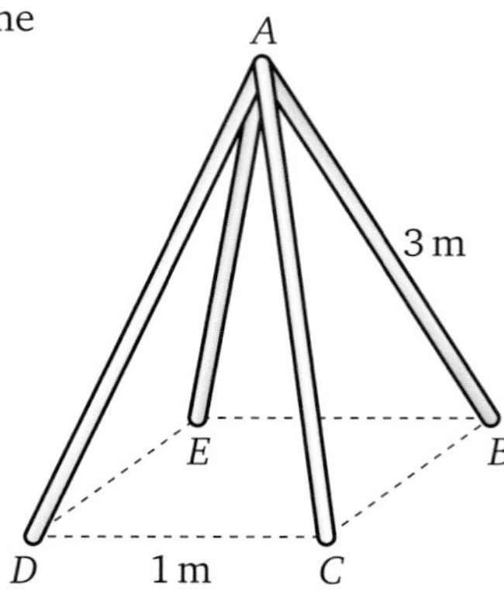

5 A shed has a base that measures 6 m by 2 m.
The roof slopes so that the front edge is 2 m tall, and the slope is 30°.

Calculate the height of the back of the shed.

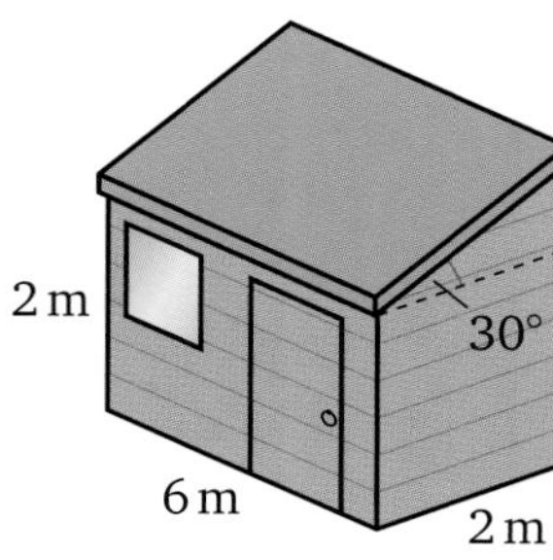

18 Assess

B

1 Calculate the sides marked x in the diagrams below.

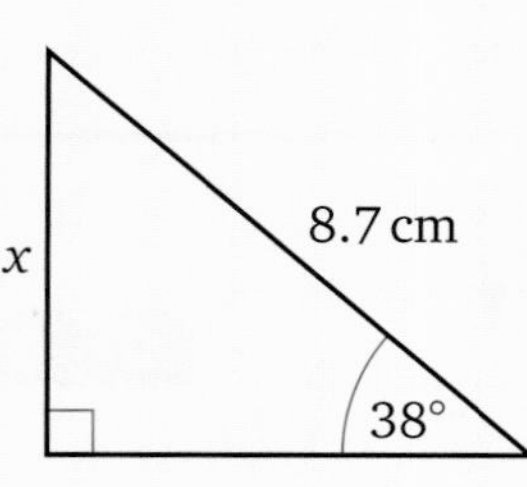

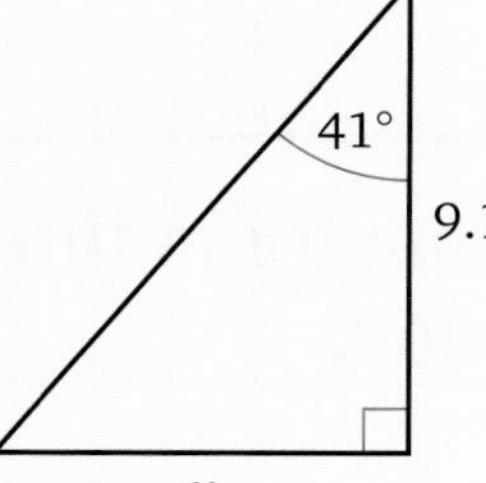

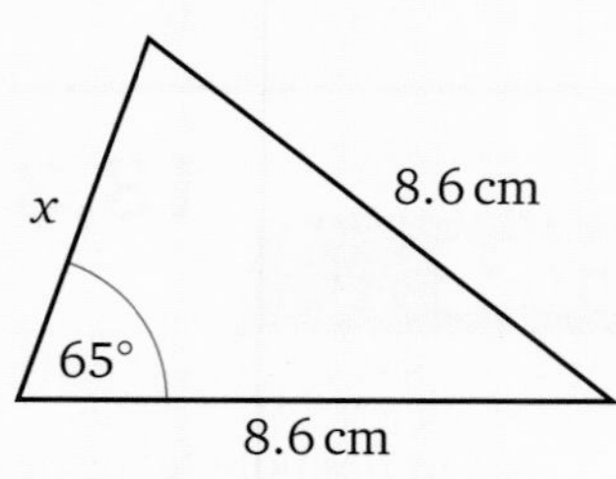

Not drawn accurately

2 Calculate the angles marked x below.

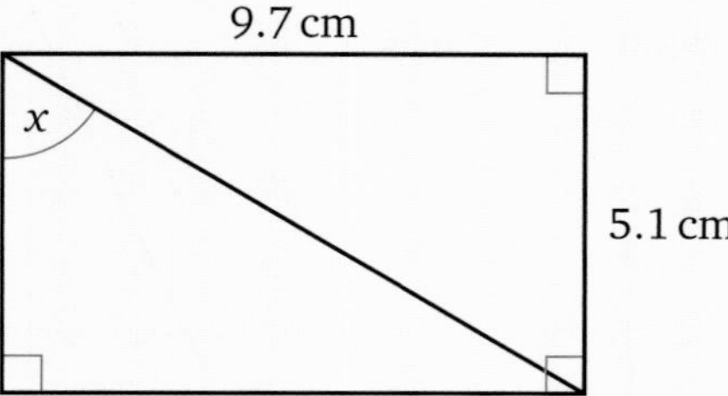

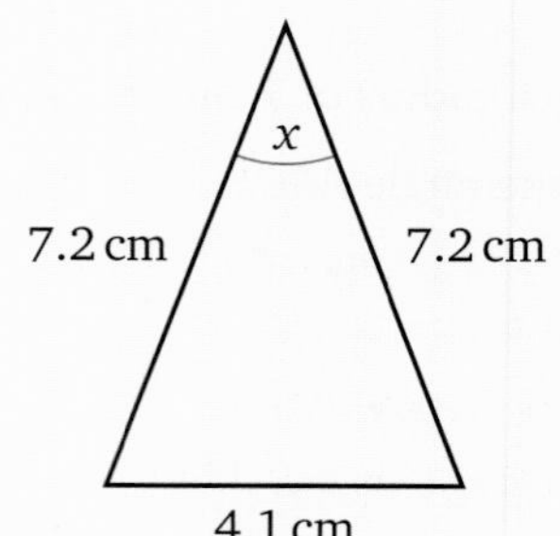

Not drawn accurately

A

3 Ware is 8.3 km from Wye on a bearing of 072°.
Hoo is due south of Ware, and on a bearing of 162° from Wye.

Calculate the distance from Wye to Hoo.

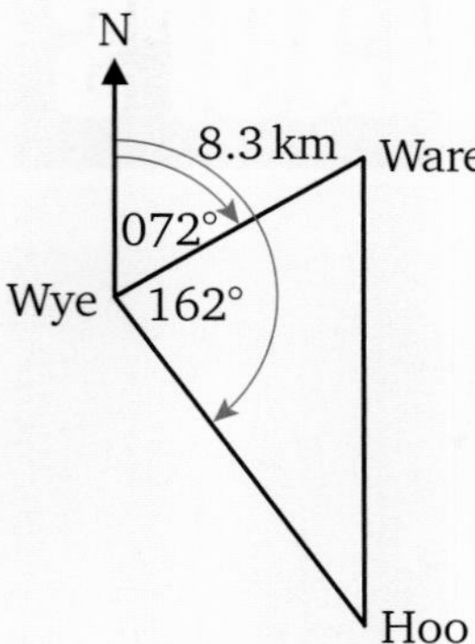

Not drawn accurately

4 A pyramid has a square base *BCDE* with sides of 8.2 cm.
X is the centre of the base.
The vertical height, *AX*, is 11.6 cm.

Calculate the angle *ACE*.

Calculate the angle *BAD*.

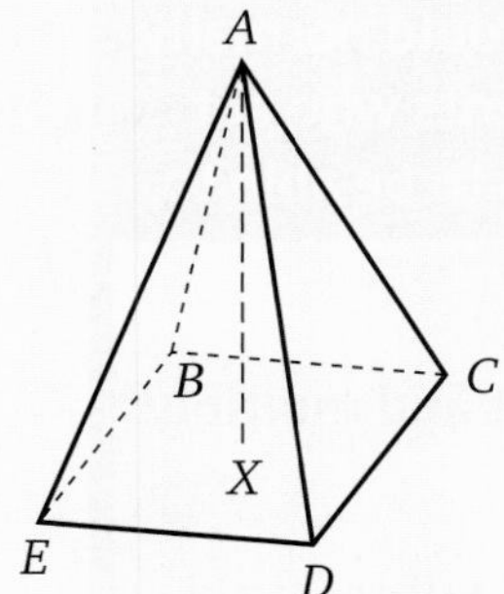

AQA Examination-style questions

1 The diagram shows two right-angled triangles.
$AD = 15$ cm.
$CD = 6$ cm.

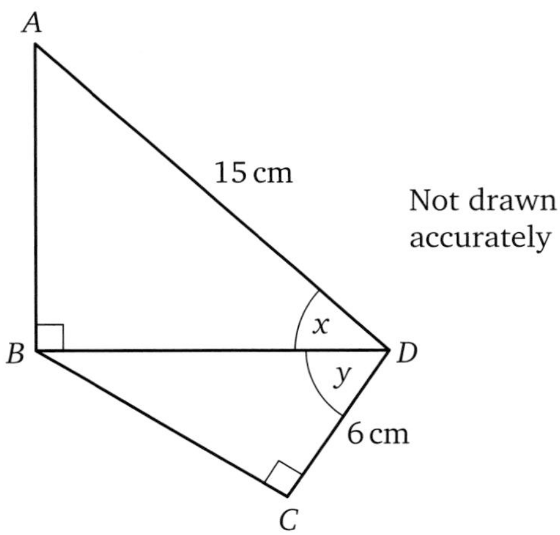

Not drawn accurately

Given that $\cos x° = \frac{2}{3}$, work out the value of $\sin y°$. *(5 marks)*

AQA 2003

2 A prism *ABCDEF* with a right-angled triangular cross section has dimensions as shown.

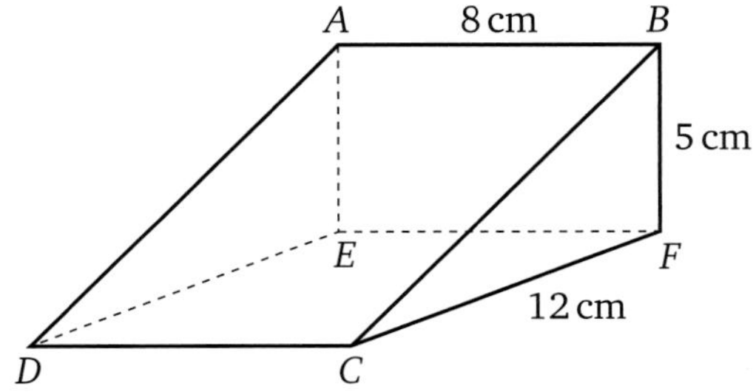

Not drawn accurately

a Calculate the length *BD*. *(3 marks)*

b Hence, or otherwise, calculate the angle *BDF*. *(2 marks)*

AQA 2008

19 Vectors

Objectives

Examiners would normally expect students who get these grades to be able to:

A

add, subtract and multiply vectors

use addition, subtraction and multiplication of vectors to solve simple geometric problems

understand the relationship between parallel vectors

A*

solve more difficult geometric problems using vectors.

Did you know?

Wind

Wind can be represented on maps using vectors. This is because wind has direction and speed.

The highest ever, officially recorded wind speed is 372 kph, or 231 mph. It was recorded on 12 April 1934, at New Hampshire's Mount Washington Observatory, USA.

Key terms

vector
magnitude
column vector
scalar
vector sum
resultant vector

You should already know:

- ✔ how to write a column vector
- ✔ how to translate a shape using column vectors
- ✔ the meaning of 'parallel'
- ✔ the relationship between similar triangles
- ✔ the properties of polygons.

Learn... 19.1 Vectors

A **vector** is a quantity that has **magnitude** (size) and direction.
A vector can be represented in a number of ways.

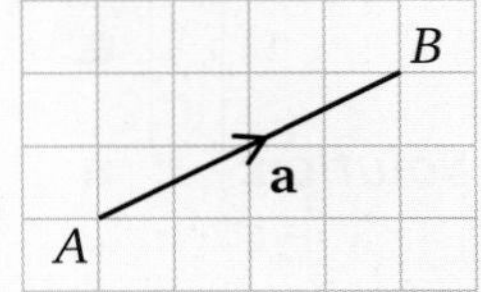

- as a **column vector** $\begin{pmatrix}4\\2\end{pmatrix}$ ← horizontal move, ← vertical move

This is a move of 4 units to the right and 2 units up.

- in type, in a book, as $\overrightarrow{AB}$ or **a**
- handwritten as $\overrightarrow{AB}$ or as $\underset{\sim}{a}$

Link

Look back to Learn 9.3 in Chapter 9 to remind yourself about column vectors.

The arrows on the vectors in the diagram shows the direction.

These vectors are the same size but in opposite directions so are two different vectors.

Compare the way the two vectors are represented.

One vector is the negative of the other.

$\begin{pmatrix}4\\2\end{pmatrix}$ $\begin{pmatrix}-4\\-2\end{pmatrix}$

$\overrightarrow{AB}$ $\overrightarrow{BA}$

a −**a** If you are handwriting this, it would be $\underset{\sim}{a}$ and $-\underset{\sim}{a}$

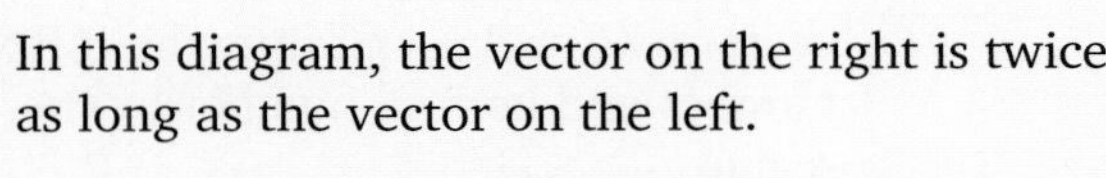

In this diagram, the vector on the right is twice as long as the vector on the left.

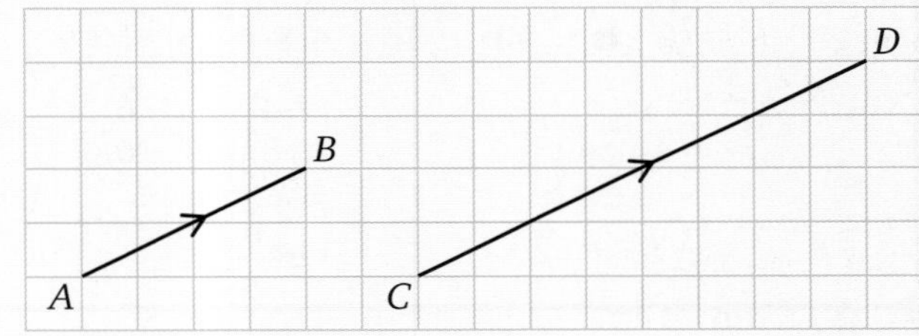

Compare the way the two vectors are represented.

The second vector $\overrightarrow{CD}$ can be represented as 2 × the first vector.

$\begin{pmatrix}4\\2\end{pmatrix}$ $\begin{pmatrix}8\\4\end{pmatrix}$

$\overrightarrow{AB}$ $2\overrightarrow{AB}$

a **2a**

When vectors are added, subtracted or multiplied the solution can be found using diagrams or column vectors.

Parallel vectors

Two vectors are parallel if their column vectors are equal or a multiple of the same column vector.

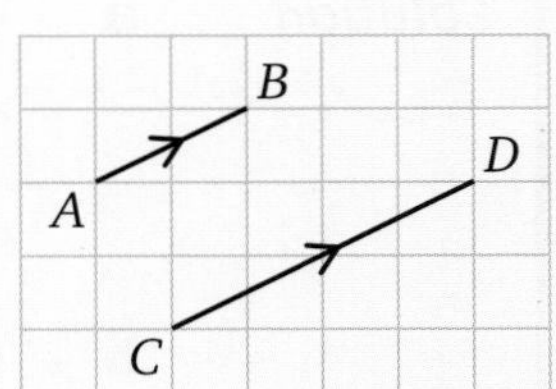

$\overrightarrow{AB} = \begin{pmatrix}2\\1\end{pmatrix}$ $\overrightarrow{CD} = \begin{pmatrix}4\\2\end{pmatrix} = 2 \times \begin{pmatrix}2\\1\end{pmatrix} = 2\overrightarrow{AB}$

$\overrightarrow{AB}$ and $\overrightarrow{CD}$ must be parallel because one is a multiple of the other.

Also, you can see from the diagram that $\overrightarrow{AB}$ is parallel to $\overrightarrow{CD}$.

Now think about the vector from D to C.

$\overrightarrow{AB} = \begin{pmatrix}2\\1\end{pmatrix}$ $\overrightarrow{DC} = \begin{pmatrix}-4\\-2\end{pmatrix} = -2 \times \begin{pmatrix}2\\1\end{pmatrix} = -2\overrightarrow{AB}$

In general, if $\overrightarrow{AB} = \begin{pmatrix}x\\y\end{pmatrix}$ and $\overrightarrow{CD} = m\begin{pmatrix}x\\y\end{pmatrix}$ then $\overrightarrow{AB}$ and $\overrightarrow{CD}$ are parallel.

m can be positive or negative.

Example: $\overrightarrow{AB} = \begin{pmatrix} 2 \\ -3 \end{pmatrix}$ $\overrightarrow{BC} = \begin{pmatrix} -4 \\ -2 \end{pmatrix}$

a Draw a diagram showing $\overrightarrow{AB}$ and $\overrightarrow{BC}$. **b** Write $\overrightarrow{AC}$ as a column vector.

Solution: **a**

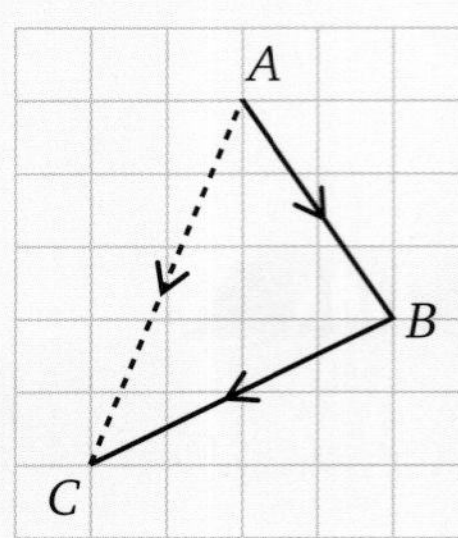

$\overrightarrow{AB} + \overrightarrow{BC}$ means $\overrightarrow{AB}$ followed by $\overrightarrow{BC}$

b $\overrightarrow{AB} + \overrightarrow{BC} = \overrightarrow{AC}$

$\begin{pmatrix} 2 \\ -3 \end{pmatrix} + \begin{pmatrix} -4 \\ -2 \end{pmatrix} = \begin{pmatrix} -2 \\ -5 \end{pmatrix}$

$\overrightarrow{AC} = \begin{pmatrix} -2 \\ -5 \end{pmatrix}$

Make sure that the vector is in the correct direction. You may be tempted to continue drawing clockwise and show the vector $\overrightarrow{CA}$ and not $\overrightarrow{AC}$.

Example: $\mathbf{a} = \begin{pmatrix} 2 \\ 3 \end{pmatrix}$ $\mathbf{b} = \begin{pmatrix} -1 \\ 2 \end{pmatrix}$ $\mathbf{c} = \begin{pmatrix} -2 \\ -2 \end{pmatrix}$

Write the following as single column vectors.

a $\mathbf{a} + \mathbf{b} + \mathbf{c}$

b $4\mathbf{b} - 3\mathbf{c}$

Solution: **a** $\mathbf{a} + \mathbf{b} + \mathbf{c} = \begin{pmatrix} 2 \\ 3 \end{pmatrix} + \begin{pmatrix} -1 \\ 2 \end{pmatrix} + \begin{pmatrix} -2 \\ -2 \end{pmatrix} = \begin{pmatrix} -1 \\ 3 \end{pmatrix}$

b $4\mathbf{b} - 3\mathbf{c} = 4 \times \begin{pmatrix} -1 \\ 2 \end{pmatrix} - 3 \times \begin{pmatrix} -2 \\ -2 \end{pmatrix}$

$= \begin{pmatrix} -4 \\ 8 \end{pmatrix} - \begin{pmatrix} -6 \\ -6 \end{pmatrix}$

$= \begin{pmatrix} 2 \\ 14 \end{pmatrix}$

You can multiply a vector by a number. This number is known as a **scalar**. To multiply a column vector by a scalar, multiply the top number and bottom number by the scalar.

Example: Use the information on the diagram to say whether each statement is true or false. If false, give the correct answer.

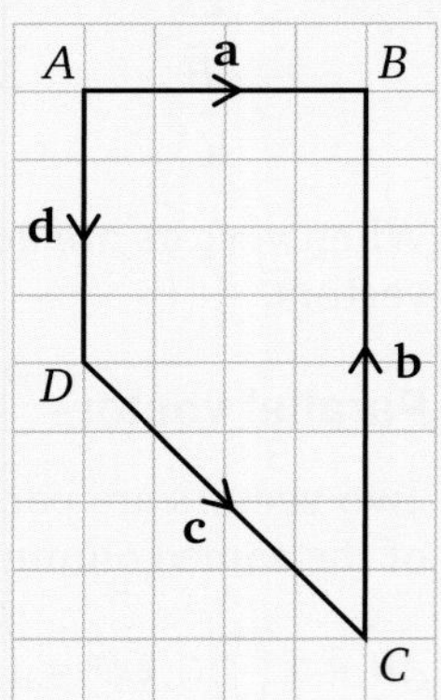

a $\mathbf{a} + \mathbf{b} = \mathbf{c} + \mathbf{d}$

b $\mathbf{b}$ is parallel to $\mathbf{d}$

c $\mathbf{b} = \mathbf{a} - \mathbf{c} + \mathbf{d}$

Solution: **a** False

$\mathbf{a} + \mathbf{b} = \begin{pmatrix} 4 \\ 0 \end{pmatrix} + \begin{pmatrix} 0 \\ 8 \end{pmatrix} = \begin{pmatrix} 4 \\ 8 \end{pmatrix}$

$\mathbf{c} + \mathbf{d} = \begin{pmatrix} 4 \\ -4 \end{pmatrix} + \begin{pmatrix} 0 \\ -4 \end{pmatrix} = \begin{pmatrix} 4 \\ -8 \end{pmatrix}$

b True

This can be seen on the diagram and also by comparing the column vectors for **b** and **d**.

$\mathbf{b} = \begin{pmatrix} 0 \\ 8 \end{pmatrix}$ and $\mathbf{d} = \begin{pmatrix} 0 \\ -4 \end{pmatrix}$

The scalar value is -2 because $\mathbf{b} = -2\mathbf{d}$

This satisfies the general rule for parallel vectors because one vector is a multiple of the other.

c False

$\mathbf{a} - \mathbf{c} + \mathbf{d} = \begin{pmatrix} 4 \\ 0 \end{pmatrix} - \begin{pmatrix} 4 \\ -4 \end{pmatrix} + \begin{pmatrix} 0 \\ -4 \end{pmatrix} = \begin{pmatrix} 0 \\ 0 \end{pmatrix}$

$\mathbf{b} = \begin{pmatrix} 0 \\ 8 \end{pmatrix}$ so does not equal $\mathbf{a} - \mathbf{c} + \mathbf{d}$.

Practise... 19.1 Vectors

A

1 In each question part, which four of the five vectors, if drawn together, would create a rectangle?

a $\begin{pmatrix}2\\0\end{pmatrix}$ $\begin{pmatrix}0\\-2\end{pmatrix}$ $\begin{pmatrix}0\\-5\end{pmatrix}$ $\begin{pmatrix}0\\5\end{pmatrix}$ $\begin{pmatrix}-2\\0\end{pmatrix}$

b $\begin{pmatrix}3\\1\end{pmatrix}$ $\begin{pmatrix}2\\-6\end{pmatrix}$ $\begin{pmatrix}-2\\6\end{pmatrix}$ $\begin{pmatrix}-3\\1\end{pmatrix}$ $\begin{pmatrix}-3\\-1\end{pmatrix}$

> **AQA Examiner's tip**
>
> Draw the vectors to help you answer this question. In an examination you can ask for squared paper to help you draw them.

2 Which of these vectors are equal? For those that are, give the column vector that represents them.

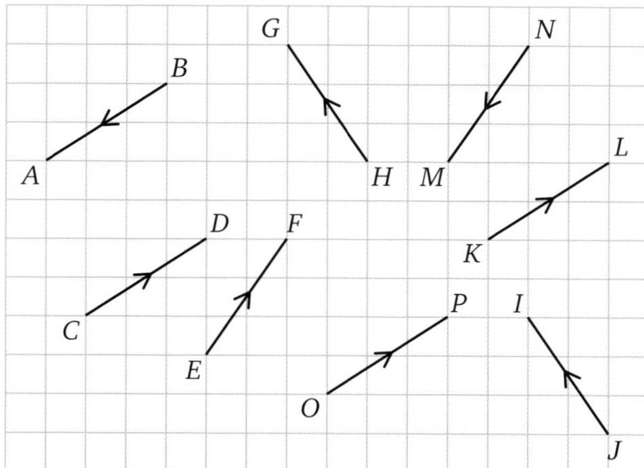

3 $\overrightarrow{AB} = \begin{pmatrix}2\\-1\end{pmatrix}$

a Which of these column vectors are parallel to $\overrightarrow{AB}$?

i $\begin{pmatrix}4\\-2\end{pmatrix}$ **ii** $\begin{pmatrix}3\\0\end{pmatrix}$ **iii** $\begin{pmatrix}-4\\2\end{pmatrix}$ **iv** $\begin{pmatrix}-10\\5\end{pmatrix}$ **v** $\begin{pmatrix}5\\4\end{pmatrix}$

b For those vectors that are parallel, write each as a multiple of $\overrightarrow{AB}$.

4 $\overrightarrow{AB} = \begin{pmatrix}3\\2\end{pmatrix}$ $\overrightarrow{BC} = \begin{pmatrix}1\\-4\end{pmatrix}$

a Draw a diagram showing $\overrightarrow{AB}$ and $\overrightarrow{BC}$.

b Write $\overrightarrow{AC}$ as a column vector.

5 Which of these statements are true and which are false? Give a reason for your answer in each case.

a $\begin{pmatrix}2x\\-4y\end{pmatrix}$ is parallel to and in the same direction as $\begin{pmatrix}5x\\-10y\end{pmatrix}$

b $\begin{pmatrix}0.2x\\1.2y\end{pmatrix}$ is parallel to and in the opposite direction to $\begin{pmatrix}-x\\-6y\end{pmatrix}$

c $\begin{pmatrix}-5x\\3y\end{pmatrix}$ is parallel to and in the same direction as $\begin{pmatrix}-15y\\25x\end{pmatrix}$

6 $\mathbf{a} = \begin{pmatrix}-1\\2\end{pmatrix}$ $\mathbf{b} = \begin{pmatrix}3\\-2\end{pmatrix}$ $\mathbf{c} = \begin{pmatrix}2\\-1\end{pmatrix}$

a Draw a vector diagram for each question part.

i $2\mathbf{b} - \mathbf{a}$ **ii** $\mathbf{a} + \mathbf{b} + \mathbf{c}$ **iii** $-\mathbf{a} - \mathbf{b} + \mathbf{a}$

b Write each of the solutions to part **a** as a single column vector.

A

7 $\mathbf{a} = \begin{pmatrix} 4 \\ 0 \end{pmatrix}$

$\mathbf{b} = \begin{pmatrix} 1 \\ -4 \end{pmatrix}$

$\mathbf{c} = \begin{pmatrix} -1 \\ -4 \end{pmatrix}$

Hint

Draw **a** then **b** then **c** then −**a** and so on, with one vector following on from the one before.

a Draw the **vector sum** $\mathbf{a} + \mathbf{b} + \mathbf{c} - \mathbf{a} - \mathbf{b} - \mathbf{c}$.

b Write down the name of the polygon in your diagram.

8 $\overrightarrow{AB} = \begin{pmatrix} -2 \\ 4 \end{pmatrix}$ $\overrightarrow{BC} = \begin{pmatrix} -3 \\ 3 \end{pmatrix}$

a Draw a diagram of $\overrightarrow{AB} + 2\overrightarrow{BC}$.

b Find the single vector equal to $\overrightarrow{AB} + 2\overrightarrow{BC}$.

9 The diagram shows a section of coastline with the sea and headland marked. There is deep water up to the land making it easy for boats to sail close in.

A small sailing boat is sailing from A to B and tacking using vectors. The sailor, Roy, plans to sail into the bay, keeping close in to the shore. However, he can only sail at 45° to the wind.

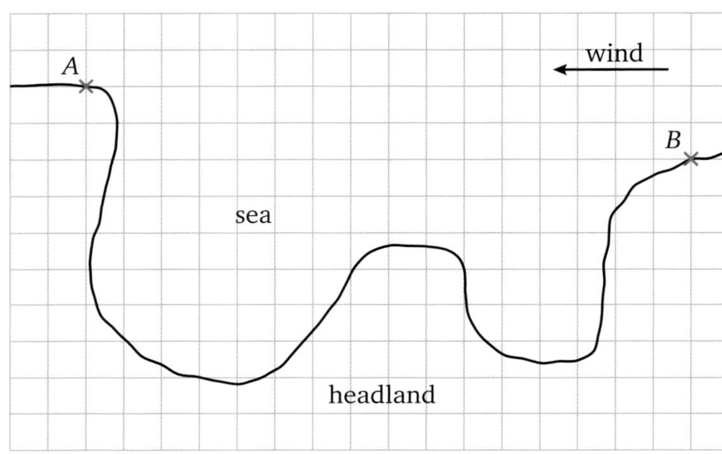

Hint

Sailing boats cannot sail directly towards the wind. They move in a series of zigzag movements, at 45° to the wind direction. This is called tacking.

a Suggest a possible journey Roy could make along the coastline from A to B using column vectors.

b Roy's friend Gordon is travelling directly from A to B in a speedboat. Describe Gordon's journey using one column vector.

c Write Roy's journey as a vector sum.

d Gordon says that Roy's vector sum is equal to his single column vector. Is Gordon right?

10 **i** On squared paper, draw each of these polygons.
ii Write vector instructions for each diagram.

a an irregular pentagon

b a right-angled triangle

c a scalene triangle

d an octagon

11 This diagram shows two sets of vectors jumbled together.
Each set of vectors has a vector sum of zero.

Can you find the two sets of vectors?

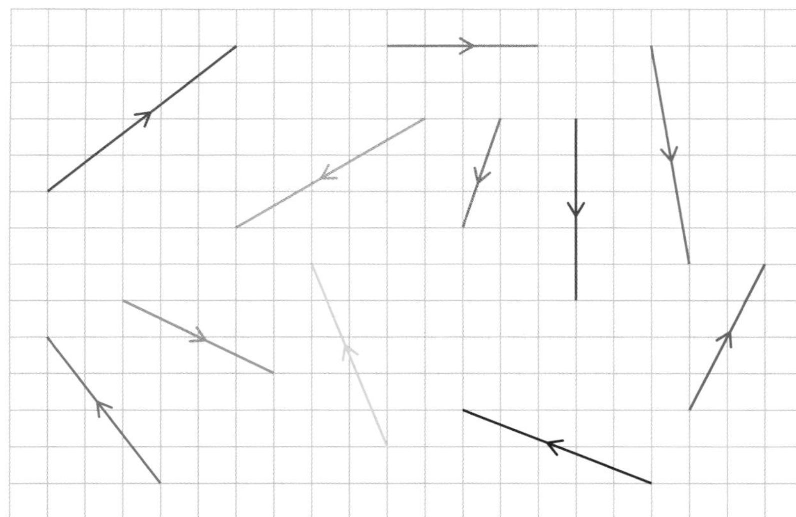

Learn... 19.2 Vector geometry

Vectors are often used in simple geometric problems.

To solve problems using vector geometry the first step is to form a vector equation.

The shapes below look the same but the directions in two of the vectors are different.

The vector equations for each diagram would be different.

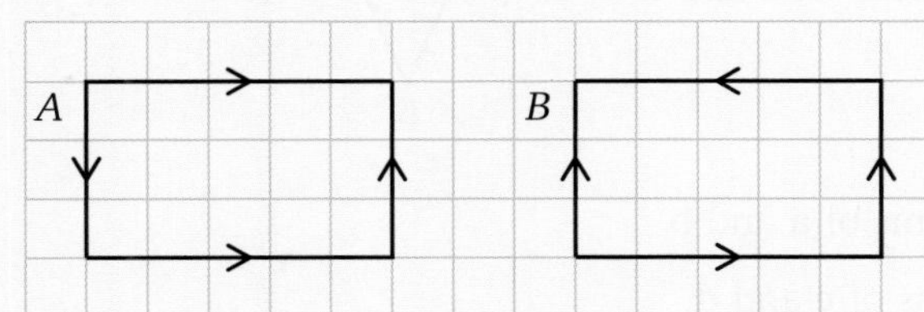

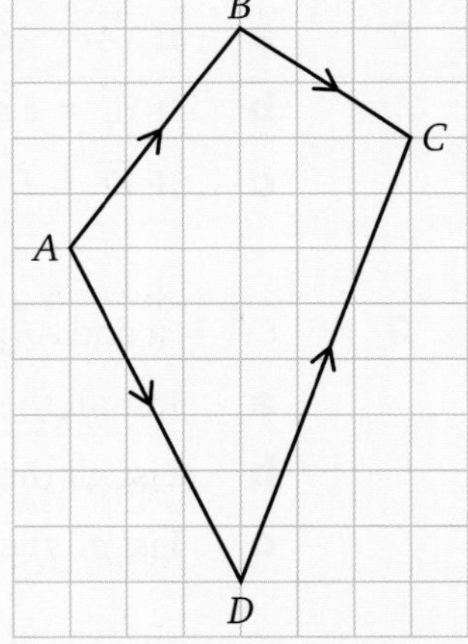

When forming equations, consider the different ways of moving from one point to another.

In the diagram, there are two ways to get from A to C using the vectors shown:

A to B then B to C $\quad \overrightarrow{AC} = \overrightarrow{AB} + \overrightarrow{BC}$

A to D then D to C $\quad \overrightarrow{AC} = -\overrightarrow{AD} + \overrightarrow{DC}$

The two routes result in the same thing, $\overrightarrow{AC}$, so can be written:

$\overrightarrow{AB} + \overrightarrow{BC} = \overrightarrow{AD} + \overrightarrow{DC}$

$\overrightarrow{AC}$ is the **resultant vector**.

Example: ABC is an isosceles triangle.
M is the midpoint of CB. $\overrightarrow{AB} = \mathbf{a}$ and $\overrightarrow{AC} = \mathbf{b}$

Even though the triangle lengths are the same, the vectors are different because their directions are different.

a Find $\overrightarrow{BC}$ in terms of **a** and **b**.

b Find $\overrightarrow{AM}$ in terms of **a** and **b**.

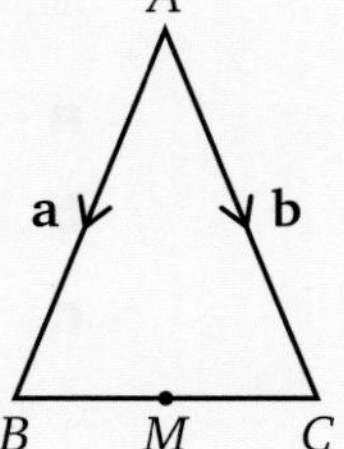

Solution:

a $\overrightarrow{BC} = \overrightarrow{BA} + \overrightarrow{AC}$ Form a vector equation.

$\overrightarrow{BC} = -\mathbf{a} + \mathbf{b}$ Note that the direction of **a** is negative because it is going from *B* to *A*.

b $\overrightarrow{AM} = \overrightarrow{AB} + \overrightarrow{BM}$

$= \overrightarrow{AB} + \frac{1}{2}\overrightarrow{BC}$

$= \mathbf{a} + \frac{1}{2}(-\mathbf{a} + \mathbf{b})$

$= \mathbf{a} - \frac{1}{2}\mathbf{a} + \frac{1}{2}\mathbf{b}$

$= \frac{1}{2}\mathbf{a} + \frac{1}{2}\mathbf{b}$

AQA *Examiner's tip*

When you form vector equations, remember to check the direction of the given vectors. Remember that the opposite of **a** is $-$**a**.

Practise... 19.2 Vector geometry

D C B A A*

A*

1 This is a kite.

a Write the vector $\overrightarrow{AD}$ in terms of **a** and **b**.

b Write the vector $\overrightarrow{AD}$ in terms of **c** and **d**.

c Write the vector $\overrightarrow{BC}$ in terms of **a** and **d**.

d Write the vector $\overrightarrow{BC}$ in terms of **b** and **c**.

e Use any of your answers to write **a** in terms of **b**, **c** and **d**.

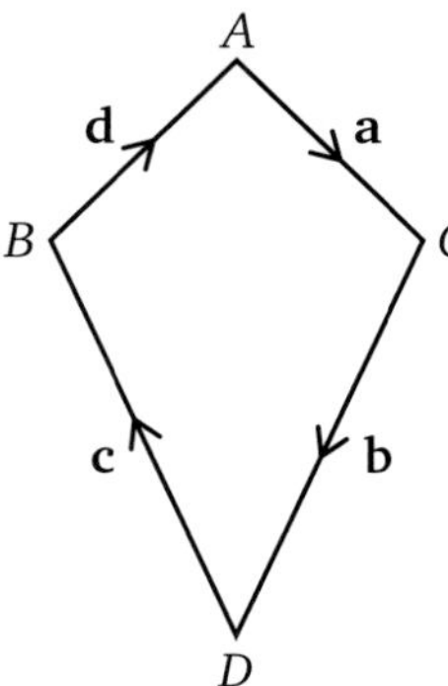

2 **a** If $\overrightarrow{XA} = 2\mathbf{a}$ and $\overrightarrow{XB} = \mathbf{a} + \mathbf{b}$, find $\overrightarrow{AB}$ in terms of **a** and **b**.

b If $\overrightarrow{YC} = \mathbf{c}$ and $\overrightarrow{YD} = \mathbf{d} - \mathbf{c}$, find $\overrightarrow{CD}$ in terms of **c** and **d**.

c IF $\overrightarrow{ZE} = \mathbf{f} - 2\mathbf{e}$ and $\overrightarrow{FZ} = 4\mathbf{f} - \mathbf{e}$, find $\overrightarrow{FE}$ in terms of **e** and **f**.

3 $\overrightarrow{OA} = \mathbf{a}$ and $\overrightarrow{AB} = \mathbf{b}$

a List all the vectors equal to **a**.

b List all the vectors equal to **b**.

c List all the remaining vectors in terms of **a** and **b**.

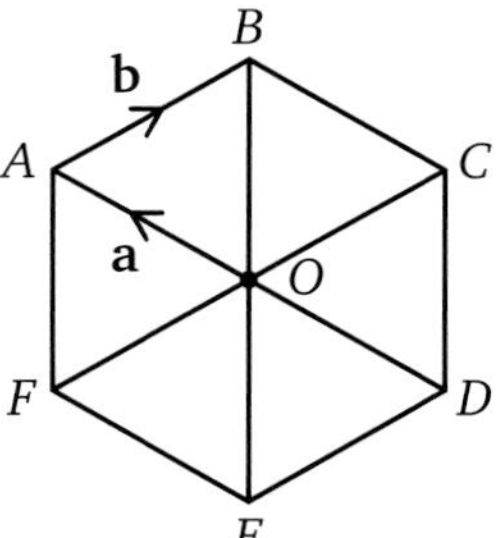

4 $\overrightarrow{AM} = 2\overrightarrow{MC} = 2\mathbf{b}$

$\overrightarrow{NC} = \frac{\overrightarrow{BC}}{3} = \mathbf{c}$

a Find $\overrightarrow{NM}$ in terms of **b** and **c**.

b Find $\overrightarrow{AB}$ in terms of **b** and **c**.

c Danny says that $\overrightarrow{AB}$ and $\overrightarrow{NM}$ are parallel. Is he right?

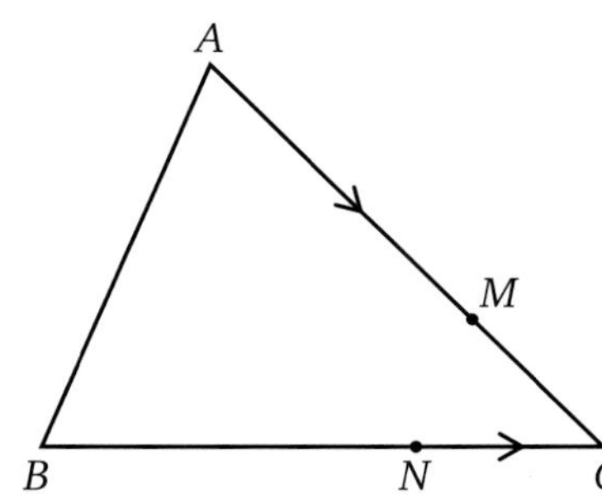

A*

5 $ABCD$ is a trapezium and $ABED$ is a parallelogram.
$\overrightarrow{AB} = \mathbf{a}$ and $\overrightarrow{AD} = \mathbf{b}$ and length $CD = 2 \times$ length AB

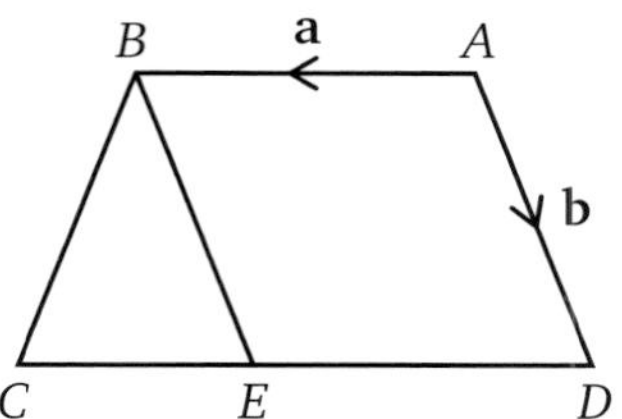

Write each of these vectors in terms of **a** and **b**.

a **i** $\overrightarrow{DC}$ **iii** $\overrightarrow{DB}$ **v** $\overrightarrow{AE}$ **vii** $\overrightarrow{BC}$

ii $\overrightarrow{AC}$ **iv** $\overrightarrow{BE}$ **vi** $\overrightarrow{EC}$

b Prove that $ABCE$ is a parallelogram.

c AC crosses BE at F. Write $\overrightarrow{AF}$ in terms of **a** and **b**.

6 $ADBO$ and $OBEC$ are both parallelograms.
$\overrightarrow{OA} = \mathbf{a}$, $\overrightarrow{OB} = \mathbf{b}$ and $\overrightarrow{OC} = \mathbf{c}$

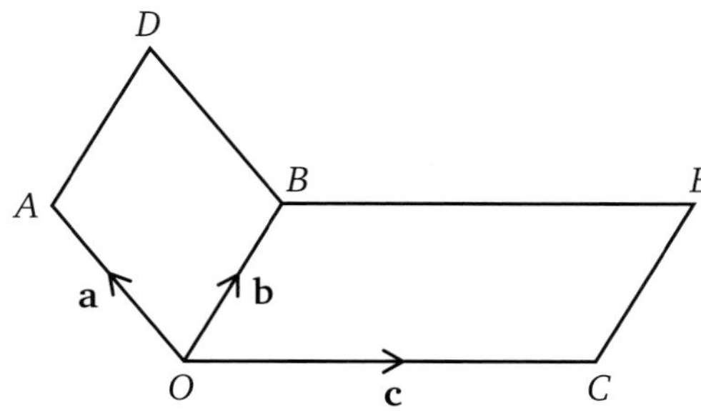

a Write each of these vectors in terms of **a**, **b** and **c**.

i $\overrightarrow{OD}$ **ii** $\overrightarrow{BC}$ **iii** $\overrightarrow{AB}$ **iv** $\overrightarrow{BE}$ **v** $\overrightarrow{DC}$

b F is a point such that $\overrightarrow{OF} = \mathbf{a} + \mathbf{b} + \mathbf{c}$
Write $\overrightarrow{CF}$ in terms of **a**, **b** and **c**.

⚠ 7 XYZ is a triangle. M is the midpoint of YZ and S is the midpoint of XZ.
YS crosses XM at R. The ratio $XR : RM = 2 : 1$
$\overrightarrow{XY} = \mathbf{a}$ and $\overrightarrow{XZ} = \mathbf{b}$

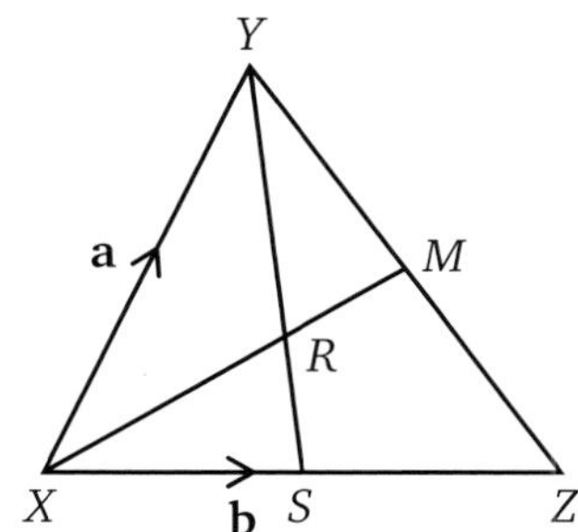

a Write the following vectors in terms of **a** and **b**.

i $\overrightarrow{YS}$ **iii** $\overrightarrow{ZM}$ **v** $\overrightarrow{XR}$ **vii** $\overrightarrow{RS}$

ii $\overrightarrow{ZY}$ **iv** $\overrightarrow{XM}$ **vi** $\overrightarrow{XS}$

b Use your answers to parts **i** and **vii** to calculate the ratio $YS : RS$

19 Assess k!

A

1 $\mathbf{a} = \begin{pmatrix} 2 \\ -1 \end{pmatrix}$ $\mathbf{b} = \begin{pmatrix} 0 \\ 2 \end{pmatrix}$ $\mathbf{c} = \begin{pmatrix} -1 \\ 0 \end{pmatrix}$

Write each of the following as a single column vector.

a $\mathbf{a} + \mathbf{b} + \mathbf{c}$

b $2\mathbf{a} + \mathbf{b}$

c $2\mathbf{b} - 2\mathbf{a} + \mathbf{c}$

A

2 Which of these vectors are equal? Give reasons for your answers.

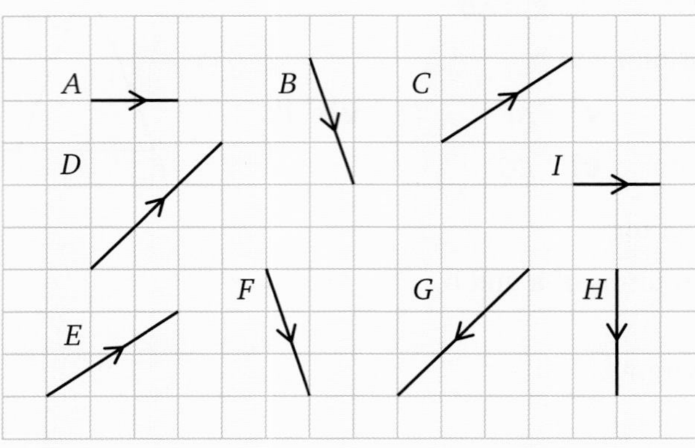

3 $\overrightarrow{AB} = \begin{pmatrix} 1 \\ 2 \end{pmatrix}$ $\overrightarrow{BC} = \begin{pmatrix} 3 \\ 4 \end{pmatrix}$

a Draw a diagram of $3\overrightarrow{AB} + \overrightarrow{BC}$.

b Find the single vector equal to $3\overrightarrow{AB} + \overrightarrow{BC}$.

4 $\overrightarrow{AB} = \begin{pmatrix} 2 \\ -3 \end{pmatrix}$

Which of these column vectors are parallel to $\overrightarrow{AB}$? For those that are, write each as a multiple of $\overrightarrow{AB}$.

a $\begin{pmatrix} 4 \\ -2 \end{pmatrix}$ **b** $\begin{pmatrix} 4 \\ -6 \end{pmatrix}$ **c** $\begin{pmatrix} 1 \\ 1.5 \end{pmatrix}$ **d** $\begin{pmatrix} -2 \\ 3 \end{pmatrix}$ **e** $\begin{pmatrix} 20 \\ -30 \end{pmatrix}$

A*

5 $ABCD$ is a parallelogram. $\overrightarrow{CD} = 3\mathbf{b}$ and $\overrightarrow{BD} = 2\mathbf{a}$

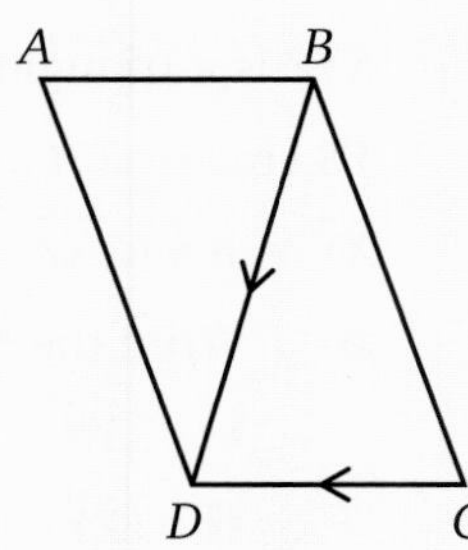

a Write the vector $\overrightarrow{AB}$ in terms of **a** and **b**.

b Write the vector $\overrightarrow{AD}$ in terms of **a** and **b**.

c Write the vector $\overrightarrow{BC}$ in terms of **a** and **b**.

d Write the vector $\overrightarrow{AC}$ in terms of **a** and **b**.

6 **a** If $\overrightarrow{XA} = \mathbf{a}$ and $\overrightarrow{XB} = \mathbf{a} + 2\mathbf{b}$, find $\overrightarrow{AB}$ in terms of **a** and **b**.

b If $\overrightarrow{YC} = 2\mathbf{c}$ and $\overrightarrow{YD} = 2(\mathbf{d} - \mathbf{c})$, find $\overrightarrow{CD}$ in terms of **c** and **d**.

7 ABC is an isosceles triangle.

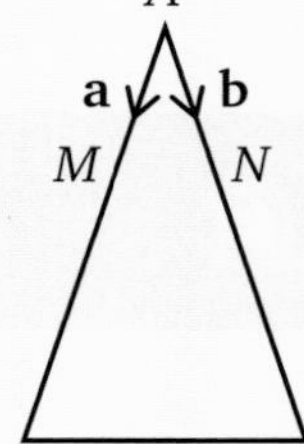

a $\overrightarrow{AM} = \mathbf{a}$ and $\overrightarrow{AN} = \mathbf{b}$

Find **MN** in terms of **a** and **b**.

b $\overrightarrow{CB} = 3(\mathbf{b} - \mathbf{a})$

Find $\overrightarrow{MC}$ and $\overrightarrow{NB}$ in terms of **a** and **b**.

c Find the ratio of AM to MC.

d Describe the relationship between triangles ABC and ANM.

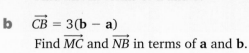

8 $ABCD$ is a rectangle. M is the midpoint of AB. $AN = 2ND$ and $BO = 3OC$

$\overrightarrow{AM} = \mathbf{a}$

$\overrightarrow{AN} = \mathbf{b}$

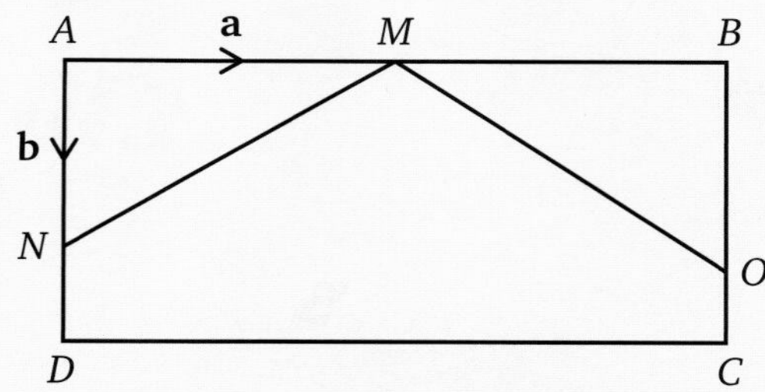

Find the following vectors in terms of **a** and **b**.

a $\overrightarrow{MB}$ **b** $\overrightarrow{BO}$ **c** $\overrightarrow{MO}$ **d** $\overrightarrow{NO}$

 Examination-style questions 

1 OAB is a triangle with P the mid point of OA and M the mid point of AB.
$\overrightarrow{OP} = \mathbf{a}$, $\overrightarrow{PA} = \mathbf{a}$ and $\overrightarrow{OB} = 2\mathbf{b}$

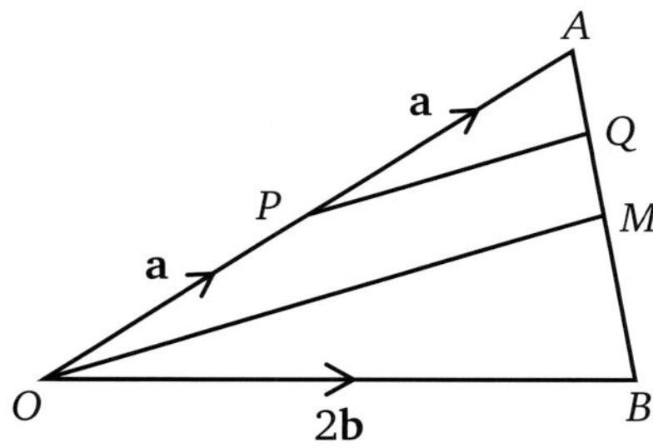

Not drawn accurately

a Write down an expression for $\overrightarrow{AB}$ in terms of **a** and **b**. *(1 mark)*

b Q lies on AB such that $\overrightarrow{AQ} = \frac{1}{4}\overrightarrow{AB}$
Show that $\overrightarrow{PQ} = \frac{1}{2}\mathbf{a} + \frac{1}{2}\mathbf{b}$
Explain your answer. *(2 marks)*

c Write down, and simplify, an expression for $\overrightarrow{OM}$ in terms of **a** and **b**. *(2 marks)*

d Explain why the answers for part **b** and part **c** show that $OPQM$ is a trapezium. *(1 mark)*

AQA 2008

20 Simultaneous equations

Objectives

Examiners would normally expect students who get these grades to be able to:

B

solve a pair of simultaneous equations such as $x + 3y = 9$ and $3x - 2y = 5$

solve a pair of linear equations graphically; identifying the point of intersection as the solution

A

solve a pair of simultaneous equations such as $y = 4x + 5$ and $y = x^2$

find the point of intersection of a linear and a quadratic equation; recognising that the solution could be found from the points of intersection of the graphs.

Did you know?

Supply and demand

The financial world uses simultaneous equations to model the balance between the supply and demand for a product or service.

When a product/service is in demand, the price will increase. However, if the price increases too much, the demand for the product/service will decrease as fewer people buy it.

The use of simultaneous equations can help foresee any problems and can try to determine where the balance between supply and demand might lie.

Key terms

simultaneous equations
coefficient
variable
eliminate
substitution

You should already know:

- ✔ how to simplify algebraic terms
- ✔ how to solve a linear equation
- ✔ how to draw linear and quadratic graphs
- ✔ how to multiply out brackets
- ✔ how to rearrange an equation to make y or x the subject
- ✔ how to solve a quadratic equation by factorisation, completing the square or by use of the quadratic formula
- ✔ the theorems associated with circles.

20.1 Solving simultaneous equations by elimination

$4x + 3y = 12$

The coefficient of x is 4.
The coefficient of y is 3.

x and y are the variables

Simultaneous equations can be solved by making the **coefficients** of one of the **variables** the same in both equations.

In these simultaneous equations, the coefficients of y are the same value. This means that you do not have to make them the same value to start with:

$4x + 3y = 12$ The coefficient of x is 4 and the coefficient of y is 3.

$3x - 3y = 2$ The coefficient of x is 3 and the coefficient of y is -3.

The coefficients of y have the same value.
Because they have different signs, the equations are ready to be **added**: $+3y$ added to $-3y$ equals zero.

In these simultaneous equations, the coefficients of y do not have the same value:

$7x - 5y = 2$ The coefficient of x is 7 and the coefficient of y is -5.

$2x - 10y = 8$ The coefficient of x is 2 and the coefficient of y is -10.

To make the coefficient of y the same value, you have to multiply the first equation by 2 throughout:

$14x - 10y = 4$

$2x - 10y = 8$

The coefficients of y now have the same value.
Because the coefficients have the same signs, the equations are ready to be **subtracted**: $-10y$ subtracted from $-10y$ equals zero.

After making the coefficients of one variable the same, you can **eliminate** that variable by adding or subtracting.

If their signs are the same, subtract the equations (remember this by s s s).

If their signs are different, add the equations.

Link

For examples and practice questions on simultaneous equations, look at the Unit 2 Higher book, Chapter 13.

Example: Solve these simultaneous equations.

$2x + 3y = 15$

$3x - 2y = 3$

Solution: Step 1: Multiply the first equation by 2 and the second equation by 3.
This makes the coefficients of y the same value.
Because the signs are different, you add the equations.

$4x + 6y = 30$

$9x - 6y = 9$

AQA Examiner's tip

Don't forget to multiply **both sides** of each equation.

Step 2: Add the equations.

$$\begin{aligned} 4x + 6y &= 30 \\ 9x - 6y &= 9 \\ \hline 13x &= 39 \\ x &= 3 \end{aligned}$$

The coefficients of y are matching numbers with different signs, so add the equations to eliminate y.

Step 3: To find the value of y, substitute $x = 3$ in the first equation.

$$\begin{aligned} 2 \times 3 + 3y &= 15 \\ 6 + 3y &= 15 \\ 3y &= 9 \\ y &= 3 \end{aligned}$$

AQA Examiner's tip

Always substitute back into the more straightforward of the two original equations. This is usually the one with positive numbers and/or smaller numbers in it.

Step 4: Use the second equation to check your answers.

$3 \times 3 - 2 \times 3 = 9 - 6$

$= 3$ ✓

AQA Examiner's tip

Make sure you do a check. Use the equation not used in Step 3. You will then know whether you have made a mistake.

Remember

- You must have a pair of terms with matching coefficients.
- If their signs are the same, subtract the equations (remember this by s s s).
- If their signs are different, add the equations.

Practise... 20.1 Solving simultaneous equations by elimination

D C B A A*

B

1 Solve these simultaneous equations by elimination.

a $4x + y = 10$, $5x - y = 8$

b $3x + 2y = 20$, $x + 2y = 9$

c $a - 3b = 2$, $3b + a = 8$

d $e + 2f = 6$, $e - f = 3$

e $3p + 2r = 3$, $8r + 3p = 21$

f $x + 2y = 8$, $x + y = 6$

2 Solve these simultaneous equations by elimination.

a $s + t = 7$, $2s + 5t = 26$

b $x + 3y = 2$, $3x - y = 26$

c $a + 2b = 4$, $3a + 5b = 9$

d $e + 2f = 1.2$, $2e - f = 6.5$

e $2p - r = 2$, $7p - 5r = 4$

f $4x - 3y = 2$, $x - y = 1$

3 Solve these simultaneous equations by elimination.

a $3e - 8f = 11$, $2e - 5f = 6$

b $2v - 7w = 57$, $3w - 11v = 6$

c $2x + 3y = 13$, $7x - 5y = -1$

d $4i + 7j = 6$, $3i - 2j = 19$

e $2.5g - h = 4.5$, $g - 2.5h = 6$

f $5x + 4y = 5$, $6y + 2x = 13$

4

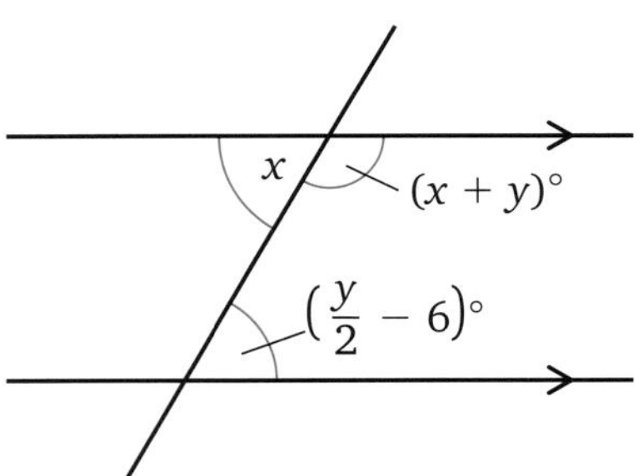

a Using the diagram above, form two equations in x and y.

b Solve these equations simultaneously to find the values for x and y.

5 Julie is thinking of a fraction $\frac{x}{y}$

If she adds 5 to the numerator and to the denominator, her fraction equals $\frac{8}{9}$

If she subtracts $\frac{1}{2}$ from the numerator and the denominator, her fraction equals $\frac{5}{7}$

What is Julie's fraction?

 6 A taxi firm charges a set amount of £x per journey plus £y for every mile travelled. A three-mile journey costs £5.70 and a four-mile journey costs £7.20.

Find the set amount of pounds per journey and the charge per mile.

 7 In a training camp, the members of staff take five flashlights with them on night expeditions. Two different brands of batteries are available.

- Brand A has an average life of x hours.
- Brand B has an average life of y hours.

Assuming that the batteries run at the average rate:

- three brand A batteries and two brand B batteries would run for 48 hours and 45 minutes in total
- four brand A batteries and one brand B battery would run for 40 hours in total.

Form a pair of simultaneous equations and solve them to find the average battery running time for each brand.

 8 Martin works in a café. At the end of each day he counts the money in the till.

- On Monday he has x £10 notes and y £5 notes. The total amount of money equals £125.
- On Tuesday he has half the number of £10 notes and five times the number of £5 notes that he had on Monday. After he counts the money, he pays the milkman £25 out of the total amount and then has £150 left.

Form a pair of simultaneous equations in x and y and solve them to find how many of the different notes he had at the end of each day.

 9 Two ice lollies and three ice-cream cones cost £3.50. Three ice lollies and two ice-cream cones cost £3.25. The cost of an ice lolly is £c and the cost of an ice-cream cone is £d.

Write down a pair of simultaneous equations and solve them to find c and d.

Learn... 20.2 Solving simultaneous equations by substitution

Simultaneous equations can also be solved by **substituting** one equation into the other.

This method is used when one of the variables is already the subject of one of the equations.

If this is not the case, you will have to rearrange one of the equations to make x or y the subject.

Example:

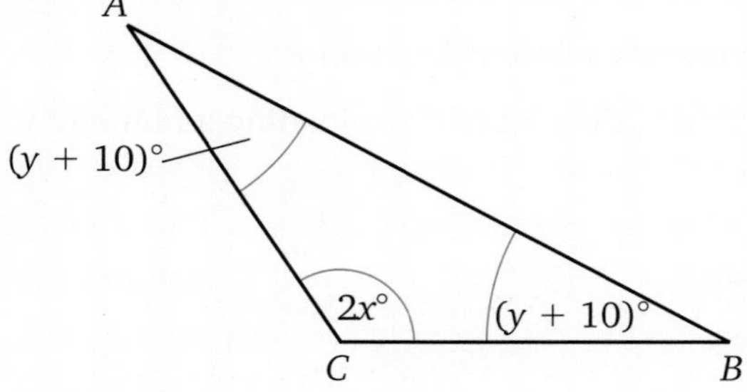

Triangle ABC is isosceles.
x is 20° more than y.

a By forming two simultaneous equations in x and y, find the values of x and y.

b Write down the value of each of the three angles in the triangle and use these to check your answers.

Solution: **a** The angles in a triangle add up to 180°.

So $2x + y + 10 + y + 10 = 180$

$2x + 2y + 20 = 180$ Collect like terms.

$2x + 2y = 180 - 20$ Subtract 20 from both sides.

$2x + 2y = 160$ Divide both sides by 2.

$x + y = 80$

As x is 20° more than y, if you add 20 to y it will equal x.

So $x = y + 20$

The two equations to be solved are:

$x + y = 80$

and $x = y + 20$

In the second equation, x is already the subject, i.e. it is in the form $x = \ldots$

Substituting for x in the first equation gives:

$y + 20 + y = 80$ Replace x with $y + 20$.

$2y + 20 = 80$ Collect the y terms.

$2y = 80 - 20$ Subtract 20 from both sides.

$2y = 60$ Divide both sides by 2.

$y = 30$

Now substitute $y = 30$ into the first equation.

$x + 30 = 80$

$x = 80 - 30$

$x = 50$

AQA Examiner's tip

Make sure you do a check. You will then know whether you have made a mistake.

Finally use the second equation to check your answers.

$50 = 30 + 20$ ✓

b The angles in the triangle are:

$2x = 2 \times 50° = 100°$

$y + 10° = 30° + 10° = 40°$ for the other two angles

Check that these angles add up to 180° because the angles in a triangle add up to 180°.

$100° + 40° + 40° = 180°$ ✓

Example: On a particular internet site, it costs £x to download a film and £y to download a track of music.
Ragini downloaded two films and nine tracks of music. It cost her £12.48.
Jamie downloaded five films and two tracks of music. It cost him £20.95.

Write down two simultaneous linear equations in x and y.

Solve these by substitution to find x, the cost of downloading a film and y, the cost of downloading a track of music.

Solution: For Ragini, £$2x$ + £$9y$ = £12.48

For Jamie, £$5x$ + £$2y$ = £20.95

The simultaneous equations are:

$2x + 9y = 12.48$

$5x + 2y = 20.95$

Rearrange the first equation to make x the subject.

$$2x = 12.48 - 9y$$
$$x = 6.24 - 4.5y$$

Substitute this expression for x into the second equation.

$$5x + 2y = 20.95$$
$$5(6.24 - 4.5y) + 2y = 20.95$$
$$31.2 - 22.5y + 2y = 20.95$$
$$31.2 - 20.5y = 20.95$$
$$10.25 = 20.5y$$
$$20.5y = 10.25$$
$$y = 0.5 \qquad £0.5 = £0.50 = 50\text{p}$$

Substitute this value for y into the formula $x = 6.24 - 4.5y$

$$x = 6.24 - 4.5(0.5)$$
$$x = 3.99 \qquad £3.99$$

Now check the answers. Substitute both the x- and y-values back into the equations to see whether the values are correct.

$2(3.99) + 9(0.5) = 7.98 + 4.50 = 12.48$ ✓

$5(3.99) + 2(0.5) = 19.95 + 1.00 = 20.95$ ✓

Practise... 20.2 Solving simultaneous equations by substitution

B

1 Solve these simultaneous linear equations by substitution.

a $x + y = 8$
$y = x + 2$

b $c - d = 1$
$c = 2d - 3$

c $r = 2p - 5$
$p + 2r = 0$

d $v = 2w - 0.5$
$4v - 5w = 1$

2 Solve these simultaneous linear equations by substitution.

a $m + n = 7$
$m = 3 + n$

b $2x + 3y = 28$
$3x + 4y = 37$

c $7x - 2y = 4$
$6x + y = 17$

d $2e - f = 22$
$3e + 2f = 40$

3 The diagram shows a circle with centre X.
The angle formed by the triangle at the circumference of the circle is $x + 3y$.
x is 30° more than y.

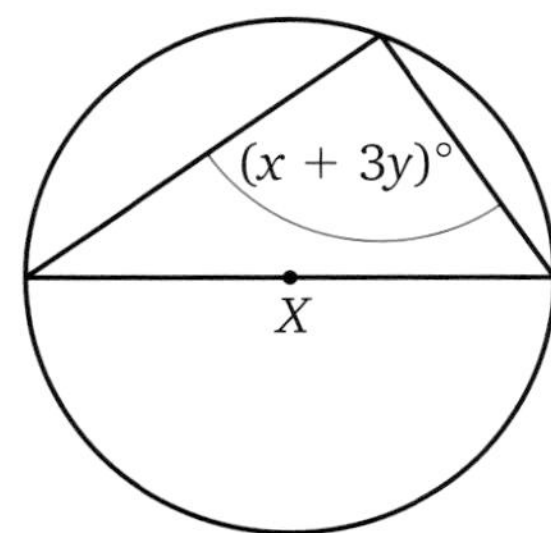

a From this, and your knowledge of circles, write down two equations in x and y.

b Solve these simultaneous equations to find the values of x and y.

A

4 Angle x is 24° larger than angle y.
The bearing of A from B is 256°.

a Find the values of x and y by forming two linear equations in x and y and solving them by substitution.

b Write down the bearing of B from A.

Not drawn accurately

5 In the diagram, the ratio of $x : y$ is $4 : 5$

O is the centre of the circle, and AB is a tangent to the circle.

a Set up and solve two simultaneous equations in x and y.

b Use these answers to find the value of y.

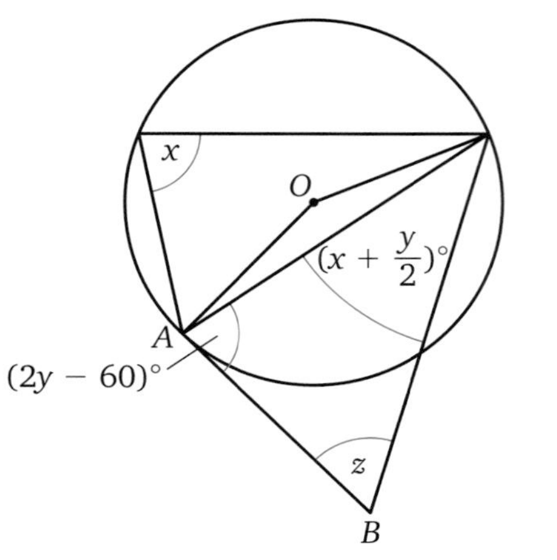

6 These are the rules for a quiz.

- You score x points for a correct answer.
- y points are taken away for an incorrect answer.
- If you correctly answer a bonus question, you score z points.
- If you do not answer, you score no points.

Three friends' results were as follows.

Adrian:	6 correct 1 incorrect 1 bonus	Total = 54 points
Beth:	5 correct 2 incorrect 2 bonus	Total = 38 points
Charlie:	7 correct 0 incorrect 4 bonus	Total = 86 points

Find the values of x, y and z.

7 John went to buy some of his favourite aftershave. It is manufactured in two sizes.

- The small size contains x millilitres.
- The large size contains y millilitres.

If he bought two small ones, he would have 30 ml less than the amount in the large size. The small size and the large size together contain a total of 135 ml.

How much aftershave is in each sized container?

8 Poppy and her three friends were going to a nightclub.
It costs less to go in before 9pm than it does to go after 9pm.
Unfortunately, one of her friends can't get there before 9pm.
If three of them go in early and one later, it will cost £15.
Her friend does not really want to go in on her own.
If two go in early and two later, it will cost £16.

If £x is the cost of early entry and £y is the cost of the later entry, find the two different entry costs.

9 Susie is x years old and her brother is y years old.
Susie is 5 years older than her brother.
They both have an aunt who is 45 years old.
The aunt is three times her brother's age plus Susie's age.

How old are Susie and her brother?

10 Two teenagers have heights of x metres and y metres.
If you add their heights together, they measure 2.53 metres in total.
The difference between their heights is 11 cm.

a Write down two simultaneous equations in x and y.

b Solve them by substitution to find each of their heights.

Learn... 20.3 Solving simultaneous linear equations graphically

Simultaneous linear equations can also be solved graphically.
Begin by finding the coordinates of some points on each straight line.
The minimum number of points needed to draw a straight line is three; two to draw it and one as a check.
This is true for linear equations but to draw quadratics, you would need more x-values.
The point of intersection of two straight lines represents the solution to the simultaneous equations.
Drawing the lines will also confirm whether or not there is a solution to the pair of simultaneous equations.

- If they are parallel, there would be no solution because the lines do not cross.
- If they are the same line, there would be an infinite number of solutions.

Example: Solve these simultaneous equations graphically by plotting both lines on the same set of axes.

$x + y = 3$

$x - y = 1$

To help you, you can copy and complete these tables of values for the two equations.

$x + y = 3$

x	0	1	3
y	3		

$x - y = 1$

x	0	1	2
y	−1		

Solution: In this question the tables have already been given to you to complete.

$x + y = 3$

x	0	1	3
y	3	2	0

$x - y = 1$

x	0	1	2
y	−1	0	1

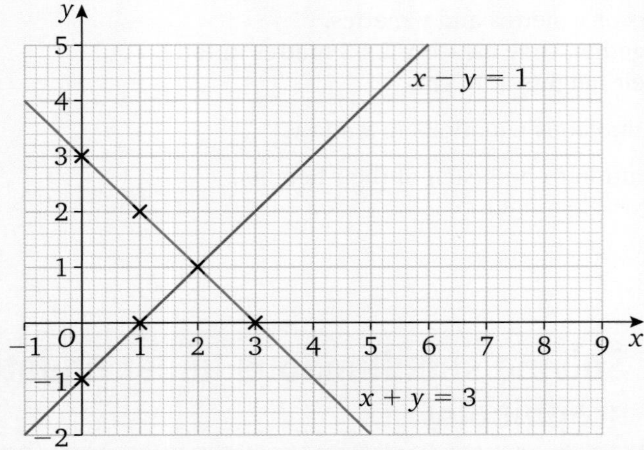

The straight lines cross at the point (2, 1).
This point represents the solution to the simultaneous equations.

$x = 2, y = 1$

Now check your solutions in both of the equations.

$2 + 1 = 3$ ✓

$2 - 1 = 1$ ✓

Example: Solve the following simultaneous equations graphically.

$y = 4 - x$

$2x - y = 2$

Solution: If you are not given the x-values that you have to use, or tables to complete, you will have to choose your own values.

One way is to begin by finding the points at which the graphs cross the x-axis and the y-axis.

This will give you a hint as to what other values to choose for each equation.

Starting with $y = 4 - x$

When $x = 0, y = 4$ (0, 4)

When $y = 0, x = 4$ (4, 0)

Choose one more value for x.

When $x = 1, y = 3$ (1, 3)

Now work out three points on the other line.

$2x - y = 2$

When $x = 0, -y = 2$ so $y = -2$ (0, −2)

When $y = 0, x = 1$ (1, 0)

Choose one more value for x.

When $x = 3, 6 - y = 2$ so $y = 4$ (3, 4)

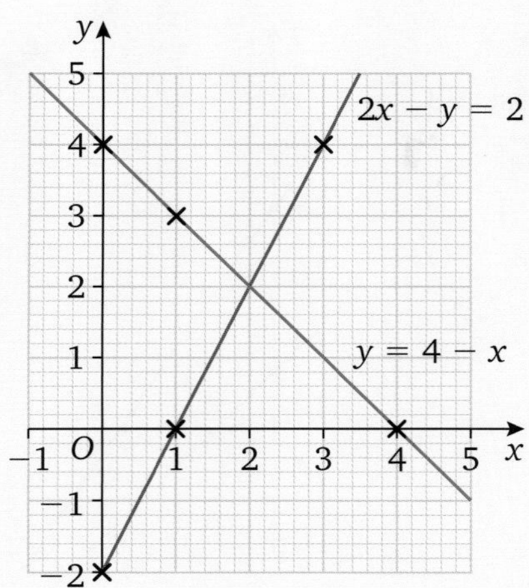

The two graphs cross at the point (2, 2). This point represents the solution to the simultaneous equations.

$x = 2, y = 2$

Now check your solutions in both equations.

$2 = 4 - 2$ ✓

$4 - 2 = 2$ ✓

Example: Cinema tickets cost £x for adults and £y for children.
Three adult tickets and one child's ticket cost £18.
Four adult tickets and two children's tickets cost £26.

Solve these equations graphically to find the prices of each type of ticket.

Solution: If 3 adult tickets and 1 child's ticket cost £18, the equation is $3x + y = 18$

If 4 adult tickets and 2 children's tickets cost £26, the equation is $4x + 2y = 26$
This can be simplified to $2x + y = 13$

So, the equations are:

$3x + y = 18$

$2x + y = 13$

Neither x nor y can be negative because they represent amounts of money.

Starting with $3x + y = 18$

When $x = 0, y = 18$ (0, 18)

When $y = 0, 3x = 18$ so $x = 6$ (6, 0)

Choose one more value for x.

When $x = 3, 9 + y = 18$ so $y = 9$ (3, 9)

Now work out three points on the other line.

$2x + y = 13$

When $x = 0, y = 13$ (0, 13)

When $y = 0, 2x = 13$ so $x = 6\frac{1}{2}$ $(6\frac{1}{2}, 0)$

Choose one more value for x.

When $x = 3, 6 + y = 13$ so $y = 7$ (3, 7)

The point of intersection is (5, 3).

An adult's ticket costs £5 and a child's costs £3.

Now check your solutions in your original equations.

$3 \times 5 + 3 = 15 + 3 = 18$ ✓

$2 \times 5 + 3 = 13$ ✓

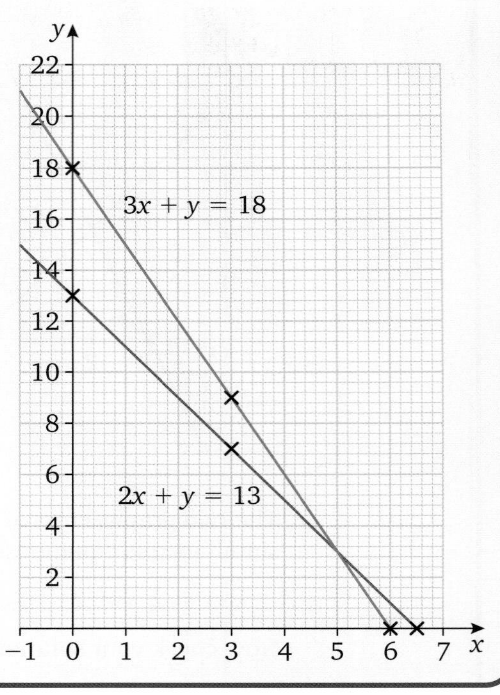

Practise... 20.3 Solving simultaneous linear equations graphically

B

1 Solve the following simultaneous equations graphically.

For each question part, consider x-values from 0 to 5.

a $x + y = 6$
$x + 2y = 8$

b $y = x + 2$
$y = 3x - 3$

c $x + 2y = 5$
$3x - y = 8$

d $3x + 2y = 10$
$2x - 2y = 5$

e $7x + 3y = 5$
$15x - 3y - 6$

f $5y + 2x = 15.5$
$3x - 5y = 4.5$

2 For each question, consider x-values from −5 to 5.

a $x + y = -1$
$2x + 5y = 1$

b $y = x - 4$
$x - 2y = 7$

c $2x + y = -8$
$x - 2y = 6$

d $2x + y = 0$
$x + 3y = 2\frac{1}{2}$

e $5x - 2y = 12$
$2x + y = 3$

f $x + y = -1$
$2x - y = -\frac{1}{2}$

3 A Chinese buffet costs £x for an adult and £y for a child.
A family of two adults and a child paid £30.
Another family of one adult and two children paid £24.

How much did the buffet cost for each adult and each child?

Solve this using a graphical method.

> **Hint**
> Consider x-values from 0 to 14.

4 A coffee and a hot chocolate cost £x and £y respectively.
Two coffees and a hot chocolate cost £9.50.
One coffee and three hot chocolates cost £13.50.

How much do you pay for each type of drink?

Solve this using a graphical method.

20.4 Solving simultaneous equations algebraically, where one is linear and one is quadratic

The substitution method is especially useful when there is one linear equation and one quadratic equation. The substitution leads to a quadratic with one unknown.

When solving these, you are finding the coordinates of any points where the straight line intersects with the curve. There are three possibilities.

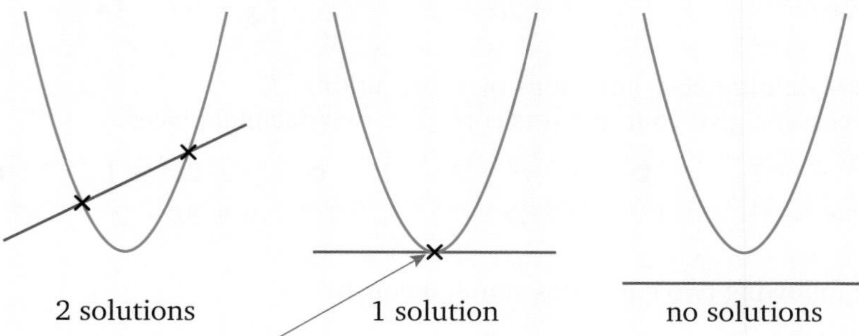

To solve these types of problem:

- Make sure that the linear equation expresses y in terms of x or x in terms of y. If the equation is not in either format, rearrange the equation so that it is.
- Substitute this expression into the quadratic equation.
- Rearrange the new equation into the quadratic form: $ax^2 + bx + c = 0$
- Solve the equation.
- Substitute each value of x into the linear equation to find the corresponding value of y.
- Check your solutions by substituting them back into the quadratic equation.

Example: Solve the simultaneous equations.

$y = 1 - 4x$ linear equation

$y = x^2 - 4$ quadratic equation

Solution: As the linear equation expresses y in terms of x, substitute this expression into the quadratic equation.

$1 - 4x = x^2 - 4$

$1 = x^2 + 4x - 4$ Collect all the x terms on one side.

$0 = x^2 + 4x - 5$ Rearrange the equation into the quadratic form $ax^2 + bx + c = 0$

$(x + 5)(x - 1) = 0$ Factorise the quadratic.

Now solve the equation.

Either $x + 5 = 0$ so $x = -5$

or $x - 1 = 0$ so $x = 1$

Substitute each value of x in the linear equation to find the corresponding value of y.

$y = 1 - 4x$

When $x = -5$ $y = 1 - 4 \times (-5) = 1 + 20 = 21$ First pair of values is $(-5, 21)$.

When $x = 1$ $y = 1 - 4 \times 1 = 1 - 4 = -3$ Second pair of values is $(1, -3)$.

To check, substitute both pairs of values into the quadratic equation, $y = x^2 - 4$

$(-5)^2 - 4 = 25 - 4 = 21$ ✓

$1^2 - 4 = 1 - 4 = -3$ ✓

Practise... 20.4 Solving simultaneous equations algebraically, where one is linear and one is quadratic

A

1 Solve these simultaneous equations by substitution.

a $y = 3x + 10$, $y = 4x^2$

b $3.5c + d = -1$, $d = 2c^2 - 2$

c $g = h^2$, $g = 9h - 14$

d $v = 4w^2$, $v = 9w - 5$

2 Solve these simultaneous equations by substitution.
Where necessary, give your answers correct to two decimal places.

a $m = n^2 + 2$, $n + m = 3$

b $f = e^2 + 3e + 1$, $f = 7e - 1$

c $y = 2x^2 - 1$, $y = 3x + 2$

d $b = 7a^2 - 4a + 3$, $b = 4a + 1$

A*

3 Solve the following two equations simultaneously.

$$(x - 2)^2 + y^2 = 4 \qquad y = x - 1$$

Give your answers correct to two decimal places.

? **4** Susie thinks of two whole numbers, x and y.

If she squares the first number and then subtracts the square of the second number, it gives the same answer as when the two numbers are added together.

The second number is also twice the first number minus 8.

What are the numbers?

? **5** The total area of the square and the rectangle is $64\,\text{cm}^2$.
The width of the rectangle (y) is five times the side length of the square (x) minus 14.

Find the dimensions of the square and the rectangle.

Not drawn accurately

x cm, x cm

y cm, $2x$ cm

? **6** The diagram shows two squares, one inside the other.
The area of the shaded part is $36\,\text{cm}^2$.
The difference between the side lengths of the squares is 2 cm.

Find the side length of each square.

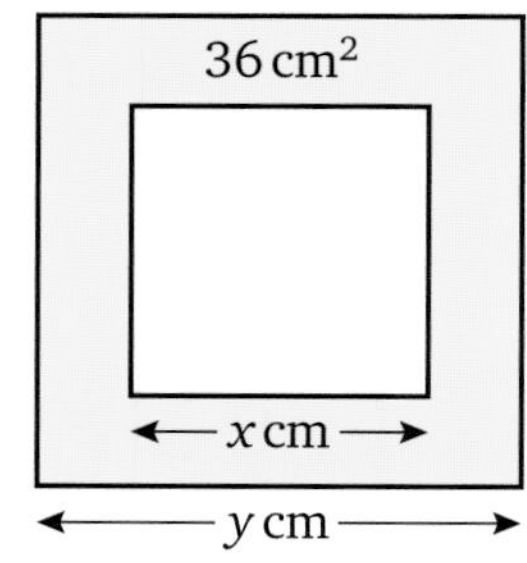

20.5 Solving simultaneous equations graphically, where one is linear and one is quadratic

In Learn 20.4, linear and quadratic equations were solved algebraically, using substitution.

For the higher grades, you also need to be able to solve them graphically.

The linear equations can be plotted in the same way as in Learn 20.3, by finding three points on the line.

As quadratic equations are curves, you will need more than three points to plot them. This usually involves the construction of a table for between six and ten points as shown in the examples.

For a recap on how to solve quadratic equations graphically look at Learn 17.5 in Chapter 17.

When solving a linear and a quadratic equation, remember that you could obtain one or two solutions, or sometimes, no solutions at all.

Sometimes straight lines are used to solve quadratic equations. In these questions, the straight line used is embedded in the equation itself and the first step is to identify the straight line to be used.

In Chapter 17, Learn 17.5, equations such as $x^2 + 2x = 0$ were solved by considering where the graph $y = x^2 + 2x$ intersected with the $y = 0$ line.

For $x^2 + 2x = 4$, the solutions were where the graph $y = x^2 + 2x$ intersected with the $y = 4$ line.

The second worked example shows you how to solve several different quadratic equations from one graph.

Example: Solve these simultaneous equations graphically by plotting both graphs on the same set of axes.

$y = -x^2 + 4x - 3$ Take values of x from 0 to 5.

$2 - 2x = y$

Solution: For the quadratic $y = -x^2 + 4x - 3$

x	0	1	2	3	4	5
$-x^2$	0	−1	−4	−9	−16	−25
$+4x$	0	4	8	12	16	20
-3	−3	−3	−3	−3	−3	−3
y	−3	0	1	0	−3	−8

$2 - 2x = y$ Rearrange the linear equation.

$y = 2 - 2x$

To draw the straight line, you will need three points.

When $x = 0, y = 2$ (0, 2)

When $y = 0, x = 1$ (1, 0)

When $x = 2, y = -2$ (2, −2)

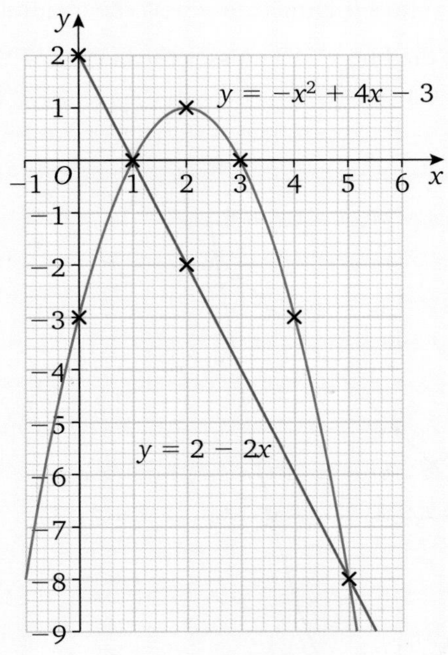

The straight line crosses the graph at (1, 0) and (5, −8).

The solutions are $x = 1$ and $x = 5$

Now check your solutions.

When $x = 1$ and $y = 0$

$y = -x^2 + 4x - 3$ $\qquad$ $-1^2 + 4 - 3 = -1 + 4 - 3 = 0$ ✓

$2 - 2x = y$ $\qquad$ $2 - 2 = 0$ ✓

When $x = 5$ and $y = -8$

$y = -x^2 + 4x - 3$ $\qquad$ $-5^2 + 4 \times 5 - 3 = -25 + 20 - 3 = -8$ ✓

$2 - 2x = y$ $\qquad$ $2 - 2 \times 5 = 2 - 10 = -8$ ✓

Example:

a Draw the graph of $y = x^2 - 6x + 8$ for values of x from $0 \leqslant x \leqslant 6$

b Use your graph and suitable straight lines to solve these equations.

i $x^2 - 6x + 5 = 0$

ii $x^2 - 7x + 6 = 0$

iii $x^2 - 4x + 4 = 0$

Solution:

For the quadratic $y = x^2 - 6x + 8$

x	0	1	2	3	4	5	6
x^2	0	1	4	9	16	25	36
$-6x$	0	−6	−12	−18	−24	−30	−36
$+8$	+8	+8	+8	+8	+8	+8	+8
y	8	3	0	−1	0	3	8

You can use the shortened table if you prefer but it is easier to make mistakes.

b i $x^2 - 6x + 5 = 0$ $\qquad$ Write down the equation you wish to solve.

$x^2 - 6x + 5 + 3 = 0 + 3$ $\qquad$ Add 3 to both sides.

$x^2 - 6x + 8 = 3$ $\qquad$ The left-hand side is now the quadratic function for y.

Solutions are found where the line $y = 3$ crosses the curve.

$x = 1$ and $x = 5$

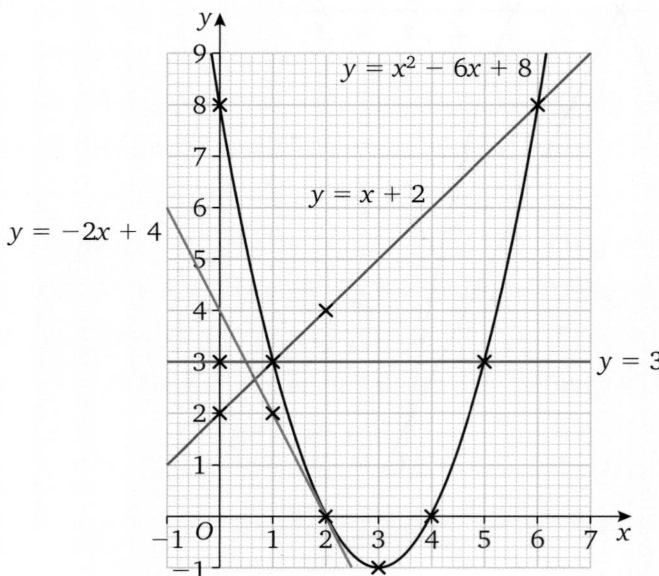

ii $x^2 - 7x + 6 = 0$ Write down the equation you wish to solve.

$x^2 - 7x + x + 6 = x$ Add x to both sides to make the term in x become $-6x$.

$x^2 - 6x + 6 + 2 = x + 2$ Add 2 to both sides to make the constant term $+8$.

$x^2 - 6x + 8 = x + 2$ The left-hand side is now the quadratic function for y.

Solutions are found where the line $y = x + 2$ crosses the curve.

To plot this straight line you need three points.

When $x = 0, y = 2$ (0, 2)

When $y = 0, x = -2$ (−2, 0) (not in the range plotted for the quadratic)

When $x = 1, y = 3$ (1, 3)

When $x = 2, y = 4$ (2, 4)

Reading from the graph, the solutions are $x = 1$ and $x = 6$

iii $x^2 - 4x + 4 = 0$ Write down the equation you wish to solve.

$x^2 - 4x - 2x + 4 = -2x$ Subtract $2x$ from both sides to make the term in x become $-6x$.

$x^2 - 6x + 4 + 4 = -2x + 4$ Add 4 to both sides to make the constant term $+4$.

$x^2 - 6x + 8 = -2x + 4$ The left-hand side is now the quadratic function for y.

Solutions are found where the line $y = -2x + 4$ crosses the curve.

To plot this you need three points.

When $x = 0, y = 4$ (0, 4)

When $y = 0, 2x = 4$ so $x = 2$ (2, 0)

When $x = 1, y = -2 + 4 = 2$ (1, 2)

This line touches the curve at (2, 0). The line is a tangent to the curve so there is only one distinct solution, $x = 2$

AQA Examiner's tip

Remember that there are three possible cases when solving a linear and a quadratic equation: you could obtain 1, 2 or 0 solutions. If the straight line and the quadratic do not intersect, there will be no solution.

Practise... 20.5 Solving simultaneous equations graphically, where one is linear and one is quadratic

D C B A A*

A

1 Solve these simultaneous equations graphically by plotting both graphs on the same set of axes.

a $y = x^2 + 3$ Take values of x from 0 to 5.
$y = 4x$

b $x + y = 4$ Take values of x from −5 to 5.
$y = x^2 - 8$

c $y = x^2 - 3x - 4$ Take values of x from −2 to 5.
$y = 1 + x$

d $y = 2x^2 - 9x + 9$ Take values of x from 0 to 5.
$x + y = 9$

e $y = 3 + 2x - x^2$ Take values of x from −2 to 4.
$2y = x + 4$

A
A*

2 **a** Draw the graph of $y = x^2 - 5x + 4$ for $0 \leqslant x \leqslant 5$

b Use your graph and suitable straight lines to solve these equations.

i $x^2 - 5x + 4 = 0$ **ii** $x^2 - 5x + 6 = 0$ **iii** $x^2 - 6x + 5 = 0$

3 **a** Draw the graph of $y = x^2 - 4x - 5$ for $-2 \leqslant x \leqslant 6$

b Use your graph and suitable straight lines to solve these equations.

i $x^2 - 4x - 5 = 0$ **ii** $x^2 - 4x + 4 = 0$ **iii** $x^2 - x - 6 = 0$

4 **a** Draw the graph of $y = 6x - x^2$ for $-1 \leqslant x \leqslant 7$

b Use your graph and suitable straight lines to solve these equations.

i $6x - x^2 = 0$ **ii** $6x - x^2 - 4 = 0$

c Use the graph to explain why you cannot solve the equation $6x - x^2 - 10 = 0$.

5 Here are the equations of a curve and a line.

$y = 3x^2 - 5x + 2$ $y = x - 1$

Show that the line is a tangent to the curve at the point (1, 0).

6 The graph shows a sketch of the curve $y = 2x^2 - 5x + 3$ and the line $y = x + 11$
They cross at the points A and B.

Show that the points A and B can be found using the quadratic equation:

$x^2 - 3x - 4 = 0$

Hence find the coordinates of A and B.

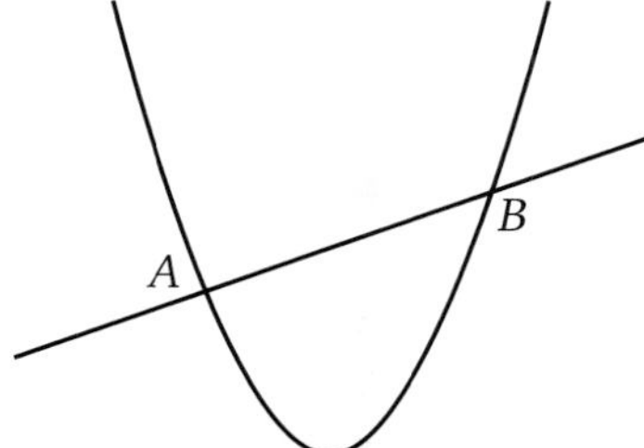

20 Assess

B

1 Solve these simultaneous equations.

a $3x + 2y = 11$
$4x - 2y = 10$

b $2x + y = 7$
$2x - y = 9$

c $3x + 5y = 5$
$4x - 3y = 26$

d $4x + 2y = 0$
$3x + y = \frac{1}{2}$

2 A vanilla smoothie costs £x and a strawberry smoothie costs £y.
Two vanilla and one strawberry smoothies cost £5.75.
One vanilla and two strawberry smoothies cost £6.25.

How much do they cost each?

3 A mother is x years old and her son is y years old.
The difference between their ages is 20 years.
Five years ago the mother was five times her son's age.

How old are they both now?

Solve this using a graphical method.

A

4 Solve these simultaneous equations.
Where necessary, give your answers to 2 decimal places

a $y = 2x^2$
$y = 7x - 3$

b $y = x^2$
$y = 4x + 5$

c $y = x^2 + 4x + 7$
$y = 3 - x$

d $y = x^2 - 2x + 5$
$2y - x = 20$

5 The diagram shows a square with an isosceles triangle removed from the corner.
The area of the shaded region is $8.5\ \text{cm}^2$.
The difference between x and y is 2 cm.

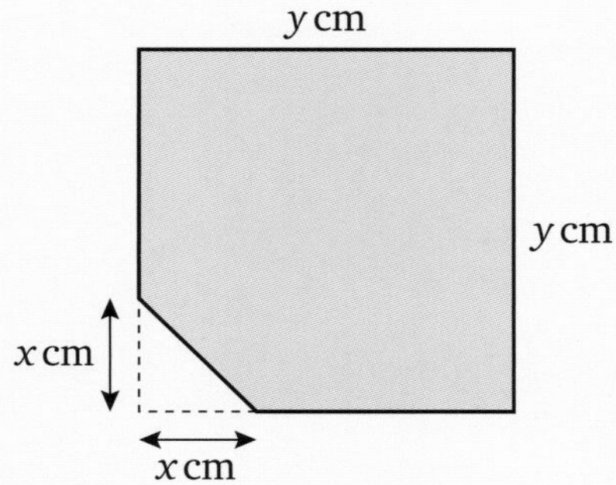

Find the values of x and y.

6 A couple are carpeting two rooms in their house with carpet tiles.
The tiles are sold in packs of four 50 cm × 50 cm square.
They have picked out two types of carpet that they like.

One is £x per pack and the other is £y per pack.
They decide to buy both carpets, using the more expensive in one room and the cheaper one in the other.
They are not sure which carpet to put in which room.

If they put the expensive carpet in the dining room and the cheaper one in the living room it will cost them £425. If they change these over, it will cost them £475.

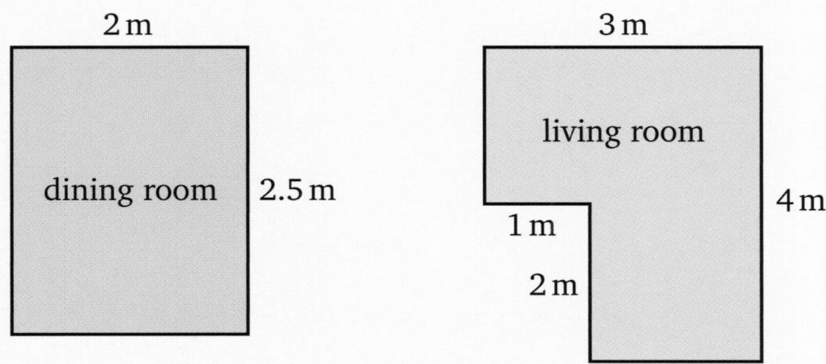

What is the cost per pack of each of the carpets?

AQA Examination-style questions

1 Solve these simultaneous equations.

$$5x + 6y = 28$$
$$x + 3y = 2$$

You **must** show your working.
Do not use trial and improvement. *(3 marks)*

AQA 2007

a The line $x + y = 5$ has been drawn on the grid.
Copy the graph and draw the line of $y = 2x - 5$ for values of x from -1 to $+5$.

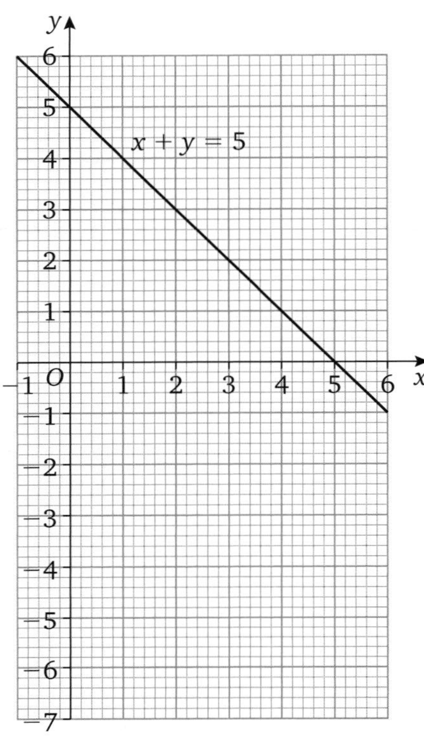

(2 marks)

b Use the graphs to find the solution to the simultaneous equations.

$$x + y = 5$$
$$y = 2x - 5$$

(3 marks)

AQA 2008

2 Solve the simultaneous equations.

$$y = 5x - 1$$
$$y = 2x^2 + 1$$

Do not use trial and improvement.
You **must** show your working. *(6 marks)*

AQA 2007

21 Cubic, circular and exponential functions

Objectives

Examiners would normally expect students who get these grades to be able to:

B

complete tables for, and draw graphs of, cubic functions and use the graphs to solve equations

A

complete tables for, and draw graphs of, reciprocal functions and use the graphs to solve equations

sketch and draw circular graphs such as $\sin x$, and $\cos x$

use the graphs to solve equations

A*

sketch and draw graphs of exponential functions and use them to solve equations

understand the graphs of circular functions for angles of any size

use symmetry of circular functions to solve equations

recognise the shapes of graphs of functions including cubic functions, reciprocal functions, circular functions and exponential functions

recognise functions when solving problems.

Did you know?

Sound waves

Sound travels in waves.

The shorter the wavelength, the higher the sound pitch.

The maximum displacement of the air molecules is called the amplitude.

The higher the amplitude, the louder the sound.

A pure tone is a tone with the shape of a sine function.

Key terms

cubic	circular functions
function	exponential
reciprocal	per annum
discontinuous	

You should already know:

- ✔ how to plot and interpret straight line and quadratic graphs
- ✔ how to use sine, cosine to calculate angles
- ✔ how to solve simultaneous equations graphically
- ✔ how to recognise and calculate proportional change.

Learn... 21.1 Cubic functions

The general form of a **cubic function** is $f(x) = ax^3 + bx^2 + cx + d$ where a, b, c and d are constants.

The highest term in a **cubic function** is a term in x^3.

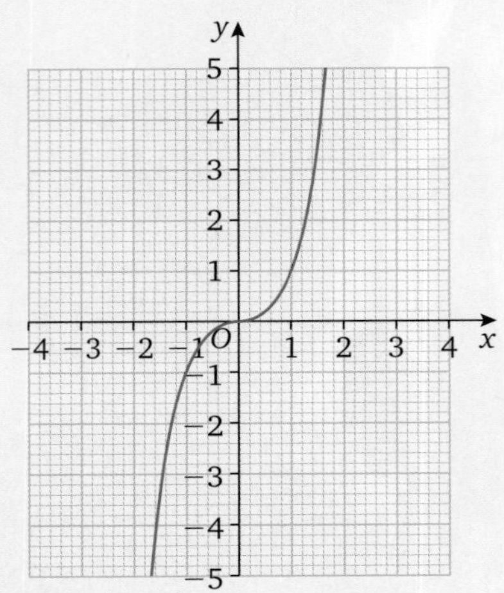

This diagram shows the graph of $y = x^3$, the simplest cubic function.

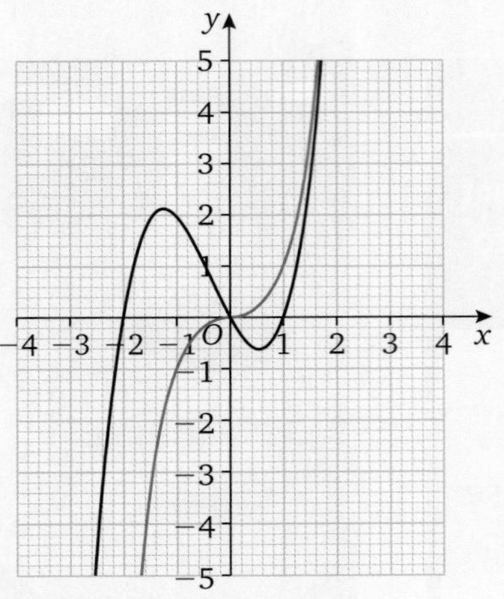

This diagram shows the graphs of $y = x^3$ in red and $y = x^3 + x^2 - 2x$ in black.

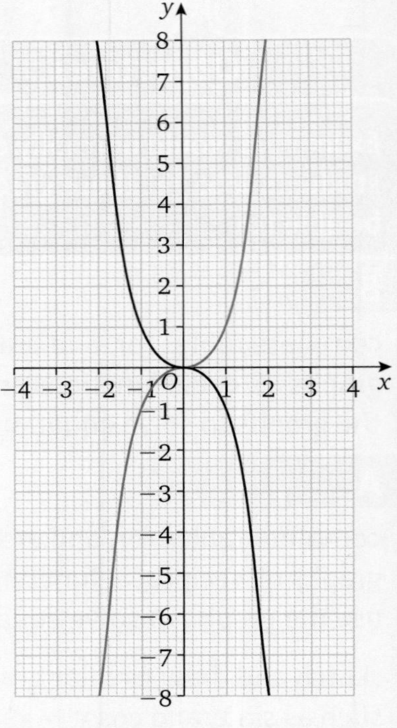

This diagram shows the graph of $y = x^3$ in red and the graph of $y = -x^3$ in black.

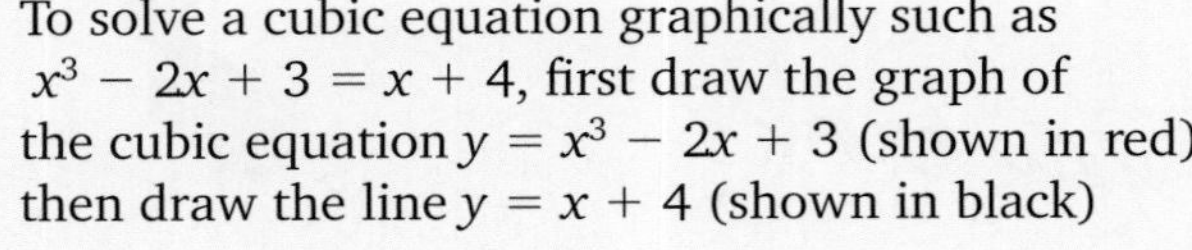

To solve a cubic equation graphically such as $x^3 - 2x + 3 = x + 4$, first draw the graph of the cubic equation $y = x^3 - 2x + 3$ (shown in red), then draw the line $y = x + 4$ (shown in black)

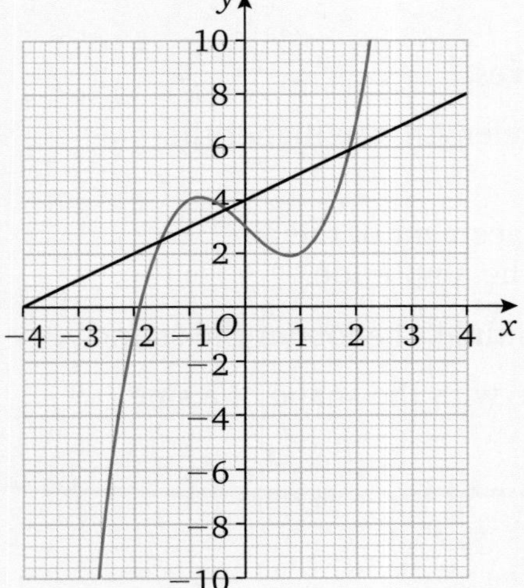

The solutions to the equation are the x-values of the points where they meet.

Example: **a** Copy and complete the table of values for $y = x^3 - 5$

x	−3	−2	−1	0	1	2	3
x^3	−27		−1	0	1		27
− 5	−5	−5	−5	−5	−5	−5	−5
y	−32		−6	−5	−4		22

b Use your table to draw the graph of $y = x^3 - 5$

c Use your graph to solve these equations.

i $x^3 - 5 = -10$

ii $x^3 - 5 = 0$

iii $x^3 - 5 = 6$

Solution: **a**

x	−3	−2	−1	0	1	2	3
x^3	−27	−8	−1	0	1	8	27
− 5	−5	−5	−5	−5	−5	−5	−5
y	−32	−13	−6	−5	−4	3	22

These values have been chosen so the graph fits the axes.

b Draw an x-axis labelled from −3 to 3 and a y-axis labelled from −40 to 30.

You need a different scale for each axis.

Plot all the points and join them with a smooth curve.

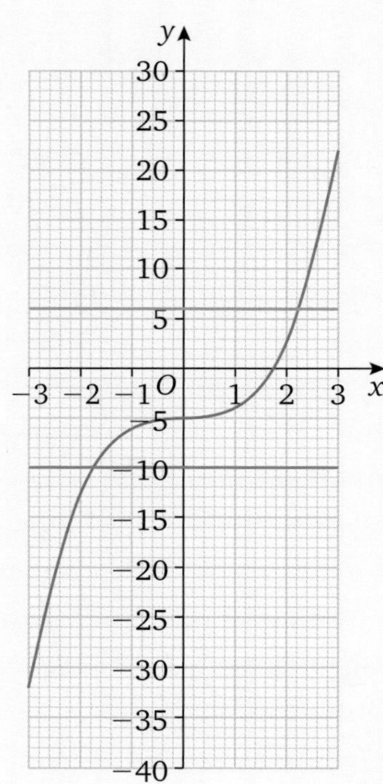

AQA Examiner's tip

Sometimes the line crosses the cubic graph more than once.

There may be one, two or three solutions to a cubic equation.

c i Draw the line $y = -10$ across the graph.
Read off the value of x where they cross.
The solution is $x = -1.7$

ii Read off where the graph crosses the line $y = 0$ (the x-axis).
The solution is $x = 1.7$

iii Draw the line $y = 6$ across the graph.
Read off the value of x where they cross.
The solution is $x = -2.2$

Practise... 21.1 Cubic functions 

D C B A A*

B

1 The diagram shows the graph of $y = x^3 - 2x^2 - 8x$

a Write down the coordinates of the points where the graph turns.

b Use the graph to solve the equation $x^3 - 2x^2 - 8x = 0$

> **Hint**
> There is more than one solution.

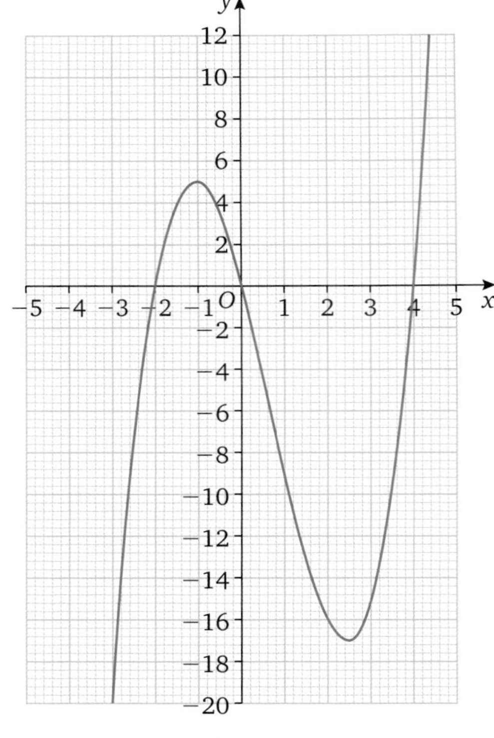

2 This diagram shows the graph of $y = x^3 - 2x^2 + x - 3$

Use the graph to solve these equations.

a $x^3 - 2x^2 + x - 3 = -4$

> **Hint**
> Read off the x-value when $y = -4$

b $x^3 - 2x^2 + x - 3 = 0$

c $x^3 - 2x^2 + x - 3 = 4$

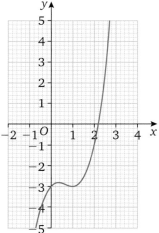

3 The diagram shows the graph of $y = x^3 + x^2 - 2x$

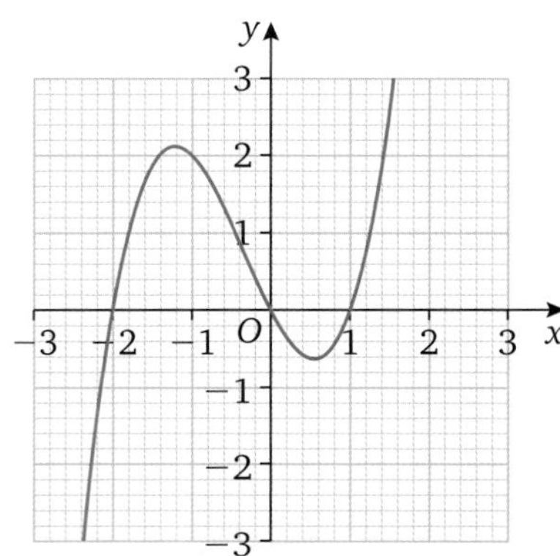

Use the graph to solve each of these equations.

a $y = x^3 + x^2 - 2x = 1$

b $y = x^3 + x^2 - 2x = 0$

c $y = x^3 + x^2 - 2x = -1$

B

4 **a** Copy and complete the table of values for $y = x^3 - 3x^2 + 4$ for values of x from -3 to $+5$.

x	−3	−2	−1	0	1	2	3	4	5
x^3		−8			1			64	
$-3x^2$		−12			−3			−48	
$+4$		+4			+4			+4	
y		−16			2			20	

b Draw the graph of $y = x^3 - 3x^2 + 4$ for values of x from -3 to $+5$.

c Use your graph to solve the following equations.

i $x^3 - 3x^2 + 4 = 0$

ii $x^3 - 3x^2 + 4 = -10$

iii $x^3 - 3x^2 + 4 = 30$

AQA Examiner's tip

You may be able to draw graphs without a table of values. However, if you do, be careful to avoid mistakes when calculating with negative values and powers.

AQA Examiner's tip

The size of the graph paper in an exam is just the right size to fit the graph.

A*

5 **a** Copy and complete the table of values for the cubic function $y = x^3 + 4x^2 - x - 4$ for values of x from -6 to $+1$.

x	−6	−5	−4	−3	−2	−1	0	1
y				8				0

b Draw the graph of $y = x^3 + 4x^2 - x - 4$

c Use the graph to solve these equations.

i $x^3 + 4x^2 - x - 4 = 0$

ii $x^3 + 4x^2 - x - 4 = -12$

6 At a theme park there is a water ride. The horizontal distance travelled from the start, x, and the vertical distance above the ground, y, are connected by the function:

$$y = \frac{-x^3}{3000} + \frac{3x^2}{40} - 5x + 150$$

where the horizontal distance is x feet and the corresponding height is y feet. The diagram shows part of the graph of this function.

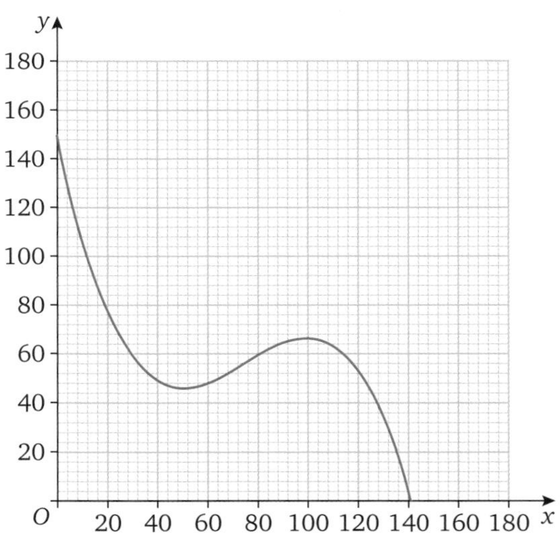

a Use the equation $y = \frac{-x^3}{3000} + \frac{3x^2}{40} - 5x + 150$ to find the value of y when $x = 50$

b Use the graph to find:

i the height of the ride when the horizontal distance is 50 feet

ii the horizontal distances when the height of the ride is 50 feet.

c Use your graph to find the horizontal distance when the ride goes below the water line.

d Malik says that a cubic function is not a good model for the whole of the theme park ride. Give your reasons why Malik is right.

Hint

Your answers to part **a** and part **b i** may not be the same but should be very close. If not, you may have made a mistake when substituting 50 into the equation.

Learn... 21.2 Reciprocal functions

The **reciprocal** of x is $\frac{1}{x}$ or x^{-1}.

$y = \frac{1}{x}$ is the simplest reciprocal function.

For this function, when $x = 0$ the value of y is undefined.

This happens because it is impossible to divide by 0.

So it is impossible to plot a point on the graph for $x = 0$

The value of y when x becomes very large is also undefined.

This happens because it is impossible to find a value of $\frac{1}{x}$ that equals 0.

This causes two breaks in the graph making it **discontinuous**, i.e. it doesn't continue.

There is a horizontal and vertical line that the graph does not cross.

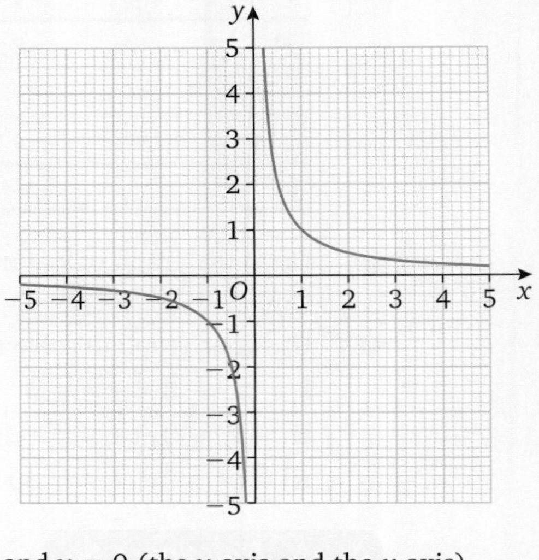

For the graph of $y = \frac{1}{x}$ the equations of these two lines are $x = 0$ and $y = 0$ (the y-axis and the x-axis).

This gives the reciprocal graph its distinctive shape.

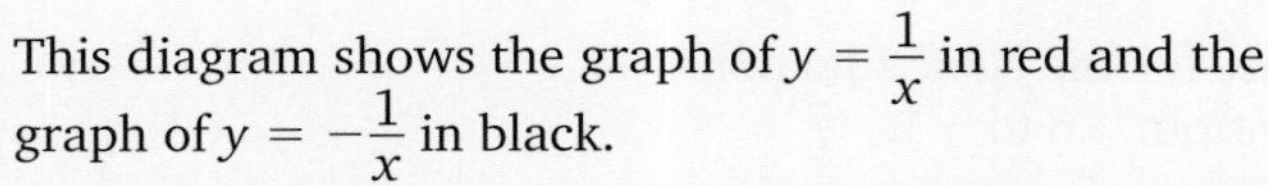

This diagram shows the graph of $y = \frac{1}{x}$ in red and the graph of $y = -\frac{1}{x}$ in black.

Notice the similarities and differences between the shapes of the two graphs.

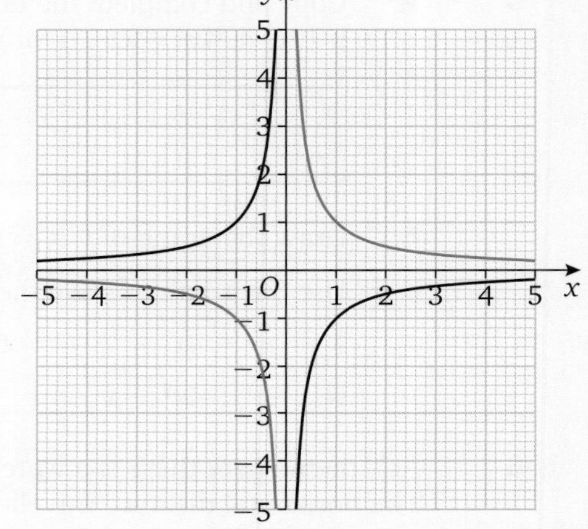

Example:

a Copy and complete the table of values for $y = \frac{10}{x}$ for $-6 \leqslant x \leqslant 6$

x	−6	−5	−4	−3	−2	−1	0	1	2	3	4	5	6
y	−1.67			−3.33				10		3.33			1.67

b Draw the graph of the reciprocal function $y = \frac{10}{x}$ for these values.

c From your graph, solve $\frac{10}{x} = 8$

Solution:

a Complete the table.

x	−6	−5	−4	−3	−2	−1	0	1	2	3	4	5	6
y	−1.67	−2	−2.5	−3.33	−5	−10	–	10	5	3.33	2.5	2	1.67

b and **c** Draw a pair of axes with values of x from -6 to 6 and values of y from -10 to 10.
Plot the points and join them with a smooth curve.
Draw the line $y = 8$ and read off the value of x at the point where they cross.
$x = 1.25$

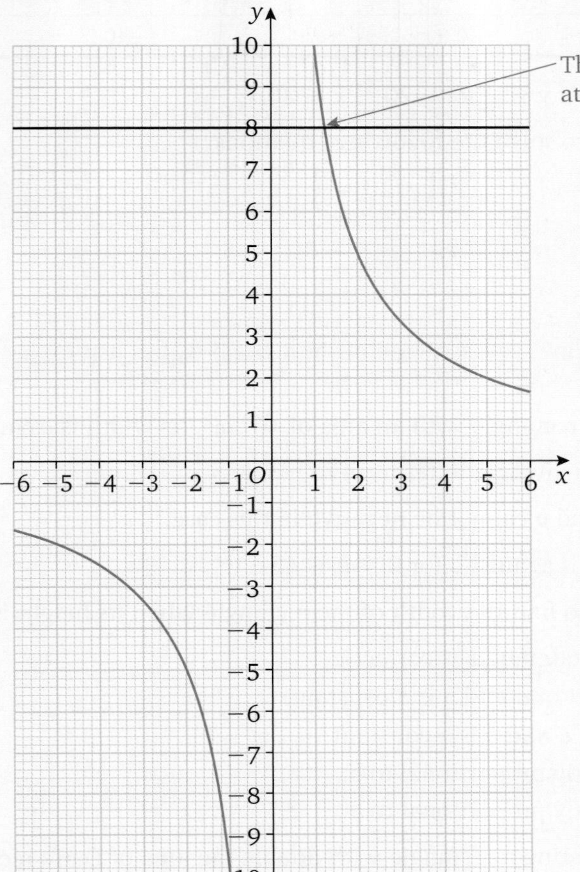

The solution is the x value at this point.

Practise... 21.2 Reciprocal functions

D C B A A*

1 This diagram shows the graph of $y = \frac{3}{x}$

a Use the graph to find solutions to these equations.

i $\frac{3}{x} = -2$

ii $\frac{3}{x} = 2$

iii $\frac{3}{x} = 0.5$

b Explain why you cannot find a solution to $\frac{3}{x} = 0$

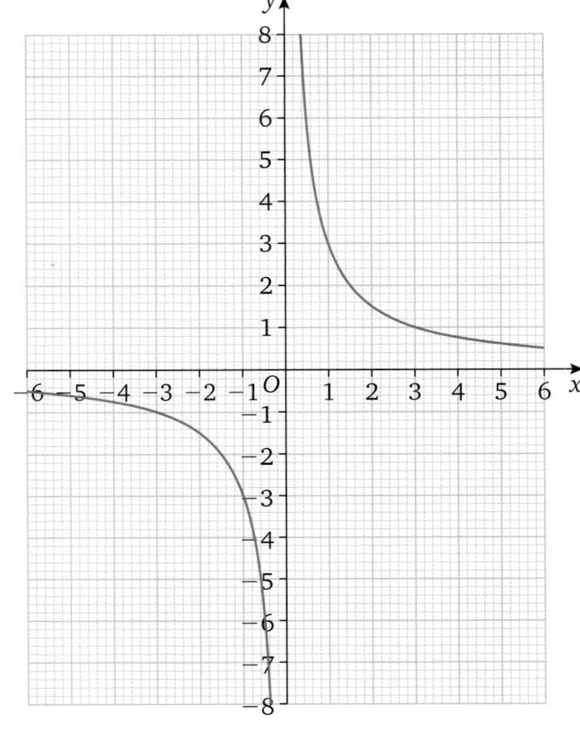

B

B

2 $y = \frac{4}{x}$ is a reciprocal function.

a Copy and complete the table of values for $y = \frac{4}{x}$

Give values to two decimal places where required.

x	−20	−15	−10	−5	−2	−1	−0.5	0	0.5	1	2	5	10	15	20
y				−0.8					40						0.2

b Draw the graph of $y = \frac{4}{x}$ for $-20 \leqslant x \leqslant 20$

c Use your graph to solve the following equations.

i $\frac{4}{x} = 4$ **iii** $20 = \frac{4}{x}$

ii $\frac{4}{x} = -3$ **iv** $\frac{4}{x} = -30$

Hint

Rather than draw a line, you can read off numerical values just for x.

3 Dan is investigating a rectangle with an area of 46 cm². Its width is a cm and its height is b cm.

a Find an equation giving a in terms of b.

b Explain why a and b cannot be negative in this case.

c Draw a graph of a against b for $0 \leqslant a \leqslant 48$

d Use your graph to find the width of the rectangle when its height is 9 cm.

e Which of these statements is correct?

i The product of a and b is a constant.

ii The sum of a and b is a constant.

iii a is indirectly proportional to b.

iv b is indirectly proportional to a.

f Becky is investigating a rectangle with double the area of Dan's rectangle.
Describe the similarities and differences between Dan's and Becky's graphs.

4 **a** On the same axes, sketch the graphs of $y = \frac{1}{x}$, $y = \frac{2}{x}$ and $y = \frac{3}{x}$

b Describe the transformation in the graph of $y = \frac{1}{x}$ when the numerator is increased.

Learn... 21.3 Circular functions

Circular functions are also known as trigonometric functions.

They can be used for angles of any size.

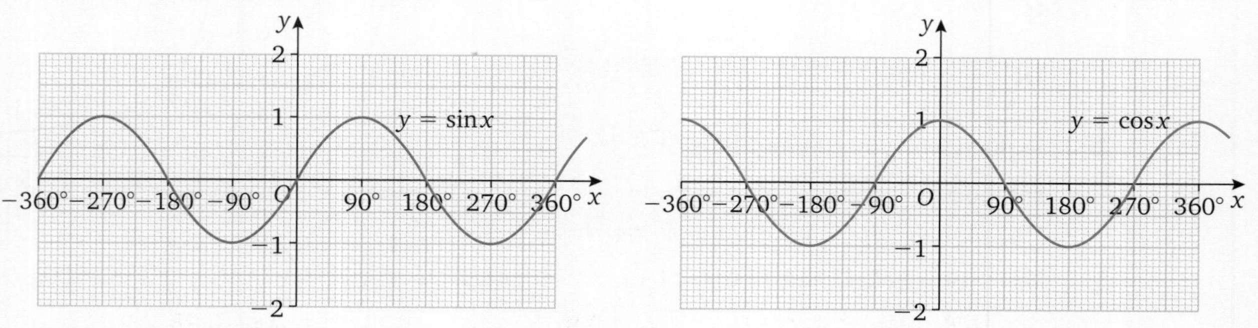

The graphs of $\sin x$ and $\cos x$ can be extended to infinity horizontally in both directions. They are wave curves as the values repeat themselves every 360°.

You can see from the diagrams that the maximum value of both $\sin x$ and $\cos x$ is +1 and the minimum value of both $\sin x$ and $\cos x$ is −1.

Example:

a Draw the graphs of $y = \sin x$ and $y = \cos x$ for $-270° \leqslant x \leqslant 270°$

b Use your graph to find all the solutions of these equations that lie within this range:

i $\sin x = 0.6$

ii $\cos x = -0.2$

Hint

You can use symmetry in all the graphs of trigonometric functions to help find solutions.

Solution:

a

x	−270°	−225°	−180°	−135°	−90°	−45°	0°	45°	90°	135°	180°	225°	270°
$\sin x$	1	0.71	0	−0.71	−1	−0.71	0	0.71	1	0.71	0	−0.71	−1

x	−270°	−225°	−180°	−135°	−90°	−45°	0°	45°	90°	135°	180°	225°	270°
$\cos x$	0	−0.71	−1	−0.71	0	0.71	1	0.71	0	−0.71	−1	−0.71	0

Join the points with a smooth curve.

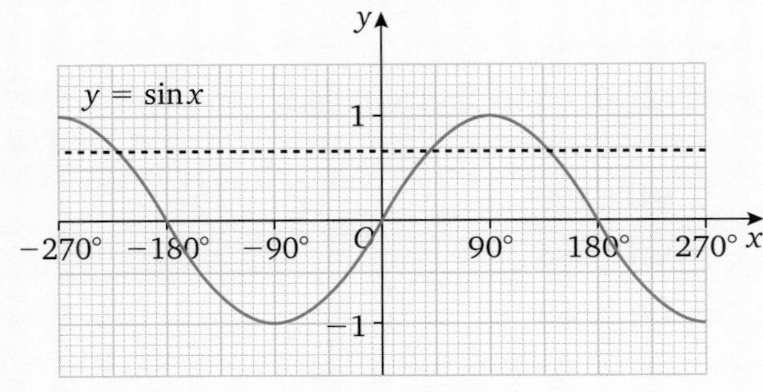

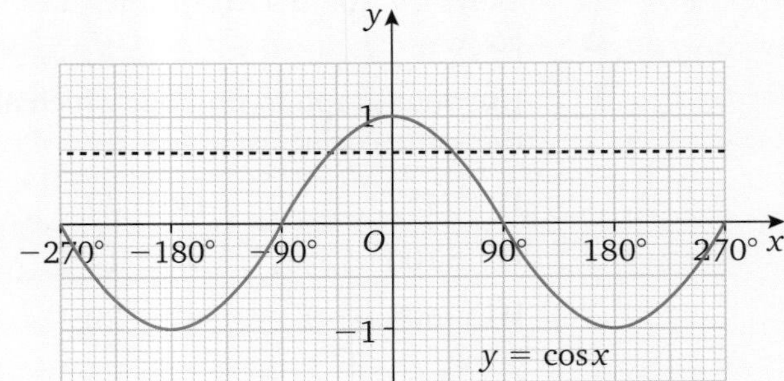

b **i** $y = 0.6$ at three points when $x = -217°$, $37°$ and $143°$

ii $y = -0.2$ at four points where $x = 102°$, $258°$, $-102°$ and $-258°$

These are accurate solutions. If you draw a graph carefully you should get readings within a few degrees of these solutions.

Practise... 21.3 Circular functions 

A

1 **a** Copy and complete the table of values for $\sin x$.

x	−360°	−270°	−180°	−90°	−45°	0°	45°	90°	180°	270°	360°
$\sin x$	0	1			−0.71				0	−1	

b Draw the graph of $\sin x$ for $-360° \leqslant x \leqslant 360°$

c Use your graph to find all the values of x for which $\sin x$ is:

i 0.5 **ii** 0.14 **iii** 0.8 **iv** 0.71 **v** 0.98

AQA Examiner's tip

Learn the shapes of the trigonometric functions so you know what they should look like before you draw them.

A

2 This diagram shows the graph of $\cos x$ for $-360° \leqslant x \leqslant 360°$

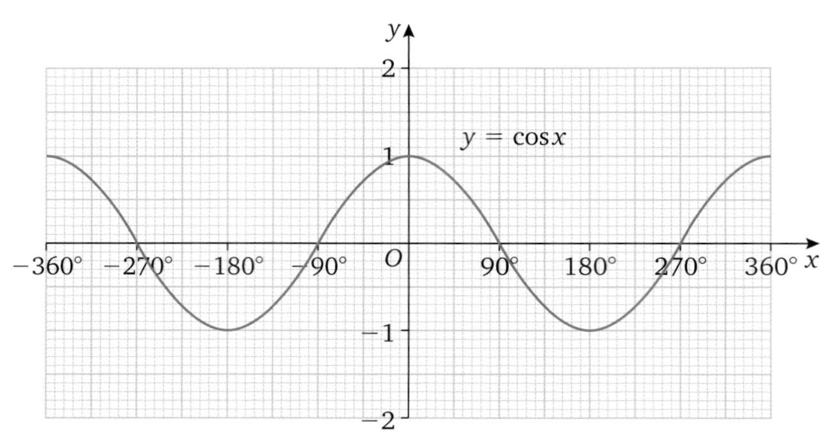

Use the diagram to solve the following equations.

a $\cos x = 0.7$

b $\cos x = -0.5$

c $\cos x = 0$

d $\cos x = 1$

e $\cos x = -1$

f Use your graph to find out which angles between $-360°$ and $360°$ have the same cosine as:

i	45°	**iv**	30°	**vii**	198°
ii	100°	**v**	14°	**viii**	286°
iii	60°	**vi**	225°	**ix**	352°

3 **a** Draw the graphs of $\cos x$ and $\sin x$ on the same axes for $0° \leqslant x \leqslant 540°$

b Use your graphs to solve the equation $\sin x = \cos x$ for $0° \leqslant x \leqslant 540°$

c Ruth says that $\sin 50° = \cos 40°$ and that $\cos 328° = \sin 58°$. She used her graphs to work it out.
Is she right? Give reasons for your answer.

d Use your graphs to decide which of these statements are true:

i	$\sin x = \sin(180° - x)$	**v**	$\sin x = -\cos x$
ii	$\cos x = \cos(180° - x)$	**vi**	$\sin x = \cos(90° + x)$
iii	$\sin x = \sin(360° - x)$	**vii**	$\sin x = \cos(180° - x)$
iv	$\cos x = \cos(360° - x)$	**viii**	$\sin x = -\sin(-x)$

In each case, give an example to illustrate your answer.

4 **a** Copy and complete the table of values for $y = \cos 2x$ for $0° \leqslant x \leqslant 360°$

x	0°	30°	60°	90°	120°	150°	180°	210°	240°	270°	300°	360°
$\cos 2x$	1	0.5		−1		0.5			−0.5		−0.5	

b Draw the graph of $y = \cos 2x$ for $0° \leqslant x \leqslant 360°$

c Compare the graph of $y = \cos 2x$ with what you know about the graph of $\cos x$. How is it different?

Learn... 21.4 Exponential functions

An **exponential** function is one where the base is a constant and the power is a variable.

The graph of $y = 2^x$ is shown below.

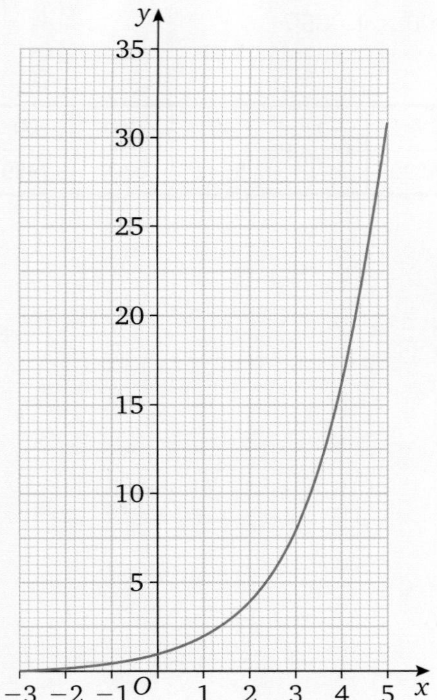

AQA Examiner's tip

If the axes are drawn for you in an examination, your graph should just fit onto them. If points go over the edge of the grid, then you have made an error in your working.

Note that $2^0 = 1$ so the graph goes through (0, 1).
This is true for any number, not just 2.
So the graph of $y = 5x$ also goes through (0, 1).

Exponential functions often occur in real situations. For example, they are used to model the growth of bacteria or to model radioactive decay. They are also relevant to the world of finance, as shown in the next example.

Example: Troy invests £4000 in a fixed interest bank account.
The interest rate is $3\frac{1}{2}\%$ **per annum**.

a Show that the value, V, of the investment after x years, can be written as $V = 4000 \times 1.035^x$

b Copy and complete this table of values showing the value of the investment over 20 years.

x (years)	0	1	2	3	4	5	10	15	20
V (£)	4000			4435		4751		6701	

c Draw a graph of V against x for $0 \leqslant x \leqslant 20$

d Explain why neither V nor x can be negative.

e Use your graph to:

i give the value of the investment after 18 years

ii find out how many years it will take for Troy's investment to grow by 50%

Solution:

a After 1 year the value of the investment is £4000 × 1.035

After 2 years the value of the investment is
£(4000 × 1.035) × 1.035 = 4000 × 1.035^2

After 3 years the value of the investment is
£((4000 × 1.035) × 1.035) × 1.035 = 4000 × 1.035^3

After y years the value of the investment is £4000 × 1.035^x

Therefore, $V = 4000 \times 1.035^x$

b

x (years)	0	1	2	3	4	5	10	15	20
V (£)	4000	4140	4285	4435	4590	4751	5642	6701	7959

c

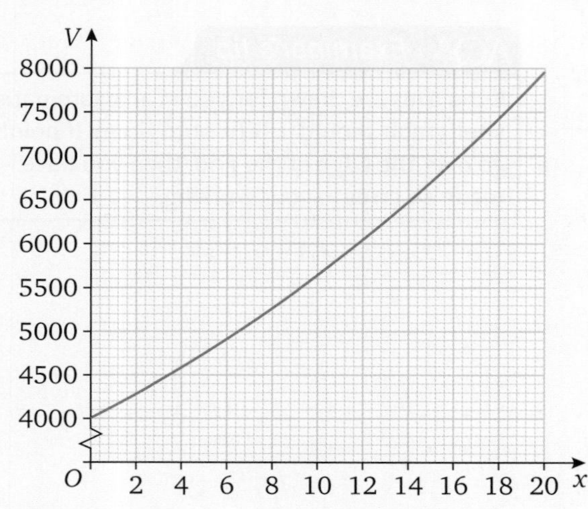

d Time cannot be negative so the value of x must always be positive.
Troy starts with £4000 and his investment increases so the value, V, will always be positive.

e **i** The value of the investment after 18 years is £7400.
ii Read off the value of x when $V = 6000$. It takes 12 years.

Practise... 21.4 Exponential functions

A*

1 This diagram shows part of the graph of $y = 5^x$

a Use the graph to estimate the value of:

i $5^{0.5}$

ii $5^{1.5}$

Remember that the powers are the x-values.

Use the graph to solve the following equations:

b **i** $5^x = 1$

ii $5^x = 10$

iii $5^x = 20$

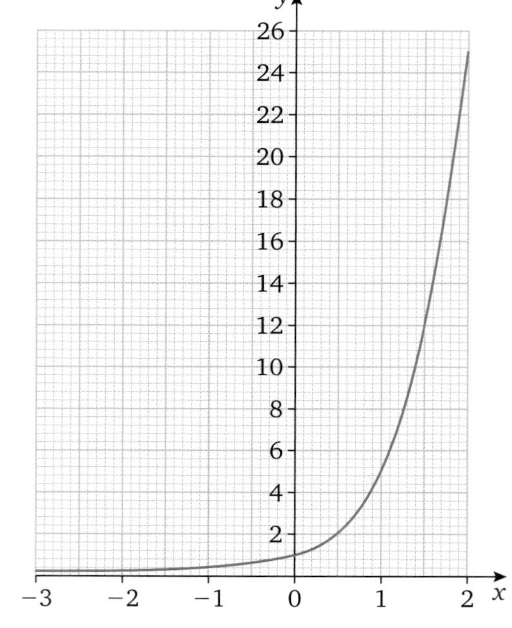

A*

2 **a** Copy and complete this table of values for the exponential function $y = 3^x$ for values of x from -3 to $+2$.

x	-2	-1	0	0.5	1	1.5	2
y	0.1			1.7	3	5.2	

b Draw the graph of $y = 3^x$ for values of x from -2 to 2.

c Use your graph to estimate the value of:

i $3^{1.2}$ **ii** $3^{0.3}$

d Use your graph to solve the equations:

i $3^x = 4$ **ii** $3^x = 2$ **iii** $3^x = 0.5$

3 **a** On a pair of axes with values of x from -3 to 4 and values of y from 0 to 10, sketch the graphs of $y = 2^x$ and the graph of $y = 5^x$

b Write down the coordinates of the point where they cross.

c Write down the solution to the equation $2^x = 5^x$

d Without sketching the graphs of any more functions, give another exponential function that passes through the same point as in **b**.

4 Claire invests £5000 in an account paying $2\frac{1}{2}$% per annum. The interest is added on at the end of each year.

a Write down an equation for v in terms of x where v is the value of the investment and x is the number of years it is invested.

b Draw a graph of the equation for values of x from 0 to 20.

c Use your graph to estimate the value of the investment after:

i 12 years

ii 17 years

Learn... 21.5 Graph recognition and graphs of loci

You should be able to recognise the sketch graphs of different functions.
This comes partly from the shape of the graph and partly from its position on the axes.
For example: from its shape you can tell that this is a quadratic graph.

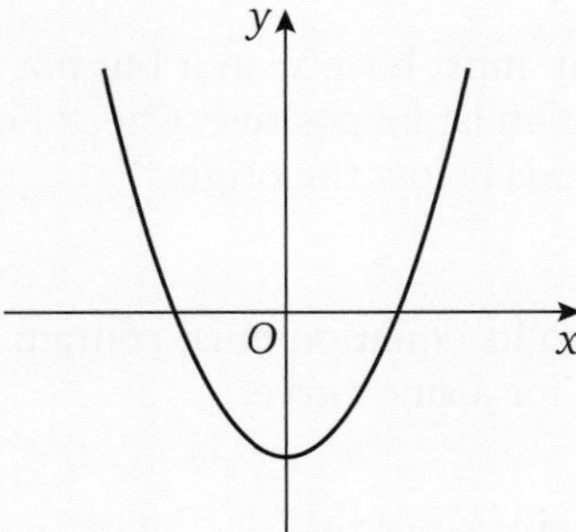

You may be asked to decide which of these could be the equation of this graph.

A $y = x^2$ B $y = 3 - x^2$ C $y = x^2 - 3$ D $y = 3x^2$

A and D both go through (0, 0) so they can be ruled out.

The graph is U-shaped so its equation contains $(+)x^2$ rather than $-x^2$.

This is the graph of C: $y = x^2 - 3$

Use your knowledge of loci to work out the equation of this graph.

If $P(x, y)$ is a point on the circle, then $x^2 + y^2 = r^2$, where r is the radius of the circle. This graph is the locus of points that are a distance r units from the origin.

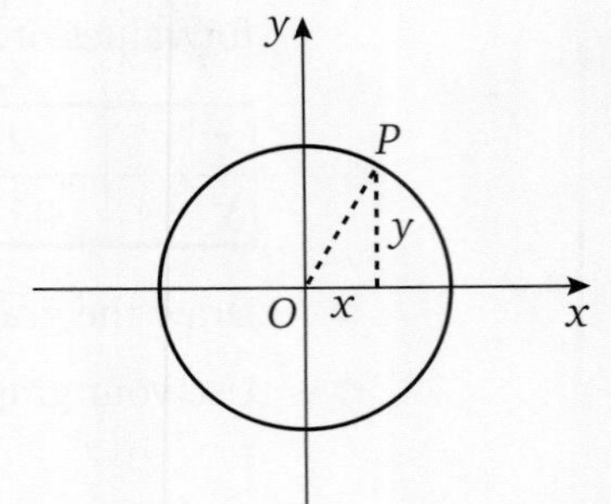

AQA Examiner's tip

You will not be tested on the equation of a circle in your exam.

Example: Here are four sketch graphs and six equations.

Match each graph to its equation, giving reasons for your choice.

Graph A

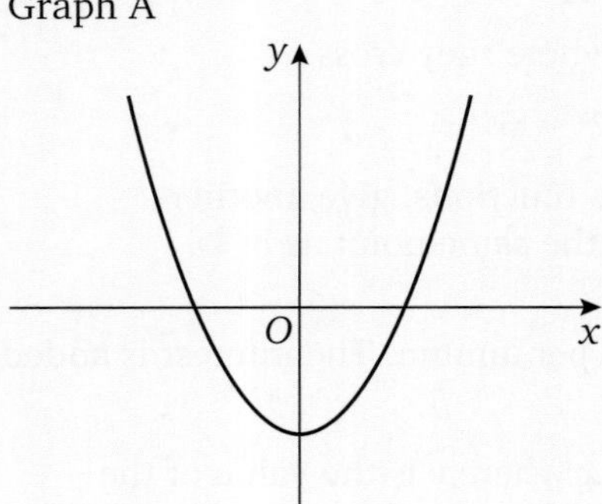

Graph B

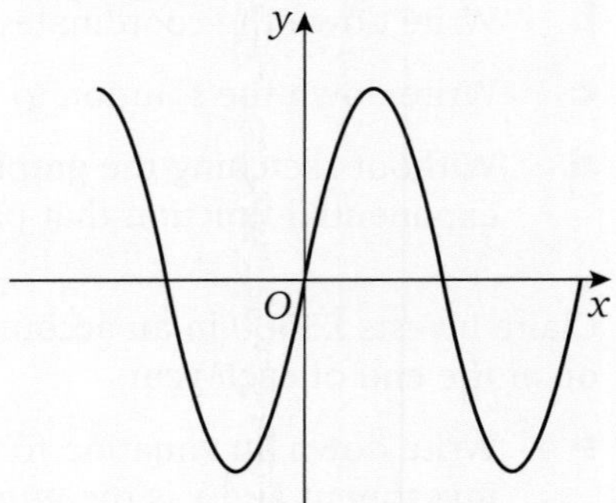

Graph C

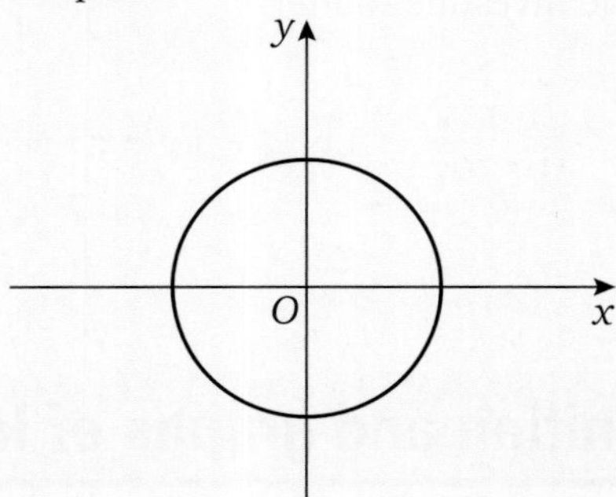

Graph D

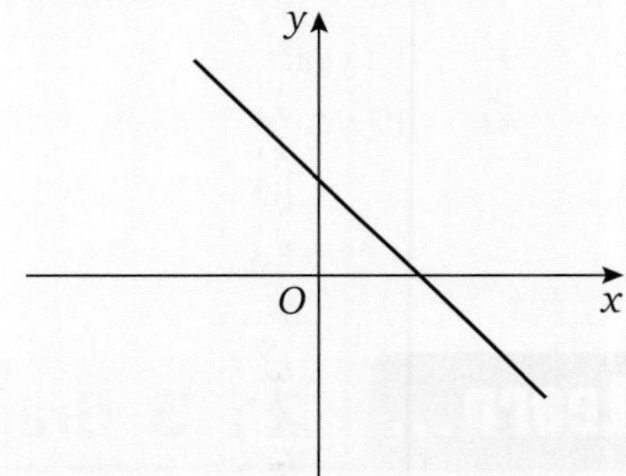

Equation 1: $y = 5 - x$ Equation 2: $y = x^2 - 5$ Equation 3: $x^2 + y^2 = 5$

Equation 4: $y = 5x$ Equation 5: $y = \frac{x^3}{5}$ Equation 6: $y = 5\sin x$

Solution: Graph A is a parabola so its equation must have x^2 in it but not any higher powers of x
It is U-shaped so the coefficient of x^2 must be positive. Check Equation 2: when $x = 0$, $y = -5$ and Graph A crosses the y-axis below the origin.
Graph A is $y = x^2 - 5$

Graph B is a repeating wave curve so its equation must contain either $\sin x$ or $\cos x$.
It goes through (0, 0) which is true for a sine curve.
Graph B is $y = 5\sin x$

Graph C is a circle with its centre at the origin. It is therefore of the form $x^2 + y^2 = r^2$
Graph C is $x^2 + y^2 = 5$

Graph D is a straight line. Equation 1 and Equation 4 are linear. Equation 4 goes through (0, 0) and has a positive gradient. Equation 1 has a negative gradient and does not go through (0, 0)
Graph D is $y = 5 - x$

Practise... 21.5 Graph recognition and graphs of loci

A*

1 Match each function with its graph.

A: $y = \frac{-2}{x}$ B: $y = 2\sin x$ C: $y = x^3 - 2$ D: $y = x^2 - 5x$

a

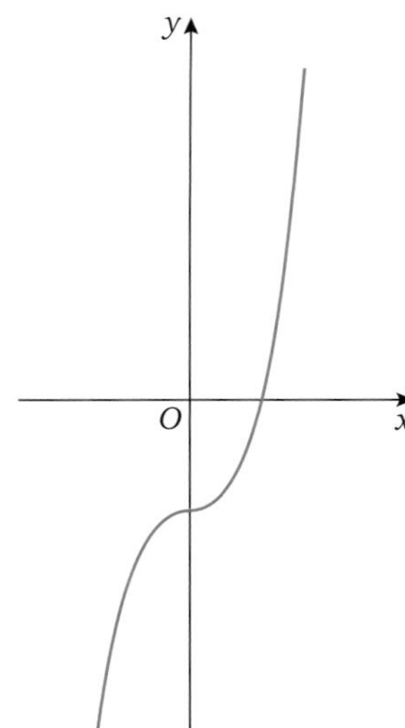

b

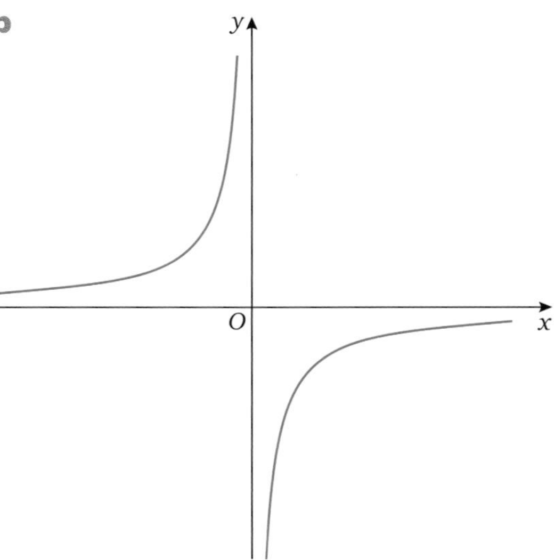

c

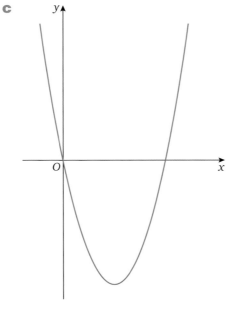

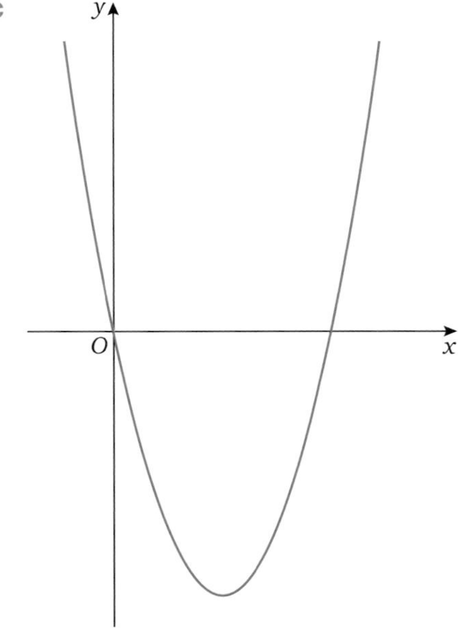

d

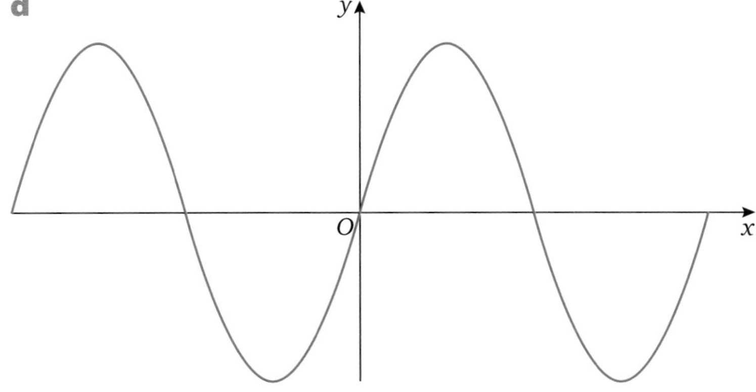

A*

2 Match each function with its graph.

A: $y = 3x^2 - 2x + 1$ B: $y = 3\cos x$ C: $y = \dfrac{3}{x}$ D: $y = 3^x$

a

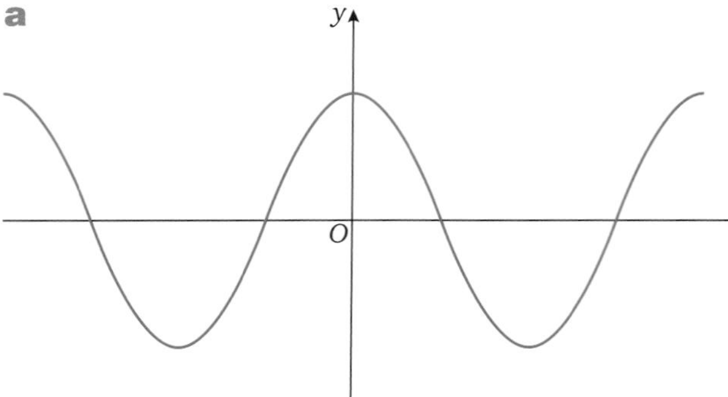

b

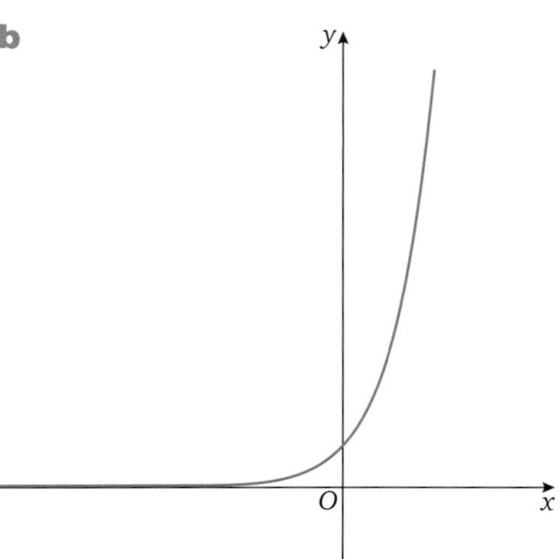

c

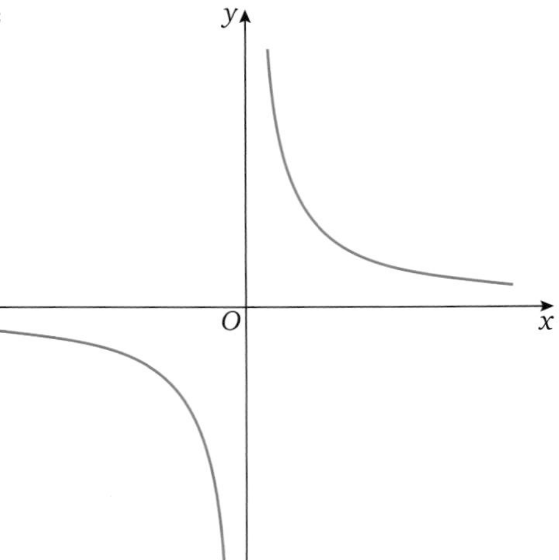

d

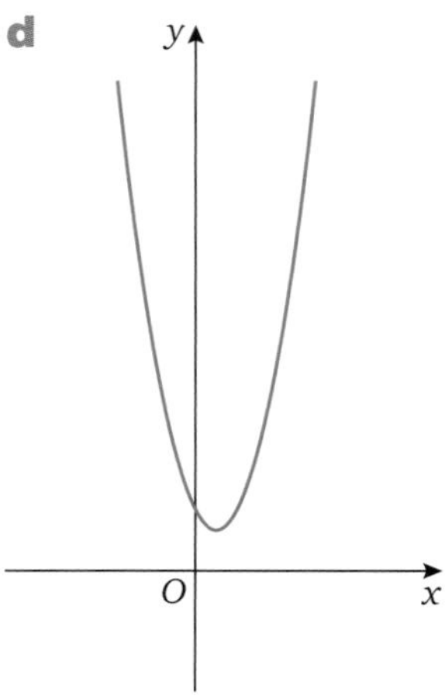

3 Anna says this is the graph of $y = 4 - x^3$

Bindia says its is the graph of $y = x^3 - 4x$

Chris says its is the graph of $y = 4x^2 - x^3$

Dave says its is the graph of $y = x^3 + 4$

Who is correct?

Explain why each of the other students is wrong.

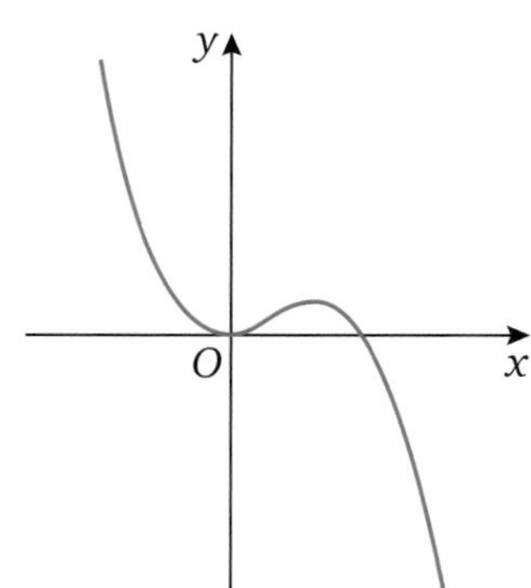

4 The sketch graph shows four curves, P, Q, R and S.

Match each curve to its equation.

a $y = (1.5)^x$ is curve …

b $y = \dfrac{1}{x}$ is curve …

c $y = 3^x$ is curve …

d $y = \dfrac{5}{x}$ is curve …

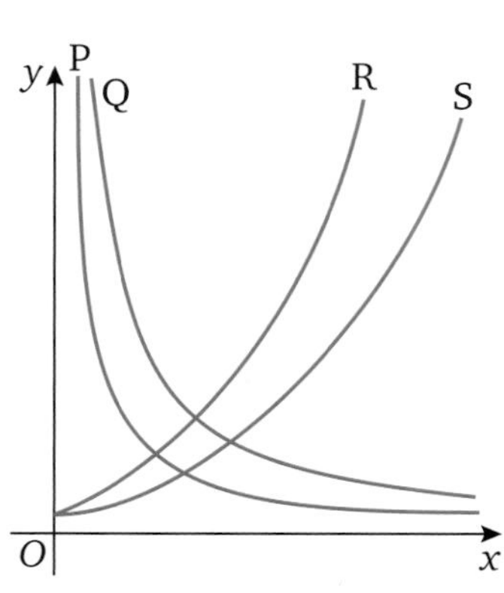

5 a Draw the graph of $x^2 + y^2 = 25$

b By drawing a line across your graph, find the points that are 5 units from the origin and satisfy the equation $y = 2x - 3$

A*

21 Assess

1 a Copy and complete this table of values for $y = -x^3 - 5$ for values of x from -3 to 3.

B

x	−3	−2	−1	0	1	2	3
$-x^3$	27		1		−1		−27
−5		−5	−5		−5		−5
y			−4		−6		−32

b Use the table to draw the graph of $y = -x^3 - 5$

c Use the graph to solve these equations.

i $-x^3 - 5 = 0$ ii $-x^3 - 5 = -10$ iii $-x^3 - 5 = 15$

2 This diagram shows the graph of $y = \frac{6}{x}$

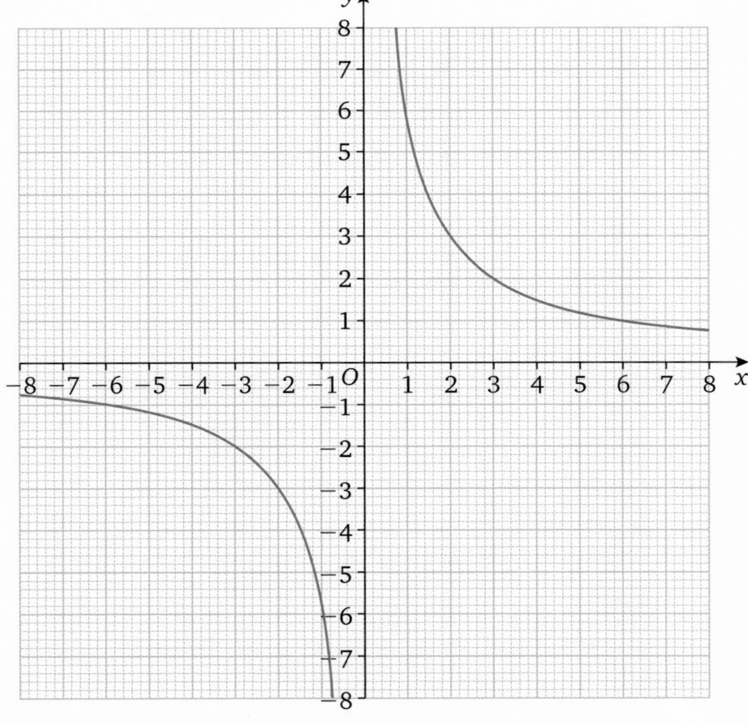

a Use the graph to find solutions to these equations:

i $\frac{6}{x} = -3$ ii $\frac{6}{x} = 5$ iii $\frac{6}{x} = 5\ -0.5$

b Explain why you cannot find a solution to $\frac{3}{x} = 0$

A

3 This diagram shows the graph of $\sin x$ for $-360° \leqslant x \leqslant 360°$

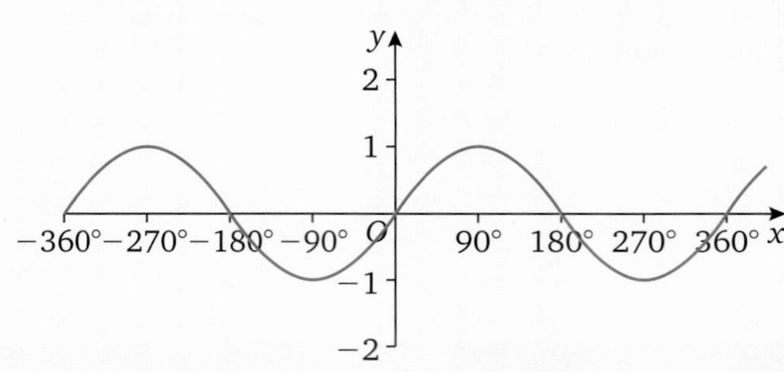

a Use the diagram, with your calculator, to find all the solutions of these equations in the range $-360° \leqslant x \leqslant 360°$.

i $\sin x = 0.9$

ii $\sin x = -0.5$

iii $\sin x = 0.2$

iv $\sin x = 1$

v $\sin x = -1$

b Use the graph to find out which angles between −360° and 360° have the same sine as:

i 45°

ii 80°

iii 20°

iv 60°

v 225°

vi 315°

A*

4 This diagram shows the graph of $y = x^3 - 3x^2 - 4x + 12$

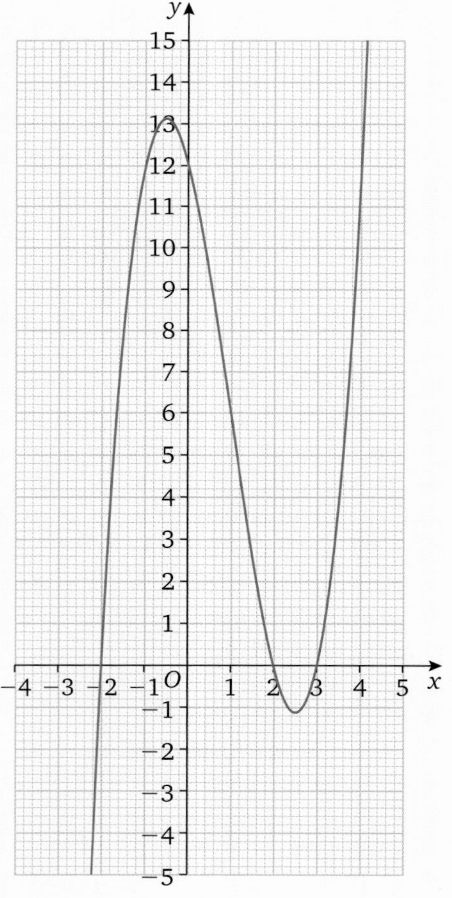

a Use the graph to solve $x^3 - 3x^2 - 4x + 12 = 0$

b $x^2 - 3x^2 - 4x + 12 = (x - a)(x - b)(x - c)$

Use the graph to find the values of a, b and c.

A*

5 This diagram shows part of the graph of $y = 4^x$

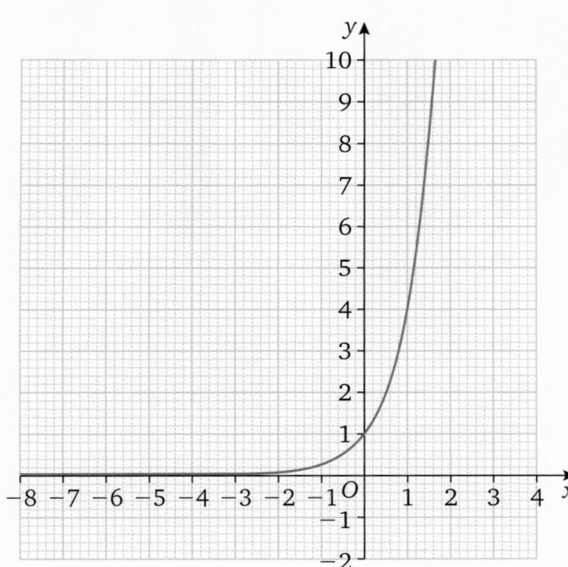

a Use your graph to estimate the value of:

i $4^{-0.5}$

ii $4^{1.5}$

iii $4^{0.75}$

b Use the graph to solve the following equations.

i $4^x = 0.5$

ii $4^x = 1$

iii $4^x = 5$

6 **a** By drawing the graphs of $y = \sin 2x$ and $y = \cos 2x$ for $-180° \leqslant x \leqslant 270°$, solve the equation $\sin 2x = \cos 2x$

b Without drawing a graph, find the next positive solution for the range of values greater than 270°.

7 **a** Complete the table of values for the cubic function $y = x^3 - 3x - 2$ for values of x from −4 to 4.

x	−4	−3	−2	−1	0	1	2	3	4
y		−20		0					50

b Draw the graph of $y = x^3 - 3x - 2$ for values of x from −4 to 4.

c Use the graph to solve the following equations.

i $x^3 - 3x - 2 = 0$

ii $x^3 - 3x - 2 = 20$

iii $x^3 - 3x - 2 = -30$

A*

8 Match each function with its graph.

A: $y = -2^x$ B: $y = \frac{-5}{x}$ C: $y = \cos 3x$ D: $y = 2x^3$

a

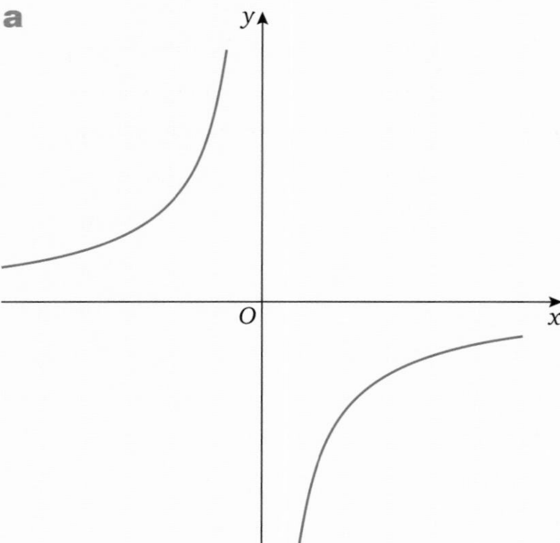

c

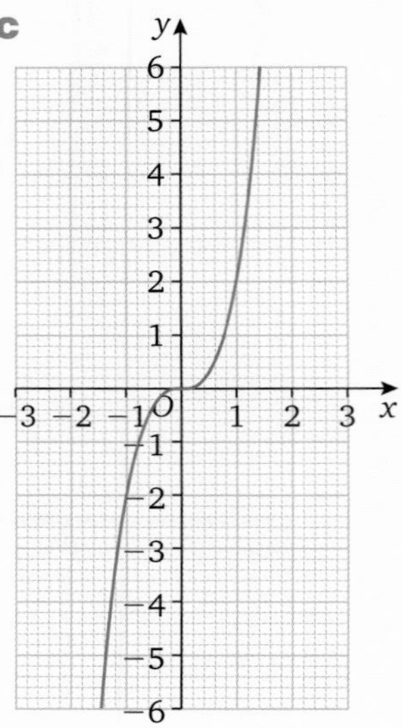

b

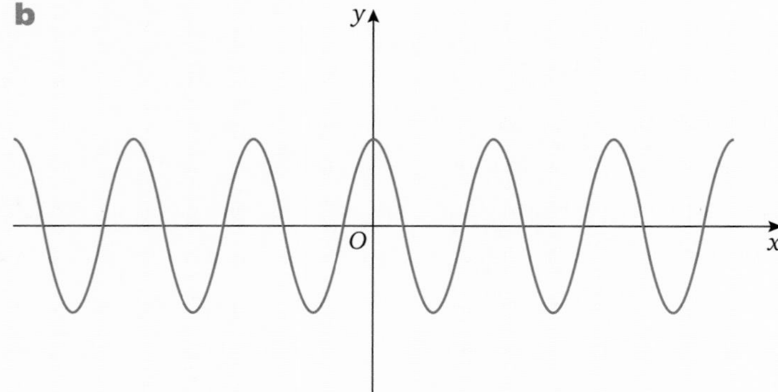

d

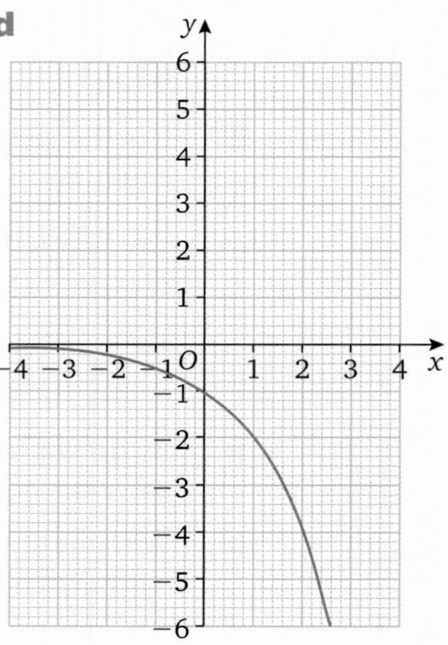

9 This table shows some of Sumaira's experimental data. Some of the values are missing.

x	0	1	2	3	4	5	6	7	8	9	10
y			14.4		20.7	24.9	29.9	35.8			61.9

a Plot these points and join them to make a smooth curve.

b Use your curve to complete the table of values.

c Describe the relationship between x and y.

d Use your graph to estimate:

i the value of y when x is 0

ii the value of y when x is 9

iii the value of x when y is 40.

AQA Examination-style questions

1 **a** Complete the table of values for $y = x^3 - x$

x	-2	-1.5	-1	-0.5	0	0.5	1	1.5	2
y		-1.875	0	0.375	0	0.375		1.875	

(2 marks)

b Copy the grid and draw the graph of $y = x^3 - x$ for values of x from -2 to -2. *(3 marks)*

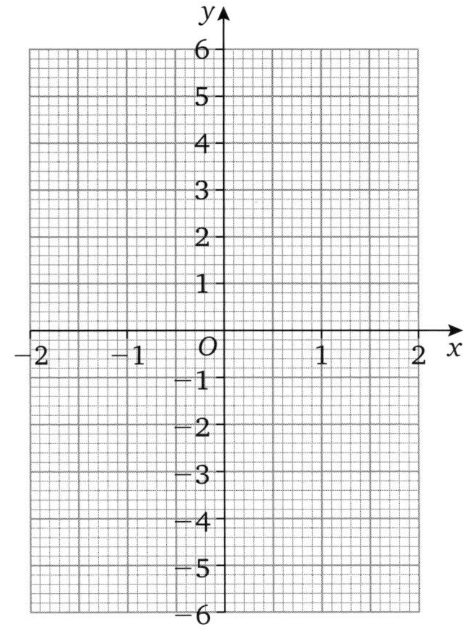

c Use the graph to solve the equation $x^3 - x = 4$ *(2 marks)*

AQA 2008

22 Trigonometry 2

Objectives

Examiners would normally expect students who get these grades to be able to:

A

use the sine and cosine rules to solve 2-D problems

calculate the area of a triangle using $\frac{1}{2}ab\sin C$

A*

use the sine and cosine rules to solve 3-D problems.

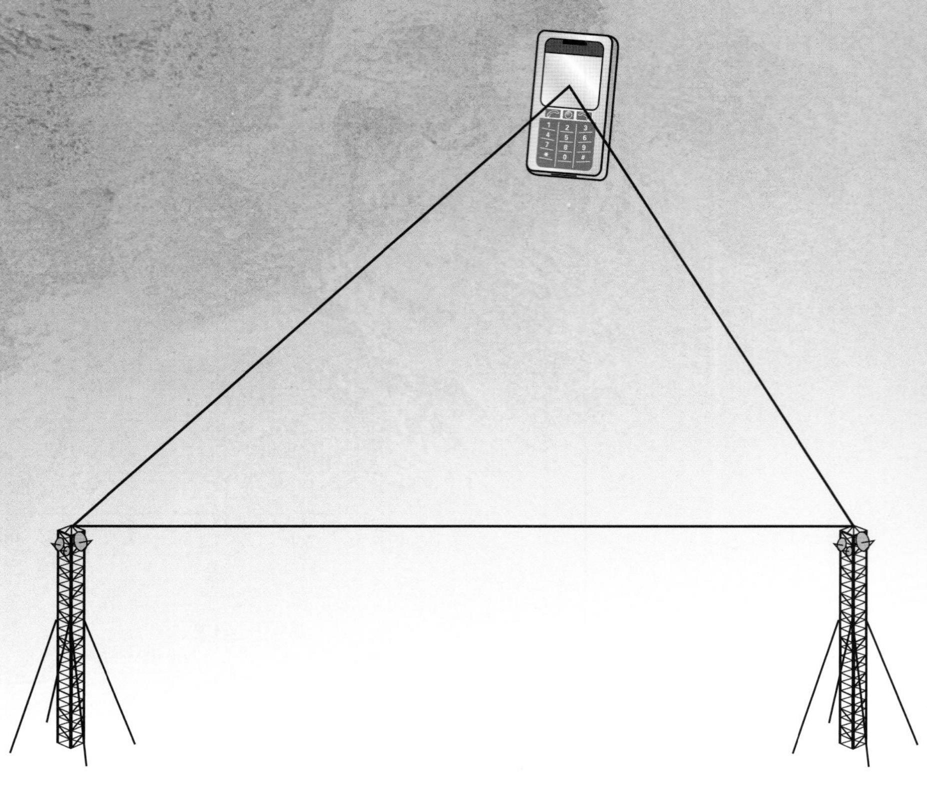

Did you know?

Finding someone in an emergency using mobile phone signals

An emergency call from a mobile phone can be used to help find the person making the call.
The distance from two masts can be calculated by the time it takes the signal to reach them.
Assuming that you know the distance between the two masts, then the cosine rule covered in this chapter allows you to calculate the angles in the triangle.
This system of triangulation is used in GPS systems as well as astronomy.

Key terms

sine rule
cosine rule

You should already know:

- ✔ how to use trigonometry in right-angled triangles
- ✔ the formula for the area of a triangle, $A = \frac{1}{2}bh$
- ✔ how to measure angle bearings.

22.1 The sine rule

It is possible to use trigonometry for triangles without a right angle.

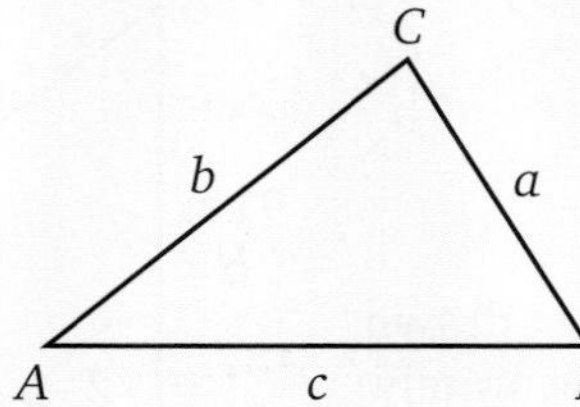

You should use upper case (capital) letters to label the angle at each vertex and lower case letters to label sides.

The sides are named after the angle they are opposite.

The perpendicular height, CD, is drawn in to create two right-angled triangles.

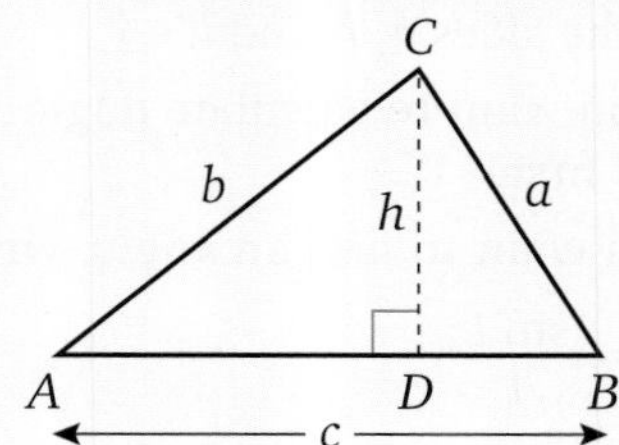

Using triangle ACD:

$\sin A = \frac{\text{opp}}{\text{hyp}}$

$\sin A = \frac{h}{b}$

or $b \sin A = h$

Using triangle BCD:

$\sin B = \frac{\text{opp}}{\text{hyp}}$

$\sin B = \frac{h}{a}$

or $a \sin B = h$

So $a \sin B = b \sin A$

Dividing both sides by $\sin A \times \sin B$,

$$\frac{a \sin B}{\sin A \times \sin B} = \frac{b \sin A}{\sin A \times \sin B}$$

$$\frac{a}{\sin A} = \frac{b}{\sin B}$$

By constructing a different perpendicular, it can be shown that:

$$\frac{a}{\sin A} = \frac{b}{\sin B} = \frac{c}{\sin C}$$

This is known as the **sine rule**.

> **AQA Examiner's tip**
>
> This formula is given to you in the formula page of exam papers.
> It can be turned upside down, giving:
> $\frac{\sin A}{a} = \frac{\sin B}{b} = \frac{\sin C}{c}$
> which is useful for finding angles, but this form is not given in the formula sheet.

Don't forget that $\sin 40° = \sin 140°$, or, generally, that $\sin a = \sin (180° - a)$

This can lead to two different answers to a question, as shown in the 'Finding an angle' example on the following page.

See Chapter 21 for the graph of $\sin x$, which shows this.

Finding a side

Example: A triangle ABC has side $AB = 8$ cm, angle $ACB = 74°$ and $CAB = 59°$

Calculate the length of BC correct to one decimal place.

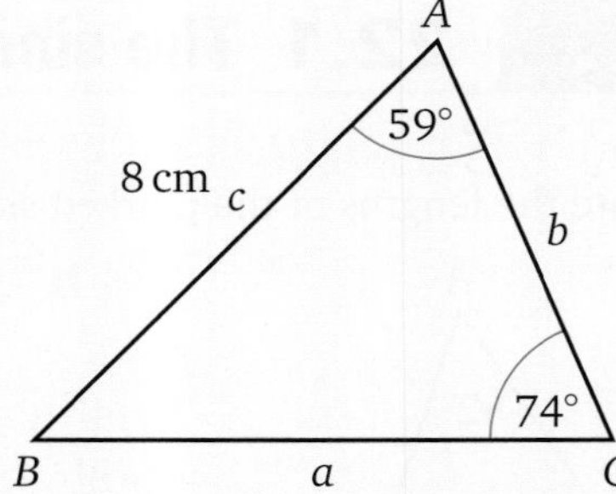

Solution: You know angles A and C, and side c. You are asked to find side a.

So use the parts of the rule containing these measurements.

$$\frac{a}{\sin A} = \frac{c}{\sin C}$$

$$\frac{a}{\sin 59°} = \frac{8}{\sin 74°}$$ Substitute values for c, A and C.

$$a = \frac{8}{\sin 74°} \times \sin 59°$$ Multiply both sides by sin 59°.

$= 7.1336...$ Calculate the answer.

$a = 7.1$ cm (1 d.p.) Round off to the required degree of accuracy.

Finding an angle

Example: Ship A is 9 km due north of a lighthouse, L.

Ship B is 11 km from ship A, and is on a bearing of 072° from the lighthouse.

Calculate the bearing of B from A.

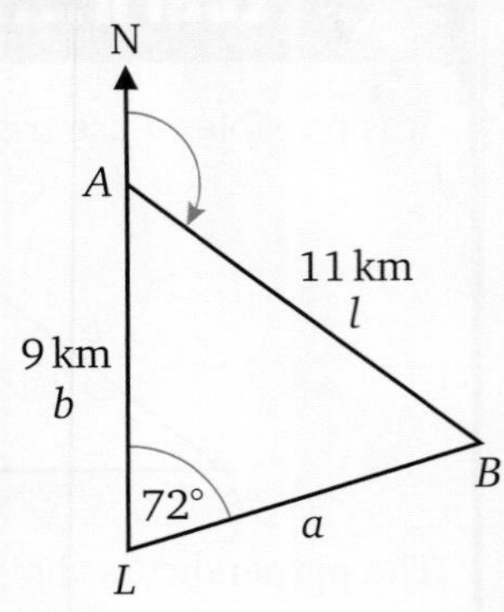

Not drawn accurately

Solution: First, label the sides a, b and l.

Although you want to calculate angle A, you know sides b, l and angle L, so you have to find angle B first.

Because you want to find an angle, write the rule with the angles as the numerator.

$$\frac{\sin B}{b} = \frac{\sin L}{l}$$

$$\frac{\sin B}{9} = \frac{\sin 72°}{11}$$

$$\sin B = \frac{\sin 72°}{11} \times 9$$

$$= 0.7781...$$

$$B = \sin^{-1} 0.7781...$$

$$B = 51° \text{ (to the nearest degree)}$$

So angle $LAB = 180° - 51° - 72° = 57°$

The bearing of B from $A = 180° - 57° = 123°$.

$\sin B = 0.7781...$ has two possible answers: $B = 51°$ or 129° (to the nearest degree). In this case, 129° is impossible as the angles in the triangle must add to 180°, but you should always check. There is another reason why, in this example, B must be an acute angle. The angle B must be less than 72° because B is opposite the side 9 km and L is opposite the side 11 km.

AQA Examiner's tip

When you use the inverse sine function on a calculator, the angle displayed is an acute angle. To get the obtuse angle which has the same sine, you need to subtract this value from 180°. This is because $\sin x = \sin (180° - x)$.

Practise... 22.1 The sine rule 

A

1 Calculate the lengths of the marked sides in the diagrams below.

a A, 52°, 9 cm, 72°, B, a, C

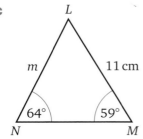

b

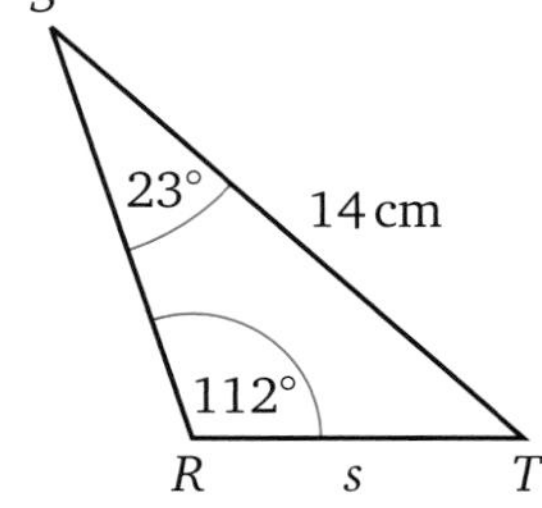

Not drawn accurately

c L, m, 11 cm, 64°, 59°, N, M

d

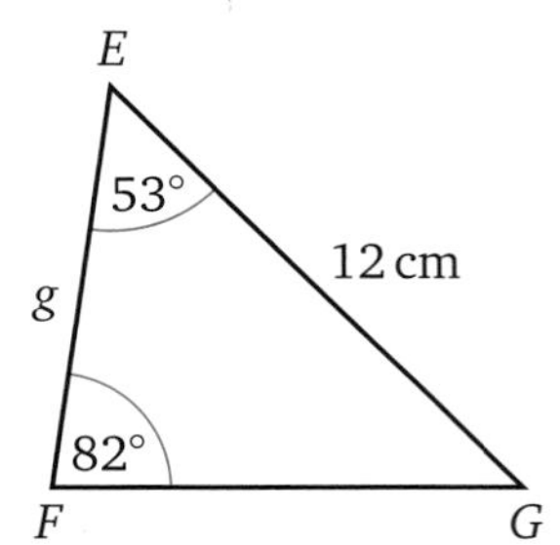

AQA Examiner's tip

If you are given two of the angles in a triangle, you can work out the third by subtraction because the sum of the three angles must be 180°.

2 Calculate the angles marked x in these diagrams.

a

A 45° 8 cm x B 6 cm C

b

D 7 cm 9 cm 68° x F E

c

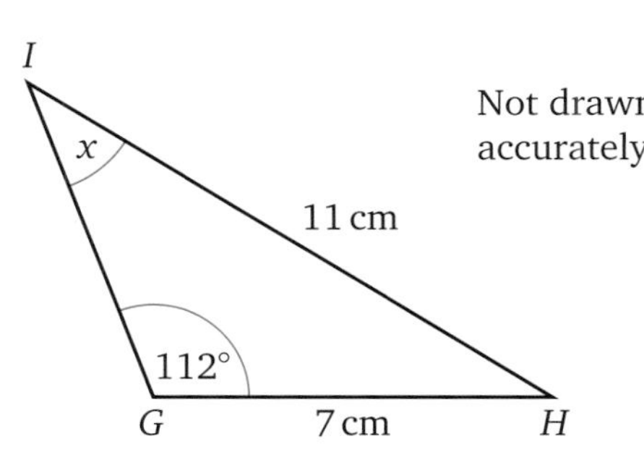

3 Triangle ABC has $AB = 9$ cm, $AC = 7$ cm and angle $ACB = 64°$.

Draw a sketch of the triangle and calculate angle ABC.

4 Triangle DEF has $DE = 7$ cm, angle $DFE = 67°$ and angle $DEF = 59°$.

Draw a sketch of the triangle and calculate the length of EF.

5 A tent is supported at A by two guy ropes, AB and AC.
AB is 1.8 m long, and AC is 2.1 m long.
The angle ABC is 72°.

Calculate angle BAC.

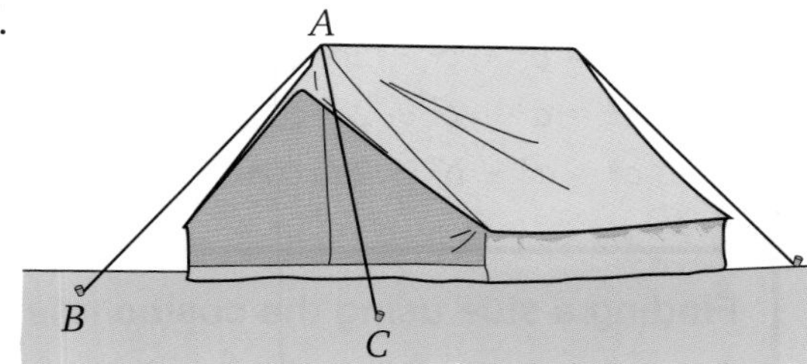

6 Two coastguards see the same boat.
Coastguard Y is 4.2 km due south of coastguard X.
The boat is on a bearing of 118° from X, and 071° from Y.

Calculate the distance of the boat from each coastguard.

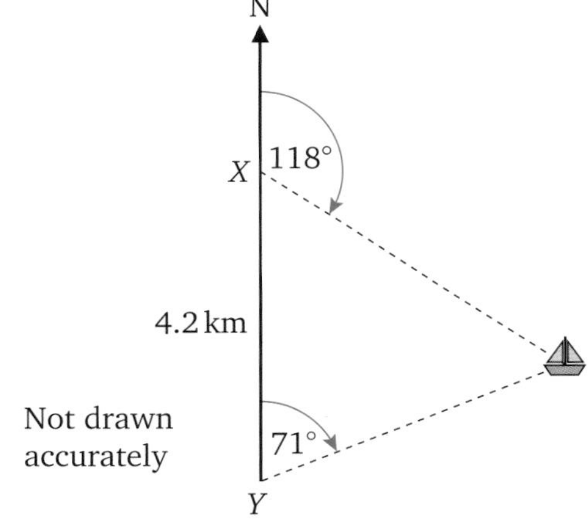

7 A child's slide has steps that are 1.6 m long.
The horizontal distance from the bottom of the steps to the end of the slide is 4 m.
The slide makes an angle of 31° with the ground.

Calculate the obtuse angle, x, between the slide and the steps.

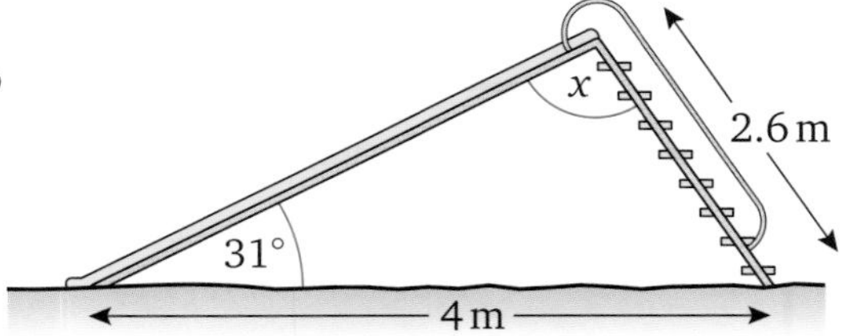

8 Mike is trying to find the height of a tree on the far side of a river.
He measures the angle of elevation of the tree to be 48°.
He walks a further 15 paces away and the angle of elevation is now 38°.
Mike's pace is about 0.8 metres long.

Use Mike's calculations to find an estimate of the height of the tree.

Learn... 22.2 The cosine rule

In the triangle ABC, CD is perpendicular to AB.

$AB = c$, which is split at D into the lengths $AD = x$ and $DB = c - x$

Using Pythagoras' theorem on the two triangles, $h^2 = a^2 - (c - x)^2$, and $h^2 = b^2 - x^2$

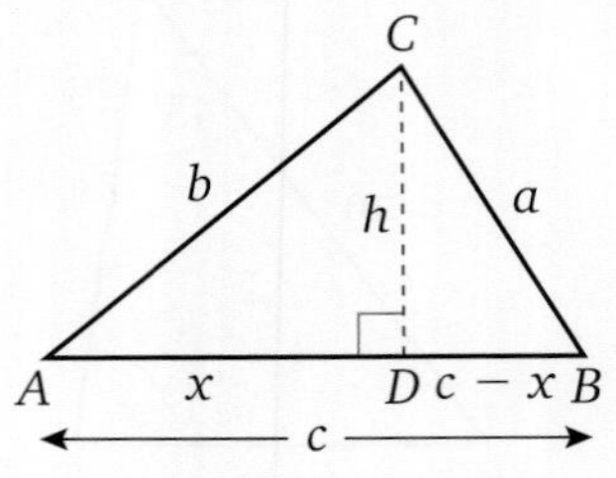

So $a^2 - (c - x)^2 = b^2 - x^2$

$a^2 = b^2 - x^2 + (c - x)^2$

$a^2 = b^2 - x^2 + c^2 - 2cx + x^2$

$a^2 = b^2 + c^2 - 2cx$

But $\cos A = \dfrac{x}{b}$

So $x = b\cos A$

So by substituting for x:

$a^2 = b^2 + c^2 - 2bc\cos A$

This is the **cosine rule**, and can be used when a question involves three sides and an angle.

It has three forms:

$a^2 = b^2 + c^2 - 2bc\cos A$

$b^2 = a^2 + c^2 - 2ac\cos B$

$c^2 = a^2 + b^2 - 2ab\cos C$

> **AQA Examiner's tip**
>
> The first version is on the formula sheet in exams, but the other two are not.
>
> Make sure that you know how to work out the other two versions from the first.

Finding a side using the cosine rule

Example: Calculate the length of the side AB in the diagram opposite.

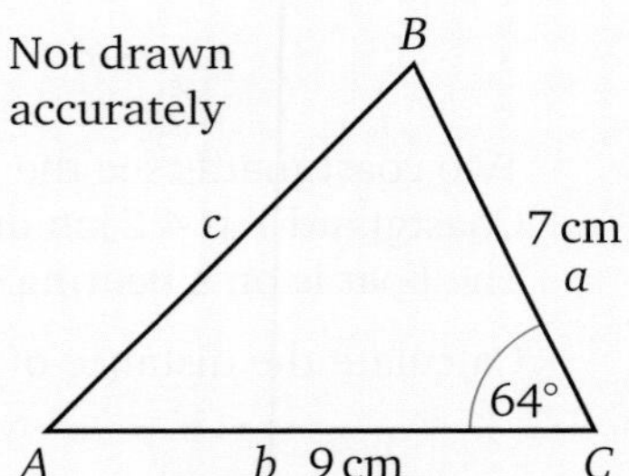

Solution: Copy and label the diagram a, b and c.

You need c, you know a, b and C.

$c^2 = a^2 + b^2 - 2ab\cos C$

$c^2 = 7^2 + 9^2 - 2 \times 7 \times 9 \times \cos 64°$

$= 74.7652\ldots$

$c = \sqrt{74.7652\ldots}$

$= 8.6466\ldots$

$c = 8.6$ cm (1 d.p.)

> **AQA Examiner's tip**
>
> Remember to take the square root at the end. Always check that your answer is reasonable.

Calculating an angle using the cosine rule

Example: Allthorpe is 14 km from Braytown on a bearing of 049°.

Crighton is 20 km from Allthorpe and 11 km from Braytown.

Find the bearing of Crighton from Braytown.

Label the sides a, b and c.

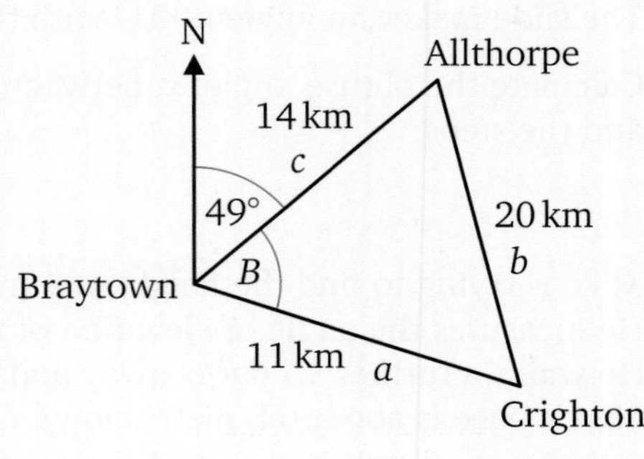

You know a, b and c and you need angle $ABC = B$

$$b^2 = a^2 + c^2 - 2ac\cos B$$

$$20^2 = 11^2 + 14^2 - 2 \times 11 \times 14\cos B$$

$$400 = 121 + 196 - 308\cos B$$

$$400 = 317 - 308\cos B$$

$$308\cos B = 317 - 400$$

$$\cos B = \frac{-83}{308}$$

AQA Examiner's tip

To find an angle using the cosine rule, you can rearrange the formula to make $\cos A$ the subject. So:

$$2bc\cos A = b^2 + c^2 - a^2$$

$$\cos A = \frac{b^2 + c^2 - a^2}{2bc}$$

You can learn this formula or practise changing the subject.

$$B = \cos^{-1}\frac{-83}{308}$$

$$= 105.6333...^\circ$$

The bearing is $49^\circ + 105.6333...^\circ = 154.6333...^\circ$

The bearing is 155° (to the nearest degree).

Link

You saw in Chapter 21 that obtuse angles have negative cosines.

AQA Examiner's tip

Many candidates round off too soon and lose marks for accuracy.
Never round off until the end.

Practise... 22.2 The cosine rule  

1 Calculate the lengths of the marked sides in the diagrams below.

a

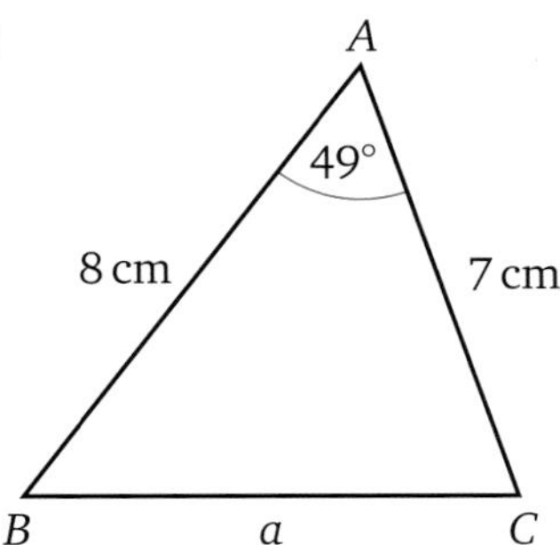

b

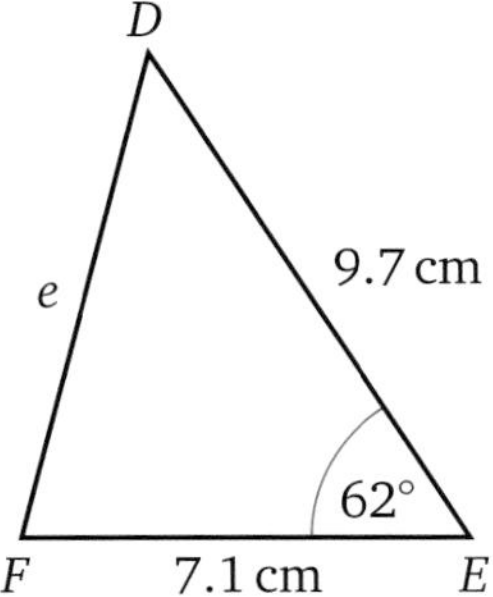

c

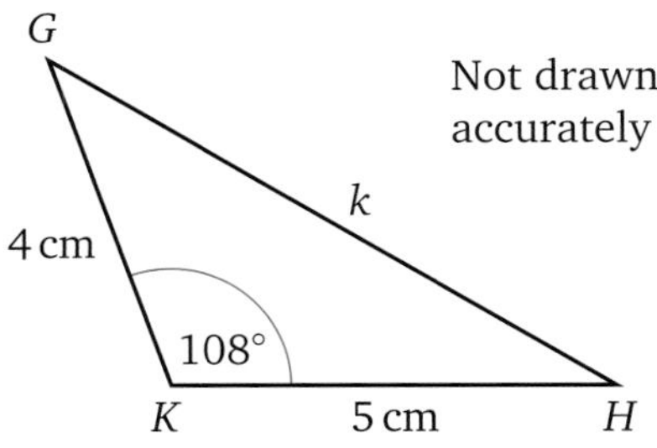

2 Calculate the angles marked x in these diagrams.

a

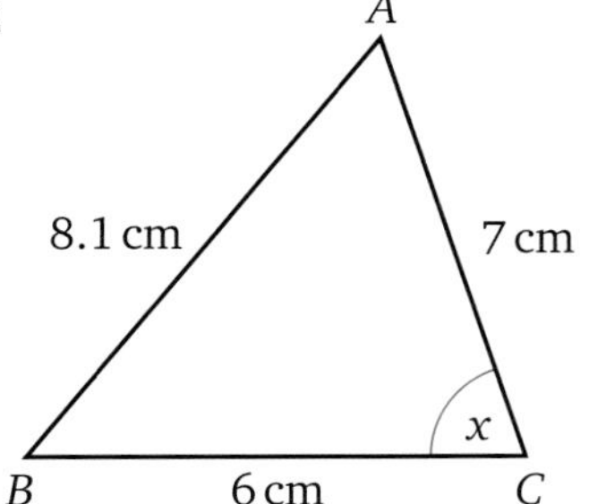

b

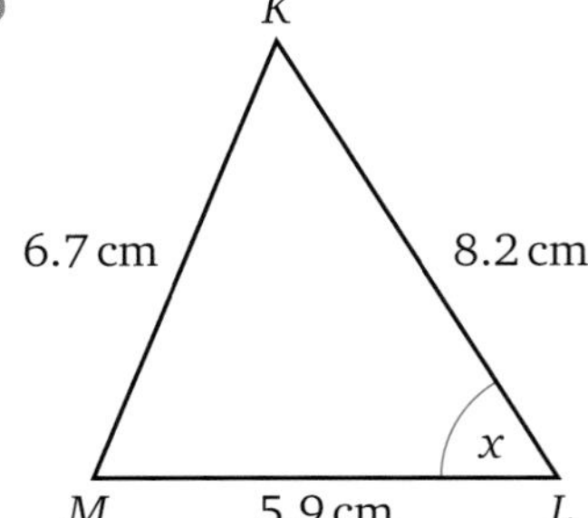

c

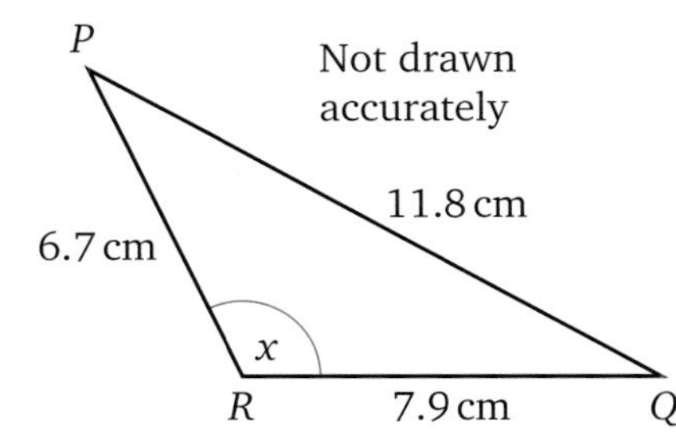

A

3 Triangle ABC has $BC = 8$ cm, $AC = 7.2$ cm and angle $ACB = 58°$.

Draw a sketch of the triangle and calculate the length of AB.

4 Triangle DEF has $DE = 8$ cm, $DF = 9.1$ cm and $EF = 6.7$ cm.

Draw a sketch of the triangle and calculate the size of angle DEF.

5 A clock has a minute hand that is 8 cm long and an hour hand that is 6 cm long.

Calculate the distance between the tips of the hands at:

a 2 o'clock

b 4.30am

6 A cuboid measures 8 cm × 6 cm × 5 cm as shown.

Calculate angle BAC.

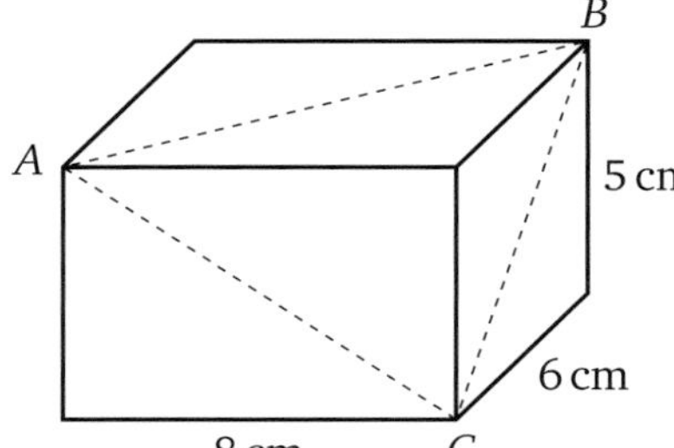

Not drawn accurately

7 Bob and Alice are walking along a straight path, AB. At point B, the path is diverted around a lake.
They can choose path BEC or path BFC.
$BE = 80$ m, $EC = 70$ m, angle $BEC = 124°$, angle $ABF = 145°$ and angle $FCD = 130°$

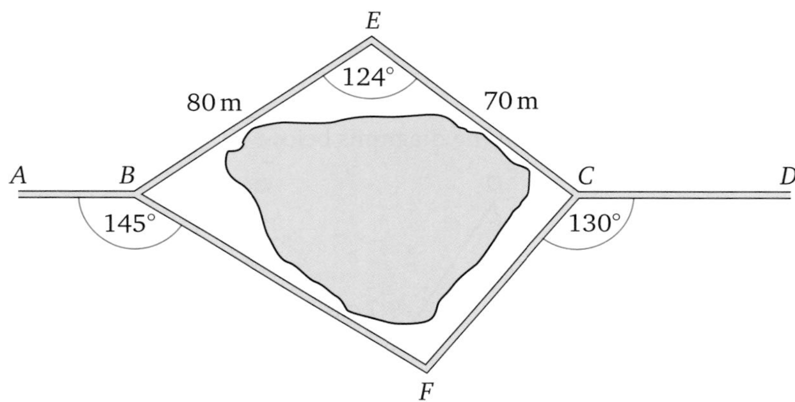

Which is the shorter route, BEC or BFC?

8 A house roof has dimensions as shown.
Angle $DCB = 90°$ and $BEC = 21°$

Calculate the length of BD.

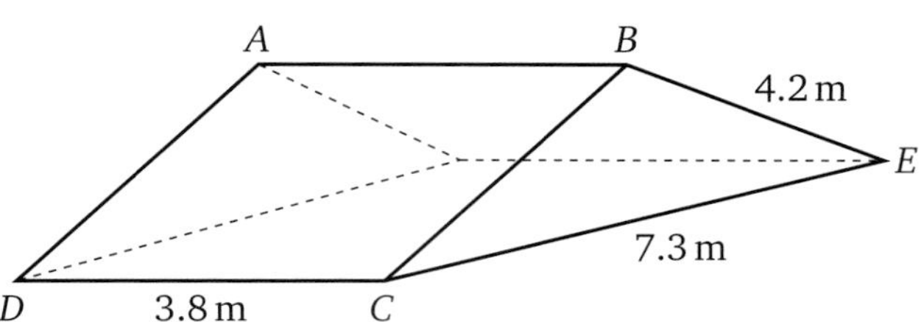

22.3 Finding the area of a triangle using trigonometry

You know that the area of a triangle can be found by the formula:

area $= \frac{1}{2} \times$ base $\times$ height

In the diagram, area $= \frac{1}{2}ch$

But $\sin A = \dfrac{h}{b}$

or $h = b\sin A$

So the area of the triangle is $\frac{1}{2}bc\sin A$

You can also use $\frac{1}{2}ac\sin B$ or $\frac{1}{2}ab\sin C$

Only this last version is given on the formula sheet in examinations.

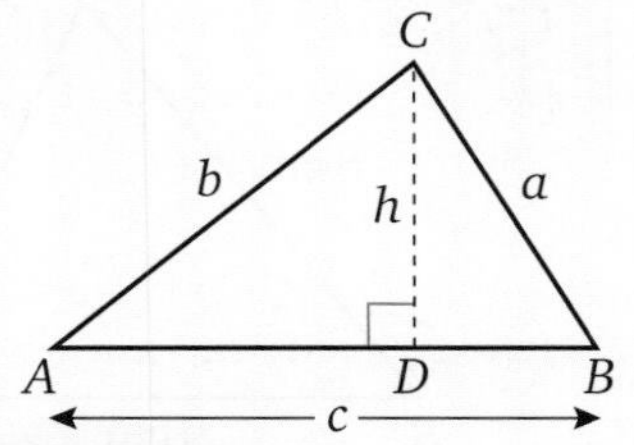

AQA Examiner's tip

You should only use this formula when the triangle has no right angle, and you do not know the height of the triangle, as $A = \frac{1}{2}b \times h$ is easier to use.

Example: Find the area of the triangle ABC, correct to 1 d.p.

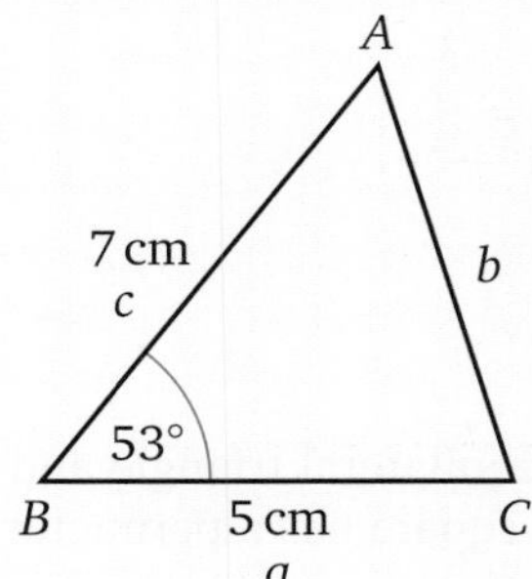

Not drawn accurately

Solution: Label the sides a, b and c.

You know a, c and B, so:

$$\text{area} = \tfrac{1}{2}ac\sin B$$
$$= \tfrac{1}{2} \times 5 \times 7 \times \sin 53°$$
$$= 14.0\ \text{cm}^2\ (1\ \text{d.p.})$$

Practise... 22.3 Finding the area of a triangle using trigonometry

A

1 Find the area of each triangle.

a

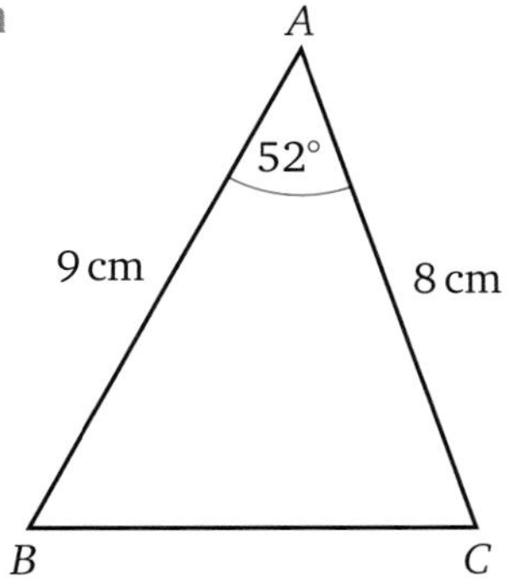

b

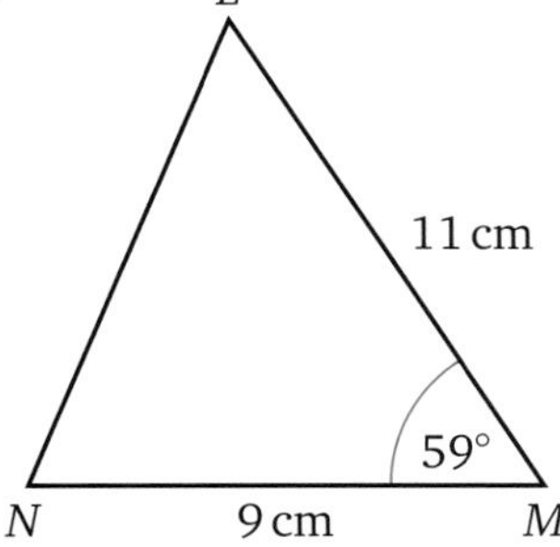

c

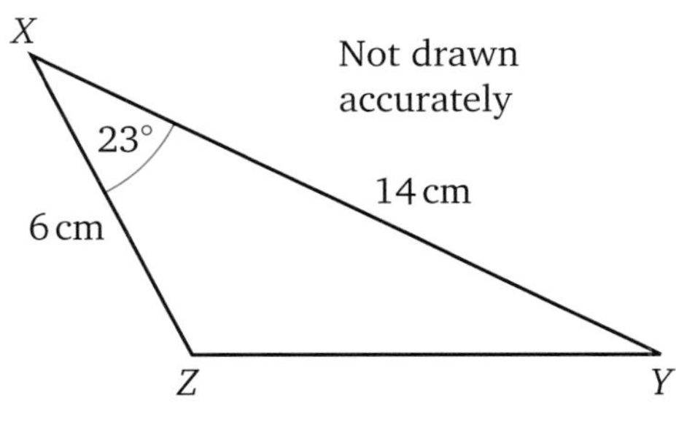

2 **a** ABC is an isosceles triangle, with $AB = BC = 8$ cm and angle $ABC = 68°$. Draw a sketch of the triangle and find its area.

b DEF is another isosceles triangle, with $DE = EF = 8$ cm and angle $EDF = 68°$. Draw a sketch of the triangle and find its area.

A

3 **a** Triangle ABC has an area of 24 cm^2. Find two possible values for angle B.

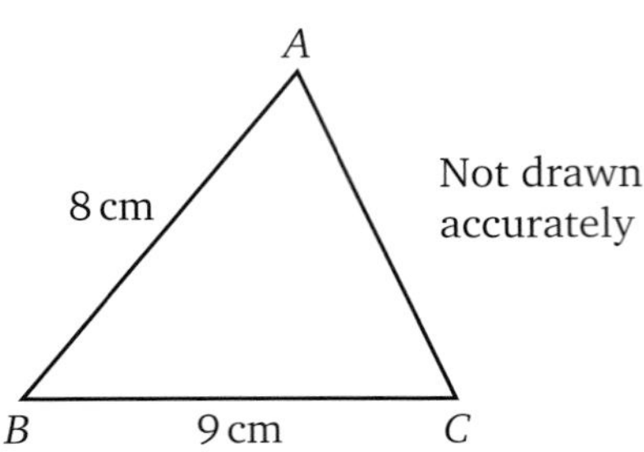

b Triangle DEF also has an area of 24 cm^2. Calculate the length of DE.

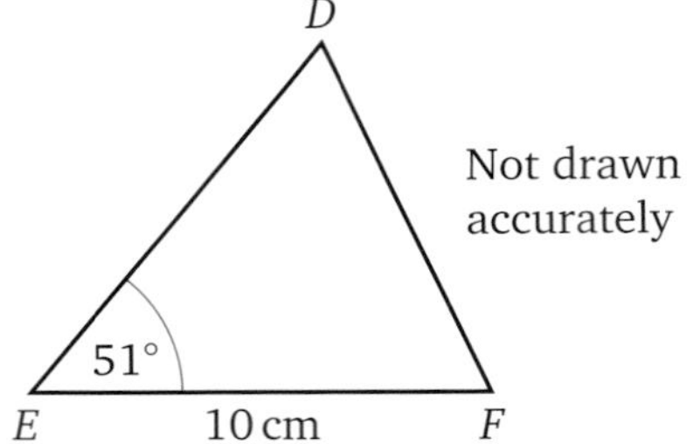

4 A farmer has a field between three roads and a hedge as shown.

Calculate the area of the field.

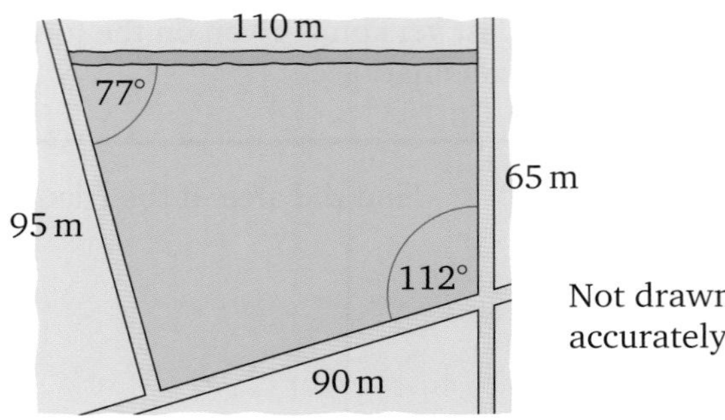

5 An equilateral triangle and a square have the same area. The square has a perimeter of 20 cm.

Find the perimeter of the triangle.

22 Assess

A

1 Calculate the sides and angles marked with letters below. The diagrams are not drawn accurately.

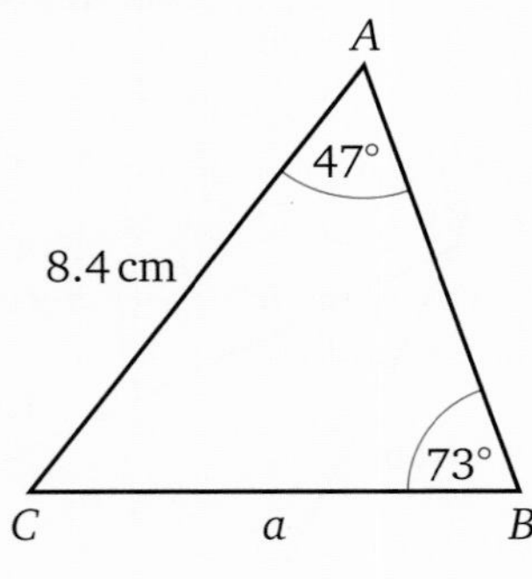

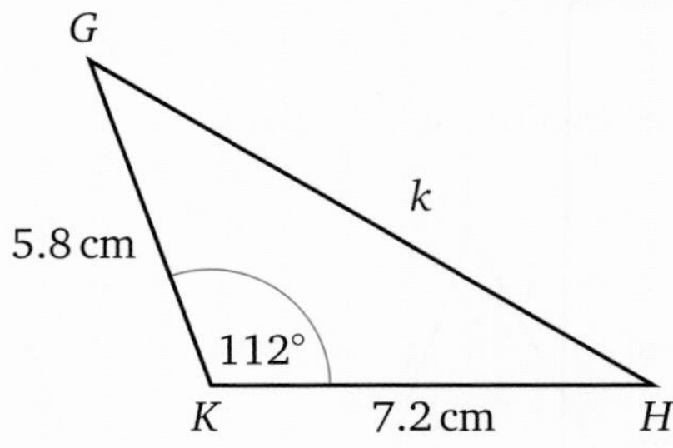

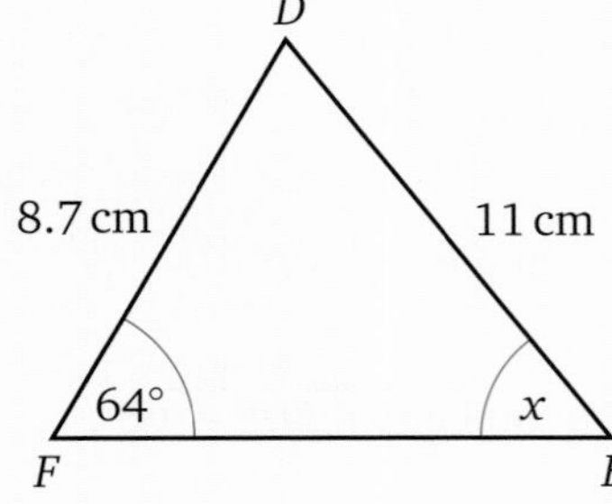

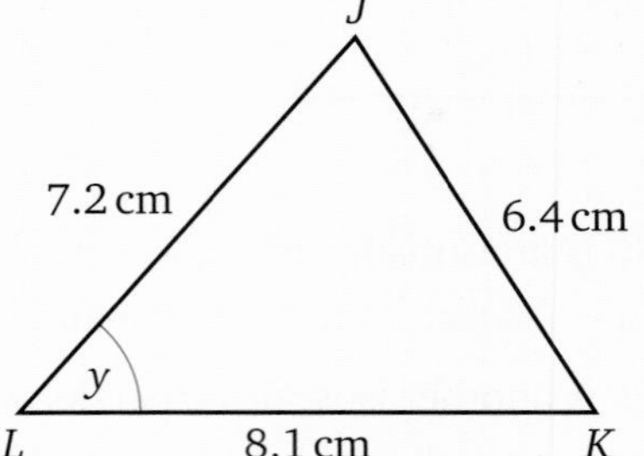

A

2 A gardener has a triangular flower bed, *ABC*.
He needs to use 7 g of fertiliser per m^2.

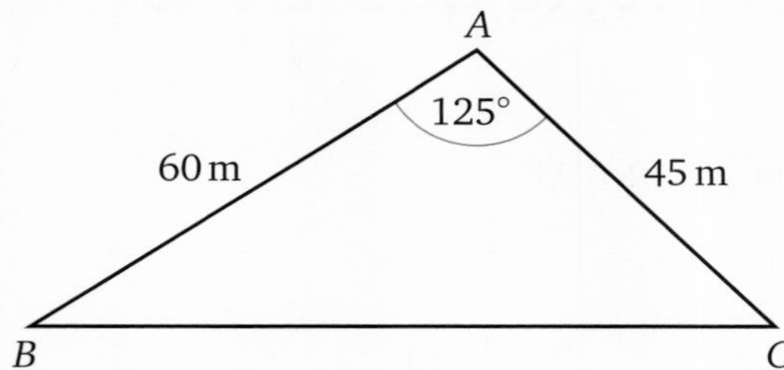

a Calculate how much fertiliser he needs.

b Calculate the perimeter of the bed.

3 Altown is 5 km from Croy on a bearing of 051°.
Broughton is 8 km from Altown and 7 km from Croy.

Calculate the bearing of Broughton from Croy.

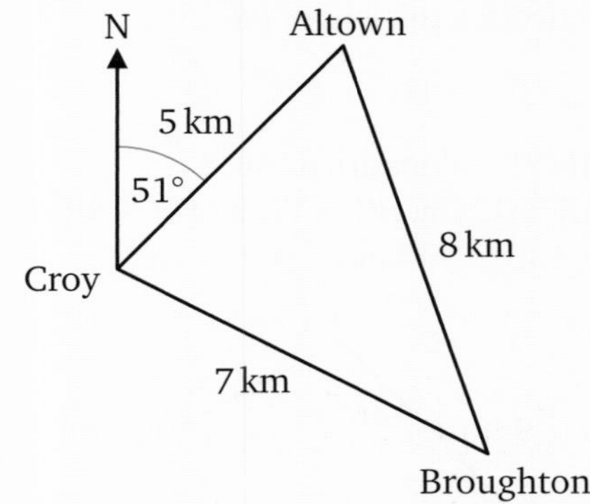

4 A parallelogram *ABCD* has $AD = 5$ cm, $AC = 8$ cm and angle $ADC = 57°$.

Calculate angle *ACD*.

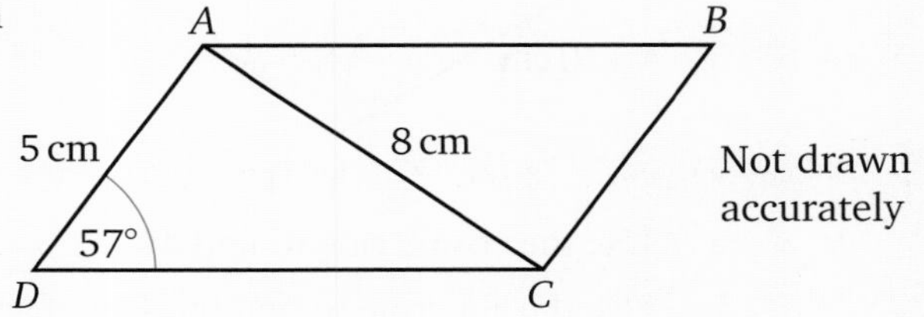

5 Calculate the area of the triangle *XYZ*.

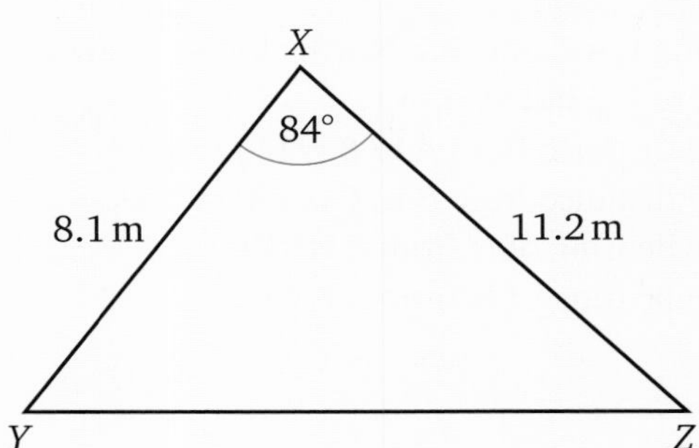

A*

6 The cuboid shown has $BC = 4$ cm, $CF = 5$ cm and $EF = 7$ cm.

Calculate angle *BDF*.

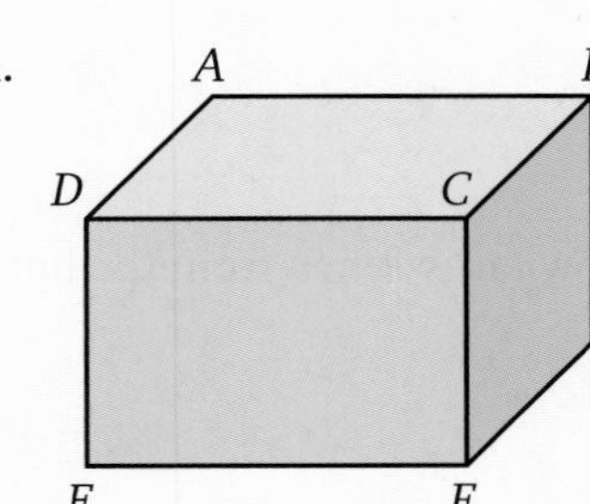

AQA Examination-style questions

1 PQR is a triangle.
$PQ = 10\,\text{cm}$, $QR = 12\,\text{cm}$ and angle $PQR = 78°$

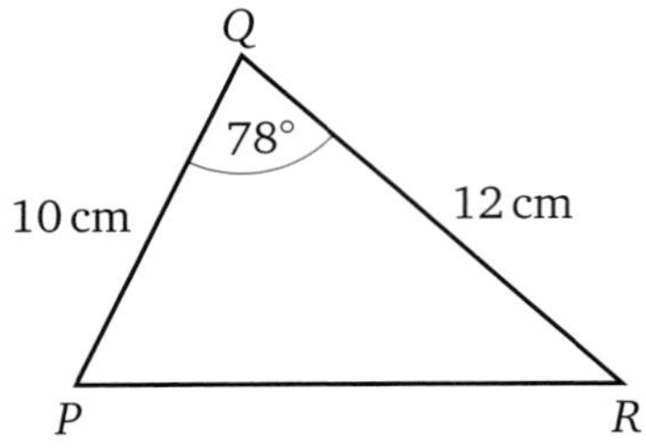

Not drawn accurately

Calculate the length PR. *(3 marks)*

AQA 2007

2 $ABCD$ is a quadrilateral.
$AB = 12\,\text{cm}$, $BC = 11\,\text{cm}$, $CD = 10\,\text{cm}$ and $DA = 9\,\text{cm}$
$\angle ABC = 74°$ and $\angle DAC = 46°$

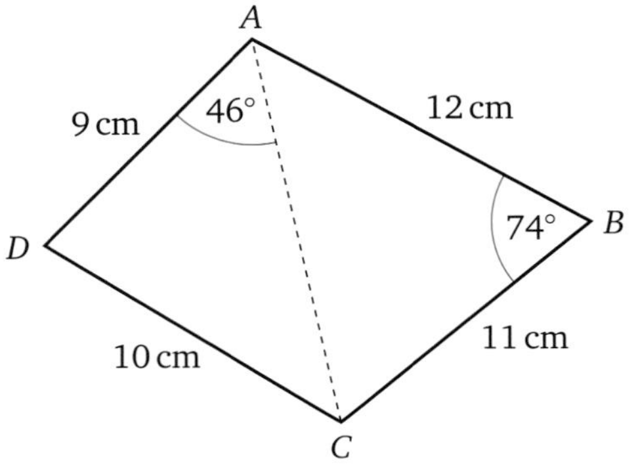

Not drawn accurately

a Use the cosine rule to find AC. *(3 marks)*

b Use the sine rule to find the size of angle ACD. *(3 marks)*

AQA 2008

3 Jenna is walking due North along a straight path, ABC.
There is a hut at H.
The distance from A to B is 140 metres.
The distance from B to C is 230 metres.
The bearing of H from A is 040°.
The bearing of H from C is 115°.

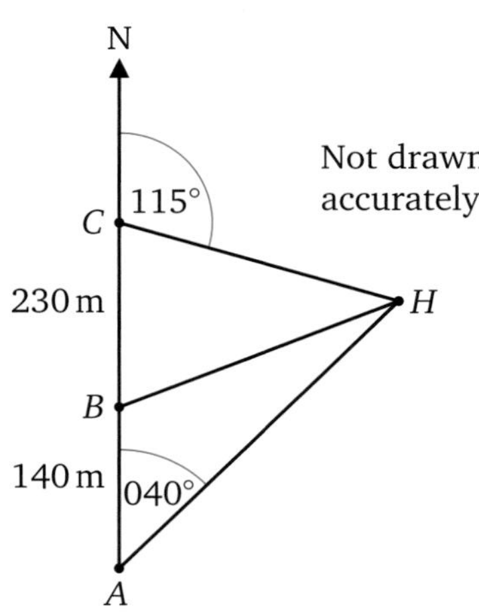

Not drawn accurately

How far is Jenna from the hut when she is at B? *(6 marks)*

AQA 2002

23 Area and volume 2

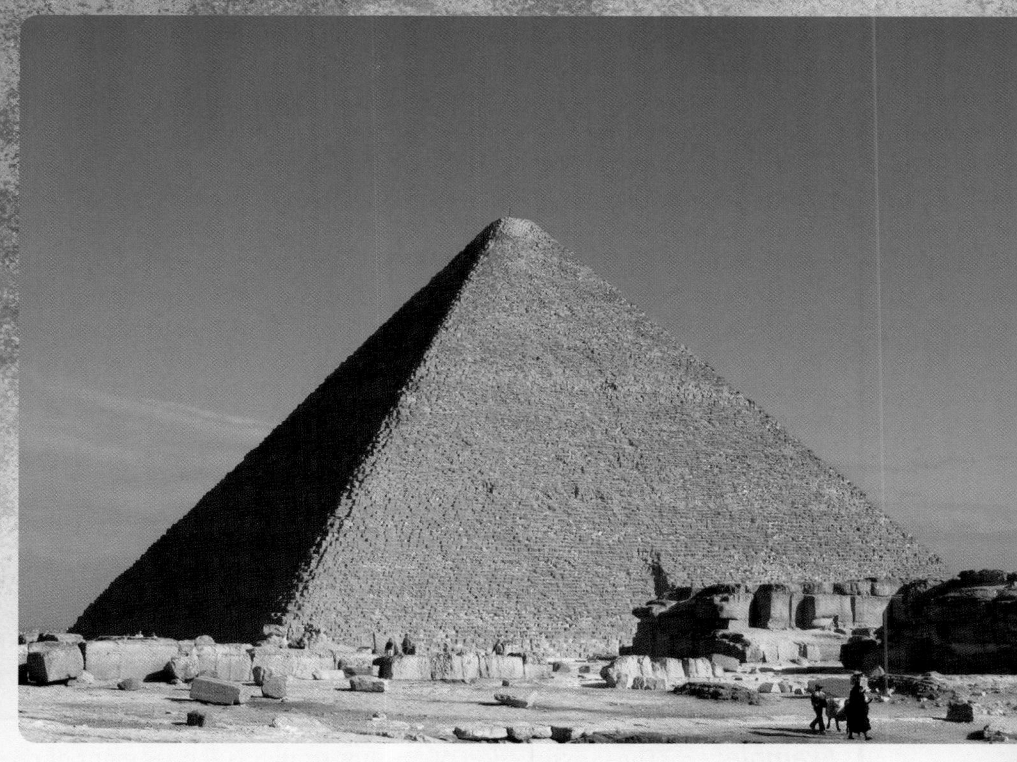

Objectives

Examiners would normally expect students who get these grades to be able to:

A

- calculate the lengths of arcs of circles
- calculate the areas of sectors of circles
- solve problems involving surface areas and volumes of pyramids, cones and spheres

A*

- solve problems involving complex shapes and solids, including segments of circles and frustums of cones.

Did you know?

The Great Pyramid

The Great Pyramid of Giza is believed to have been built as a tomb for the Egyptian King Khufu. It was completed approximately five and a half thousand years ago and was the world's tallest building for 3800 years. It is the only one of the Seven Wonders of the Ancient World still remaining.

The pyramid has a square base with sides 231 metres long and is approximately 147 metres high, taller than 33 double-decker buses. Its volume is more than a quarter of a million cubic metres, the volume of 50 Olympic swimming pools.

Key terms

arc
sector
segment
tetrahedron
frustum

You should already know:

- ✔ how to work out the circumference and area of a circle
- ✔ how to work out the volume and surface area of prisms and cylinders.

23.1 Arcs and sectors

An **arc** is part of the circumference of a circle.

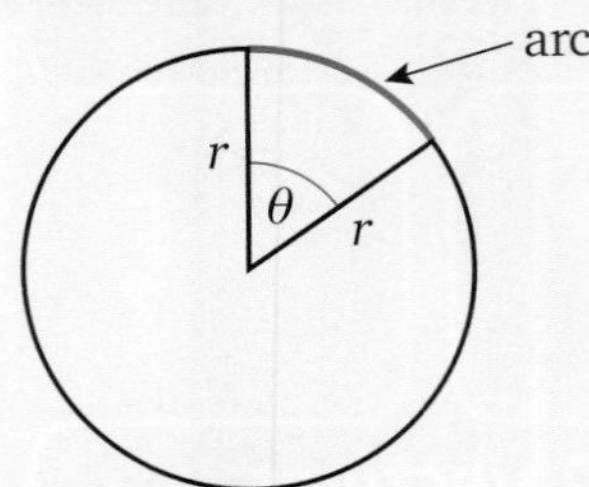

The length of an arc of a circle is proportional to the angle that the arc makes at the centre of the circle.

The whole circumference of a circle, $2\pi r$, makes an angle of 360°.

You can use the unitary method to find the length of any arc.

Length of arc with an angle of 360° is $2\pi r$.

Length of arc with an angle of 1° is $\frac{1}{360} \times 2\pi r$.

Length of arc with an angle of $\theta°$ is $\frac{\theta}{360} \times 2\pi r$.

The Greek letter θ, called theta, is often used to represent an angle.

A **sector** of a circle is a wedge-shaped piece of the area of a circle, bounded by two radii and an arc.

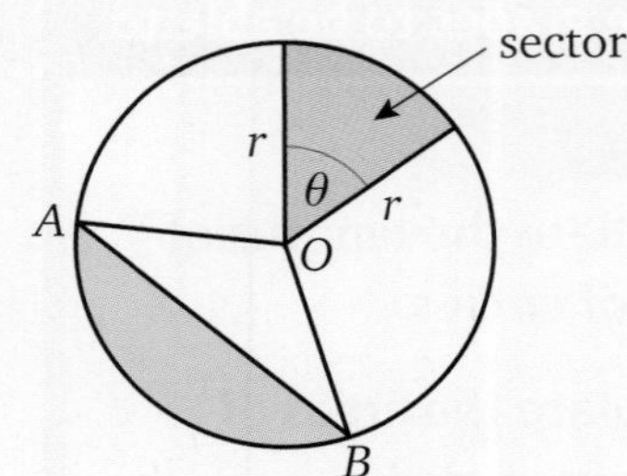

The whole area of a circle is πr^2.

The area of a sector with an angle of $\theta°$ is $\frac{\theta}{360} \times \pi r^2$.

A **segment** of a circle is a region bounded by a chord and an arc.

The area of the shaded segment in the diagram is the area of the sector OAB minus the area of the triangle OAB.

Example: The cross-section of a water trough is the segment of the circle shown shaded in the following diagram. The length of the trough is 3 metres.

Find the volume of the trough.

First find the area of the segment.

$$\text{Area of sector } OAB = \frac{125}{360} \times \pi \times 48^2\,\text{cm}^2$$

$$\text{Area of triangle } OAB = \tfrac{1}{2}ab\sin C = \tfrac{1}{2} \times 48^2 \sin 125°\,\text{cm}^2$$

$$\text{So area of segment} = \left(\frac{125}{360} \times \pi \times 48^2 - \tfrac{1}{2} \times 48^2 \times \sin 125°\right)\text{cm}^2$$

$$= 1569.6109\ldots\,\text{cm}^2$$

$$= 0.1569\ldots\,\text{m}^2$$

The radius of the trough is given in centimetres, but the length is in metres. To find the volume, these must be in the same units, so either convert the length to cm or convert the area to m².

The area of the segment = the area of cross-section of the water trough

$$\text{Volume of water trough} = \text{area of cross-section} \times \text{length}$$

$$= 0.1569\ldots\,\text{m}^2 \times 3\,\text{m}$$

$$= 0.471\,\text{m}^3 \text{ (3 s.f.)}$$

When the cross-section of a 3-D shape is uniform, the volume is given by the formula:

$$V = A \times h$$

where V is volume, A is area of the cross-section and h is the length.

Practise... 23.1 Arcs and sectors 

A

1 Find the length of the arcs indicated. O is the centre of the circle in each case.

a

5 cm
O 100°
5 cm

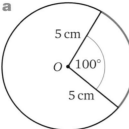

b

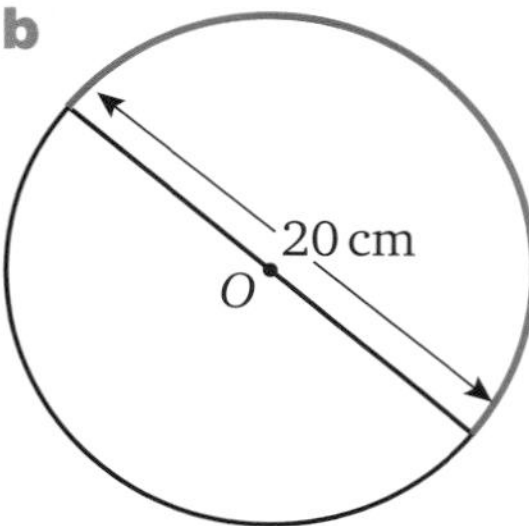

c

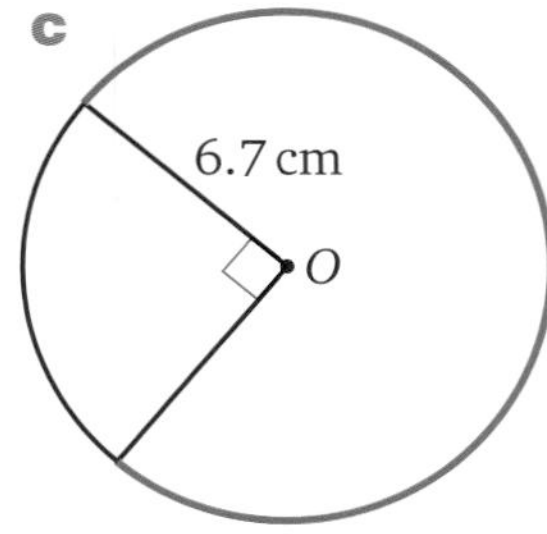

Not drawn accurately

2 Find the areas of the shaded sectors. O is the centre of the circle in each case.

a

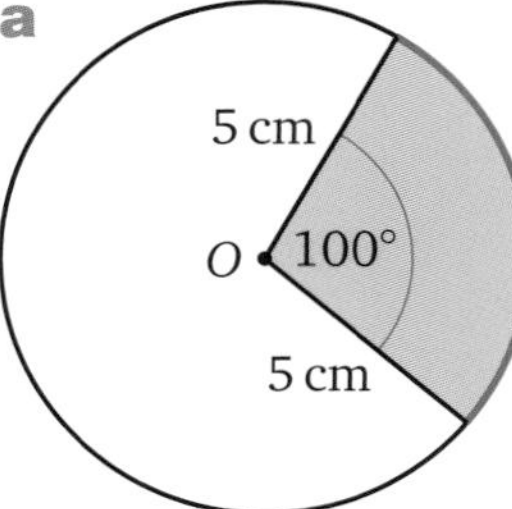

b

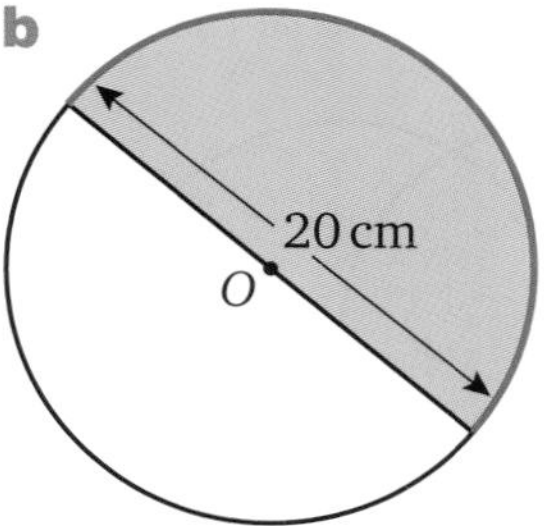

c

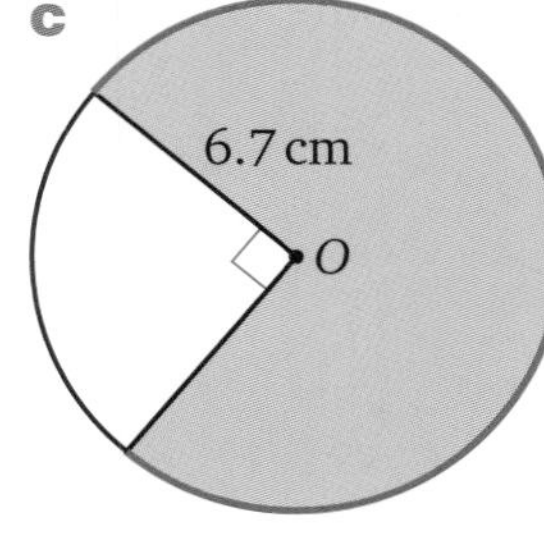

Not drawn accurately

3 Find the length of the arc of a circle that makes:

a an angle of 90° at the centre of a circle with radius 10 cm

b an angle of 150° at the centre of a circle with radius 4 cm

c an angle of 53° at the centre of a circle with radius 12.5 cm.

4 An arc makes an angle of 30° in a circle with radius 2 cm.
What angle does an arc of the same length make in a circle with radius 10 cm?

5 Find the area of the sector of a circle that makes:

a an angle of 120° at the centre of a circle with radius 5 cm

b an angle of 57° at the centre of a circle with radius 22 cm

c an angle of 210° at the centre of a circle with radius 1 m.

A*

6 Find the shaded areas.

a Radius = 12 cm

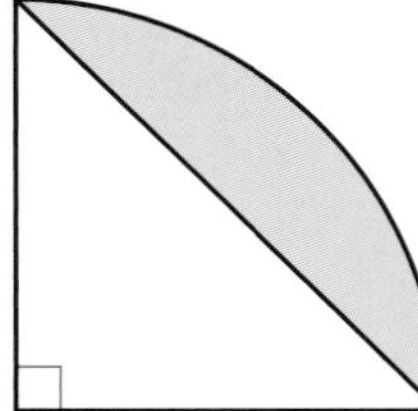

b Radius = 4.5 cm

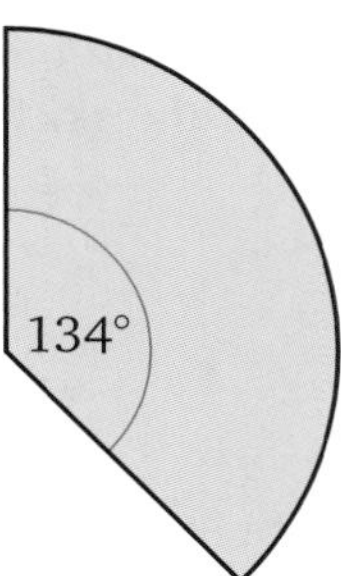

c Radius = 62 cm

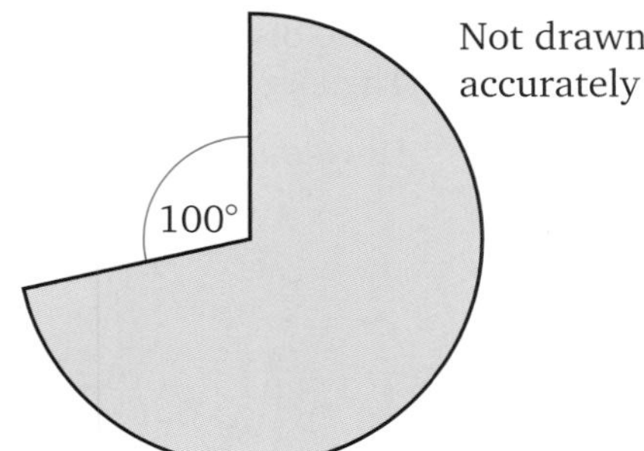

Not drawn accurately

A*

7 Find the shaded areas. The arcs are quarter circles and the squares have sides 10 cm long.

a

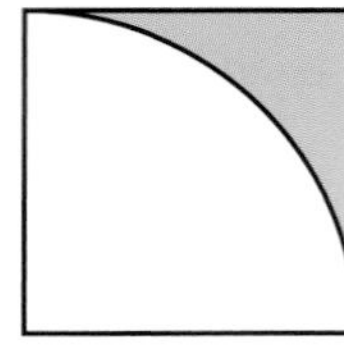

b

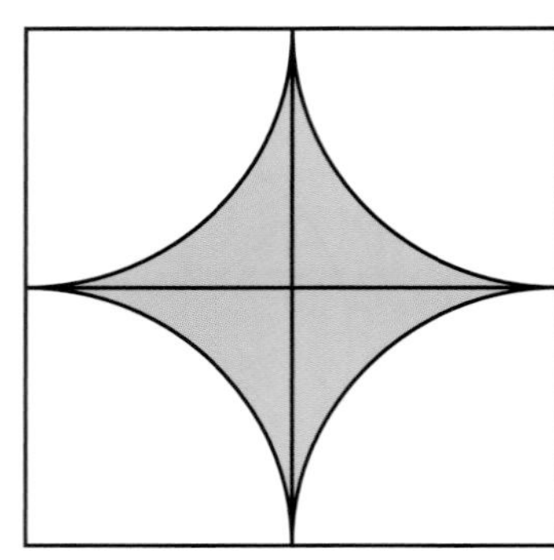

c

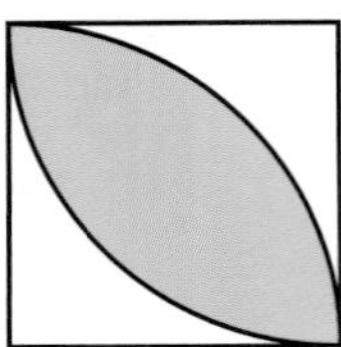

8 Two circles with radius 5 cm overlap so that the centre of one is on the circumference of the other. Find the perimeter and area of the overlap region.

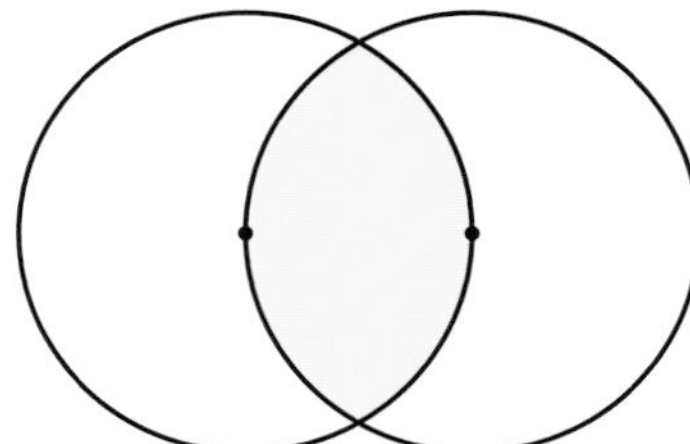

9 A semicircular piece of card of radius R is folded into a cone with height h.

Prove that $h = \frac{R}{2}\sqrt{3}$

 **10** The diagram shows the layout of four flower beds, each in the shape of a quarter circle of radius 6 m.

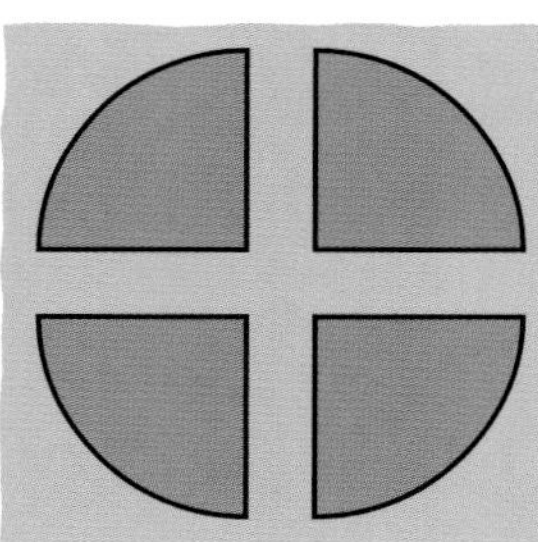

Each bed is to be surrounded with edging strip costing £1.23 per metre. How much will this cost?

 **11** The diagram shows the pattern for a skirt consisting of three quarters of a circle of radius 55 cm (with a small circle cut out of the centre for the waist). The large circular arc forms the hem of the skirt, to be finished with ribbon.

How long is this circular arc?

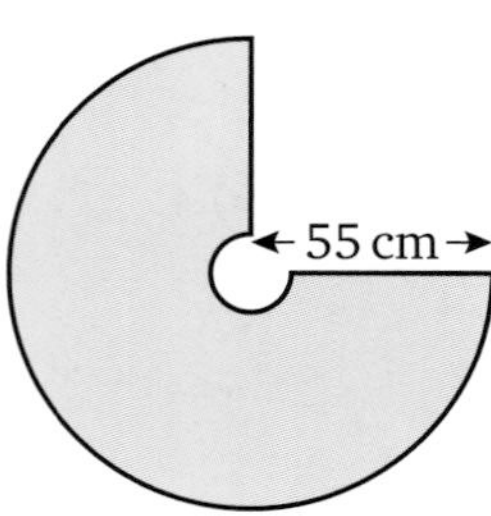

23.2 Volumes and surface areas of pyramids, cones and spheres

Volume of a pyramid

If a triangle is fitted exactly inside a rectangle with a side in common with the rectangle, then the area of the triangle is half the area of the rectangle.

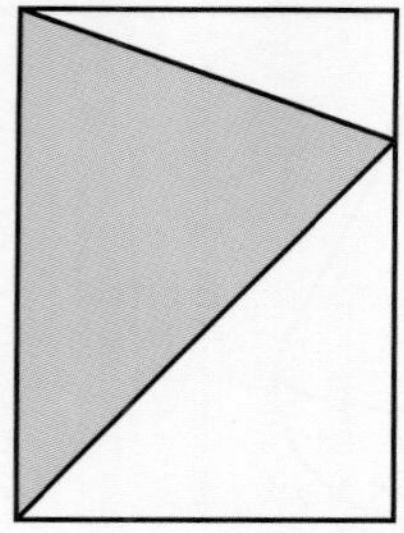

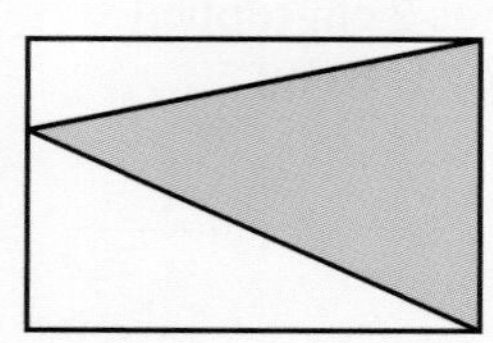

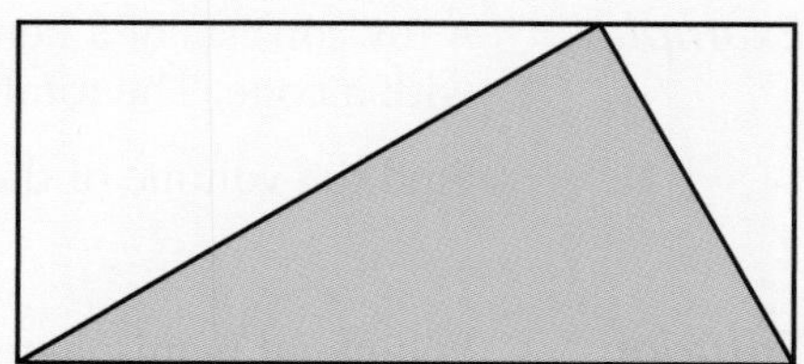

Similarly, a pyramid can be fitted inside a prism sharing the base and height.

The volume of the pyramid is one-third of the volume of the prism.

The volume of a prism is: area of base × height.

So the volume of the pyramid is $\frac{1}{3}$ × area of base × height.

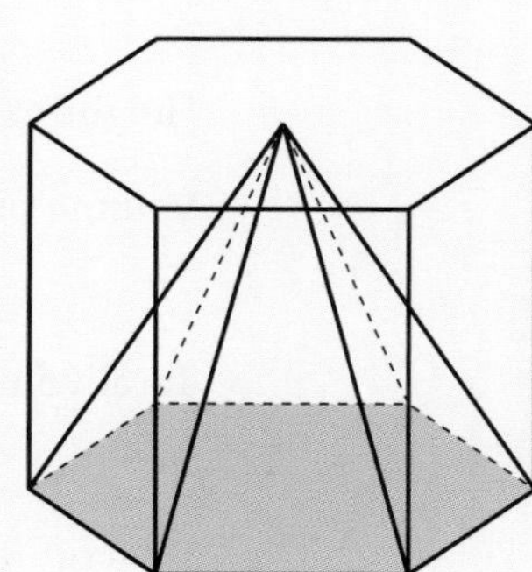

AQA Examiner's tip

A prism has a uniform cross-section, so $V = A \times h$

For solids that come to a vertex (cone and pyramid),

$V = \frac{1}{3}A \times h$

Example: A pyramid with height 5 cm and a square base with sides of 4 cm fits inside a cuboid measuring 4 cm by 4 cm by 5 cm.

Calculate the volume of the pyramid.

Solution: The volume of the cuboid is $4 \times 4 \times 5\ \text{cm}^3$.

The volume of the pyramid is $\frac{1}{3}$ × area of base × height.

$= \frac{1}{3} \times 4 \times 4 \times 5\ \text{cm}^3$

$= 26.7\ \text{cm}^3$ (3 s.f.)

Volume of a cone

A cone can be fitted exactly into a cylinder.

The volume of the cone is one third of the volume of the cylinder.

A cone with perpendicular height h cm and base radius r fits into a cylinder with height h and radius r.

The volume of the cylinder is $\pi r^2 h$.

So the volume of the cone is $\frac{1}{3}\pi r^2 h$.

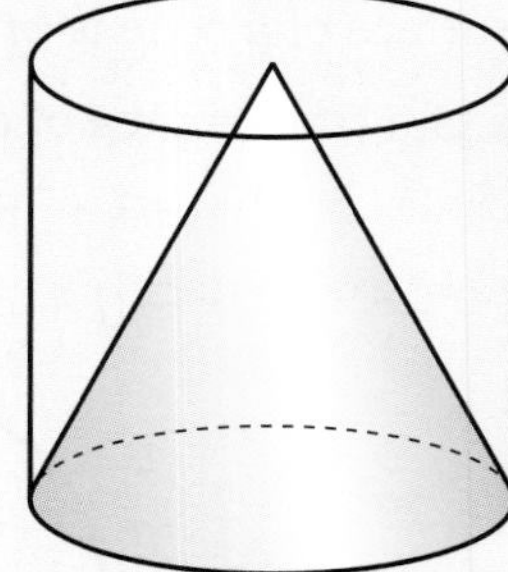

A cone is a pyramid with a circular base.

AQA Examiner's tip

This formula is given on the exam paper.

Volume of a sphere

The volume of a sphere with radius r is $\frac{4}{3}\pi r^3$.

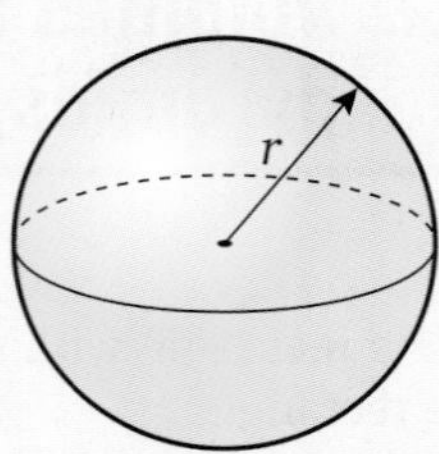

Example: A toy consists of a hemisphere of radius 3 cm topped with a cone. The total height of the toy is 8 cm.

Find the volume of the toy.

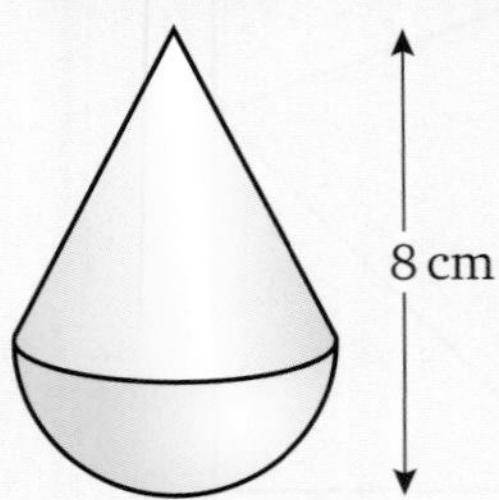

Not drawn accurately

Solution:

Volume of hemisphere $=$ half volume of sphere

$= \frac{1}{2} \times \frac{4}{3}\pi \times 3^3\ \text{cm}^3 = \frac{2}{3}\pi \times 3^3\ \text{cm}^3$

$= 56.5486\ldots\ \text{cm}^3$

Height of cone $= 8 - 3 = 5\ \text{cm}$

Volume of cone $= \frac{1}{3}\pi \times 3^2 \times 5\ \text{cm}^3$

$= 47.1238\ldots\ \text{cm}^3$

Total volume of toy $=$ volume of hemisphere $+$ volume of cone

$= 103.6725\ldots\ \text{cm}^3$

$= 104\ \text{cm}^3$ (3 s.f.)

Surface areas

The surface area of any solid is the sum of the areas of all the faces.

A **tetrahedron** is a pyramid with a triangular base. A regular tetrahedron is a pyramid whose faces are all equilateral triangles.

The surface area of a regular tetrahedron is the sum of the areas of the four equilateral triangles that form its faces.

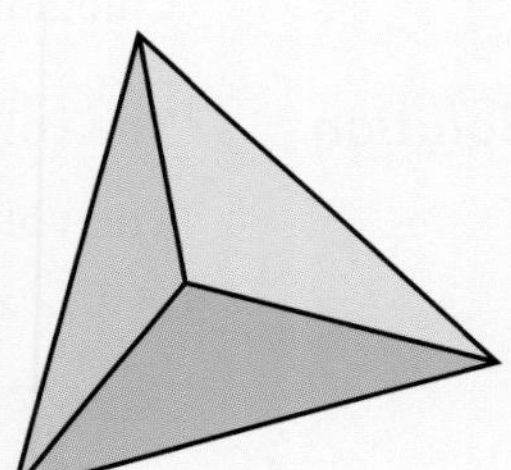

If the edges of the tetrahedron have length x, the area of each triangle is:

$\frac{1}{2}x^2 \sin 60°$

So the surface area of the tetrahedron $= 4 \times \frac{1}{2}x^2 \sin 60°$

$= 2x^2 \sin 60°$

The surface area of a cone is the sum of the area of its circular base and the area of its curved surface.

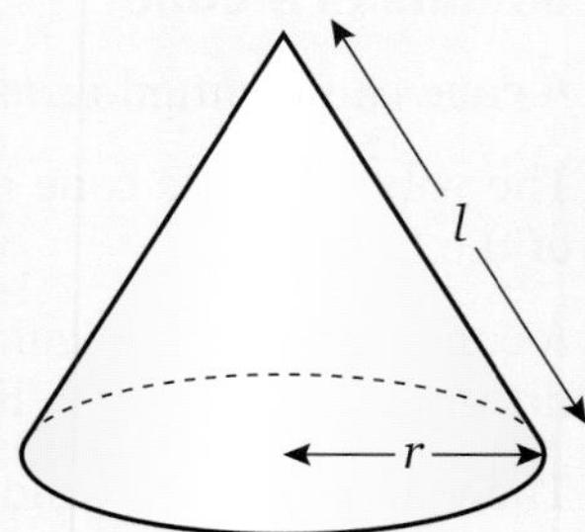

The area of the curved surface of a cone is πrl.

So the total surface area of a cone $=$ area of curved surface $+$ area of base

$= \pi rl + \pi r^2$

$= \pi r(l + r)$

The surface area of a sphere is $4\pi r^2$.

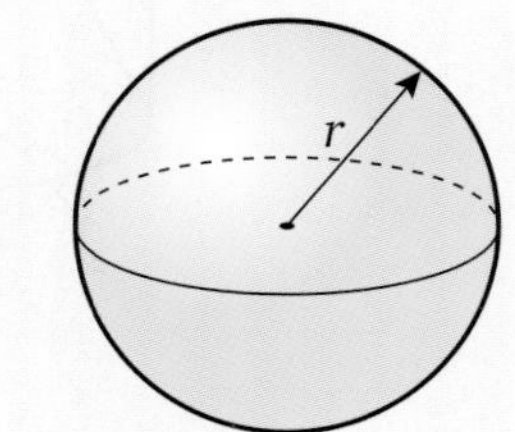

Example: The **frustum** of a cone has base radius 6 cm and top radius 3 cm. The distance between the top and the base is 3 cm. Find the volume of the frustum.

Not drawn accurately

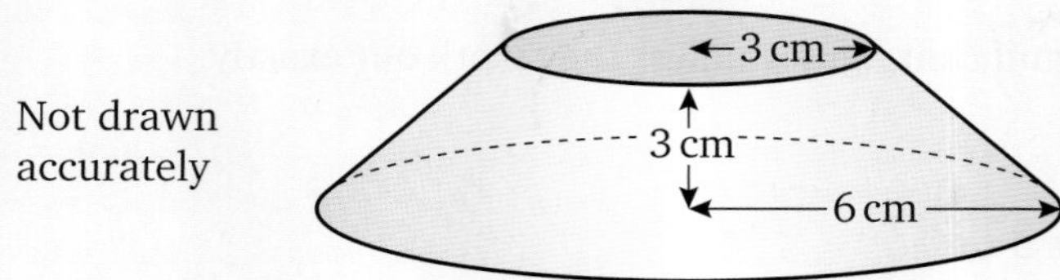

The frustum of a cone is the part remaining when the top of a cone is removed.

Solution: Let the height of the missing part of the cone be h cm.

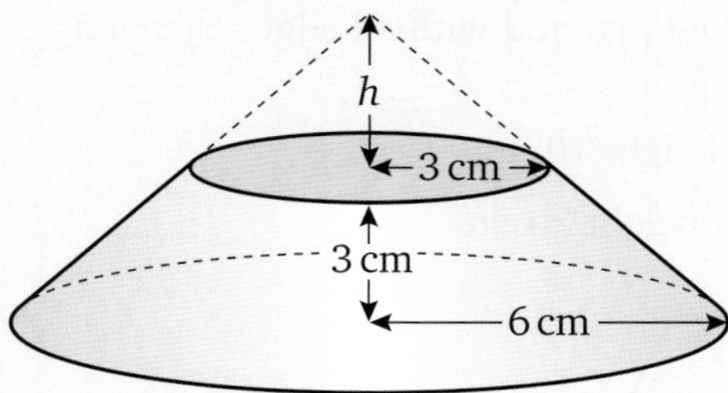

By similar triangles, $\dfrac{h+3}{h} = \dfrac{6}{3}$

$h + 3 = 2h$

$h = 3$

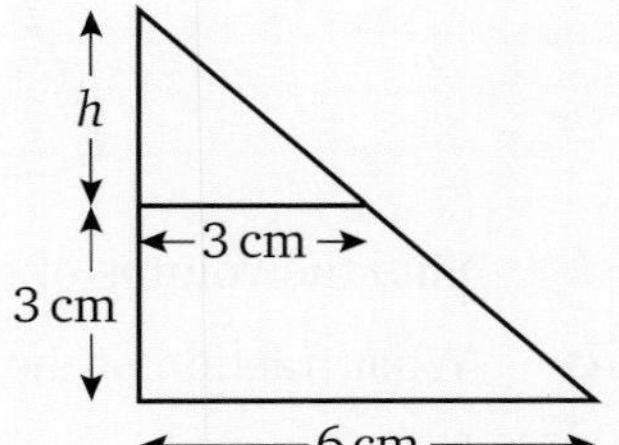

(This is obvious from the diagram, but the similar triangle method above applies to any frustum, not just this special case.)

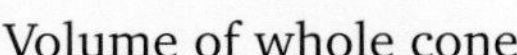

Volume of whole cone $= \frac{1}{3}\pi \times 6^2 \times 6\,\text{cm}^3$

$= 72\pi\,\text{cm}^3$

Volume of missing part of cone $= \frac{1}{3}\pi \times 3^2 \times 3\,\text{cm}^3$

$= 9\pi\,\text{cm}^3$

So volume of frustum $= 72\pi\,\text{cm}^3 - 9\pi\,\text{cm}^3$

$= 63\pi\,\text{cm}^3$

$= 198\,\text{cm}^3$ (3 s.f.)

Summary of formulae

Volume of a pyramid $= \frac{1}{3} \times$ area of base $\times$ perpendicular height

Volume of cone $= \frac{1}{3} \times$ area of base $\times$ perpendicular height

$= \frac{1}{3}\pi r^2 h$

Area of curved surface of cone $= \pi r l$

Volume of sphere $= \frac{4}{3}\pi r^3$

Surface area of sphere $= 4\pi r^2$

AQA Examiner's tip

All of these formulae are given on the exam paper.

Practise... 23.2 Volumes and surface areas of pyramids, cones and spheres

Give your answers correct to three significant figures unless they work out exactly.

A

1 Find the volume of:

- **a** a sphere with radius 5 cm
- **b** a hemisphere with radius 10 cm
- **c** a pyramid with height 12 cm and a hexagonal base with area 36 cm^2.

2 Find the total surface area of a square-based pyramid with all edges 5 cm long.

3 The small cone has base radius 7 cm and height 10 cm.

The large cone has base radius 14 cm and height 20 cm.

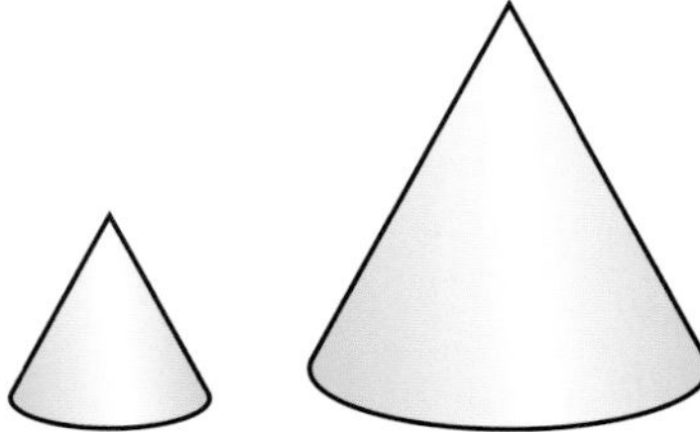

- **a** Find the volume of each cone.
- **b** What fraction of the volume of the large cone is the volume of the small cone?

4 Two spheres, *A* and *B*, have radii 2 cm and 6 cm.

Find the ratio of:

- **a** radius of A : radius of B
- **b** surface area of A : surface area of B
- **c** volume of A : volume of B

5 **a** Find the volume of:

- **i** a pyramid with height 15 cm and square base with sides of length 10 cm
- **ii** a cone with base radius 3.5 cm and height 15 cm.

b What is the radius of the base of a cone of height 15 cm that has the same volume as the pyramid in part **a i**?

A*

6 What is the height of a cone of radius 6 cm that has the same surface area as a hemisphere of radius 6 cm?

7 A measuring cylinder with radius 3 cm contains water to a depth of 10 cm.
A small metal sphere is dropped into the cylinder and the water level rises by 2 cm.

What is the radius of the sphere?

8 A semicircular piece of card with radius 10 cm is curved to form a cone.

What is the volume of the cone?

9 Prove that the area of the curved surface of a cone with radius r and slant height l is πrl.

 **10** The shape of most buckets is an approximation to a frustum of a cone. One such bucket has top diameter 24 cm, base diameter 18 cm and height 26 cm. These are all internal measurements.

How many litres will the bucket hold?

 **11** A grain hopper is in the form of a cylinder on top of the frustum of a cone. Its measurements are as shown in the diagram.

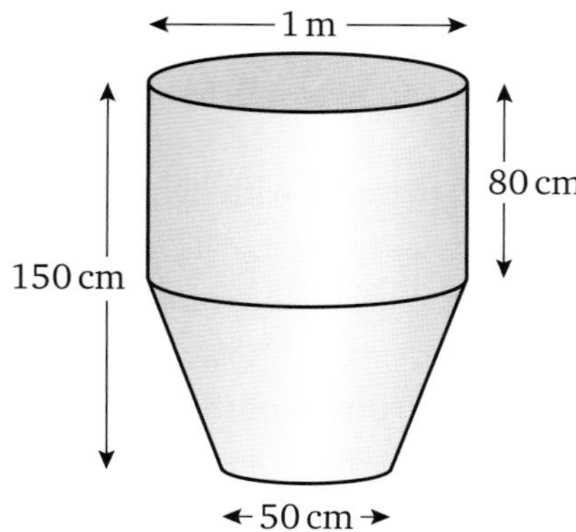

a What volume of grain will the hopper hold?

b What area of sheet metal is needed to make the hopper?

23 Assess 

Give your answers to three significant figures.

A

1 Find the area of each sector.

a Radius 4.9 cm
Angle 58°

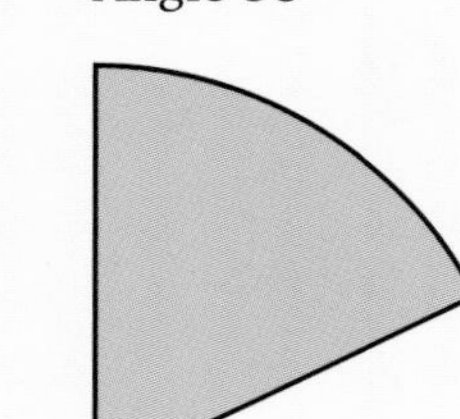

b Radius 15 cm
Angle 134°

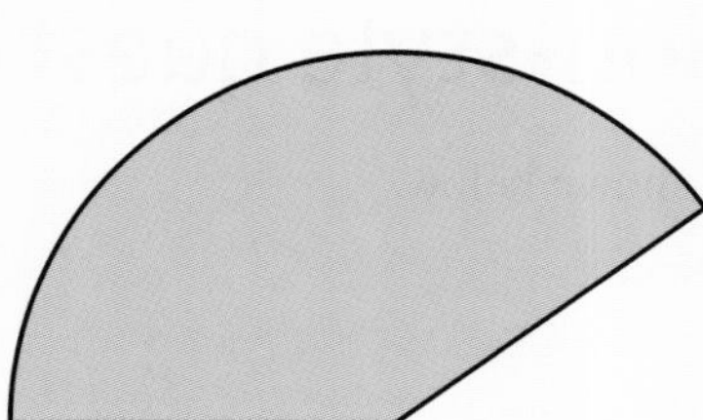

c

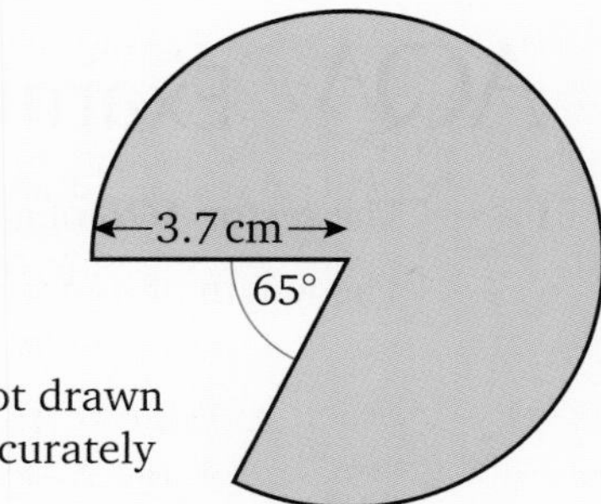

Not drawn accurately

2 An arc of a circle of radius 10 cm is 4π cm long.

What angle does the arc make at the centre of the circle?

3 Which has the bigger volume, a sphere with diameter 24 cm or a cone with diameter 24 cm and height 48 cm?

A*

4 A goat is tethered to the corner of a square field measuring 20 m by 20 m. The goat can graze half the field.

How long is the goat's tethering rope?

5 The Pyramid of Menkaure is the smallest of the three main pyramids at Giza. Its measurements are approximately half those of the Great Pyramid (see start of chapter).

What is its approximate volume?

A*

6 The diagram shows an equilateral triangle with sides 3 cm long. A sector of a circle with radius 1 cm has been cut from each vertex to make the shaded shape.

Find the area and perimeter of this shaded shape.

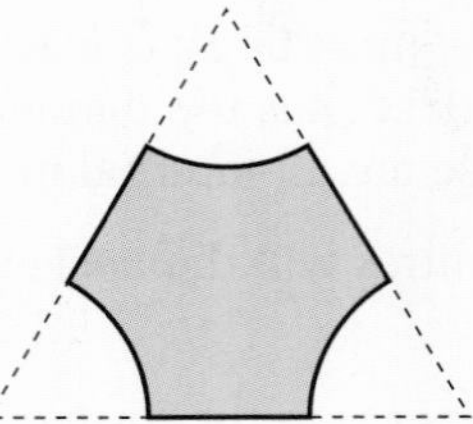

7 A rectangular block of metal measuring 6 cm by 8 cm by 5 cm is melted down. It is recast as spherical ball bearings of diameter 0.9 cm.

How many ball bearings can be made and how much metal is left over?

8 From the top of an 8 cm square-based pyramid with height 10 cm, a pyramid of height 6 cm is removed.

a What fraction of the volume of the original pyramid is the volume of the removed pyramid?

b Find the volume of the remaining frustum of the pyramid.

9 The cross-section of a tunnel is the major segment of a circle with radius 3.2 m. The tunnel is to be 0.75 km long.

How many cubic metres of rock must be removed to make the tunnel?

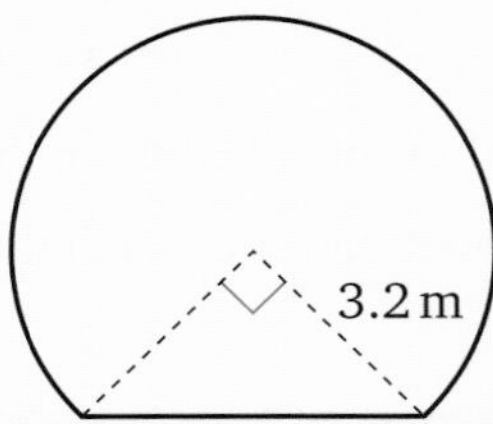

10 An oil storage tank consists of a cylinder with a hemisphere on each end. The cylinder is 1 m long and its diameter is 75 cm. The metal used to make the tank is 1 cm thick.

How many litres of oil will the tank hold?

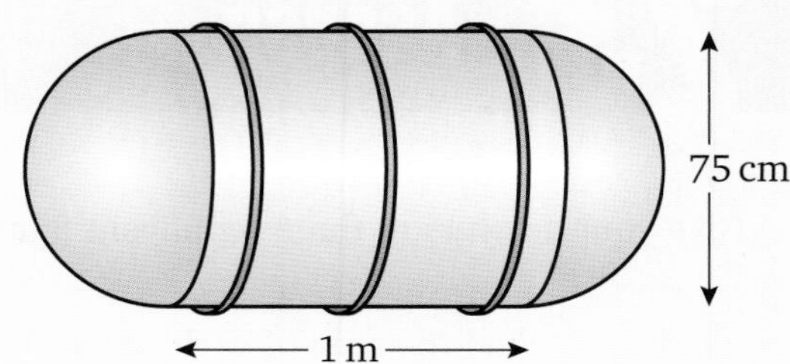

AQA Examination-style questions

1 The sector AOB of a circle is shown below.

The length of its arc AB is 10π cm.

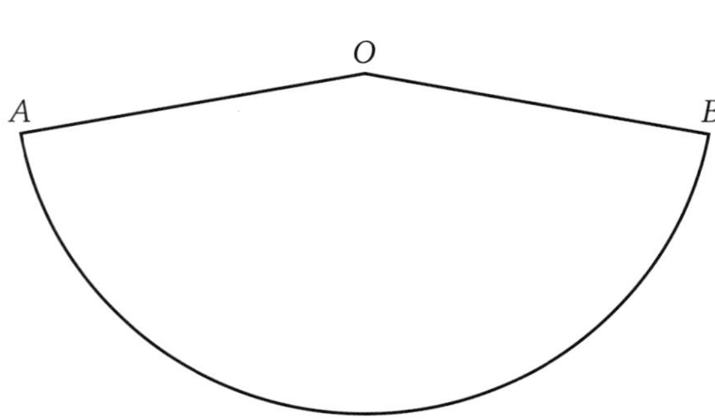

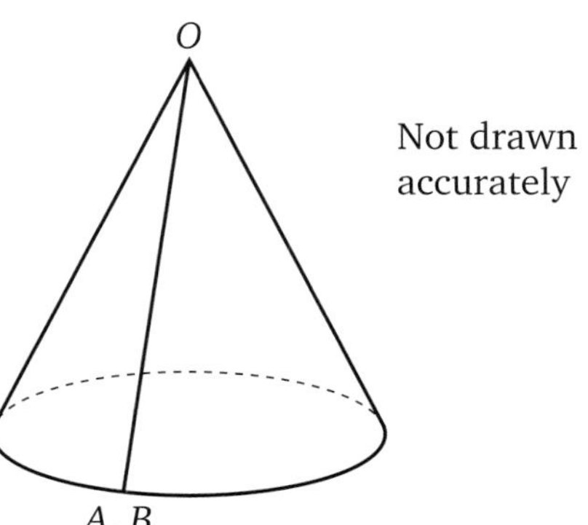

Not drawn accurately

The sector is folded so that the straight edges meet and form a cone as shown.

a Calculate the radius of the base of the cone. *(3 marks)*

b The volume of the cone is $80\pi\,\text{cm}^3$.
Work out the perpendicular height of the cone. *(3 marks)*

AQA 2009

24 Transforming functions

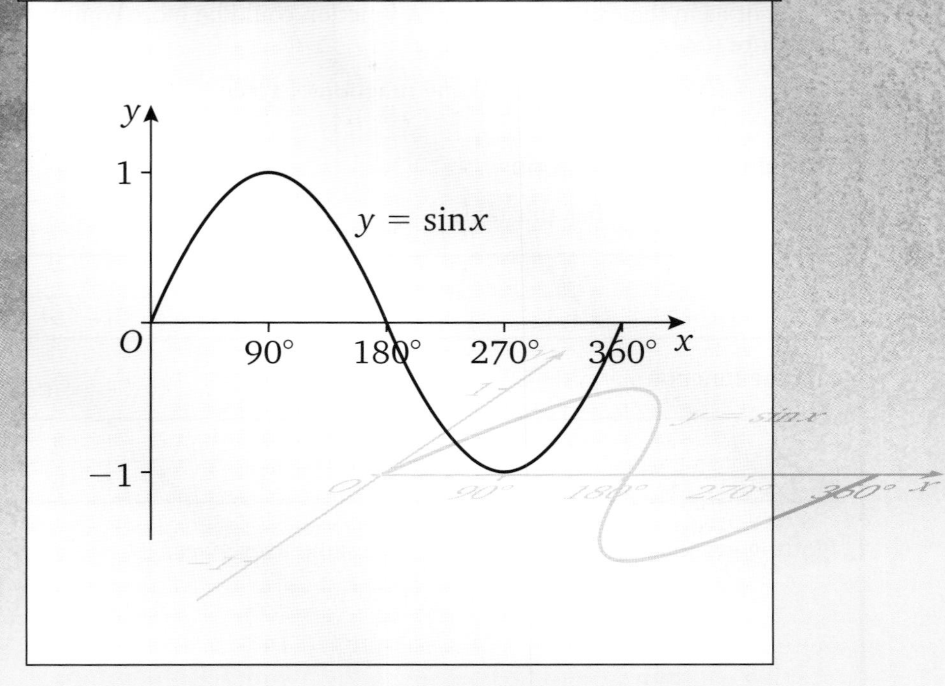

Objectives

Examiners would normally expect students who get these grades to be able to:

A/A*

understand and apply function notation

given $y = \mathrm{f}(x)$ or a sketch of $y = \mathrm{f}(x)$, draw transformations of $\mathrm{f}(x)$

understand that $\frac{y}{a} = \mathrm{f}(x)$ and $y = \mathrm{f}\left(\frac{x}{a}\right)$ represent a one-way stretch with a scale factor a parallel to the y- and x-axes respectively of $y = \mathrm{f}(x)$

understand that $y = \mathrm{f}(x) + a$ and $y = \mathrm{f}(x - a)$ represent translations of $y = \mathrm{f}(x)$.

Did you know?

Just as you can transform shapes, you can also transform graphs

You can apply transformations such as translations and reflections to graphs just as you can to shapes.

The effects can be quite interesting, and can be used to help solve some difficult problems. Given a complex function it is often possible to use transformations to simplify it and then solve simple problems. These solutions can then be 'mapped back'.

This type of work is developed in courses which go beyond GCSE.

Key terms

transformation
function
translation
one-way stretch

You should already know:

- ✔ how to plot graphs
- ✔ the graphs of standard functions including trigonometric functions
- ✔ how to describe translations using vectors.

Learn... 24.1 Transforming functions

You need to be able to use **transformations** on graphs.

If y is a **function** of x, then it can be written as $y = f(x)$. A function of x can be very simple, such as $y = x$, or as in the example, $y = x^2$. A function could be more complex such as $y = 3x^2 - 4x + 7$, and we can write $f(x) = 3x^2 - 4x + 7$

Here $y = f(x)$ is used to refer to any function of x.

Translations of functions

The function $y = x^2$ has this table:

x	−3	−2	−1	0	1	2	3
y	9	4	1	0	1	4	9

and the function $y = x^2 + 3$ has the table:

x	−3	−2	−1	0	1	2	3
y	12	7	4	3	4	7	12

Plotting graphs for both functions on the same axes gives:

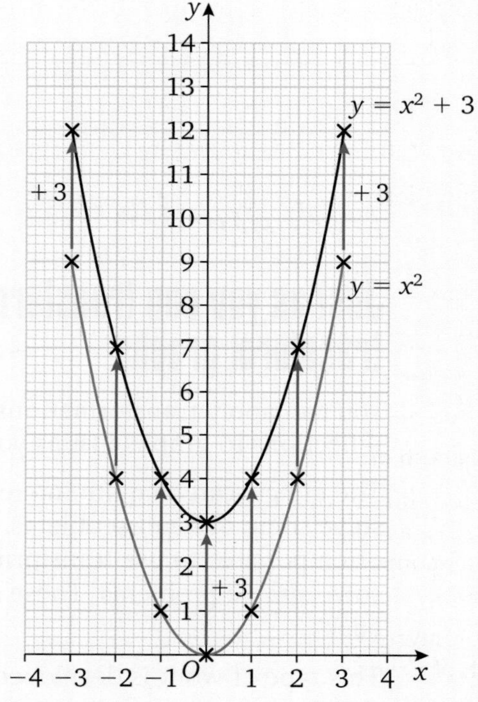

The graph of $y = x^2 + 3$, drawn in black, has exactly the same shape as the graph of $y = x^2$ but is 3 units above it. This is the effect of the '+ 3'.

In general, $y = x^2 + a$ will be the same shape as the graph of $y = x^2$, but will be a units above it.

It is a **translation** with vector $\begin{pmatrix} 0 \\ a \end{pmatrix}$

This result applies to graphs of all functions, so the graph of $y = f(x) + a$ is a translation of the graph of $y = f(x)$ by a vector $\begin{pmatrix} 0 \\ a \end{pmatrix}$

This is true whether a is positive or negative, for example:

- If a is 4 then the translation vector is $\begin{pmatrix} 0 \\ 4 \end{pmatrix}$
- If a is −2 then the translation vector is $\begin{pmatrix} 0 \\ -2 \end{pmatrix}$

It is important to note that $y - a = f(x)$ is the same as $y = f(x) + a$

Now consider $y = (x - 2)^2$

x	−1	0	1	2	3	4	5
y	9	4	1	0	1	4	9

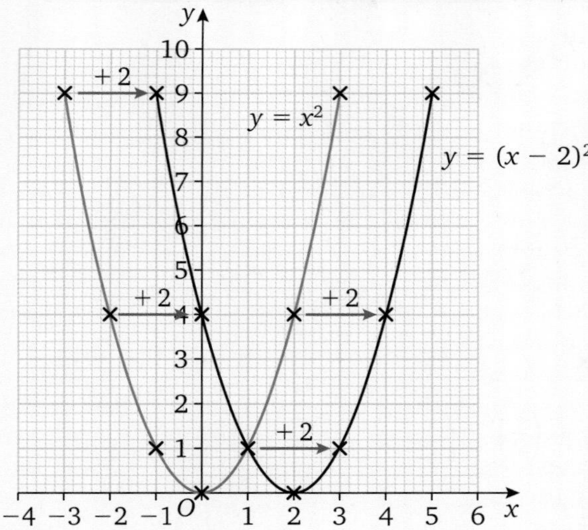

Every point on the graph of $y = x^2$ has 'moved' 2 units to the right.

The graph of $y = (x - 2)^2$ is a translation of the graph of $y = x^2$ by the vector $\begin{pmatrix} 2 \\ 0 \end{pmatrix}$

Similarly, the graph of $y = (x + 3)^2$ is a translation of the graph of $y = x^2$ by the vector $\begin{pmatrix} -3 \\ 0 \end{pmatrix}$

It may help to think of $y = (x + 3)^2$ as $y = (x - (-3))^2$ so a is -3.

We conclude that $y = (x - a)^2$ is a translation of the graph of $y = x^2$ by the vector $\begin{pmatrix} a \\ 0 \end{pmatrix}$

In general, $y = \mathrm{f}(x - a)$ is a translation of the graph of $y = \mathrm{f}(x)$ by the vector $\begin{pmatrix} a \\ 0 \end{pmatrix}$

Stretches

Now consider the function $y = 2x^2$

x	−3	−2	−1	0	1	2	3
y	18	8	2	0	2	8	18

These points are plotted on the graph of $y = x^2$.

The graph shows that all the y-coordinates of $y = x^2$ have been multiplied by 2. This effect is called a stretch in the y-direction, with scale factor 2. Note that any points on the x-axis remain unchanged (because $0 \times 2 = 0$).

The graph of $y = 3x^2$ would be a stretch in the y-direction, with scale factor 3, of the graph of $y = x^2$

In general, the graph of $y = ax^2$ would be a stretch in the y-direction, with scale factor a, of the graph of $y = x^2$

As before, this result applies to graphs of all functions, so the graph of $y = a\mathrm{f}(x)$ is a **one-way stretch** in the y-direction, with scale factor a, of the graph of $y = \mathrm{f}(x)$

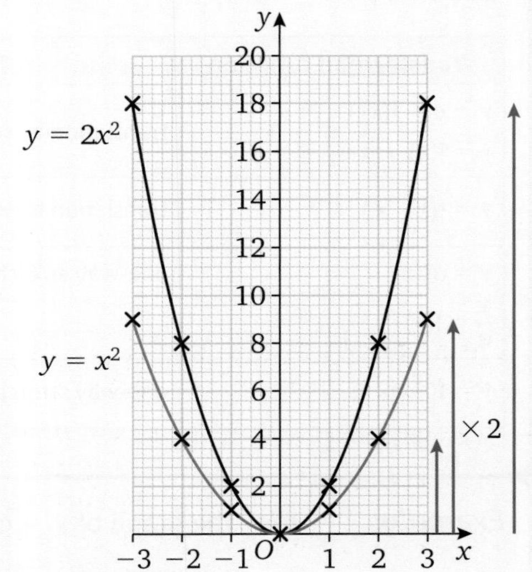

If a is negative, for example, -2, then all the y-coordinates are multiplied by this and the table of values becomes:

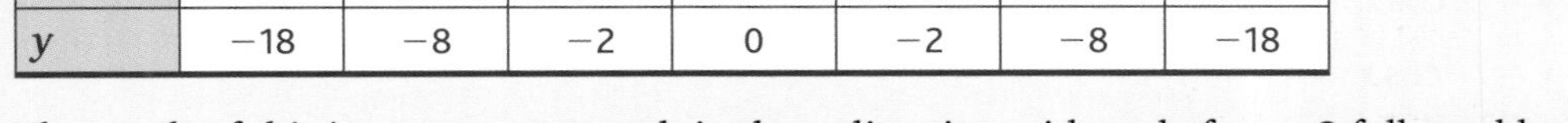

x	−3	−2	−1	0	1	2	3
y	−18	−8	−2	0	−2	−8	−18

The result of this is a one-way stretch in the y-direction with scale factor 2 followed by a reflection in the x-axis. This can be expressed as a single transformation, a one-way stretch in the y-direction, scale factor -2.

It is important to note that $y = a\mathrm{f}(x)$ can be written as $\frac{y}{a} = \mathrm{f}(x)$

Now consider $y = \left(\frac{x}{2}\right)^2$

x	−4	−3	−2	−1	0	1	2	3	4
y	4	2.25	1	0.25	0	0.25	1	2.25	4

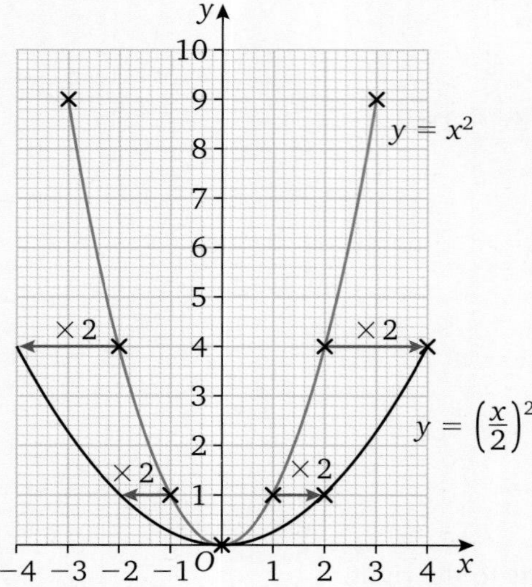

The graph shows that the x-coordinates of $y = \left(\frac{x}{2}\right)^2$ are exactly double the x-coordinates of $y = x^2$
This is a one-way stretch in the x-direction, scale factor 2.

Similarly, $y = (2x)^2$ is a one-way stretch in the x-direction, scale factor $\frac{1}{2}$, of the graph of $y = x^2$
It may help to think of $2x$ as being the same as $\frac{x}{1/2}$

We conclude that the graph of $y = \left(\frac{x}{a}\right)^2$ is a one-way stretch in the x-direction, scale factor a, of the graph of $y = x^2$

In general, the graph of $y = \mathrm{f}\left(\frac{x}{a}\right)$ is a one-way stretch in the x-direction, scale factor a, of the graph of $y = \mathrm{f}(x)$

Summary

Transformed function	Transformation
$y - a = \mathrm{f}(x)$ $y = \mathrm{f}(x) + a$	translation by vector $\begin{pmatrix} 0 \\ a \end{pmatrix}$
$y = \mathrm{f}(x - a)$	translation by vector $\begin{pmatrix} a \\ 0 \end{pmatrix}$
$y = a\mathrm{f}(x)$ $\frac{y}{a} = \mathrm{f}(x)$	one-way stretch in y-direction, scale factor a
$y = \mathrm{f}\left(\frac{x}{a}\right)$	one-way stretch in x-direction, scale factor a

Example: Draw the graph of $y = \cos x$. Use your graph to plot the graphs of:

a $y = \cos x - 3$

b $y = 2\cos x$

c $y = -\cos x$

In each case, state the transformation used.

Solution: Plot the graph of $y = \cos x$

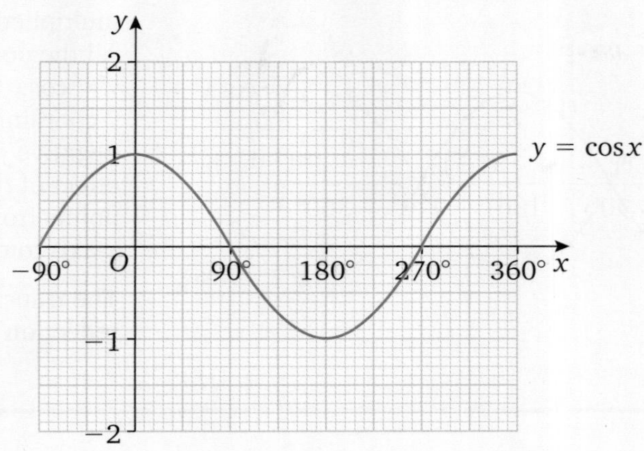

a

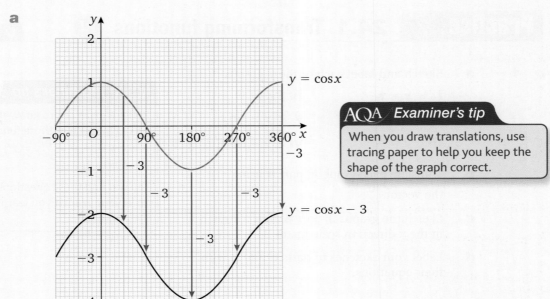

y = cos x
−3
y = cos x − 3

All the y-coordinates have 3 subtracted from them.

The transformation is: a translation with vector $\begin{pmatrix} 0 \\ -3 \end{pmatrix}$ or a translation three units in the negative y-direction.

AQA Examiner's tip

When you draw translations, use tracing paper to help you keep the shape of the graph correct.

b

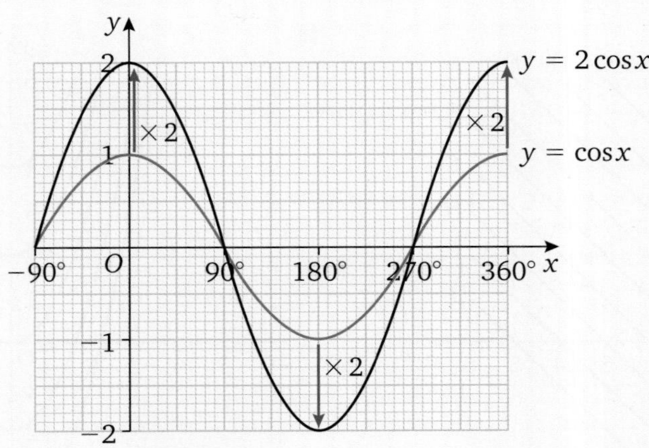

For the graph of $y = 2\cos x$ all the y-coordinates will be multiplied by 2.

The transformation is: a stretch in the y-direction, factor 2.

c

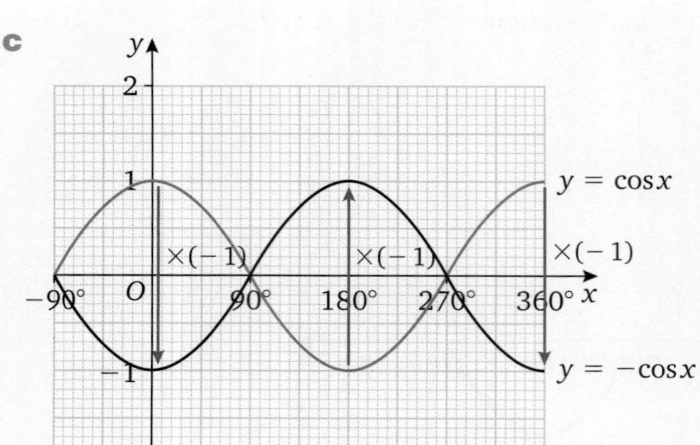

To get the graph of $y = -\cos x$ all the y-coordinates have been multiplied by -1.

All the points on the graph $y = \cos x$ that had positive y-coordinates now have a negative y-coordinate. The effect of this is to move all the points from one side of the x-axis to the other side.

The transformation is: a reflection in the x-axis.

Practise... 24.1 Transforming functions 

A*

1 **a** Sketch and label each of these graphs.

i	$y = x$	**iv**	$y = x^3$
ii	$y = x^2$	**v**	$y = \sin x$
iii	$y = \frac{1}{x}$	**vi**	$y = \cos x$

b Translate each graph in part **a** using the vector $\begin{pmatrix} 0 \\ 4 \end{pmatrix}$

c Sketch the graphs in part **a** after a stretch in the y-direction scale factor 3.

d Label your sketches in parts **b** and **c** with their equations.

> **AQA Examiner's tip**
>
> You are expected to know the shape of the curves for each of these functions: $y = x, y = x^2, y = \frac{1}{x}, y = x^3, y = \sin x$, $y = \cos x$.
>
> You will find these questions much easier to do in your exam if you learn these curves and their equations.

2 For each set of graphs, do the following.

a Write down the equation of the red graph.

b By transforming the red graph, write the equation of each black graph.

c Write the transformation that produces each family of graphs.

i

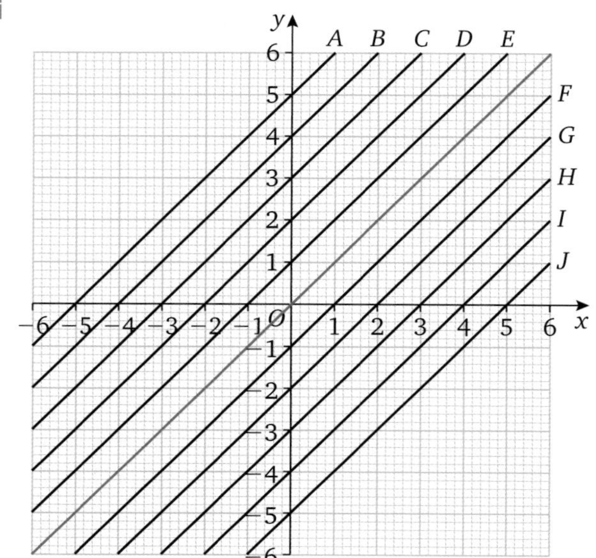

ii

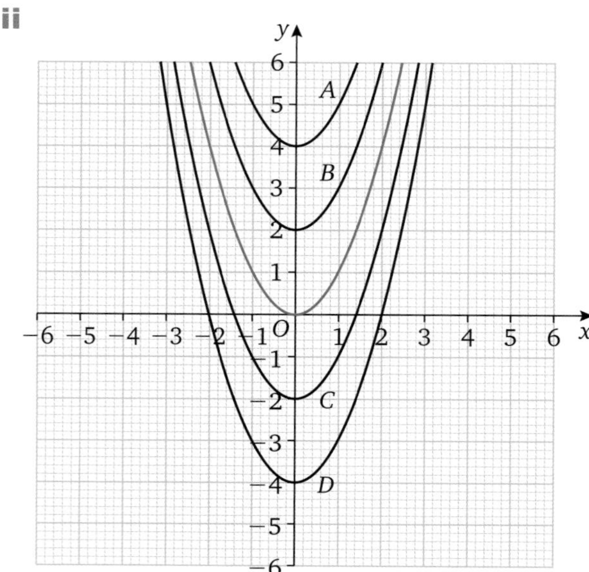

3 This diagram shows seven transformations of the graph of $y = \sin x$ A*

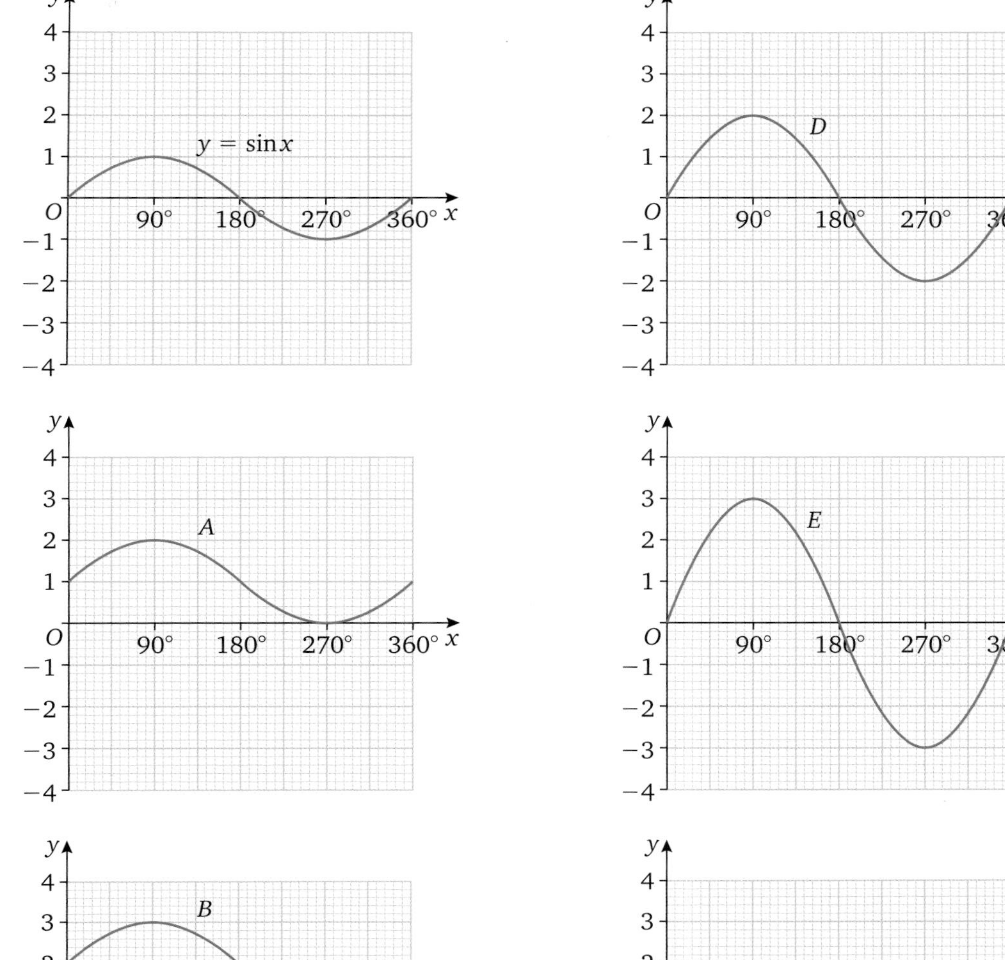

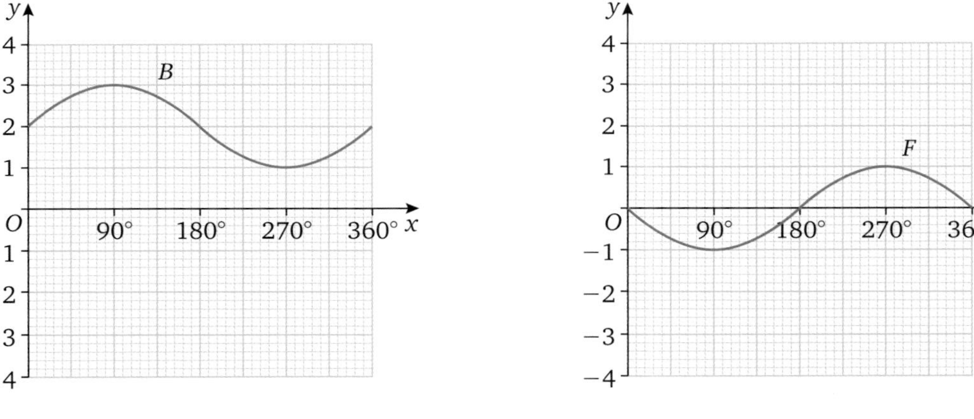

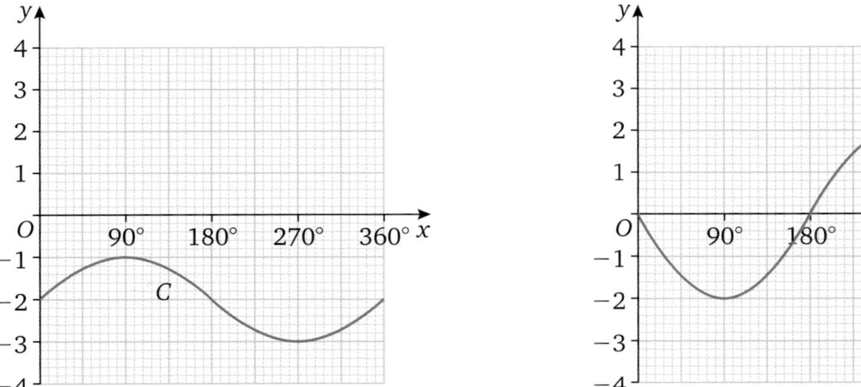

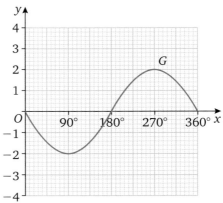

a Match each graph with one of the following functions.

i	$y = 2\sin x$	**iii**	$y = -\sin x$	**v**	$y = \sin x + 1$	**vii**	$y = \sin x - 2$
ii	$y = 3\sin x$	**iv**	$y = -2\sin x$	**vi**	$y = \sin x + 2$		

b Describe the transformation that needs to be carried out to get the graph of each function from the graph of $y = \sin x$

A*

4 This is the graph of $y = x^3 - 2x^2$

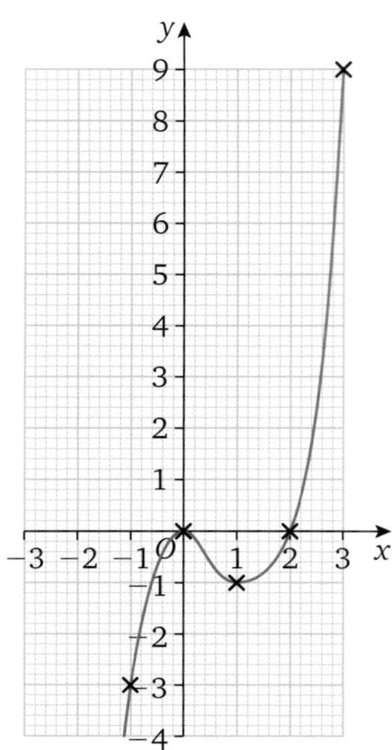

Hint

Think of the graph of $y = x^3 - 2x^2$ as the graph of $y = f(x)$

Sketch the graphs of:

a $y = x^3 - 2x^2 + 2$

b $y = x^3 - 2x^2 - 1$

c $y = 2x^3 - 4x^2$

d $y = 2x^2 - x^3$

5 This is the graph of $y = \frac{1}{x}$

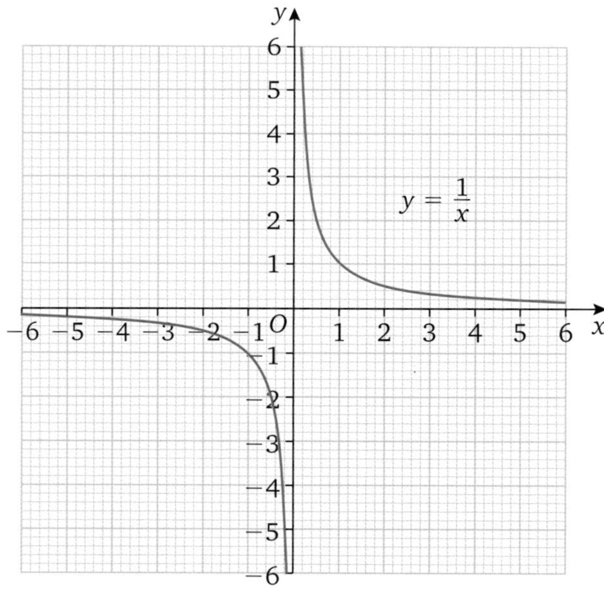

a Sketch each of these graphs.

i $y = \frac{1}{x} + 1$

ii $y = \frac{1}{x} + 2$

iii $y = \frac{1}{x} - 1$

iv $y = \frac{2}{x}$

v $y = \frac{1}{2x}$

vi $y = -\frac{1}{x}$

b Describe the transformation needed to get each graph you have drawn from the graph of $y = \frac{1}{x}$

A*

6 For each of the graphs **a–c**, sketch the following transformations and describe the transformation in each case.

i $y = -f(x)$ **ii** $y = 2f(x)$ **iii** $y = f(x - 1)$

a $y = f(x)$

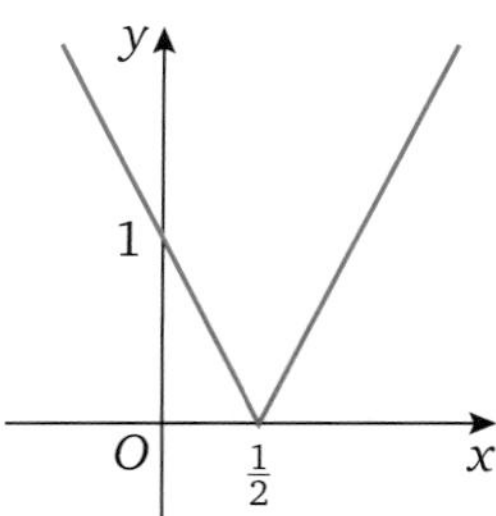

b $y = f(x)$

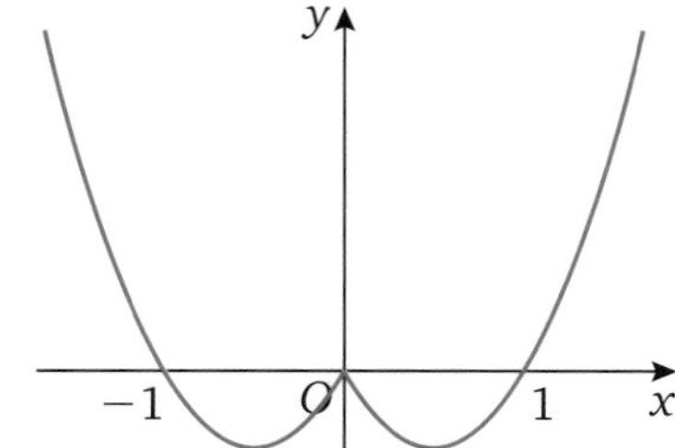

c $y = f(x)$

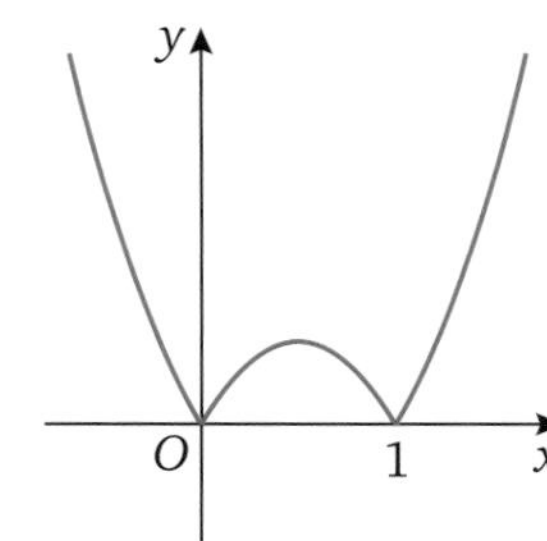

7 James, Ken, Hamish and Faisal have been given this graph by their teacher.

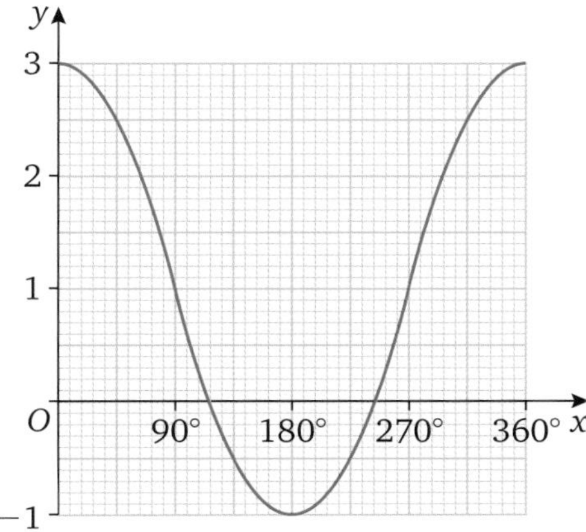

James says it is the graph of $y = 3\cos x$ because it starts at 3 on the y-axis.
Ken says it is the graph of $y = \cos x + 2$ because it goes up to 3 at its highest.
Hamish says it is the graph of $y = 2\cos x + 1$ because it is stretched and translated.
Faisal thinks that one of them is likely to be correct, but is not sure which one.

Who should Faisal choose? Explain your answer, and justify why he should not choose either of the others.

8 Lucy has been given the graph of $y = f(x)$ by her teacher.
He asks her to draw the graph of $y = af(x) + b$

a What transformations should she carry out and in which order should she do them?

b Check your answer on some examples using $f(x) = x^2$

9 What are the effects of changing the values of a, b and c in the function $y = a(x - b)^2 + c$?

Summarise your results in terms of simple transformations of the graph of $y = x^2$

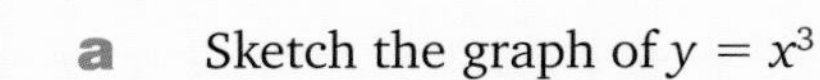

1 **a** Sketch the graph of $y = x^3$

b Sketch the graphs of:

i $y = x^3 + 2$

ii $y = x^3 - 2$

iii $y = 2x^3$

iv $y = \frac{1}{2}x^3$

v $y = -2x^3$

c Describe the transformation needed to obtain each graph you have drawn from the graph of $y = x^3$

2 Look at these graphs.

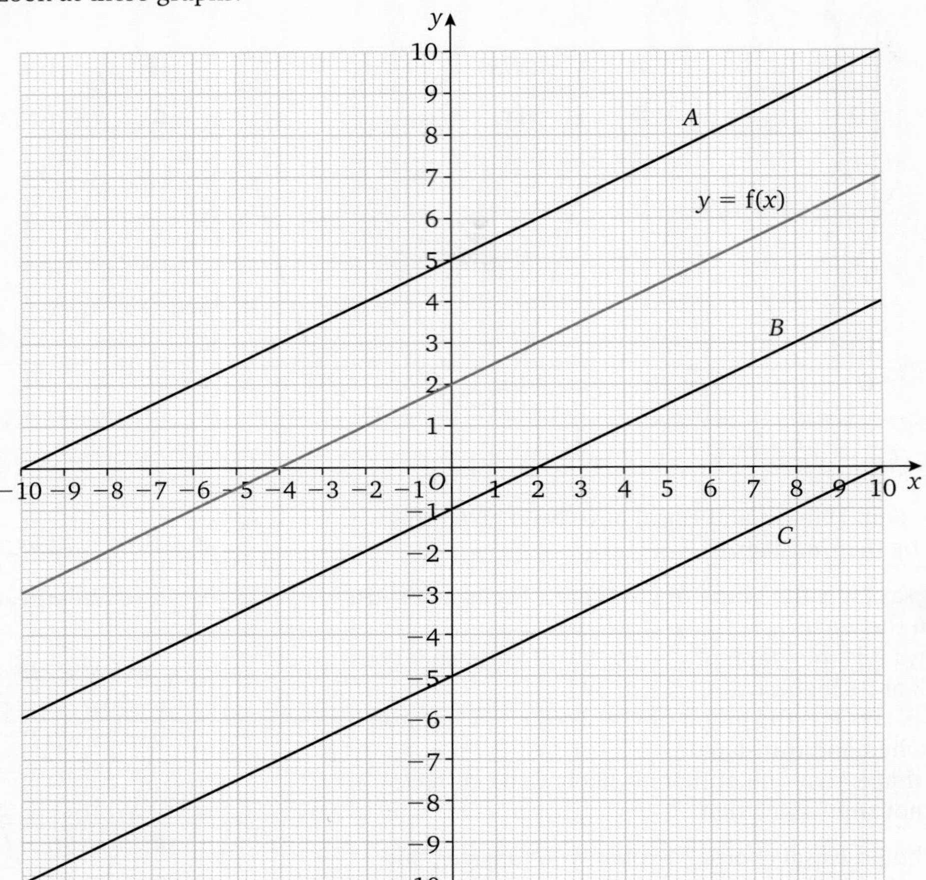

a What is the equation of the line labelled $y = f(x)$?

b For each of the graphs labelled A, B and C, write their equations in the form $y = f(x) + a$

c Work out the equations of lines A, B and C in the form $y = mx + c$ and show that you get the same answers as you did in part **b**.

3 Hamilton is given the graph of $y = f(x)$. He knows that the graph of $y = f(x) - 1$ is a translation of the graph of $y = f(x)$ by vector $\begin{pmatrix} 0 \\ -1 \end{pmatrix}$. He says that the graph of $y = f(x - 1)$ must be a translation by vector $\begin{pmatrix} -1 \\ 0 \end{pmatrix}$

Is Hamilton correct? Explain your answer.

AQA Examination-style questions

1 The graph of $y = 3 - x^2$ is sketched below.

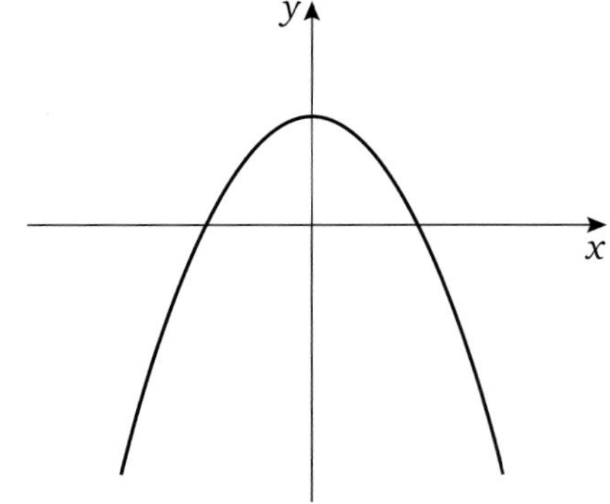

Not drawn accurately

Copy the sketch and on the same axes, sketch the graph of $y = 7 - x^2$ *(1 mark)*

AQA 2009

Consolidation 3

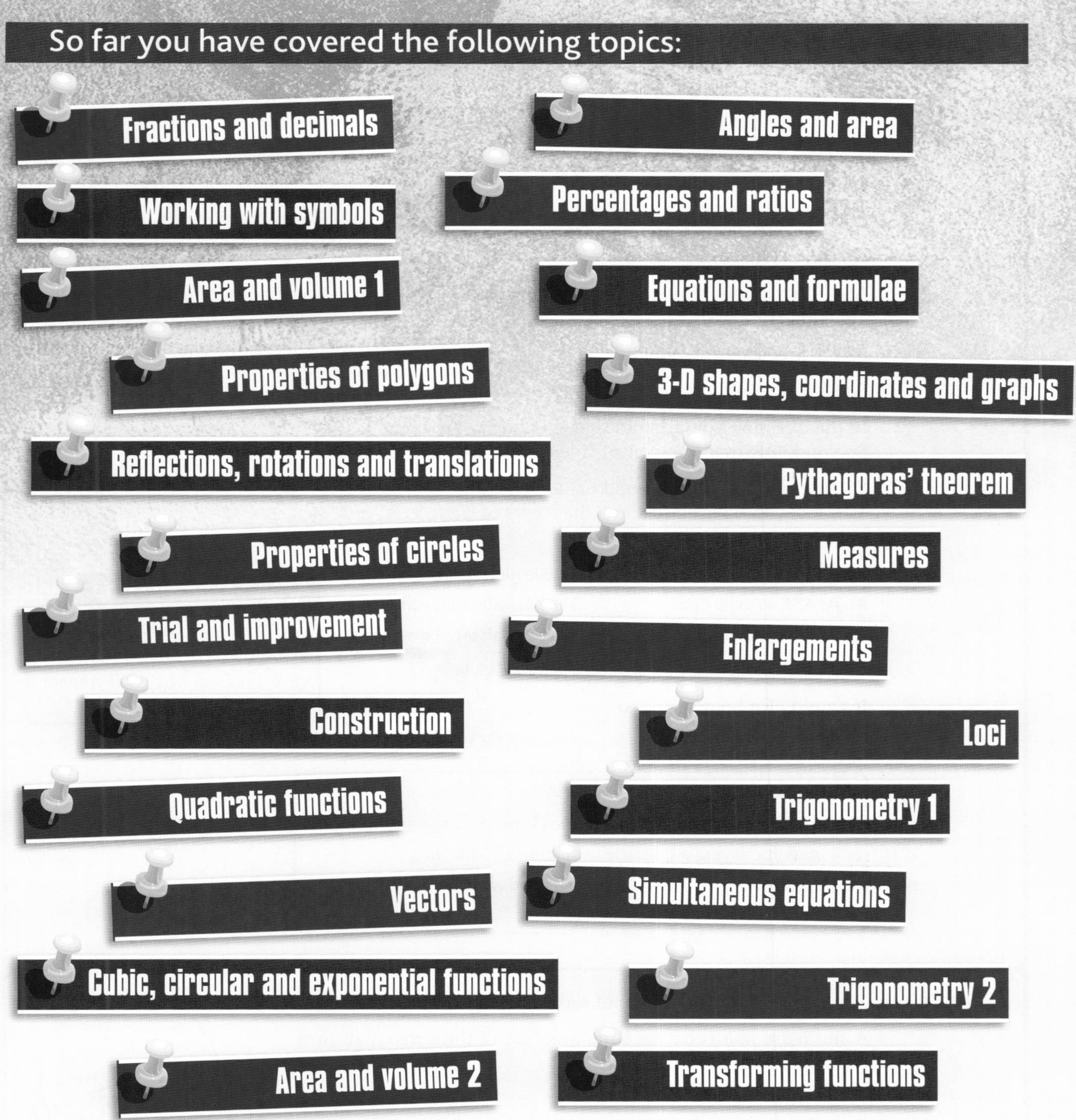

All these topics will be tested in this chapter and you will find a mixture of problem solving and functional questions. You won't always be told which bit of mathematics to use or what type a question is, so you will have to decide on the best method, just like in your exam.

Example: In the diagram, the lines *AC* and *BD* intersect at *E*.

AB and *DC* are parallel and $AB = DC$

Prove that triangles *ABE* and *CDE* are congruent.

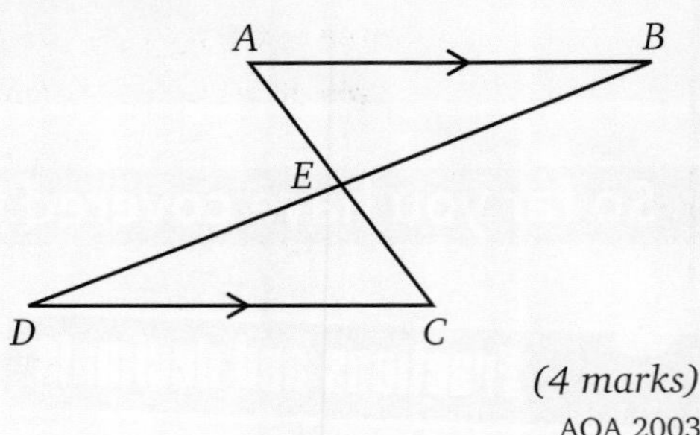

(4 marks)

AQA 2003

Solution: To prove that a pair of non-right angled triangles are congruent, one of these conditions has to be met:

ASA (Two angles and a corresponding side are equal.)

SSS (Three sides are equal.)

SAS (Two sides and the angle between them.)

In the question you are told that *AB* and *CD* are equal. So you know that one pair of sides are equal.

You are also told that *AB* and *CD* are parallel. This fact can be used to show that pairs of angles in the triangles are equal.

> **AQA Examiner's tip**
>
> A useful first step in trying to prove that two triangles are congruent is to write down the conditions for any two triangles to be congruent.
>
> Then try to decide the one that is most likely to apply to the question.

$AB = CD$	This is given in the question.
Angle ABE = Angle CDE	They are alternate angles.
Angle EAB = Angle ECD	They are alternate angles.

So triangles *ABE* and *CDE* are congruent by ASA.

It would also be possible to state that:

Angle AEB = Angle CED	They are vertically opposite.

> **AQA Examiner's tip**
>
> Try to set your proof out a step at a time.
>
> Give a reason why each step is true.

> **Mark scheme**
>
> 1 mark for noting that $AB = CD$ and choosing a correct method.
> - 1 mark for saying angle ABE = angle CDE and why.
> - 1 mark for angle EAB = angle ECD and why.
> - 1 mark for the final correct statement.

Example: The diagram shows a circle of radius 8 centimetres, centre *O*.

A and *B* are points on the circumference of the circle that form a triangle with *O*.

Angle *AOB* is 120°.

Work out the area of the shaded segment.

Give your answer to an appropriate degree of accuracy.

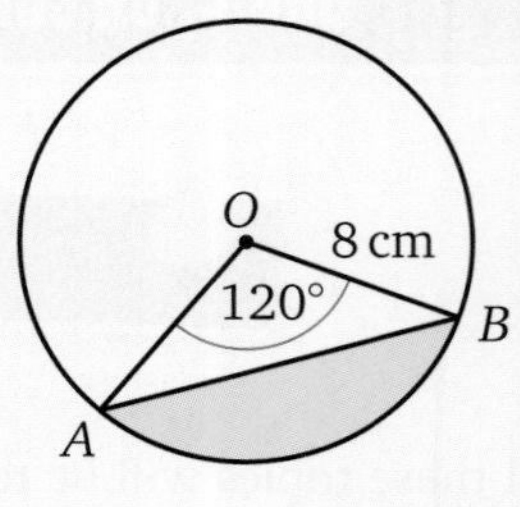

(4 marks)

Solution: The shaded area = area of sector AOB − area of triangle AOB

The angle AOB is 120° so the sector is $\frac{120}{360}$ $(= \frac{1}{3})$ of the circle.

The area of a sector with an angle of $\theta°$ is $\frac{\theta}{360} \times \pi r^2$

Area of sector $= \frac{120}{360} \times \pi \times 8^2 = 67.0206\ldots \text{ cm}^2$

> **AQA Examiner's tip**
>
> Do not round your answer too soon as this might affect your final answer. Work out the area of the sector to at least six significant figures.

Area of triangle = $\frac{1}{2}ab\sin C$

Area of triangle = $\frac{1}{2} \times 8 \times 8 \times \sin 120°$

= 27.7128...cm^2

Shaded area = 67.0206... cm^2 − 27.7128... cm^2

= 39.3078... cm^2

= 39.3cm^2 (3 s.f.)

AQA Examiner's tip

In the question you were asked to give your answer to an appropriate degree of accuracy. Here three significant figures are appropriate. 39 cm^3 (rounded to the nearest whole number) would also gain a mark.

Mark scheme

- 1 mark for $\frac{120}{360} \times \pi \times 8 \times 8$
- 1 mark for $\frac{1}{2} \times 8 \times 8 \times \sin 120°$
- 1 mark for subtracting the triangle's area from the sector's area.
- 1 mark for the correct final answer.

Consolidation

D

1 A rectangle has a perimeter of $14x$ cm and an area of $12x^2$ cm^2.

Work out the length and width of the rectangle in terms of x.

2 This diagram represents an acre.
The acre is used to measure large areas of land.

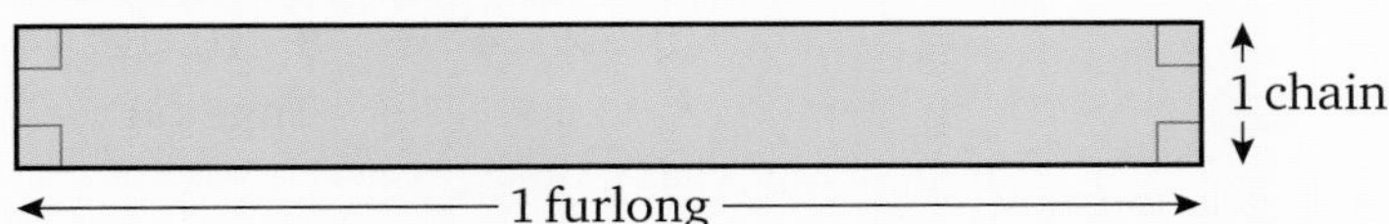

The metric unit for measuring large areas is the hectare.

1 hectare is approximately 2.5 acres.

1 furlong = 10 chains

1 chain = 22 yards

Use these values to estimate the number of square yards in 1 hectare.

3 The price of tickets for a boat trip to the Farne Islands is:

Adults	£12
Children	£8

On one of the trips, there are x adults and y children.

£T is the total price of their tickets.

a Write a formula for T in terms of x and y.

b On one boat trip, the total price of the tickets is £672.
The number of children's tickets sold is 18.
How many adult tickets are sold on this trip?

D

4 *ABCDEF* is a regular hexagon with centre *O*.
The hexagon is divided into congruent triangles.
G is the midpoint of *AB*.
H is the midpoint of *ED*.
Two triangles, *X* and *Y*, are labelled.

Triangle *X* maps onto triangle *Z* by a rotation of 180° about *O* followed by a reflection in the line *BD*.
Triangle *X* maps onto triangle *Y* by a translation onto triangle *T* followed by a reflection.

a On a copy of the hexagon, mark the position of triangles *Z* and *T*.

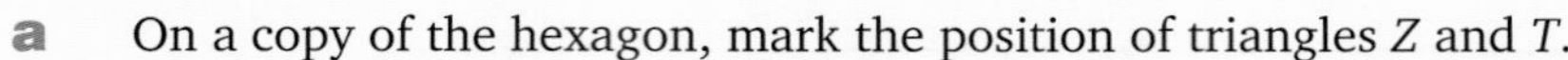

b Describe fully the single transformation that maps triangle *T* onto triangle *Y*.

5 Orange squash is made by mixing concentrated orange juice with water.

Holly is making some orange squash.

She has poured 50 millilitres of concentrated orange juice into a cylindrical glass.

The graph shows what happens next.

a Describe fully what is happening during the first three seconds.

b Write the ratio of concentrated orange juice to water in its simplest form when the orange squash has been made.

c For how many seconds was Holly drinking squash from the glass?

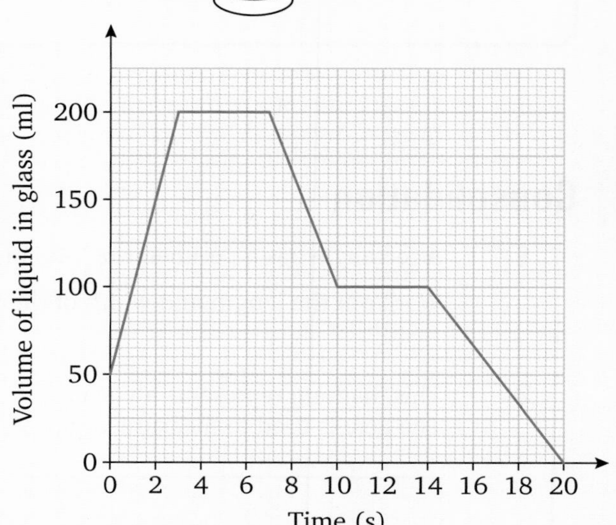

6 Here are details of Sudhir's bicycle journey.

Stage 1: After the start he cycles at a speed of 12 km/h for $2\frac{1}{2}$ hours.

Stage 2: He stops for 30 minutes.

Stage 3: He cycles back towards the start for 1 hour, travelling 10 km.

Stage 4: He stops for another 30 minutes.

Stage 5: He cycles back to the start at a speed of 8 km/h.

On a copy of these axes, draw a distance–time graph to represent Sudhir's journey.

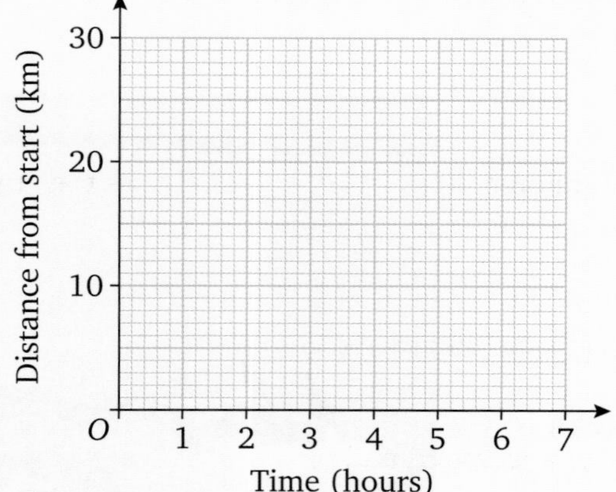

D C

7 There are some red counters, white counters and blue counters in a bag.
There are 50% more white counters than red counters.
The number of blue counters is three-fifths of the number of white counters.

Write the ratio of the numbers of red to white to blue counters in its simplest form.

AQA Examiner's tip

Although you can use an algebraic method to solve a question like this, sometimes it can be easier to solve it by choosing a particular number (100 is best) for the red counters and then working out the others from this number.

8 **a** The perimeter of this rectangle is $8x + 4$.

Write down an expression for the area of this rectangle.

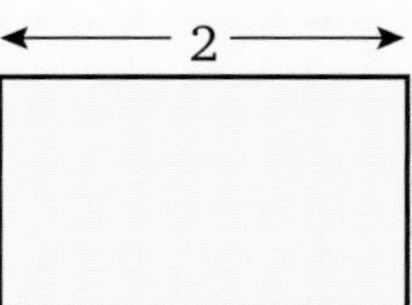

b Write down an expression for the area of a square with perimeter $8x + 4$.

9 In a test there are 10 questions.
If you attempt a question and get it right, you score 5 marks.
If you attempt a question and get it wrong, you lose 2 marks.
If you do **not** attempt a question, you lose 10 marks.

a The table shows how three friends, Andrew, Bill and Clare, do in the test.

Name	Number of questions attempted	Number of questions correct
Andrew	10	6
Bill	9	8
Clare	8	7

In the test the mark for a Merit is 30.
The mark for a Pass is 20.

How well do the three friends do in the test?
You must show working to justify your answer.

b Tim takes the test.
He scores −7.

How many questions did Tim **not** attempt?
You must show working to justify your answer.

10 **a** Rachel tries to draw a quadrilateral with exactly three interior right angles.

Explain why she finds this impossible.

b Tom says that he can draw a hexagon with exactly five interior right angles.

Is this possible?
Show working to justify your answer.

C

11 Work out the coordinates of the point where the graph of $y = 2x - 3$ intersects the graph of

$x + 2y = 4$

12 Plot the points $A(1, 2)$, $B(4, 1)$ and $C(2, 0)$ on a centimetre grid.

a Show that $AC = BC$

b Draw the locus of points that are equidistant from A and B.

C

13 In box A there are x beads.
In box B there are two more beads than in box A.
In box C there are three times as many beads as there are in box B.
In box D there are five times as many beads as there are in box A.
In box E there are four more beads than there are in box D.

Show that the total number of beads in boxes A, B, C and D is double the number of beads in box E.

14 Use trial and improvement to find a solution to the equation $x^3 - 2x = 45$

The table shows the first trial.

x	$x^3 - 2x$	Conclusion
3	$3^3 - 2 \times 3 = 21$	too small

Continue the table to find a solution to the equation.
Give your answer to one decimal place.

15 Adil and James are planning a mountain walk.

They find this rule to help them estimate how long the walk will take.

> Estimating the time of a mountain walk
>
> On a mountain walk it takes:
>
> 1 hour for every 3 miles travelled horizontally
>
> *plus*
>
> 1 hour for every 2000 feet climbed
>
> Add 10 minutes of resting time for each hour you walk

When planning their mountain walk, Adil and James estimate they will:

- travel 24 kilometres horizontally
- climb for 900 metres.

They plan to start their walk at 09:00.

At what time are they likely to complete their walk?

16 David has a large number of 2p coins.

He puts them into four piles: A, B, C and D.

In pile B there are 30 more coins than in pile A.

In pile C there are 12 times as many coins as there are in pile A.

In pile D there are 8 times as many coins as there are in pile A.

David notices that the total number of coins in piles A, B and D is equal to the number of coins in pile C.

What is the total number of 2 pence coins that David has?

C

17 This formula gives the stopping distance of a car travelling on a dry road in terms of its speed.

$$d = \frac{v^2}{150} + \frac{v}{5}$$

d is the stopping distance in metres

v is the speed of the car in km/h

a **i** Copy and complete this table of values for $d = \frac{v^2}{150} + \frac{v}{5}$

v (km/h)	0	15	30	45	60	75	90	105	120
d (metres)					36			94.5	

ii Draw the graph of $d = \frac{v^2}{150} + \frac{v}{5}$ for values of v from 0 to 120.

b The stopping distance on a wet road is double that on a dry road.
Use your graph to answer these questions.

i A car is travelling at a speed of 95 km/h on a wet road.
Estimate its stopping distance.

ii The speed limit on a motorway in the UK is 112 km/h.
A car is forced to brake suddenly in wet conditions on a motorway.
Its stopping distance is 240 metres.
Was the car travelling within the speed limit?

C B

18 The volume V of this square-based pyramid is given by the formula $V = \frac{1}{3}x^2h$ where x and h are measured in centimetres.

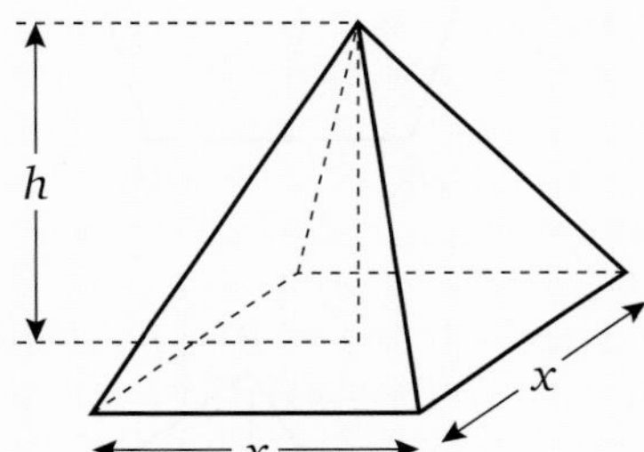

a Work out the volume of the pyramid when $x = 6$ and $h = 8$

b Aaron has been asked to make a square-based pyramid.
The height must be 2 cm bigger than the length of the base.
The volume of the pyramid must be 125 cm³.
He uses trial and improvement to work out the length of the base, x.

Complete his worksheet to find x to one decimal place.

Length of base, x (cm)	Height, h	x^2	Volume (cm³)	
7	9	49	147	too big

C B

19 **a** Work out the length x in this triangle.

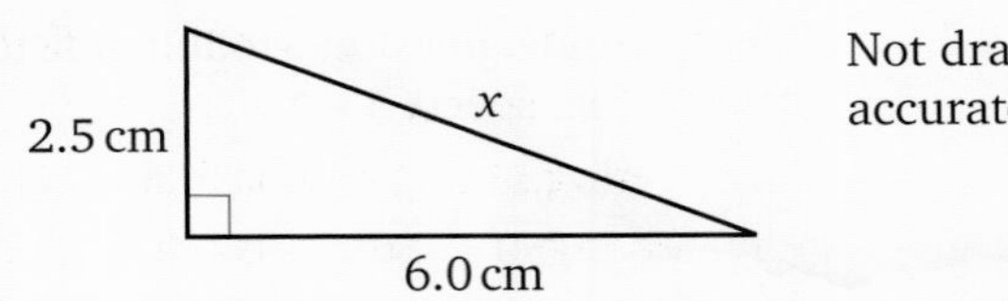

b The diagram shows a 2-metre long ladder leaning against a vertical wall.

The ladder reaches 1.8 metres up the wall from the ground.
To use a ladder safely, the ladder should be inclined with a gradient of 4.

Is the ladder safe to use in the position shown?

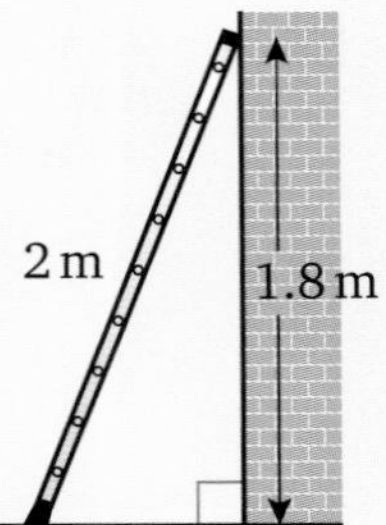

c The sides of an equilateral triangle are 12 centimetres long.

Work out the height, h, of the equilateral triangle.

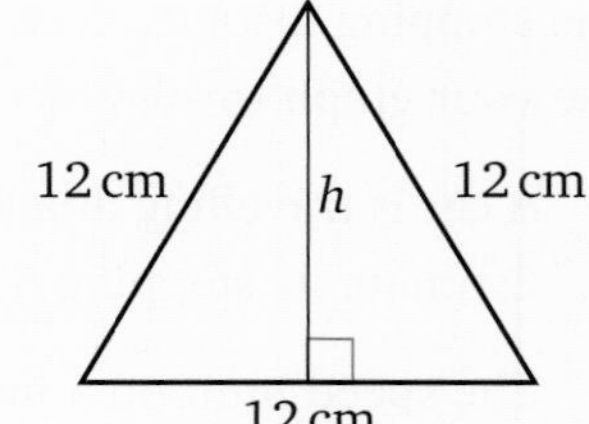

B

20 $ABCDE$ is a regular pentagon.

Prove that $ACDE$ is a trapezium.

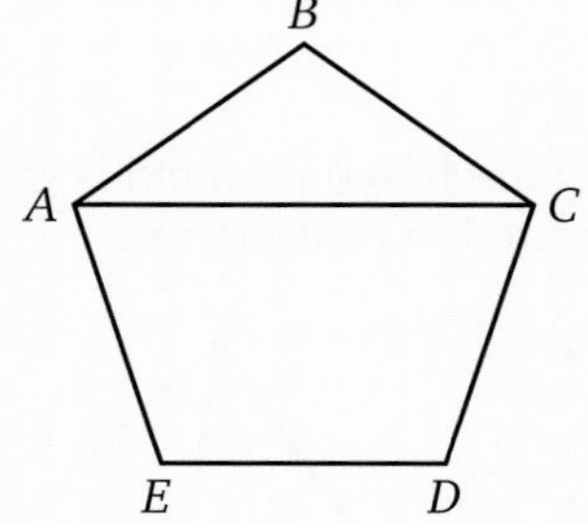

21 Work out the length of CD.

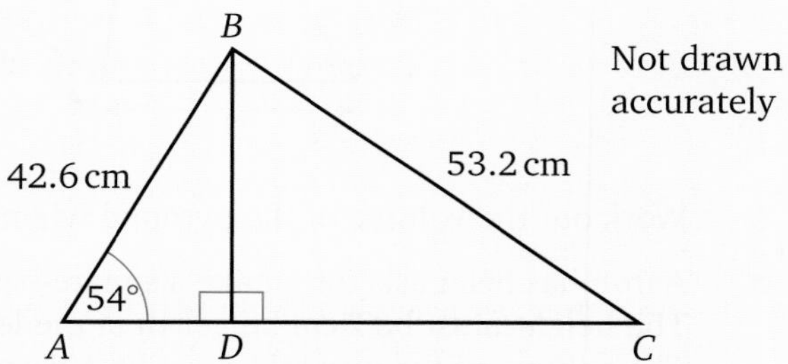

22 Mrs Sim has this photo frame.

The photo frame is designed for a photograph of length 24 cm and width 17.5 cm.
Mrs Sim has a photograph of her daughter stored on her digital camera.
The print of this photograph has length 8 cm and width 5 cm.

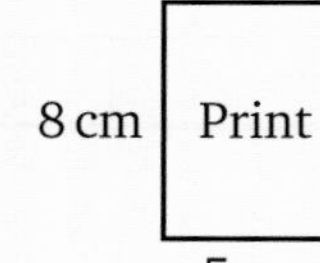

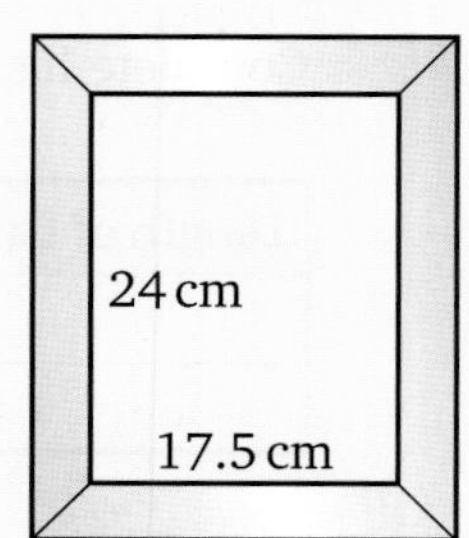

Not drawn accurately

a Show that an enlargement of the print will **not** fit in the photo frame exactly.

b Mrs Sim decides to crop the 8 cm side of her print before enlarging.
What length should she make this side so that the enlargement will fit in the photo frame exactly?

23 The diagram shows a sketch of a field, $ABCD$.
$AB = 64$ metres and $BC = 102$ metres.
The bearing of B from A is 348°.
The bearing of C from A is 040°.
The bearing of D from C is 162°.
D is due east of A.

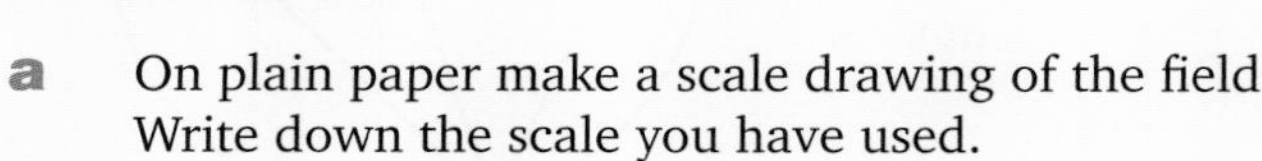

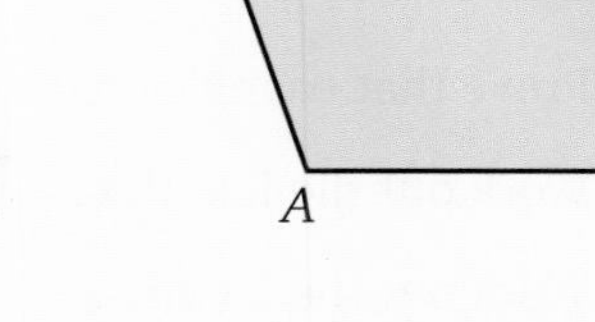

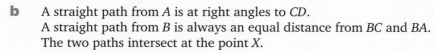

a On plain paper make a scale drawing of the field.
Write down the scale you have used.

b A straight path from A is at right angles to CD.
A straight path from B is always an equal distance from BC and BA.
The two paths intersect at the point X.

Using ruler and compasses only, construct both of these paths on your scale drawing of the field.

c Billy runs along the path from A to X, then along the path from X to B in 30 seconds.
Work out his average speed in metres per second.

24 **a** Copy and complete the table for $y = x^3 - 3x - 5$

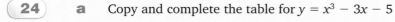

x	−2	−1	0	1	2	3
y		−3	−5			13

b Draw axes with values of x from −2 to 3 and values of y from −8 to +14.
On these axes, draw the graph of $y = x^3 - 3x - 5$ for values of x

c Use your graph to solve $x^3 - 3x - 5 = 8$

25 Mrs Grey can buy four cakes and three drinks for £7.

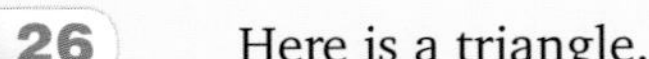

If she buys two more cakes and one more drink the cost will be £10.20.

Work out the cost of one cake.

26 Here is a triangle.

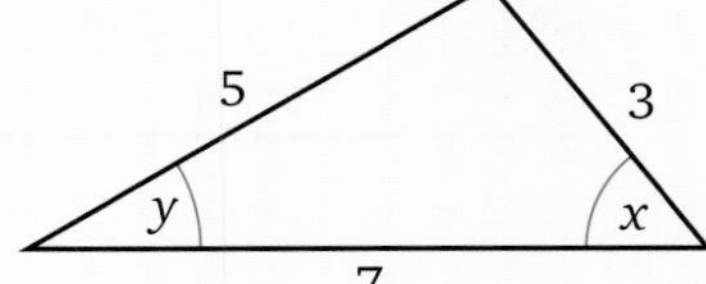

a Show that the triangle cannot be right angled.

b Explain why angle x is greater than angle y.

c Which one of these six statements is correct?

i $\dfrac{\sin x}{7} = \dfrac{\sin y}{5}$ **iv** $\dfrac{\sin x}{5} = \dfrac{\sin y}{7}$

ii $\dfrac{\sin x}{3} = \dfrac{\sin y}{7}$ **v** $\dfrac{\sin x}{3} = \dfrac{\sin y}{5}$

iii $\dfrac{\sin x}{5} = \dfrac{\sin y}{3}$ **vi** $\dfrac{\sin x}{7} = \dfrac{\sin y}{3}$

27 **a** Factorise fully $w \times w \times 2 + w + w + w + w + 2$

b Simplify $(x + 2)^2 - (x - 2)^2$

c Prove that $\dfrac{1}{x - 1} + \dfrac{1}{x + 1} = \dfrac{2x}{x^2 - 1}$

B A

28 $ABCD$ is a cyclic quadrilateral.
The diagonals AC and BD intersect at X.
$BX = 7$ cm, $CX = 4$ cm and $DX = 6$ cm
Angle $BXC = 138°$

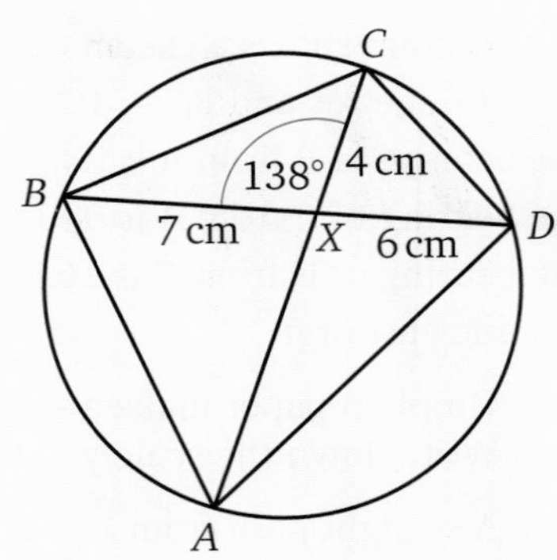

Not drawn accurately

a Prove that triangles BXC and AXD are similar.

b Work out the length of AX.

c Work out the length of BC.

A

29 The volume of a sphere is $288\pi \text{ mm}^3$.

Calculate the radius of the sphere.

30 Here is a triangle.

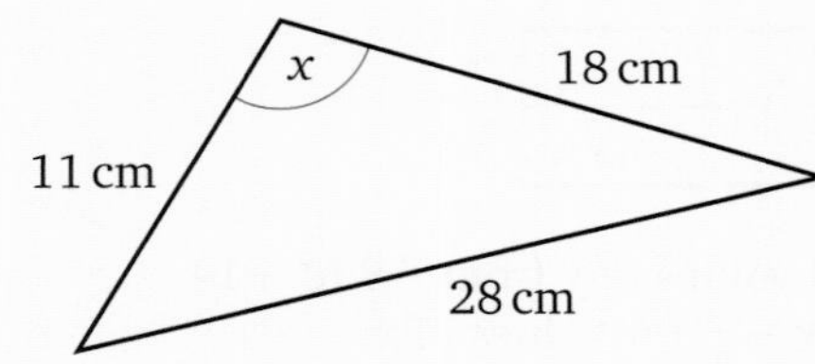

Not drawn accurately

Work out angle x.

31 $ABCD$ is a quadrilateral.

$\overrightarrow{AB} = \begin{pmatrix} 3 \\ 1 \end{pmatrix}$ $\overrightarrow{BC} = \begin{pmatrix} 2 \\ -5 \end{pmatrix}$ $\overrightarrow{CD} = \begin{pmatrix} -5 \\ -2 \end{pmatrix}$

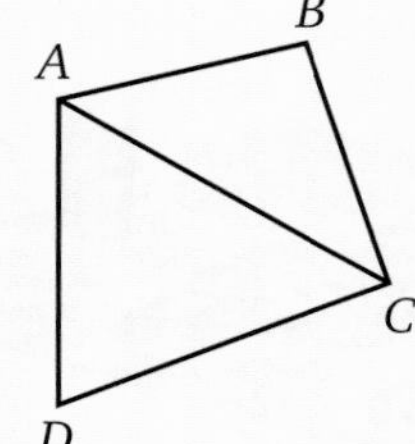

Not drawn accurately

Prove that D is vertically below A.

32 Given that y is inversely proportional to x^2, copy and complete the table.

x	5	10		
y	5			500

A A*

33 $\overrightarrow{OP} = -2\mathbf{a} + 4\mathbf{b}$ and $\overrightarrow{OQ} = 4\mathbf{a} - 2\mathbf{b}$

a Express $\overrightarrow{PQ}$ in terms of $\mathbf{a}$ and $\mathbf{b}$.

b R is the midpoint of $\overrightarrow{PQ}$.
Express $\overrightarrow{OR}$ in terms of $\mathbf{a}$ and $\mathbf{b}$.

c $\overrightarrow{PS} = 7\mathbf{a} + \mathbf{b}$
Express $\overrightarrow{OS}$ in terms of $\mathbf{a}$ and $\mathbf{b}$.

d What **two** facts do $\overrightarrow{OR}$ and $\overrightarrow{OS}$ indicate about the points O, S and R.

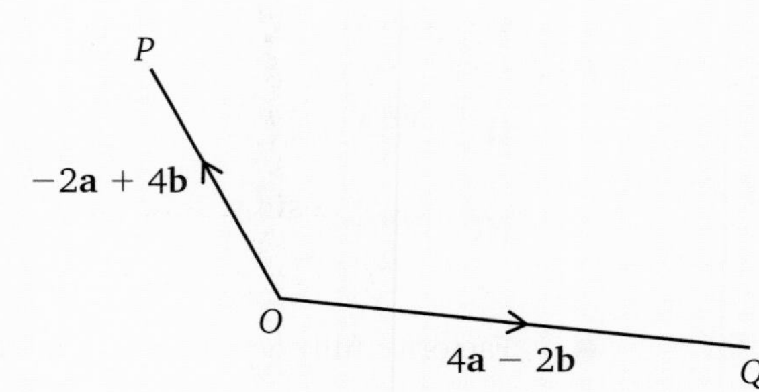

A A*

34 **a** Here are four equations.

A $y = \dfrac{3}{x}$ **B** $y = 3^x$ **C** $y = 3x$ **D** $y = x^3$

Match each graph to its equation.

i

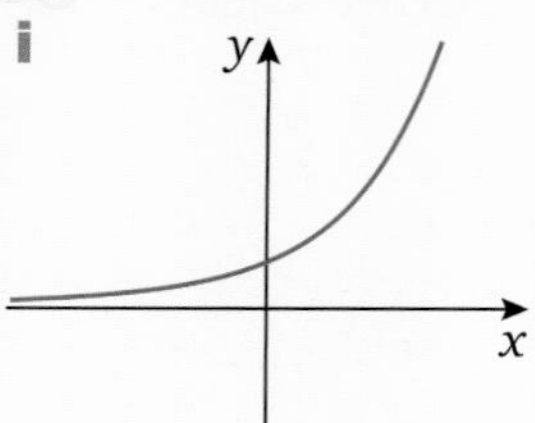

ii

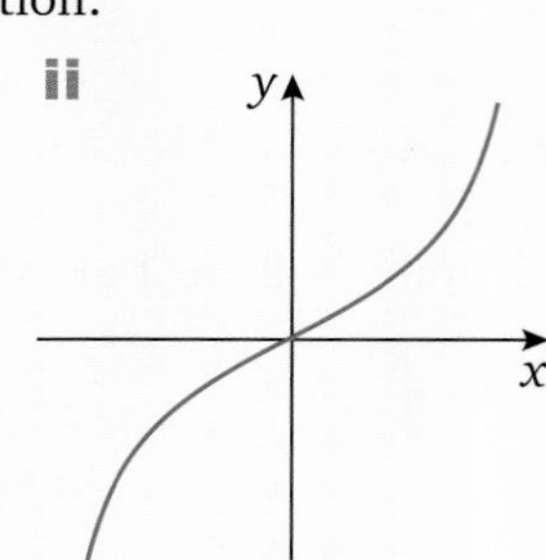

iii

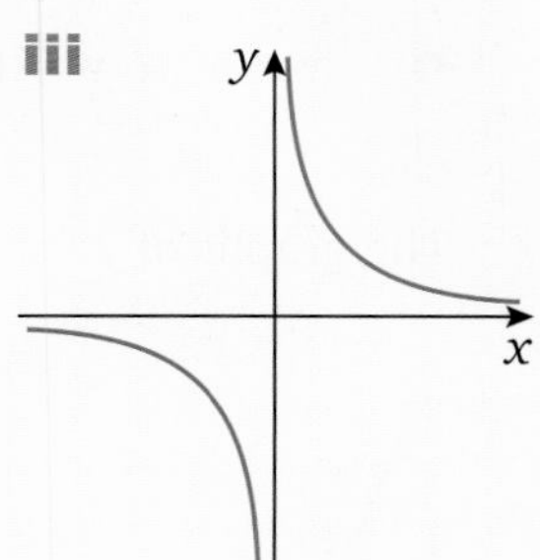

b Here is a sketch graph.

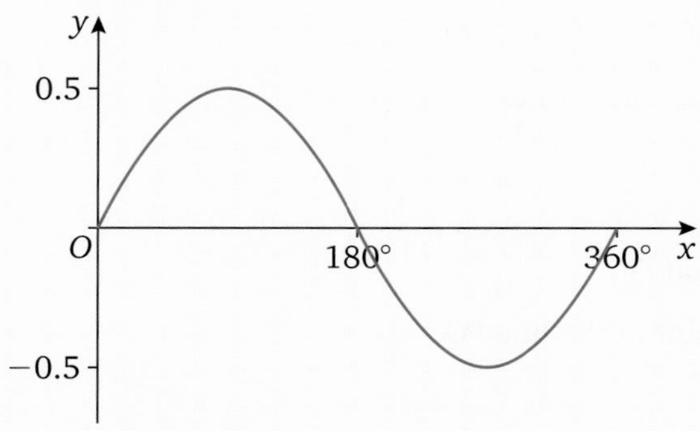

i Explain why this **cannot** be a sketch of the graph of $y = \sin x$ for values of x from 0° to 360°.

ii Write down the equation of the sketch graph.

A*

35 In triangle ABC, $AB = 5$ cm, $AC = 7$ cm and angle $ABC = 105°$

Work out the area of the triangle.

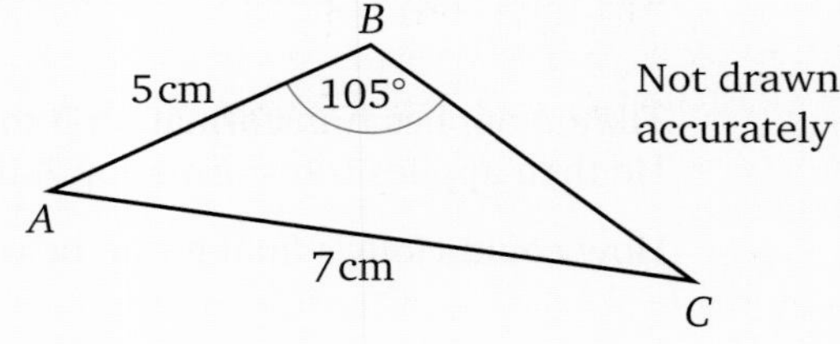

36 An oil container consists of a cylinder on top of a cone.
The diameter of the cylinder is 200 cm.
The height of the cone and the cylinder are both 100 cm.

Initially the container is full.
Oil flows from the container until it reaches a level 40 cm from the bottom of the container.

Work out the volume of oil that has flowed from the container.

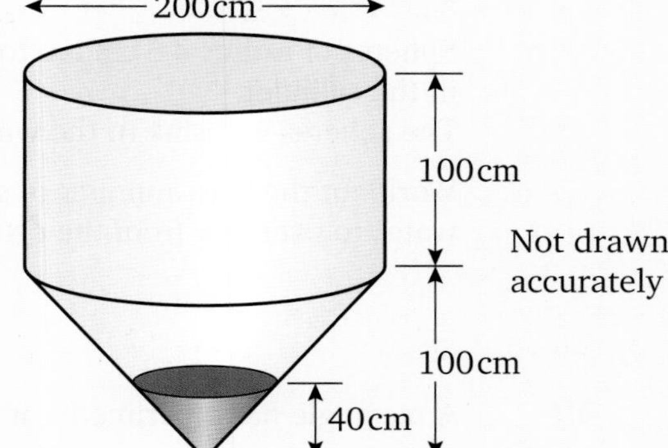

37 AB, BC and CD are straight lines.

$\overrightarrow{AB} = \begin{pmatrix} 3 \\ 5 \end{pmatrix}$ $\overrightarrow{CD} = \begin{pmatrix} 12 \\ k \end{pmatrix}$ angle ABC = angle BCD

Work out k.

Give reasons for your answer.

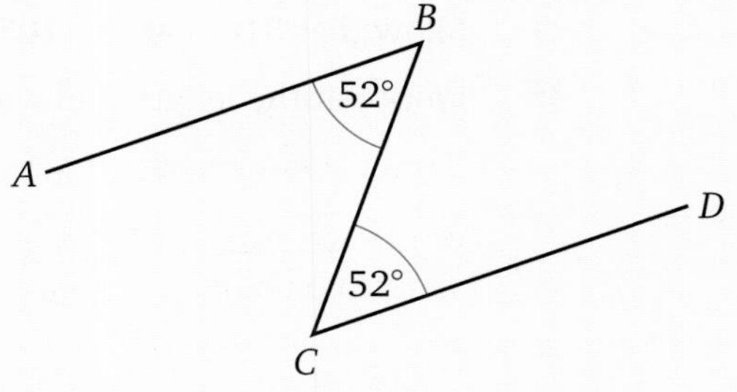

A*

38 Describe fully the transformations that map the first graph to the second graph.

a $y = x^3$ to $y = x^3 - 4$

b $y = \sin x$ to $y = 2\sin x$

c $y = x^2$ to $y = (x + 1)^2$

39 The graphs of $y = x^2 - x - 3$ and $y = 2x + 1$ are shown.

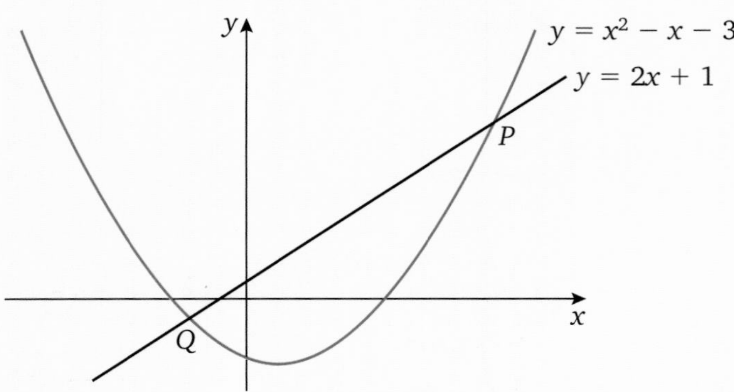

The graphs intersect at points P and Q.

Find the coordinates of the midpoint of the line PQ.

40 Here are two transformations.
Transformation A: reflect in the x-axis
Transformation B: translate 4 units vertically downwards

Holly applies transformation A to the graph $y = x^2$
She then applies transformation B to the graph she has obtained.

Tikram applies transformation B to the graph $y = x^2$
He then applies transformation A to the graph he has obtained.

How could Holly's final graph be transformed to Tikram's final graph?

41 A cylinder is filled with water to a depth of 3 cm from the top as shown.

Spheres of radius 4.5 cm are to be placed, one at a time, in the cylinder.
The spheres will sink in the water.

Work out the least number of spheres needed for the water to overflow from the cylinder.

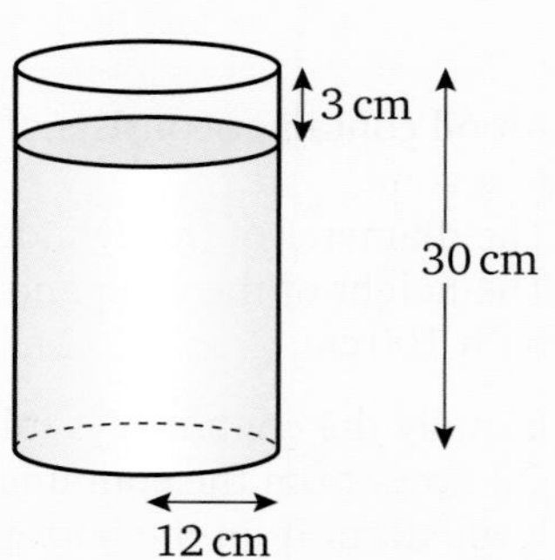

Not drawn accurately

42 A rectangle has a perimeter of 18.4 cm and an area of 21 cm^2.

Let the length of the rectangle be x cm.

a Show that $5x^2 - 46x + 105 = 0$.

b Work out the length and width of the rectangle.

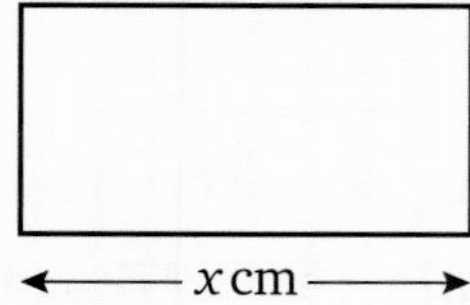

AQA Examination-style questions

1 A triangle has angles of 80°, x and $4x$.
Show that the triangle is isosceles.

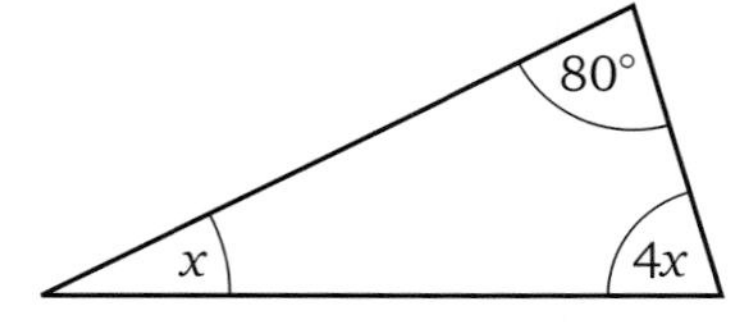

Not drawn accurately

(4 marks)
AQA 2008

2 A and B are two similar cylinders.
The height of cylinder A is 10 cm and its volume is 625 cm^3.
The volume of cylinder B is 5000 cm^3.

Calculate the height of cylinder B.

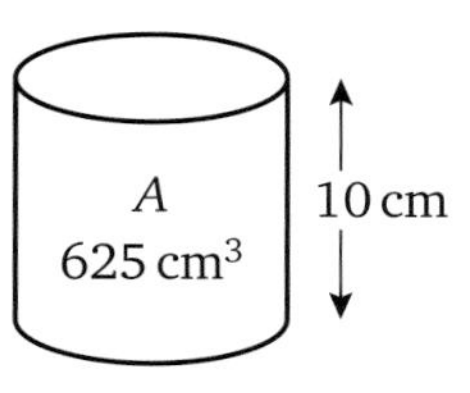

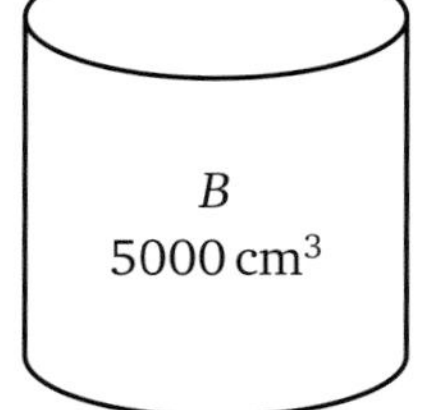

Not drawn accurately

(3 marks)
AQA 2000

3 $ABCD$ is a quadrilateral.
$AB = 4.6$ cm
$BC = 6.9$ cm
$AD = 3.8$ cm
Angle $ABC = 90°$
Angle $CAD = 48°$

Work out the perimeter of the quadrilateral.

Give your answer to an appropriate degree of accuracy.

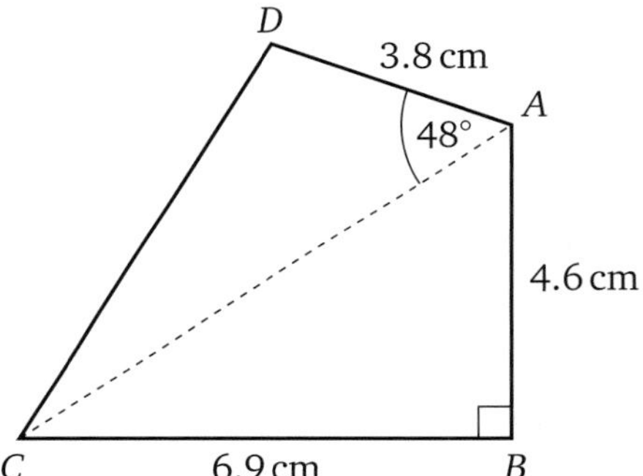

Not drawn accurately

(5 marks)
AQA 2009

4 Sam has made some wooden play blocks for a nursery class.
Each block is a prism with an L-shaped cross-section.

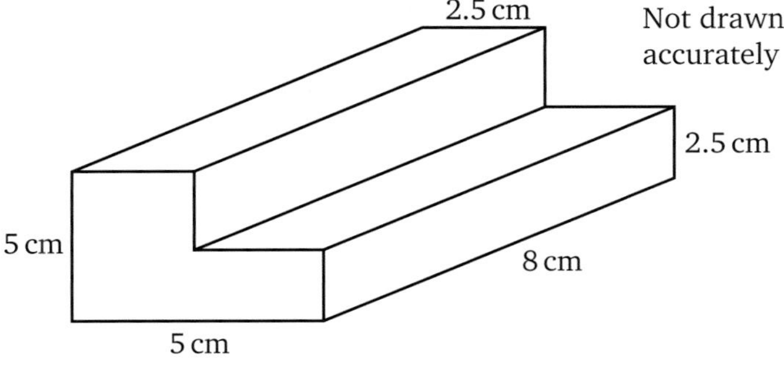

Not drawn accurately

Sam paints the entire surface of the blocks with one coat of red paint.
Sam has enough red paint to cover 3000 square centimetres.

What is the maximum number of blocks that he can paint red? *(4 marks)*
AQA 2009

Glossary

adjacent side – in a right-angled triangle, the side adjacent to a known or required angle (but not the hypotenuse).

alternate angles – angles formed by parallel lines and a transversal that are on opposite sides of the transversal. For example, the angles marked a, which are on opposite sides of the transversal.

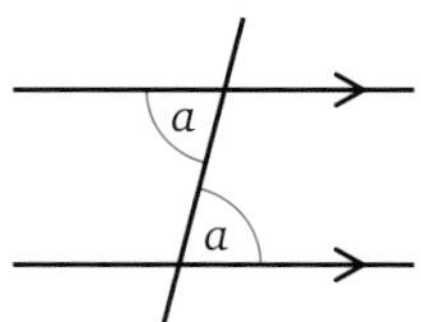

alternate segment – the angle between a tangent and a chord is equal to the angle in the alternate segment. For example, the red angles in the diagram are equal.

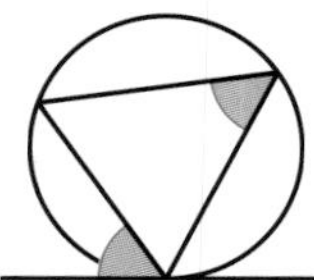

angle of elevation, angle of depression – the angle between a given line and the horizontal.

angle of rotation – the angle by which the object is rotated.

arc (of a circle) – part of the circumference of a circle; a minor arc is less than half the circumference and a major arc is greater than half the circumference.

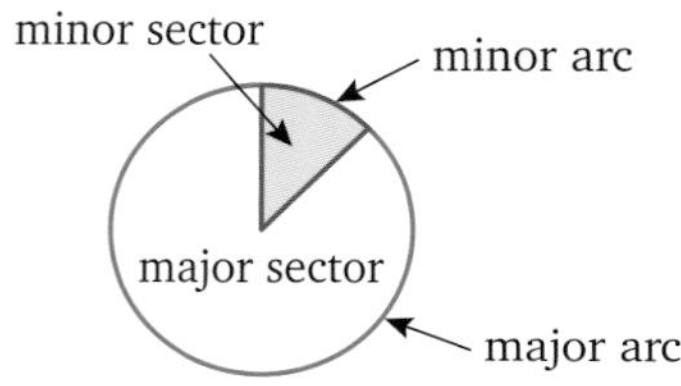

base – the lowest part of a 2-D or 3-D object (i.e. the side that it stands on).

bearing – an angle that denotes a direction. A bearing is measured clockwise from north.

bisect – to divide into two equal parts.

bisector – a bisector is a line that cuts either an angle or a line into two equal parts.

centre of enlargement – the fixed point from which the enlargement is made.

centre of rotation – the fixed point around which the object is rotated.

chord – a straight line joining two points on the circumference of a circle.

circular functions – functions of the form $y = \sin x$ and $y = \cos x$ and $y = \tan x$

coefficient – the number (with its sign) in front of the letter representing the unknown, for example $4p$ (4 is the coefficient).

column vector – a way of writing a vector, for example $\begin{pmatrix}4\\3\end{pmatrix}$ which means a move of 4 units to the right and 3 units up.

compound measure – a measure formed from two or more measures. For example, speed $= \dfrac{\text{distance}}{\text{time}}$

congruent – exactly the same size and shape; one of the shapes might be rotated or flipped over.

consecutive – next to each other in a sequence. For example, the numbers 3.4 and 5 are consecutive.

construct – this is the process of drawing a diagram accurately with a 'straight edge' and compasses only.

coordinates – a system used to identify a point; an x-coordinate and a y-coordinate give the horizontal and vertical positions relative to the origin.

corresponding angles – angles in the similar position between parallel lines and a transversal. For example, the angles marked b, which are on the same side of the transversal.

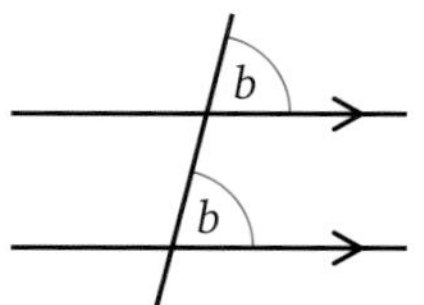

cosine – in a right-angled triangle, the length of the adjacent side divided by the length of the hypotenuse. Abbreviated to cos.

cross-section – a cut parallel to a face of, and usually at right angles to the length of, a prism.

cubic – an equation, expression or function in which the highest power of x is 3. It may also include x^2 or x terms and constants, for example $5x^3 + 2x^2 + 5x - 4 = 0$

cyclic quadrilateral – a quadrilateral with all four vertices on a circle.

cylinder – a prism with a circle as a cross-sectional face.

decagon – a polygon with ten sides.

decimal places – the number of digits after the decimal point. For example, the number 23.456 has three decimal places (4 tenths, 5 hundredths and 6 thousandths). Numbers can be rounded to different numbers of decimal places; 23.456 to 1 d.p. is 23.5

denominator – the bottom number of a fraction, indicating how many fractional parts the unit has been split into. In the fractions $\frac{4}{7}$, $\frac{23}{100}$, $\frac{6}{9}$ the denominators are 7 (indicating that the unit has been split into 7 parts, which are sevenths) 100 and 9

density – to find the density of an object, divide the mass by the volume. The units for density are usually grams per cubic centimetre (g/cm^3) or kilograms per cubic metre (kg/m^3).

depreciation – a reduction in value over time (of used cars, for example).

diagonal – a line joining two vertices of a polygon (that are not next to each other).

dimension – the measurement between two points on the edge of a shape, for example length.

direct proportion – if two variables are in direct proportion, one is equal to a constant multiple of the other. If one of the variables doubles then so does the other; if one of the variables halves then so does the other. In general, it means that $y = kx$ where k is the constant of proportionality.

discontinuous – if a graph has breaks in it, it is discontinuous. For example, the graph of $y = \frac{1}{x}$ does not cross the point $x = 0$

discount – a reduction in the price. Sometimes this is for paying in cash or paying early.

elevation – this is the view of an object when viewed from the front or side. Sometimes called front elevation (view of the front), or side elevation (view of a side).

eliminate – to remove one of the unknowns from a pair of simultaneous equations by adding or subtracting like terms. The unknown being eliminated must have a matching coefficient in both equations.

enlargement – an enlargement changes the size of an object according to a certain scale factor.

equation – a statement showing that two expressions are equal, for example, $2y - 17 = 15$

equidistant – to be equidistant from two points is to be the same distance from both points.

equilateral – having all sides of equal length.

equilateral triangle – a triangle that has all three sides equal in length.

equivalent fractions – two or more fractions that have the same value. Equivalent fractions can be made by multiplying or dividing the numerator and denominator of any fraction by the same number. For example, the fractions $\frac{4}{7}$ and $\frac{8}{14}$ are equivalent and have the same value.

expand – to remove brackets by multiplying to create an equivalent expression (expanding is the opposite of factorising).

exponential functions – functions of the form $y = Ak^{mx}$ where A, k and m are constants. For example, if £1500 is invested at 2.5% compound interest, the amount of money, £y, in the account after x years is given by the formula $y = 1500 \times 1.025^x$, which is an exponential function with $A = 1500$, $k = 1.025$ and $m = 1$. The amount of money in the account is said to 'grow exponentially'. This does not necessarily mean that it grows very rapidly, but that its rate of growth is proportional to the amount in the account.

exponential growth – this occurs when the rate of growth is proportional to the quantity present. Repeated percentage increases leads to exponential growth.

expression – a mathematical statement written in symbols, for example, $3x + 1$ or $x^2 + 2x$

exterior angle – the angle between one side of a polygon and the extension of the adjacent side.

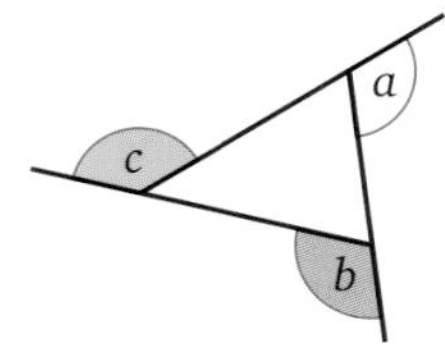

a, *b* and *c* are exterior angles

face – one of the flat surfaces of a solid. For example, a cube (such as a dice) has six flat faces.

factorise – (in algebra) to include brackets by taking common factors (factorising is the opposite of expanding).

formula – a formula shows the relationship between two or more variables, for example, in a rectangle area = length × width, or $A = lw$

front elevation – see **elevation**.

frustum (of a cone) – a cone with the top part cut off.

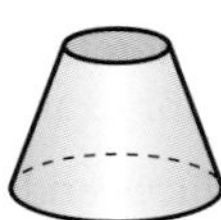

function – a function tells you what to do to the value of a variable to work out the value of a second variable. For example, if y is a function of x, then the function will tell you what to do to the value of x to work out the value of y. If y is a function of x, then it can be written as $y = f(x)$

hexagon – a polygon with six sides.

hypotenuse – in a right-angled triangle, the longest side, opposite the right angle.

identity – two expressions linked by the ≡ sign are true for all values of the variable, for example, $3x + 3 \equiv 3(x + 1)$.

image – the shape following a transformation of the object, for example, reflection, rotation, translation or enlargement.

integer – any positive or negative whole number or zero, for example, −2, −1, 0, 1, 2

interior angle – an angle inside a polygon.

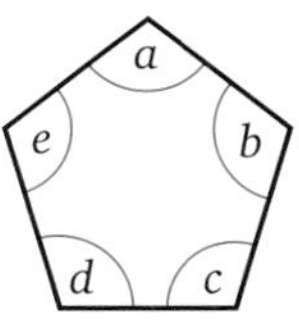

a, *b*, *c*, *d* and *e* are interior angles

interior (or **allied**) – angles between two parallel lines and a transversal, which are on the same side of the transversal and between the parallel lines. For example, the angles marked *a* and *b*.

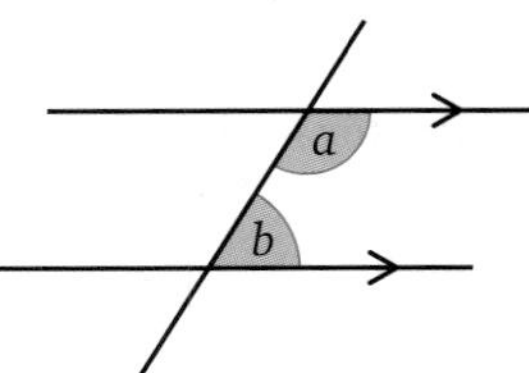

inverse proportion – if two variables are in inverse proportion, their product is a constant. If one of the variables increases, then the other decreases (and vice versa). In general, $y \propto \frac{1}{x}$ means that $y = k \times \frac{1}{x}$ and $xy = k$ where k is the constant of proportionality.

isosceles triangle – a triangle with two equal sides and two equal angles.

line of symmetry – a shape has reflection symmetry about a line through its centre if reflecting it in that line gives an identical-looking shape.

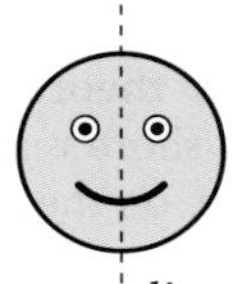

line of symmetry

loci – plural of locus – see **locus**.

locus – a locus is the path followed by a moving point. It is also a set of points that meet a condition.

lower bound (or limit) – the value of a rounded quantity lies between two limits, the upper and lower bounds. The smallest possible value is called the lower bound (or limit).

magnitude – size.

mass – the weight of an object, measured in tonnes (t), kilograms (kg), grams (g) or milligrams (mg).

mixed number – a fraction that has both a whole number and a fraction part, for example, $1\frac{4}{7}$, $3\frac{1}{2}$, $5\frac{3}{4}$

net – a net shows the faces and edges of an object. When the net is folded up it makes a 3-D object. For example, the net of a cube when folded up makes a cube.

nonagon – a polygon with nine sides.

numerator – the top number of a fraction, indicating how many parts there are in the fraction, for example, in the fractions $\frac{4}{7}$, $\frac{23}{100}$, $\frac{6}{9}$ the numerators are 4, 23 and 6.

object – the shape before it undergoes a transformation, for example, translation, reflection, rotation or enlargement.

octagon – a polygon with eight sides.

one-way stretch – a one-way stretch in the x-direction increases all the x-coordinates by a scale factor. Likewise, a one-way stretch in the y-direction increases all the y-coordinates by a scale factor.

opposite side – in a right-angled triangle, the side opposite a known or required angle.

parabola – the locus of a point that moves so that it is always the same distance from a fixed point and a given line.

parallel – two lines that stay the same perpendicular distance apart.

pentagon – a polygon with five sides.

per annum – means 'per year'.

percentage – the number of parts per hundred. For example, 15% means $\frac{15}{100}$

perpendicular – at right angles to; two lines at right angles to each other are perpendicular lines.

perpendicular bisector – a line drawn at right angles to a line segment, cutting the segment into two equal parts.

perpendicular height – the height of a shape that is 90° to the base.

plan – this is the view when an object is seen from above. Sometimes called the plan view.

plane of symmetry – a 3-D object has a plane of symmetry, whereas a 2-D shape has a line of symmetry.

polygon – a closed two-dimensional shape made from straight lines.

prism – a solid that has the same cross-section all the way through.

Pythagoras' theorem – in words 'the sum of squares on the two shorter sides of a right-angled triangle is equal to square on the hypotenuse' or $c^2 = a^2 + b^2$

quadrant – one quarter of a circle.

quadratic equation – an equation in which the highest power of x is 2. It may also include x terms and constants, for example $5x^2 + 2x - 4 = 0$

quadratic expression – an expression containing terms where the highest power of the variable is 2.

quadrilateral – a polygon with four sides.

ratio – a ratio is a means of comparing numbers or quantities. If two numbers or quantities are in the ratio 1 : 2, the second is always twice as big as the first. If two quantities are in the ratio 2 : 5, for every 2 parts of the first there are 5 parts of the second.

reciprocal – the reciprocal of a number is 1 divided by that number. Any number multiplied by its reciprocal equals 1. For example, the reciprocal of 6 is $\frac{1}{6}$ because $6 \times \frac{1}{6} = 1$ and $1 \div 6 = \frac{1}{6}$. The number 1 is its own reciprocal and the number zero has no reciprocal.

recurring decimal – a decimal whose digits after the point eventually form a repeating pattern. A dot over the digits indicates the repeating sequence, for example, $\frac{2}{7} = 0.\dot{2}8571\dot{4}$

reflection – a transformation involving a mirror line (or axis of symmetry), in which the line from the shape to its image is perpendicular to the mirror line. To describe a reflection fully, you must describe the position or give the equation of its mirror line, for example, the triangle A is reflected in the mirror line $y = 1$ to give the image B.

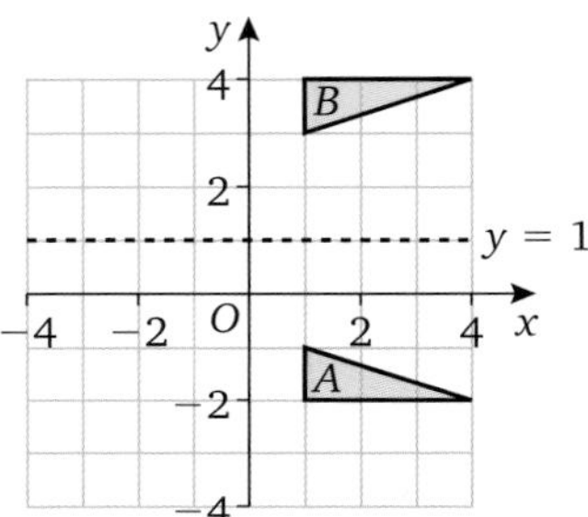

rotation – a transformation in which the shape is turned about a fixed point called the centre of rotation. To describe a rotation fully, you must give the centre, angle and direction (a positive angle is anticlockwise and a negative angle is clockwise), for example, the triangle A is rotated about the origin through 90° anticlockwise to give the image C.

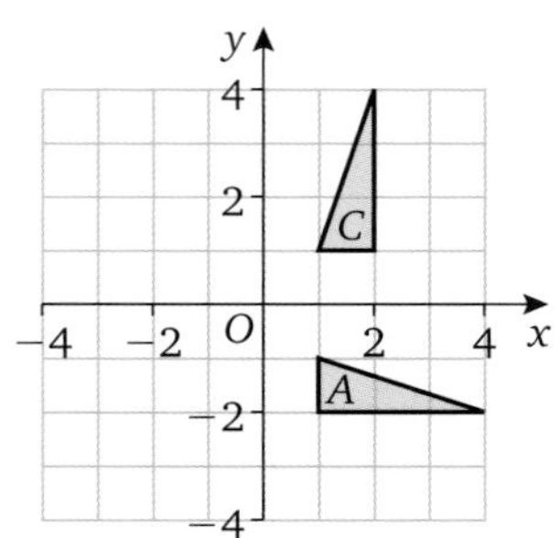

rounding – a number can be expressed in an approximate form rather than exactly. For example, it may be written to the nearest integer or to the nearest thousand. This process is called rounding. The number 36 754 rounded to the nearest thousand is 37 000.

scalar – a quantity with size but not direction, for example, the number 2.

scale factor – the scale factor of an enlargement is the ratio of the corresponding sides on an object and its image.

sector – the area bounded by two radii and an arc in a circle.

segment – the region bounded by an arc and a chord in a circle.

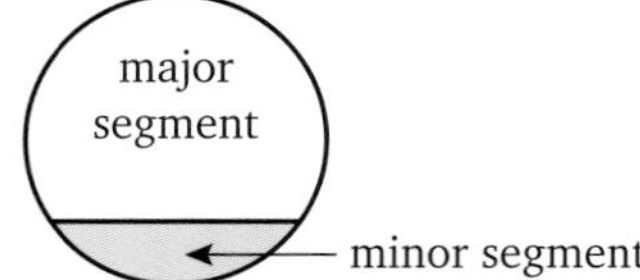

side elevation – see **elevation**.

significant figures – the closer a digit is to the beginning of a number then the more important or significant it is. For example, in the number 23.657, 2 is the most significant figure. 23.657 = 20 correct to 1 s.f.

similar – shapes are similar (mathematically similar) if they have the same shape but different sizes. That is, one is an enlargement of the other.

simplify – (in algebra) to make simpler by collecting like terms.

simultaneous equations – a pair of equations containing two unknowns where both equations are true at the same time.

sine – in a right-angled triangle, the length of the opposite side divided by the length of the hypotenuse. Abbreviated to sin.

solid – a three-dimensional shape.

substitution – in order to use a formula to work out the value of one of the variables, you replace the letters by numbers. This is called substitution.

subtend – when the end points of an arc are joined to a point on the circumference of a circle the angle formed is subtended by the arc.

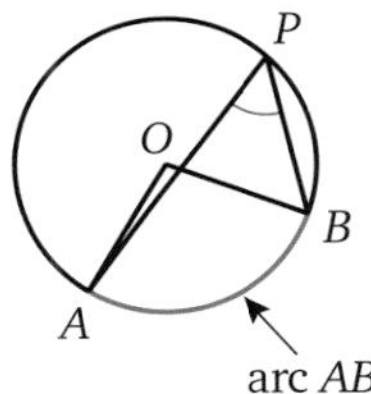

arc AB subtends $\angle AOB$ at the centre

arc AB subtends $\angle APB$ at the circumference

supplementary angles – two angles are supplementary if their sum is 180°.

surd – a number containing an irrational root, for example $2\sqrt{7}$.

surface area – the exposed area of a solid object, often measured in square centimetres (cm^2) or square metres (m^2).

tangent – in a right-angled triangle, the length of the opposite side divided by the length of the adjacent side. Abbreviated to tan.

tangent (to a circle) – a straight line that touches the circle at only one point.

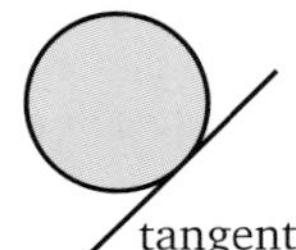

terminating decimal – a decimal that has a finite number of digits after the decimal point for example, $\frac{1}{64} = 0.015625$

tetrahedron – a solid made up of four triangular faces.

three-dimensional (3-D) – a shape that has three dimensions: length, width and depth.

transformation – reflections, rotations, translations and enlargements are examples of transformations as they transform the position, orientation or size of a shape.

translation – a transformation where every point moves the same distance in the same direction so that the object and the image are congruent.

transversal – a line that crosses two or more parallel lines.

trial and improvement – a method for solving algebraic equations by making an informed guess, then refining this to get closer and closer to the solution.

triangle – a polygon with three sides.

triangular prism – a prism with a triangular cross-section.

trigonometry – the study of the relationship between the length of sides and the size of angles in a triangle.

unitary method – a way of calculating quantities that are in proportion. For example, if 6 items cost £30 and you want to know the cost of 10 items, you can first find the cost of 1 item by dividing by 6, then find the cost of 10 by multiplying by 10.

unitary ratio – this is a ratio in the form $1 : n$ or $n : 1$. This form of ratio is helpful for comparison, as it shows clearly how much of one quantity there is for one unit of the other.

upper bound (or limit) – the value of a rounded quantity lies between two limits, the upper and lower bounds. The upper limit is called the upper bound. The actual value of the quantity must lie below this bound.

Value Added Tax (VAT) – this tax is added on to the price of some goods or services.

variable – a symbol such as x, y or z representing a quantity that can take different values.

vector – a quantity with magnitude (size) and direction. In this diagram, the arrow represents the direction and the length of the line represents the magnitude.

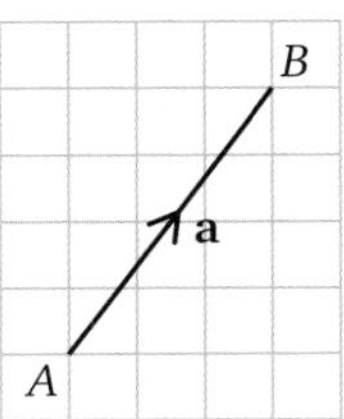

In print, this vector can be written as $\overrightarrow{AB}$ or **a.** In handwriting, this vector is usually written as $\overrightarrow{\text{AB}}$ or $\underset{\sim}{a}$. The vector can also be described as a column vector:

where $\begin{pmatrix} x \\ y \end{pmatrix}$ ← x is the horizontal displacement ← y is the vertical displacement

vector sum – the sum of two or more vectors.

vertices – plural of vertex – see **vertex**.

vertex – the point where two or more edges meet.

vertically opposite angles – the opposite angles formed when two lines cross.

volume – the amount of space a solid takes up, often measured in cubic centimetres (cm^3) or cubic metres (m^3).

Index